Food Color
and
Appearance

Second

John B. Hutchings, FInstP, FIFST

Fellow of the Institute of Physics
Fellow of the Institute of Food Science and Technology
Consultant
Colmworth, Bedford
United Kingdom

A Chapman & Hall Food Science Book

AN ASPEN PUBLICATION®
Aspen Publishers, Inc.
Gaithersburg, Maryland
1999

Aspen Publishers, Inc., is not affiliated with the American Society of Parenteral and Enteral Nutrition.

Library of Congress Cataloging-in-Publication Data

Hutchings, John B.
[Food colour and appearance]
Food color and appearance/John B. Hutchings.—2nd ed.
p.   cm.
Rev. ed. of: Food colour and appearance. 1994.
Includes bibliographical references.

1. Coloring matter in food.   2. Color of food.
3. Food—Sensory evaluation.   4. Food presentation.   I. Title.
TX571.C7H88   1999
664'.062—dc21
99-20527
CIP

Orders: (800) 638-8437
Customer Service: (800) 234-1660

About Aspen Publishers • For more than 35 years, Aspen has been a leading professional publisher in a variety of disciplines. Aspen's vast information resources are available in both print and electronic formats. We are committed to providing the highest quality information available in the most appropriate format for our customers. Visit Aspen's Internet site for more information resources, directories, articles, and a searchable version of Aspen's full catalog, including the most recent publications: **http://www.aspenpublishers.com**
Aspen Publishers, Inc. • The hallmark of quality in publishing
Member of the worldwide Wolters Kluwer group.

Editorial Services: Jane Colilla

ISBN 978-1-4419-5193-9

Printed in the United States of America

1   2   3   4   5

Table of Contents

Preface

Our behavior is controlled to a large extent by appearance, but the appearance of our food is of paramount importance to health and well-being. In day-to-day survival and marketing situations, we can tell from their optical properties whether most foods are fit to eat. Although vision and color perception are the means by which we appreciate our surroundings, visual acceptance depends on more than just color. It depends on *Total Appearance*. The need continues for food technologists to increase understanding of how the appearance of raw materials and products behave under processing, and how the customer behaves toward product appearance in a growing, more discriminating, and worldwide market.

The chapters that follow describe the philosophy of total appearance, the factors comprising it, and its application to the food industry. Included are considerations of the evolutionary, historical, and cultural aspects of food appearance; the science of food color and appearance; the principles of sensory appearance assessment and appearance profile analysis, as well as instrumental measurement; and the interaction of product appearance, control, and acceptance in the varied environments in which food is prepared, manufactured, and consumed.

This work provides grounding in the science and methodology of food appearance. It is a basic and reference text, suitable for those scientists, technologists, designers, manufacturers, and product developers in industry and academia whose concern is for how foods look. It will also serve as a text for students of food science, technology, and marketing, and for those concerned with food presentation and food appearance specification. The message to take away is that total appearance in foods can be understood, described, quantified, and optimized.

THE SECOND EDITION

Apparent in the literature is a growing concern for food product appearance, and, since the first edition, there have been changes in broad approach as well as detail. Advances have been made in extending the philosophical approach into the *Information Transfer Process*, aiming to understand links between material properties, Basic Perceptions, and the Derived Perceptions, those of Visual Expectation. An extra chapter has been included giving examples of Appearance Profile Analysis, a disciplined method for the study of these perceptions.

There have also been welcome advances in instrumentation. The advent of calibratable image capture devices provides a basis for an instrument that will allow us to examine the whole product without destruction of the sample. We are getting closer to making measurements related with a wide range of visual properties significant to the customer. This technique has the potential to revolutionize food appearance science, whether the product is in the laboratory, on the production line, on the shelf, in the pan, or on the table.

Also in this new edition, the coloring bases of foods are presented in terms of driving forces existing within specific food systems. This is a more logical and useful approach in a volume widely used within industry, academic research, and postgraduate study. This edition is concerned solely with food itself and the environment of sensory assessment.

I am grateful for continued support from my family and from Unilever Colworth House Research Laboratory staff Peter Lillford and Kathey Parkinson, with Glenda Scott and Keith Plater, who have permitted me to keep abreast with new methodology.

CHAPTER 1

Food Color and Appearance in Perspective

We regard the appearance of our food with interest. The look of a meal provokes expectations, stimulates or depresses the appetite, and can engender joy or melancholy. When we eat to savor and enjoy, rather than merely to survive, those extra pains taken to use color and appearance to increase temptation and appetite prior to and during consumption are worthwhile. Traditional meals around the world look different. The browns and greens of the West; the paler, more subtle shades of the Chinese dish; and the brightly colored, delicately sculptured components of the Thai meal are examples. We respond to the aesthetic nature of color, pattern, and design. However, within each foodstuff, appearance has deeper meanings and associations.

We eat to gain energy, but processing of some kind is normally called for to increase edibility. The orange merely has to be peeled and separated into segments. Red beans must be boiled for 20 minutes to reduce toxicity and increase tenderness. The preparation of a gourmet meal involves a great deal more sophistication in material handling and cooking technique. Cooking and processing make natural ingredients more palatable and tender enough to eat quickly and lengthen their shelf life. Edible ingredients are manipulated to create sweet products such as ice cream and impulse foods, succulent pastries, and sweets, as well as the more savory pickles and sauces that occupy the same plate as traditional meals. Most of these products may be colored already or may be deliberately pigmented using natural or artificial colorants.

1

Processing alters pigments and product structure, resulting in changes to color and appearance. These changes are used as indicators of processes occurring within the product. Color indicates stage of ripening, and change of brown color indicates stage of bread baking and toasting, meat roasting, and coffee creaminess. Change of translucency indicates the depth of cooking of fish and meat, and change of skin condition is symptomatic of fruit cooking. In summary, appearance indicates ripeness, freshness, and cooking stage.

In this chapter, an attempt is made to put the color of food into some form of evolutionary context and to give an overview of its significance at the domestic level. Included also are accounts of how our foods came to be the colors they are, the biological importance of food pigments, the use of color additives, some of the rules governing food appearance, and the relative importance of appearance and other sensory properties. In other words, an account is given of the place occupied by food color in our everyday lives as human beings.

EVOLUTION OF FOOD COLOR

Some hues occur frequently in foods provided by nature. Green, red, pink, orange, yellow, and purple are common. Blue green is rare, and no blues exist at all. It is instructive to consider the fundamental, natural rules governing the colors of the world around us and the food we eat (Hutchings 1998). From this point of view, raw natural foods may be divided into four groups: those derived from green leaves, colored fruit and vegetables, mammal flesh, and fish flesh. Although the colors of these foods attract and excite us, most do not owe their existence to us. Some colors developed independently of humans; others evolved with us. We have adapted to appreciate and enjoy them, and we have learned the color codes indicating when each food can be eaten with safety.

Plant leaves evolved to be green by default. Their color arises from the reflection of unwanted radiant energy by energy-absorbing photosynthetic systems. When life began on earth in an aqueous environment, the first colored living organism probably was the halobacterium. This contains a purple photosynthetic pigment that absorbs a broad band of radiation between the overenergetic ultraviolet and the heat-inducing infrared. This radiation band is in the middle of what became, when humans arrived in the world, the visible part of the spectrum. These wavelengths possess the energy useful for photosynthesis. The halobacterium could not fix the carbon dioxide needed for growth and obtained its nutrients from the primeval broth of chemicals in which it lived. Its purple color permitted the maximum absorption of energy penetrating into the broth. Probably at this time, the much more efficient green photosynthetic pigment chlorophyll, able to convert carbon dioxide to produce growth-promoting glucose, was also starting to evolve. Purple bacteria absorb much of the energy available in the middle of the wavelength band. Hence, the long (red) and short (blue) wavelengths present in daylight were available for absorption by the chlorophyll. As the broth chemicals were consumed, the halobacterium population diminished, and the green system flourished (Goldsworthy 1987).

Leaves contain a complex mixture of many forms of chlorophyll and carotenoid pigments. The latter have evolved to assist photosynthesis and protect the leaf during this process. The different shades of green of the leaf food we eat probably result from the energy absorption optimization tactics of the particular plant growing in its own environment. Whatever the origins and functions of the leaf energy absorption spectrum, it is the energy not required to drive photosynthesis that is the basis of our leaf food color.

The driving force for the evolution of coloration in flowering plants and fruits has arisen from coevolution of predator vision and the reflectance and transmission characteristics of the flower or fruit. Zoologists are satisfied that the mechanism of color vision can develop or diminish in various species according to the requirements of survival (Walls 1963). Humans (Hill and Stevenson 1976) and fish (Muntz 1975) still retain the capacity of changing photopigments in response to their environment.

Fruit maturation involves pigment changes due to chlorophyll destruction, revelation and synthesis of carotenoids (yellows and oranges), and synthesis of anthocyanins (reds and purples). Yellow and orange tropical fruits appear to have coevolved with the trichromatic color vision of Old World monkeys (Mollon 1989). Fruit of these colors is consistently taken and forms a substantial part of their diet. The monkey needs trichromatic color vision to find the fruit among the foliage. In contrast, fruits predominantly taken by birds are red and purple, while fruits taken by ruminants, squirrels, and rodents are dull colored, green, or brown. Each animal species has evolved its own retinal cone pigments, which enable it to search for food efficiently. As well as providing food for animals, the plant is also helped. Fruit is usually carried away to be eaten, and the seeds are dispersed where they are spat out or excreted. This coevolution argument is similar to that used to link the visual characteristics of pollinators to the color of flowers they pollinate. The insect gets the pollen and the nectar that goes with it, and the plant is able to produce fertile seeds.

The third driving force for the color of our food is coloration by coincidence. Into this class fall those optimizations that happen to result in a biochemical that is colored. Meat flesh color may be an example. Early in the evolution of life on this planet, nature found the basic porphyrin structure. Variation has occurred from time to time, but porphyrins have always been involved in the respiration function of plant and animal organisms. Examples are green chlorophyll in leaves, brown pinnaglobin in bivalves, blue hemocyanin in some molluscs, and red hemoglobin in mammals. Blood pigments appear within the organism and are not a primary factor in outward appearance. Hence, meat may be red, not as a result of any energy absorption or vision/coloration survival mechanism, but because hemoglobin is a very efficient respirator. Although blood color does not affect our appearance in a primary way, the fact that the blood is a strong contrast color does assist the diagnosis of injury and illness.

The color of fish flesh is governed by blood or by melanin derivatives produced for skin coloration or arises directly from its diet. Pink flesh color arises from the carotenoid astaxanthin contained in the diet of crustacea. Carotenoids enter this food chain originally from photosynthesis.

Blue fruit is rare. This is not because nature cannot synthesize blue colors, for there are blue flowers. As already indicated, there are convincing arguments linking the color of food with the visual characteristics of the gatherer. It is also evident that in real life, colors do not exist by themselves. They are always seen in contrast to other colors. Just as monkeys must isolate yellow and orange fruits against a background of green, they must isolate them against a background of the sky. Blue fruit would be difficult to pick out against such a background, and possible driving forces for a mature blue would be reduced.

It is tempting to extend the argument linking visual pigments to the bulk of our food. Most fruits change color from green to yellow to orange and may progress to red. On the other hand, red meat ages in the reverse order. It moves from red to brown to green. In both cases, an ability to detect the yellow (or brown)-versus-green boundary would be a survival benefit. As the movement of the change is in spectral order, it would be convenient if these critical changes had a common mechanism for detection. Work on large color contrasts indicates that perceived contrast will be greatest if the colors concerned are opponent pairs—for example, red versus green (Neri et al. 1986). In evolutionary terms, the need for a visual alarm system to detect poorly colored natural foods is debatable. Overage vegetation shows patches of easily visible high-contrast dark color of enzymic browning. Mold exhibits a change in surface texture, and overage meat and fish smells, so color vision may not be required for detection.

A convenient extension of this argument links our color vision characteristics and our settled agrarian society. Food production, storage, preservation, and distribution have become more organized and controlled. As we have generally ceased to become gatherers, the need for a developed red-versus-green detection mechanism may decline, and the possible need for a visual alarm function may decline. This is the most common form of color vision deficiency, and it occurs in industrialized Western Europe and North America. The lowest occurrence is found in the tribal regions of Central Africa and among the Inuit in the far north. These peoples depend for their survival on close contact with the natural world. So it appears that food's availability and selection provide the driving forces for both normal and defective color vision in humans.

CULTURAL ANTHROPOLOGY, ORAL TRADITION, AND FOOD APPEARANCE

Food taste and appearance are normally regionally based through local produce and economic and political events. In North America, for example, all culinary regions are traceable to the specific populations of American Indians, pioneers, missionaries, cowboys, and immigrants (Cousminer and Hartman 1996). Oysters provide an example of preference linked to availability. These bivalves sometimes develop green gills. In areas like New Brunswick, where this condition is uncommon, they can be rejected, but in regions such as Virginia, where they are more common, they tend to be preferred, and in Marennes in France they are highly prized (Boon 1977). Even in a small country such as England, there are regional food differences based on color. Darker foods are preferred in the north. Treacle toffee and licorice are preferred sweets, and icing sugar is dyed colors that are darker and more saturated. Proportionally higher sales of darker beers, brown sherry, and Demerara rum occur there. In the southwest counties of Devon and Cornwall, there is a preference for white in foods. Fresh milk and cream are used not only in great proportion with their famed cream teas but on fruit and vegetables also (Allen 1968). The color of products varies with crop variety particular to a locality. The color of tortillas derives from the color of the maize used. In Mexico they are turquoise, in the United States bluish gray.

There are also racial differences. In North America, individuals with a Mexican food background tend to regard appearance as a less important attribute, while those with a Chinese background regard it more highly (Schutz and Wahl 1981). In a supermarket trial in which yellow and white cheeses were displayed, individuals from the black and other ethnic groups tended to select yellow cheeses, whereas white individuals tended to select white (Scanlon 1985). There are also individually based preferences. Again in the United States, significant differences in preference occur between those who are over-weight and underweight, between males and females, and between blacks and whites (Meiselman and Wyant 1981). Health-conscious panelists give snack foods significantly higher *Looks Too Greasy* scores than tasters who are flavor orientated (Moskowitz 1985).

Food names signify different things to different peoples. This is because the name stands not for a physical object but for a concept or a sign. Rice and potatoes are both white in color, but rice can be also ethnically white. In the Andes, a neighbor may be seen as pretentious for cooking it. An expensive food stands for wealth, a cheap one for poverty (Weismantel 1988). In the normal eating situation, foods must be within the color range the eater has been brought up to regard as proper. Although potatoes and potato products may be presented in a great variety of sizes and shapes, their color must be within the correct range. Although we may accept color changes in dress and decor, we are reluctant to accept inappropriately colored food.

In the Western world, *naturalness* is a quality to which we can become conditioned. In Britain, white-shelled eggs are associated with battery production; those with brown shells are seen as more natural. Brown eggs are common in southern New England, but white are preferred in New York (Lawless 1995, cited in Francis 1995) and the Middle East (King 1979). Such preferences may be founded on product quality, as brown eggs contain a higher yolk weight and yolk/albumin ratio than tinted- or white-shelled eggs (Campo 1995). Meat products such as paté have a non-natural appearance and score low on *naturalness, perceived quality,* and *preference* (Steenkamp 1986). However, we often demand that the natural appearance of meat and fish be disguised or hidden from sight by a crumb coating. Perhaps we have an aversion to eating parts of what were living creatures. Feelings may be different on festive occasions, when the natural forms of the sucking pig, whole salmon, or roast turkey may even be enhanced, possibly through a subconscious link with the hunt (Lyman 1989).

The debate over whether meat ought to be bloody and rare or well done probably has as much to do with appearance and the sight of blood as with texture. In France, all meat, except chicken, is preferred rare. The reason for the exception is the high risk of salmonella. Anthropologically, to eat bloody meat is making a potent statement of power over other highly evolved animals. In Britain, rare cooked meat is interpreted by the French as overcooked (Blythman 1992). Among Australians, there is wide disagreement when ordering beefsteaks. Out of a number of diners, 4.5 percent ordered **rare**, 19.8 percent **medium rare**, 33.6 percent **medium**, 14.9 percent **medium/ well done**, and 27.2 percent **well done** (Cox et al. 1997).

Existence of migrant workers encourages marketing of foreign foods. For those who feel alienated, a sense of identity and reassurance is provided by a diet of their own regional food. This in turn increases the choice available to the indigenous population. Similarly, food not eaten since childhood can bring back fond memories of upbringing and security. The back-to-nature trends of cosmetics and toiletries, the successful and lucrative marketing of regional dishes, rural-fresh vegetables, and foods made from old recipes result in appearance images and feelings of warmth, comfort, and good health (Newall 1987). An identification of some North Americans with Ireland, and therefore with green, leads them to coloring the Chicago River green and to drinking green beer annually on St Patrick's Day (Hutchings 1997).

Oral tradition and folklore play a large part in our view of foods. Although over a great area of Europe and the East, many colors are used to dye eggs, red is by far the most popular. In 2,900 BP they were exchanged by the Chinese at spring festivals. Today in Britain, they are used in Easter games such as egg rolling. In Romania, where red-painted eggs are called "love apples," and in Slovenia, they represent love and health. In Hungary, where the name for Easter eggs is *piros tojas* or red eggs, and in Russia, Yugoslavia, and Greece, they represent blood lost by Christ on the cross (Newall 1991).

Aspects of appearance as well as color are important in traditional foods. On Lammas ("Loaf Mass") Day, celebrations in medieval England included the baking of loaves colored red (with rose petal), golden orange (with saffron), yellow (with lemon), green (with parsley), blue (with thistle), indigo (with plum), and purple (with violet). Architectural shapes of castles and warships, depictions of animals, depictions of Eve in the Garden of Eden, and star and moon shapes were part of the traditional fare (Cosman 1984). There is a fine tradition of shaped breads in central Europe. These are used to reinforce the Christian belief of the transubstantiation to the Body of Christ. *Christstollen* is a sweet yeast loaf with fruit and nuts swaddled in white icing to represent Jesus. This was first documented in 1329. Traditions involving shaped breads occur in eastern and western Europe, Scandinavia, and Mexico. Ceremonies concerning them range from reinforcement of religious beliefs, to celebrating the arrival of seasons of the year, to increasing fertility of the fields, to bringing good luck. Pretzels were believed to drive witches away. Iced fruit cakes, optionally tiered, are customary for weddings and christenings in many countries. Gloss plays a part in food folklore also. The Japanese ceremony of "breaking the mirror" is carried out when a new wooden cask of sake is broken open with a mallet. The name refers to the fact that the surface of the drink is so clear it looks like glass. People of all religions and races ceremonially share foods during celebrations. Table settings and room decoration also play their part in festive meals.

Dance, song, and ceremony are relevant parts of the food production process. Dances were used to greet the return of the growing season of spring, to accompany sowing, to make rain, to celebrate the end of harvest, and to accompany the stirring of the mincemeat and the Christmas pudding mix. Customs vary around the world, but in no part are all these traditions and rites obsolete. The cooking of a flambé at the table and the preparation of a pizza or a sushi in front of a restaurant customer are theatrical performances. A well-lighted and displayed supermarket bakery department is part of a performance-orientated personal relationship between the store management and the customer.

On every continent, rice has long been the staple diet of half the human population. It is not surprising, therefore, that it has become the object of folklore and tradition, much of which concerns the use of color. In China, rice is called "white jade beads," and the phrase "having a bowl of rice" means having a steady job. Although the most common variety is white, it does occur naturally in a number of other colors. Black rice is used as a tonic for the kidneys and red for modulating one's spirit, and there are also light-yellow and light-violet varieties. Rice dyed red is given to wedding guests. In some provinces, cooked black rice is eaten in the lunar calendar of April to commemorate the birth of an ox or to mark the struggle against miscellaneous levies. White rice forms one part of the collection of bright colors that symbolize happiness in ceremonies practiced by the Canchis Indians in southern Peru. Rice colored yellow with turmeric or saffron is widely used in customs in India and Pakistan and by Indians in Malaysia. It is used to scare away demons affecting fertility, it forms an invitation to wedding feasts, and it is sprinkled on the couple afterwards. White rice is used as a contrast color in celebratory dishes. In Japan, red rice balls (steamed rice mixed with red beans) form an essential part of the celebration of happy occasions such as weddings, New Year's Day, and anniversaries (Hutchings et al. 1996).

PHYSIOLOGICAL ROLE OF PIGMENTS

Natural food pigments are important to our well-being. There are seven major groups found in biological materials. Some play a significant physiological role in mammals, but there is still doubt as to whether others have any such function. Table 1–1 lists the groups, together with their function or suspected function in plants and an indication of their value to the herbivores and carnivores further along the food chain. Animals must obtain vitamins A, B_2, K_1, and K_2 from plants, as they cannot synthesize them.

A large part of current nutritional advice is that we should eat at least five varieties of fruit and vegetables each day. Most of these contain natural pigments evolved to work by chemical action and designed, among other functions, as antioxidants to dispose of free radicals. Hence, the protective function of this group of pigments within plants appears to extend to herbivores that consume them, possibly performing similar functions. The value of a variety of natural food pigments to human diet seems indisputable (Schweitzer 1997).

Table 1–1 Major Organic Plant Pigment Groups and Their Value to Herbivores and Carnivores

Pigment Group	Function in Plants	Value to Herbivores and Carnivores
Betalaines	Pollination? Virus protection?	No physiological function known
Carotenoids	Photosynthesis Pollination	Vital; precursors of vitamin A that cannot be synthesized
Chlorophylls/ Porphyrins	Photosynthesis	Not essential? Hemoglobin can be synthesized
Flavins	Quench molecular oxygen	Vital; vitamin B_2 cannot be synthesized
Flavonoids	Growth control Pollination	Antioxidants
Indoles	Unknown	Melanins can be synthesized
Quinones	Respiratory enzymes	Vital; includes vitamins K_1 and K_2, which cannot be synthesized

Source: Reprinted from J. Hutchings, Color in Plants, Animals and Man, in *Color for Science, Art and Technology*, K. Nassau, ed., pp. 222–246, with permission from Elsevier Science.

FOOD COLOR AND USE OF COLORANTS

Added colorants in foods date from at least 3,700 BP when Egyptians were coloring their candy. Sugar imported into Europe from Alexandria in the twelfth century was colored with madder and kermes, but the use of cochineal probably predates this (McLaren 1983). Awareness of browning would have predated the use of color additives. For an extended shelf life in fresh fruits and vegetables, enzymic browning must be eliminated. Foods depending on enzymic browning for their quality include black tea, dates, prunes, and raisins. Nonenzymic browning provides many of the flavors of baking and cooking.

In the home, we are used to seeing a wide range of colors and used to eating food containing color additives. This has a comparatively recent origin for the bulk of the population. Colors were not commonly available for decorating the home, and many colored foods were not eaten. The "slender meles" of the "poure widewe" of the thirteenth century are described in *Piers Plowman:*

> No win ne dranke she, neyther white ne red:
> Hire bord was served most with white and black.
> Milk and broun break, in which she fond no lack,
> Seinde [singed] bacon, and somtime an ey or twey.

The "white and black" refers to milk and coarse black bread, the general diet being completed with cheese, eggs, and occasionally bacon or fowl.

Among natural colors available to the better-off fifteenth-century British household were (Mead 1967)

- yellow and orange—from saffron ("For hen in broth, colour it with saffron for God's sake!") and egg yolks
- black and red—from blood (when killing a swan, "keep the blood to colour the chaudron") and saunders, from sandalwood (when making gingerbread, "if you will have it red, colour it with saunders enough")
- green—from mint and parsley (for glazing rissoles "with some green thing, parsley or yolks of eggs together, that they be green")
- brown—from bread (when cooking Nombles, "colour it with brown bread") and ginger

These colorants could be blended. *Braun ryalle* ("royal brown") was made with saffron, green leaves, and saunders (Mead 1967).

In Europe, annatto (yellow/orange) and chocolate were in use in the seventeenth century. Europeans were suspicious about the harm that might arise from eating vegetables and fresh or dried fruit, so their consumption was not widespread before the nineteenth century. The Scots, however, were using dried fruits in 1700 to achieve the color of their black buns (Hartley 1954). Although tomatoes arrived in Spain from the New World in the sixteenth century, the red varieties did not become widely used in Europe or America until the twentieth (Tannahill 1988). Colors could be purchased. Tomlinson's butter color, which "does not colour the butter milk," was on the market in the 1880s (Opie 1991).

A number of low-flavored colors may be obtained from botanical sources for use in the home. Table 1–2 contains a list of several that can be dried (Rietz 1961).

Additives have been used to change or restore the appearance of foods in three ways. First, we feel more comfortable if the

Table 1–2 Low-Flavored Colors Obtainable from Botanical Sources (data from Rietz 1961)

Color	Source
Blue	Indigo
	Campeachy
	Violet
	Cudbear
Brown	Caramel
	Catechu
	Walnut hull
	Coffee
Red	Huckleberry
	Alkanet
	Pokeberry
	Mallow
	Carotene
Pink	Heather
Yellow	Safflower
	Turmeric
	Buckthorn berry
	Saffron
	Annatto
Green	Tea
	Chlorophyll
	Clover

food we are eating is the appropriate color. In the United States, a tax of 10 cents per pound was levied on the cost of yellow oleomargarine to avoid unfair competition with butter. To overcome the tax, margarine was sold in the white form, together with a capsule of yellow dye. This was added later in the home (Judd and Wyszecki 1975). Mint-flavored ice cream is white before green coloring is added, and white chocolate is reported to taste less like chocolate than the customary brown product (Dunker 1939). Product color also influences our ability to identify a flavor and to estimate its strength and quality (see Chapter 5).

Second, colors are added for numerous manufacturing reasons:

- to alleviate damage to the appearance caused by processing and to preserve product identity
- to ensure color uniformity in the marketplace of products that naturally vary in color
- to intensify the colors of manufactured foods, such as sauces and soft drinks, and to make colorless foods, such as gelatin-based jelly, more attractive
- to help protect flavor- and light-sensitive vitamins during shelf storage by a sunscreen effect
- to serve as a visual indication of quality (Newsome 1986)

Third, colors are added to dilute more expensive food materials. In the past, pepper was diluted with papaya seed, cocoa with brick dust, coffee with chicory, and ginger with pea flour. Alum and plaster of Paris were used to whiten flour, and flour and chalk to whiten milk (Tannahill 1988). In more recent adulterations, cherry concentrates have been found to contain a grape and beetroot juice mixture, salmon has been replaced by cheaper trout, goat's milk has contained cow's milk, appleless apple juice has been marketed (Watson 1997), and betalaines from pokeberry juice have been used to give wine a more desirable red color (Pasch et al. 1975). In yogurt and milk products, cherry has been replaced by a grape plus beetroot combination, and apricot mixtures have been replaced by peaches and apricot aroma.

Historically, food manufacturers used for their products much of the range of inorganic dyes and colorants then commonly used in paint and wallpaper. Copper salts gave added greenness to canned peas and preserved pickles, and red lead was used to color cayenne pepper and the rind of Gloucester cheese. Dried thorn leaves colored with verdigris were sold as green China tea. Port was colored red with Brazil wood or elderberry husks, and French wine was colored with fuscine, a triphenylmethane dye. Spent leaves of black tea were stiffened with gum, colored with black lead, and resold as black Indian tea (Tannahill 1988). The bright colors of early–nineteenth-century boiled sweets owed their origins to the use of highly poisonous lead, mercury, arsenic, and copper salts. These particular additives were prohibited in British law in 1860. However, the range of colors and dyestuffs available to the world blossomed with the discovery in 1856 of the first aniline dye. By 1925, at least 90 of these colorants could be found in foods.

Edward Abbott, who wrote Australia's first cookery book in 1864, showed concern at adulteration, advising readers to grind their own flour, brew their own port, and avoid bright-colored peppers, spices, sauces, anchovies, herrings, and green pickles. Also to be avoided were colored confections, especially those green, blue, or red, since the presence of poisonous salts could be expected (Symons 1982).

Early legislation was restricted to lists of prohibited ingredients. No permission or clearance was required for other materials. The present positive rules permitting use of specific additives were laid down in 1965. For permission to be gained for a proposed additive, not only must data demonstrating its safety in use be provided, but also the need for it must be demonstrated.

The process of freshening up the appearance of raw vegetables by spraying them with water is a common sight in the warmer countries, such as those lining the Mediterranean Sea. Although this does conveniently add weight to the produce, its main purpose is to restore turgor pressure lost through evaporation. The appearance of older fish can also be thus freshened. This practice was common, for example, in medieval Venice, but it is now prohibited (Tannahill 1988).

In some parts of the world, there has been a marked change in consumer attitudes to additives. In Europe, an initial view that additives were "bad" changed to "E- (for Europe) numbered additives are safe to eat." Many customers checked packs to ensure that additives possessed E numbers. Subsequently, however, attitudes changed, and E numbers are now looked upon with suspicion. Products are being rejected simply because E numbers are listed among the ingredients. A legislative device intended to be a safety signal for consumers has become an alarm bell. Large and intensely competitive food manufacturers and supermarket chains are, under this pressure, tending to market products containing the minimum additives deemed necessary to produce food to fit their marketing strategies. They are excluding colorants that, although permitted, are receiving adverse publicity. Negative advertising is tending to replace positive advertising. This draws the purchaser's attention away from other ingredients to what is not there and consequently to what is inconsequential. Consumer concern about colorants has led to the wider availability of colorings from natural sources or colorings manufactured to be molecularly identical to natural pigments. Some natural as well as synthetic colorants have been linked with adverse reactions in some people. In the generally safe area of processed foods, concern for food intolerance is probably now greater than that for food safety.

Although artificial colorants are used in staple vegetable, fruit, and meat meal products, they must be used with discretion. This is because we are sensitive to the color-versus-quality relationship. If a food presented in a natural form is not the color expected, it may well be seen as a product of lower quality. The situation changes dramatically with nonstaple foods. Existence of artificial colors allows us to have fun with food. It seems that we can accept impulse desserts and sweets of almost any color. Even named-flavor ice creams can be given quite outrageous colors, yet still be accepted as tasting of that flavor. An example is to be found in strawberry ice cream, which can range from orange brown through pink to bright purple. Orange desserts that have an intense, almost fluorescent, color are eaten without reservation. For these product types, there appear to be few constraints for the color to match closely that normally associated with the flavor.

Although not a color existing in natural food, even blue is now accepted. In the late nineteenth and early twentieth centuries, blue was a warning color through its widespread use for medicine bottles. An exception was wedding cake decoration, for which blue represented good luck. The medical use of blue has ceased, and with it blue's warning connotations. Blue is now available for use in sweets and ice cream, acting as a dark contrast color for milk products.

Attempts to omit color additives do not always meet with success. A possibly apocryphal story circulating in Sweden in the early 1980s concerned colorants omitted from orange drink. Because cashiers at self-service restaurants could not tell the difference between the orange drink and plain water, colorants had to be reintroduced. Another colorant omission concerned "clear" colas. This was also unsuccessful in spite of its appeal to those with increasing fears of additives (Cardello 1993).

Some colors are unwanted and are bleached out to improve product appearance. Salted fish darkens with time. The darkening and any blood stains that happen to be present can be bleached out using hydrogen peroxide. Similarly, marinated herring, tripe, and in-shell walnuts are lightened (McNeillie and Wetmur 1994).

Products have their appearance enhanced by means other than color. For example, a brand of Polish vodka contains small pieces of gold flake to enliven the look of the otherwise clear drink. Mexican mezcal tequila contains worms. When the bottle is passed round a group of drinkers, the person having the last mouthful, including the worms, is regarded as "manly."

Manufacturers are governed by food additive legislation of their own country and the countries to which they are exporting their products. Only those colorants listed in the country of consumption may be used, and within this framework manufacturers are free to use them as they wish. However, this can lead to ethical considerations not specifically included in statutory regulations. One of these has been highlighted by Francis and Clydesdale (1970):

> The degrading effect of ascorbic acid on anthocyanin pigments is a well-known phenomenon and constitutes a dilemma for a processor. For product stability

reasons, he would like to leave out the added ascorbic acid. However, the nutritionists have spent many years attempting to educate the public that fruit juices are a good source of vitamin C (ascorbic acid). In the opinion of the authors, it's not professionally ethical to market fruit juices as breakfast drinks when they contain little or no ascorbic acid. This constitutes a deliberate undermining of good nutritional education. An even worse practice is to fortify fruit juices with riboflavin and thiamine and label them "fortified." The public has come to accept the word as meaning fortified with vitamin C, and sometimes the fine print is difficult to read, particularly in a supermarket.

Other ethical problems have also been noted. At one time, yellow additive was used in cake products to give the impression of a greater egg yolk content. The manufacturer may logically argue that differences in yellowness can be caused by some yolks' being a stronger color than others or that it is an unfounded assumption that the consumer is concerned about the inclusion of eggs anyway. As in the fruit juice example, we may ask if the manufacturer is deliberately trying to produce in the consumer's mind an impression about the product that the manufacturer knows is patently not true. After all, egg yolks contain nutrients that are essential for our well-being. In addition, what suspicions do consumers put into the back of their minds in the hope that others are looking after their welfare? How much is the manufacturer consciously or subconsciously relying on this?

An ethical question bordering on fraud concerns the presentation of meat. No single lighting regime is optimum for all foods. Red meat can be made to appear redder by illuminating it with pink or "warm" light or by placing a red refelctor near the display. The effect is short lived, as the true state of the surface pigments is revealed when the meat is unpacked at home (Francis 1995).

Products allegedly targeted at children include "Alcopops." These are sugar drinks, some brightly colored, containing alcohol and marketed in brightly colored bottles painted with cartoon characters or portraying someone who could be under 18 years old. Under public pressure, some have been withdrawn from the supermarket shelves. If these products have led to an increase in alcohol abuse by children, who is to blame? Perhaps the designer of the bottle label for not behaving professionally, or perhaps the directors of the companies producing these products for unethically filling this "market niche."

The science of labeling, informing, and understanding the thoughts of purchasers in an age when they personally prepare perhaps only a very small part of their diet is woefully incomplete.

OTHER APPEARANCE ATTRIBUTES

Translucency

A translucent material is one that both transmits and scatters light. As a phenomenon, translucency occurs between the extremes of transparency and opaqueness. Transparency or clarity occurs when there is no visually apparent light scattering. It is a property greatly sought by many processors. Ancient Egyptians used layers of stones and shellfish shells, and wool and cotton fibers have long been used as filters to clarify water and drinks. The haze in spirits, wines, and beers normally leads to their rejection. However, in Germany and Belgium, there is a preference for the whitish-yellow hazy wheat beer, wheat lacking the husks that create a natural filter (Jackson 1996). Anisette-based drinks are clear when poured from the bottle, but when water is added, they turn white. For their flavor, these drinks rely on aromatic terpines, which are soluble in alcohol but not in water. When water is added, they are forced out of solution to form a milky haze. Addition of milk to coffee or tea decreases clarity and increases opacity.

Gelatin-based jellies have greater visual impact if they are clear. Product identity can depend on the extent of the clarity. Clearer orange drinks can be seen as refreshing, and probably containing artificial colorant and are more likely to be appreciated by younger people. More opaque, lighter-colored fruit drinks, on the other hand, are often regarded as health-giving breakfast beverages obtained from real fruit and are more likely to be appreciated by adults. Turbidity in fruit juices can be a positive or a negative attribute depending on the expectation of the consumer. In origin, hazes are biological, caused by growth of microorganisms, or chemical, caused by physicochemical instability and the coagulation of particles that tend to form a sludge. Opalescent apple juice contains a stable cloud of soluble as well as insoluble pectin-stabilized particles.

Transparency is destroyed when light scattering occurs. This is largely controlled by particle or cell size (see Chapter 3). Fish flesh provides an example. The color of raw salmon flesh is a translucent deep pink red that, on cooking, turns a more opaque light pink. The path length of light in the raw fish is greater because the beam penetrates farther. There is, therefore, greater opportunity for selective absorption of the incident light to take place. As a result, a deeper color is seen. However, the sarcoplasmic proteins that precipitate on cooking do not allow such deep penetration. Light is scattered back to the viewer, who sees a much lighter, paler color. The albumin of fresh duck eggs develops into a turbid gel, but immersion of the eggs in sodium hydroxide and salt results in an increase in transparency (Su and Lin 1993).

The color of green vegetables immediately changes when they are plunged into boiling water. The main reason for this is that water replaces air in cells just beneath the outer layers of the structure. This reduces light scattering, and the color appears a deeper green.

Much of the cooking process is aimed at changing food texture through a change in structure. Protopectin is solubilized, cellulose is softened, starch takes up moisture, and proteins are precipitated. All these factors contribute to the perceived color, mainly through changing the translucency. Light-scattering phenomena also contribute to powder reflectance properties, such that color changes with particle size.

Gloss

A wide range of gloss occurs in foods, but extremes are rare. Gloss is symptomatic of production method in yellow spreads, some of which achieve a relatively high gloss. Gloss differences help us to distinguish between components of mixtures, as in the case of fat floating on hot gravy. Lack of gloss in chocolate is symptomatic of the state of crystallization and granularity. Glossiness of moist surfaces such as fish reinforces perceptions of freshness, but only if there is sufficient directional light.

Waxy layers and deposits are components of the surface cuticle tissue of many fruits. As well as preventing fluid loss, their natural gloss reduces penetration and damage from visible and ultraviolet light; shaded plants rarely possess surface gloss (Barber and Jackson 1957). Different varieties of apples exhibit different levels of gloss. Glossy surfaces look attractive. Hence, waxes designed to reduce gas exchange, weight loss, and fungal growth are also designed to be glossy. These include most fruit waxes, such as carnuba wax, polyethylene wax, beeswax rice wax, and shellac (Kaplan 1986).

Gloss or *glaze* has three meanings in the kitchen: to finish vegetables glacé; to color under a salamander, as with fish in a sauce; and to coat with a jam or fruit purée, as with a flan or tart. Dried foods and baked foods tend to be matte if not deliberately made shiny with perhaps a sugar glaze. Freshly cooked vegetables are coated with butter so that their shininess is preserved while the meal is being eaten. Sweet or savory aspic glazes are used to decorate a wide range of foods; in Spain, these include the succulent suckling pig for the table.

Structure, Uniformity, and Pattern

Structural patterns in natural products are characteristic of a food and are among the means whereby it is recognized. Patterns are controlled by optimization of biological and biochemical properties of growth, nutrient transfer, and physical strength. The presence of color patterns in, for example, older varieties of apple is governed by principles of natural evolution. These include optimization of reproduction and seed dispersal properties. There are regional preferences for apple color pattern: Asian markets prefer striped apples, North Americans tend toward blushed, and Europeans tend toward green. Color patterns in modern varieties are controlled by selective breeding, perhaps for properties other than appearance.

In composite natural foods, we demand the correct proportion of each element. For example, for greater acceptability, lamb chops cut for display must be trimmed to optimum lean-to-fat-to-bone ratios (Jeremiah et al. 1993). Significantly more beefsteaks and roasts are purchased when fat thickness is less than 8 mm (Savell et al. 1989). In composite manufactured foods, such as pizza, a complex surface pattern of ingredients is possible. However, the appearance of each constituent of the matrix must fulfill the criteria of freshness or goodness normal to that constituent. Very precise and regular patterning tends to give the impression that the product has been manufactured by machine. The artisan look of a calculated randomness leads to a more homemade appearance. Similar evidence of hand making in decorated chocolates is also welcome.

Perfection of appearance is the shopper's main criterion for selection of fruit and vegetables in the supermarket. For example, for wholesale and retail markets, it is desirable that apples in a box be presented in the same way and look as similar as possible. Our preoccupation with appearance has lead to the rigorous standardization of size and shape and a reduction in the number of varieties on sale. Items of imperfect appearance are restricted to "organic" produce.

Although there are restrictions to the colors appearing on our plate, there is no restriction to shape. We can carve, circle, dice, strip, or bunch vegetables to increase perceived variability.

RELATIVE IMPORTANCE OF APPEARANCE, FLAVOR, AND TEXTURE AS PRODUCT ATTRIBUTES

Color is all-important in the survival sense. An inappropriately colored food, such as green meat or immature green fruit, is rejected. In the store, 37 percent of meat purchases from self-service counters are made on impulse, primarily on the basis of an attractive color (Nelson 1964). Appearance, particularly gloss, is the most significant factor in determining the choice of confection at the point of sale (Musser 1973). The most annoying feature of a savory sauce was that it was "too oily looking" (Moskowitz 1995).

Can these observations be reconciled with the results of studies that conclude that flavor and texture are more important than color as product attributes? Szczesniak and Kahn (1971) used word association tests to determine the relative awareness of all food attributes. They concluded that form and color occupied a rather poor third and fourth place respectively to flavor and texture. Hamilton and Bennett (1983), in an investigation to determine the relative importance of sensory attributes for a selection of fresh white fish, found that flavor and texture were the most important and appearance the least. Tuorila-Ollikainen et al. (1984) found that sucrose concentration and the presence of a fruity flavor in soft drinks significantly influenced overall liking, whereas color had little effect.

The situations of survival and modern everyday eating are different. The survival sense is needed today in the initial selection of food. We use our eyes when buying or harvesting. Even though flavor and texture are important, we cannot taste everything before we buy. At these times, sight acts as the major screening sense. When we get the food home, the sample we have purchased is not in competition with others. We use the eyes differently. Their task of screening has largely finished, and they are now directed toward monitoring what has been purchased during its preparation and presentation for eating. In the taste panel situation, we trust the organizers to give us wholesome samples that will not do us harm, so the visual screening sense is not necessary. The in-mouth enjoyment and nutritional benefit of a food are in the eating, and once it has passed the test of acceptance by the visual sense, what it looks like seems irrelevant. Once the food has been judged safe to eat, it can be enjoyed through the greater warmth and intimacy of the flavor and texture contact senses. Appearance affects in-mouth response, but the appearance itself becomes less important.

Further insight into this question can be gained from the extensive work of Schutz and Wahl (1981). They determined "the perceived relative contribution of appearance, flavor and texture to the overall acceptability of the food while it is being eaten"—that is, not during selection, purchasing, or preparation. They used questionnaires with the names of foods acting as stimuli. The authors pointed out that the results from this type of study may differ considerably from one in which actual physical samples of the food are used. However, this study has laid the groundwork for any future such investigation. Ninety-four foods were included, and 420 valid responses were obtained from adults living in the greater metropolitan area of Sacramento, California. Foods included were 20 types of meat and fish, 29 vegetables, 14 fruits, 16 breads and cereals, 5 dairy products, and 10 miscellaneous foods.

The results of this study were that flavor ranked as the most important attribute for all foods, but the relative importance of appearance and texture varied from food to food. Fifty-one products (54 percent) received higher scores for appearance than for texture. The 10 highest scores for appearance were obtained for pimientos, iceberg lettuce, cream puffs, blueberries, radishes, cherry-flavored gelatin, mushrooms, strawberries, coffeecake, and orange marmalade. This list included the brightly colored fruits and vegetables. The authors commented that "perhaps appearance is especially important for this group as an indicator of quality and enters in the subsequent decision to accept or reject, or it may be that the more colorful aggressive appearance demands more attention." Shewfelt et al. (1986) confirmed this conclusion in finding appearance more important than texture in distinguishing between fresh snap beans of high and intermediate price groups. Similarly, visual cues of apple skin color are greater drivers of perceived apple ripeness than tactile cues (Richardson-Harman et al. 1998).

In the Schutz and Wahl (1981) survey, the 10 lowest appearance scores were given to soy sauce, sunflower seeds, lemon juice, cola soft drink, cheddar cheese, apricot nectar, coffee, beer, barley, and barbecue sauce. This group consists primarily of liquids for which "flavor seems to overshadow all other attributes." Taken overall, appearance received significantly higher scores than texture for fruit, dairy, and miscellaneous groups. The reverse was the case for the bread and cereals group, for which judgment of freshness relied primarily on the assessment of texture.

Demographic analysis indicated that there were no significant differences for appearance among the various educational levels or among age groups of respondents. There was some indication, however, that those with a Mexican food background regarded appearance as less important, while those having a Chinese background regarded appearance as rather more important than other racial groups. Yoshida (1981) has studied the responses of Japanese consumers to a wide range of foods. In agreement with Schutz and Wahl, he found that color and appearance were highly desirable attributes of fruits, vegetables, and cakes. They were less important in condiments and convenience foods.

Color plays a dominant role in influencing the purchase of yogurt-based desserts, perhaps through the association of a high fruit content and color intensity (Ulberth et al. 1993). Color affects intake also. Flavored yogurts were used in an experiment in which color and flavor were compared as competing cues during consumption. Two sets of three samples were used. The first set of yogurts, of different colors, consisted of hazelnut, orange, and black currant flavors. The second set, cherry, raspberry, and strawberry, were the same color. No peripheral visual cues were present that could give a clue to the flavor. Each of the three flavors in any one set was presented to subjects in succession. Each test consisted of three 10-minute periods. At the start of each period, the subject was given a fresh pot of yogurt. In the first period, there was no difference in intake for any of the conditions. When variety was introduced in the subsequent two periods, intake was lower if the same flavor was given than if a different flavor was introduced. Total intake was increased by 19.5 percent in the variety condition. In the test involving the second set of similarly colored samples, there was no such increase in consumption when variety was introduced. This occurred even though the tastes were distinctive and the subjects could tell what flavors they were eating. Hence, it appears that color can influence intake (Rolls et al. 1982).

Different attributes of appearance may also assume different degrees of importance to the look of a product. For example, *Darkness, Shininess, Size* (area), and *Thickness* were found to govern the *Appearance Liking* of a series of chocolates (Moskowitz 1985). Each of these attributes assumes criticality if the product is deficient in that attribute. For example, if the chocolate is too thick, the *Thickness* will be the cause for rejection irrespective of the other attributes. If the *Thickness* is then brought within the bounds that the consumer sees as reasonable for the product, another of the attributes may lead to rejection. That is, these properties form a hierarchy.

It is more difficult to sell a customer a badly colored product than to sell him or her a badly flavored product because the in-mouth quality may not be discovered until the food reaches home. Itinerant salespeople take advantage of this principle. However, this also occurs in supermarkets, which tend to display perfect-looking fresh foods at the expense of varieties that may have better flavor but poorer appearance. One might postulate that if these studies had been carried out before the advent of modern methods of production and supply, a different result may have been obtained. Similarly, different conclusions might have been reached if the tests cited at the beginning of this section had been carried out at the point of supply—for example, in the supermarket or on the fishing trawler. The importance of appearance might then have been found to be paramount. Color catches the eye, directs attention, and induces expectation.

Some confirmation of these observations on the relative importance of the senses may be obtained from studies on primates. The mere sight of food (without odor) is not sufficient to influence the flow rate or amylase content of saliva (Steiner et al. 1977). Pictures of food result in no increase in saliva flow, but the sight of abstract pictures results in a lower flow rate (Birnbaum et al. 1974). However, the amount of salivation in humans increases when subjects are deprived of food and then faced with palatable food (Wooley and Wooley 1973). In agreement with these observations, neurons that respond when a monkey sees its favorite foods are not activated by smell, touch, or eating in the dark, nor are they associated with salivation. This response occurs only when the monkey is hungry; it decreases as the monkey approaches satiety but is revived at the sight of another food. This also occurs with humans, for whom the pleasantness of a particular food normally decreases as eating progresses. The response is clearly an important factor in food selection (Rolls et al. 1982). These studies indicate the independence of operation of the appearance sense. Hence, appearance governs flavor expectation as a result of experience as well as by saliva release. Once the food is in the mouth, the fact that it looked especially good does not dominate over the opinion of goodness of flavor. If the taste of a food is poor, a good appearance will not improve it.

The Colour Society of Australia organized an eating seminar lunch as part of its 10th anniversary celebrations at Edith Cowan University of Western Australia at Perth in 1997. This consisted of white foods only for soup, main course, and dessert. Interviews with diners revealed that many had felt very uncomfortable during the meal because the choice between food identification and conversation had to be made. Although a wide variety of foods were available, some diners found the occasion "boring" because of the lack of color. Chef John O'Connor stated that it was the most depressing meal he had ever created. During a meal, color is used as the primary identifier of food. For equally colored dishes, the eater is driven to using texture, while for reduced foods, flavor is the arbiter. The greater concentration required for this can be annoying and disturbing.

Appearance provides the initial definition of the product. It provides a state of expectation, a mental image of what the in-mouth properties should be. If a food cannot be recognized, it is the maker's word as to what the product is. When the visually produced image is that the food is uniformly "good," appearance becomes less important. If the lights are out, we must wait for the generally slower processes of texture and flavor to provide product identification. In doing so, we eat without the visual confirmation that the product is wholesome.

REFERENCES

Allen DE. (1968). *British tastes*. London: Hutchinson.

Barber HN, Jackson WD. (1957). Natural selection in action in eucalyptus. *Nature* 179:1267–1269.

Birnbaum D, Steiner JE, Karmell F, Ilsar M. (1974). Effect of visual stimuli on human salivary secretion. *Psychophysiology* 11:288–293.

Blythman J. (1992). And how would madam like it cooked? *Independent*, 11 July, 35.

Boon DD. (1977). Coloration in bivalves. *J Food Sci* 42:1008–1012, 1015.

Campo JL. (1995). Relationship between shell colour and compositional characteristics in brown, tinted or white eggs. *Arch Gefluegelkunde* 59:161–164.

Cardello AV. (1993). New product trends: sense or nonsense? *Cereal Foods World* 38:870–871.

Cosman MP. (1984). *Medieval holidays and festivals*. London: Judith Piatkus.

Cousminer J, Hartman G. (1996). Understanding America's regional taste preferences. *Food Technol* 50(7):73–77.

Cox RJ, Thompson JM, Cunial CM, Winter S, Gordon AJ. (1997). The effect of degree of doneness of beef steaks on consumer acceptability of meals in restaurants. *Meat Sci* 45:75–85.

Dunker K. (1939). The influence of past experience upon perceptual properties. *Am J Psychol* 52:255–265.

Francis FJ. (1995). Quality as influenced by color. *Food Qual Pref* 6:149–155.

Francis FJ, Clydesdale FM. (1970). Color measurement of foods: cranberry products. *Food Prod Dev* 4(2):56–60.

Goldsworthy A. (1987). Why trees are green. *New Sci* 116(1590):48–51.

Hamilton M, Bennett R. (1983). An investigation into consumer preferences for nine fresh white fish species and the sensory attributes which determine acceptability. *J Food Technol* 18:75–84.

Hartley D. (1954). *Food in England*. London: Macdonald.

Hill AR, Stevenson RWW. (1976). Long term adaptation to ophthalmic tinted lenses. *Mod Prob Ophthalmol* 17:264–272.

Hutchings J. (1997). Folklore and symbolism of green. *Folklore* 108:55–63.

Hutchings J. (1998). Color in plants, animals and man. In *Color for science, art and technology*, ed. K Nassau, 222–246. Amsterdam: Elsevier.

Hutchings J, Akita M, Yoshida N, Twilley G. (1996). *Colour in folklore with particular reference to Japan, Britain and rice*. London: Folklore Society.

Jackson M. (1996). A pint of cloudy please. *Independent*, 25 May, 41.

Jeremiah LE, Gibson LL, Tong AKW. (1993). Retail acceptability of lamb as influenced by gender and slaughter weight. *Food Res Int* 26:115–118.

Judd DB, Wyszecki G. (1975). *Color in business, science and industry*, 3rd ed. New York: John Wiley.

Kaplan HJ. (1986). Washing, waxing and colour adding. In *Fresh citrus fruits*, ed. WF Wardowski, S Nagy, W Grierson, 379–395. New York: Van Nostrand.

King S. (1979). Presentation and the choice of food. In *Nutrition and life styles*, ed. M Turner, 67–78. London: Applied Science.

Lyman B. (1989). *A psychology of food: more than a matter of taste.* New York: Van Nostrand Reinhold.

McLaren K. (1983). *The colour science of dyes and pigments.* London: Hilger.

McNeillie A, Wetmur K. (1994). Lighten up. *Food Processing* 55(11):120–122.

Mead WE. (1967). *The English medieval feast.* London: Allen & Unwin.

Meiselman HL, Wyant KW. (1981). Food preferences and flavor experiences. In *Criteria of Food Acceptance Symposium proceedings,* ed. J Solms, RL Hall, 144–152. Zurich: Forster.

Mollon JD. (1989). "Tho she kneel'd in that place where they grew . . .": the uses and origins of primate colour vision. *J Exp Biol* 146:21–39.

Moskowitz HR. (1985). *New directions for product testing and sensory analysis of foods.* Westport, Conn: Food and Nutrition Press.

Moskowitz HR. (1995). Food quality: conceptual and sensory aspects. *Food Qual Pref* 6:157–162.

Muntz WBA. (1975). Visual pigments in the environment. In *Vision in fishes,* ed. MM Ali, 565–578. New York: Plenum.

Musser JC. (1973). Gloss on chocolate and confectionery coatings. In *Proceedings of the 27th PMCA Production Conference,* 46–50.

Nelson P. (1964). Set-up seen in five years on central meat cutting. *Supermarket News* 13:48.

Neri DF, Luria SM, Kobus DA. (1986). The detection of various color combinations under different chromatic ambient illuminations. *Aviation Space Environ Med* 57:555–560.

Newall VJ. (1987). The adaptation of folklore and tradition (folklorismus). *Folklore* 98:131–151.

Newall VJ. (1991). Colour in traditional Easter egg decoration. In *Colour and appearance in folklore,* ed. J Hutchings, J Wood, 36–39. London: Folklore Society.

Newsome RL, ed. (1986). Food colours. *Food Technol* 40(7):49–56.

Opie R. (1991). An exhibit in the Museum of Advertising and Packaging, Gloucester, UK.

Pasch JH, von Elbe JH, Dinesen N. (1975). Betalaines as natural food colorants. *Food Prod Dev* 9 (Nov):38, 42, 45.

Richardson-Harman N, Phelps T, McDermott S, Gunson A. (1998). Use of tactile and visual cues in consumer judgments of apple ripeness. *J Sensory Stud* 13:121–132.

Rietz CA. (1961). *A guide to the selection, combination and cooking of foods.* Westport, Conn: Avi.

Rolls BJ, Rolls ET, Rowe EA. (1982). The influence of variety on human food selection and intake. In *The psychobiology of human food selection,* ed. LM Barker, 101–125. Westport, Conn: Avi.

Savell JW, Cross HR, Francis JJ, Wise JW, Hale DS, Wilkes DL, Smith GC. (1989). National consumer retail beef study: interaction of trim level, price and grade on consumer acceptance of beef steaks and roasts. *J Food Qual* 12:251–274.

Scanlon BA. (1985). Race differences in selection of cheese color. *Percept Motor Skills* 61:314.

Schutz HG, Wahl OL. (1981). Consumer perception of the relative importance of appearance, flavor and texture to food acceptance. In *Criteria of Food Acceptance Symposium proceedings,* ed. J Solms, RL Hall, 97–116. Zurich: Forster.

Schweitzer C. (1997). Health benefits of natural mixed carotenoids. *Food Technol Eur* 4(4):24–28.

Shewfelt RL, Resurreccion AVA, Jordan JL. (1986). Quality characteristics of fresh snap beans in different price categories. *J Food Qual* 9:77–78.

Steenkamp J-BEM. (1986). Perceived quality of food products and its relationship to consumer preferences, theory and measurement. *J Food Qual* 9:373–386.

Steiner JE, Konijn AM, Ackermann R. (1977). Human salivary secretion in response to food-related stimuli. *Isr J Med Sci* 13:545.

Su H-P, Lin C-W. (1993). A new process for preparing transparent alkalised duck egg and its quality. *J Sci Food Agric* 61:117–120.

Symons M. (1982). *One continuous picnic.* Adelaide: Duck Press.

Szczesniak A, Kahn EL. (1971). Consumer awareness of and attitudes to food texture. *J Texture Stud* 2:280–295.

Tannahill R. (1988). *Food in history.* London: Penguin.

Tuorila-Ollikainen H, Mahlamaki-Kultanen S, Kurkela R. (1984). Relative importance of colour, fruity flavour and sweetness in the overall liking of soft drinks. *J Food Sci* 49:1598–1600, 1603.

Ulberth F, Kneifel W, Schaffer E. (1993). Colour intensity preferences observed with selective fruit yogurts. *Milchwissenschaft* 48:15–17.

Walls GL. (1963). *The vertebrate eye and its adaptive radiation.* New York: Hafner.

Watson A. (1997). Good food costs you more. *Food Processing* 66(6):19–20.

Weismantel MJ. (1988). *Food, gender and poverty in the Equadorian Andes.* Philadelphia: University of Pennsylvania Press.

Wooley SC, Wooley OW. (1973). Salivation to the sight and thought of food: a new measure of appetite. *Psychosom Med* 35:136–142.

Yoshida M. (1981). Trends in international and Japanese food consumption and desirable attributes of foods as assessed by Japanese consumers. In *Criteria of Food Acceptance Symposium Proceedings,* ed. J Solms, RL Hall, 117–137. Zurich: Forster.

CHAPTER 2

Appearance of Objects and Scenes: The Information Transfer Process

All visually perceived scenes result in appearance images. Based upon these images, judgments are made, expectations are formed, and actions are undertaken. Appearance of people, products, packages, advertisements, and retail spaces is assuming ever-greater importance in daily life, yet the science controlling the formation of such images has received little attention. Marketing success relies upon creating in potential customers appropriate positive images of goods and services, especially in competitive situations.

The food technologist is under pressure to increase the understanding of materials used in product development and processing. Marketed are ever-increasing numbers of novelty snacks, low-calorie, and fat/alcohol/caffeine/sugar/starch-free foods, and products of greater sophistication. Demand is increasing for raw material upgrading, high and consistent standards of produce, and additive reduction. The opening of wider and wider markets has exposed a lack of understanding of the ways and lives of potential customers. Factors such as culture and climate are assuming greater importance for the producer and manufacturer. Such factors, as well as the individual's state of well-being, make up the *Total Appearance* of the scene and influence the customer's decision to buy. Images and expectations ought to be under control.

Product design relies upon manipulation of the material physics of the design elements. Thus, an understanding of appearance science will allow us not only to improve the performance of products directly in the marketplace but also to use effectively the growing techniques for judging product and service effectiveness via direct observation, computer simulation, and virtual reality. The *Information Transfer Process* model describes the route from product or scene to appearance images, together with the methodology for analysis of the route.

INFORMATION TRANSFER PROCESS: AN OVERVIEW

The *Information Transfer Process* model provides methods for studying and understanding the formation of appearance images relevant to a wide range of situations. These include the product, the package, the pack in the freezer and the store, and the design of the restaurant and supermarket—in fact, all appearance images, however they arise.

When we scan the contents of the plate of food in front of us or look into a freezer cabinet, we form appearance images. Some of these images attract us favorably; others drive us away. The *Information Transfer Process* model traces the ways in which images are created from the scene physics by the product designer or manufacturer and contains the rules by which effective transfer can take place.

Using this model, we can examine the factors affecting *Total Appearance* images, we can use *Appearance Profile Analysis* to formally assess images of the scene, and we can attempt to link the scene's material properties with *Total Appearance* images. *Total Appearance* is discussed in this chapter; *Appearance Profile Analysis*, a disciplined method of analysis of the scene, is described in Chapter 7; and, for specific populations, examples of links of material properties to images are given in Chapters 8 and 9.

The scene in view can be described in terms of the physics of the elements of the scene and the way the elements have been assembled—that is, the design. The scene physics and design working together contribute to the stimulus, which is converted into appearance images in the brain of the perceiver of the scene. There are two broad types of image: *basic perceptions*, such as size and shape, and *derived perceptions* (or *visual expectations*), such as creaminess and value. Thus, in the *Information Transfer*

Process, developers/manufacturers manipulate material properties to create a scene consisting of scene physics. This results in the viewer's *basic perceptions of the scene* and *derived perceptions of the scene.*

A product or scene possesses *physics properties,* which can be summarized as *spatial properties* of dimension; *spectral properties* dependent on wavelength of light reflected or transmitted; *goniophotometric properties* dependent on angles of illumination and viewing; and *temporal properties* dependent on movement and time. When the product is viewed under *illumination,* which itself can be defined, it results in the two types of perception. *Basic perceptions* are of size, shape, surface texture, color, translucency, gloss, and the uniformity of such qualities. *Derived perceptions,* formed through repeated eating experience, make up *visual expectations* such as *visually assessed identification, visually assessed flavor, visually assessed texture,* and *visually assessed satisfaction. Derived perceptions* and *visual expectations* include perceptions of a food as a "dessert," as "lemony," as "tough," or as "filling." The extent to which these expectations are subsequently confirmed or disconfirmed can have a profound effect on acceptance (Cardello 1994).

When we view a scene, our images are normally the gestalt *derived perceptions.* Nevertheless, they are linked with the *basic perceptions* through the specific properties of the individual visual mechanisms. The derivation of these perceptions and images can be called *Total Appearance,* a model of the analysis and derivation of a personal opinion of quality incorporating all visually perceived information concerning the product, its situation, and its environment. The *Total Appearance* model includes consideration of appearance images, what they are, how they arise, how they can be measured, and how they can be manipulated. It can be applied to any situation in which the individual may find him- or herself, but applications described are confined to foods. This chapter describes the place of *Total Appearance* within perception as a whole, then considers the factors composing it within the context of the *Information Transfer Process.*

TOTAL APPEARANCE—HISTORY OF RESEARCH

The first detailed consideration of appearance within perception as a whole came from food industry workers. The sensory properties of food can be depicted in the form of a circle, the perimeter of which is divided into three zones defined by the major senses (see Figure 2–1).

Although the senses may be regarded as individual and separate, the perceptions to which they give rise are not. Some properties are perceived in different ways by more than one sense. Also, individual attributes can arise from combinations of signals from different senses. Two regions of overlap concern *mouth-feel,* on the kinesthetics/flavor boundary, and *visually assessed viscosity* or *visually assessed texture* on the appearance/ kinesthetics boundary. The third region of overlap is *flavor ex-*

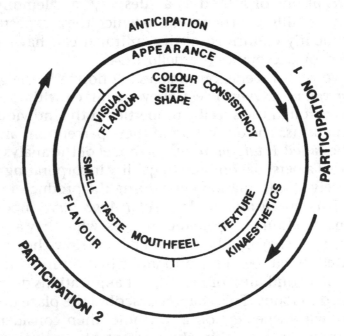

Figure 2–1 The Modified Kramer Circle (following Kramer and Szczesniak 1973, via Hutchings 1977, with permission).

pectation. Product sensory properties form a continuum ordered according to perception and method of perception.

Product attributes can be different for each perceiver, each product perceived, and each situation in which the product is perceived. Thus, product attributes can be divided into groups according to time of perception. One has a basis of expectation, i.e., *anticipatory,* and two, both *participatory,* have their basis in experience. The anticipatory group consists of the visually assessed flavor, visually assessed texture, appearance sector perceptions, and smell. The second stage, when the food is handled, the anticipatory and the first participation properties are combined. The latter include peripheral tactile cues such as cutting. At the last stage the in-mouth, later participatory group of attributes, come to the fore. Again, these groupings are not isolated and individual. As consumption progresses, there is a continual feedback of information. This gives a running total impression of the product. Whatever the comparative importance of these groupings may be, they do occur in order, and those judgments formed during the earlier stages can influence those made later (Hutchings 1977). Over the period of appreciation, the importance attached to the appearance attributes declines, and flavor becomes more influential on the judgment of quality and preference (see Chapter 1).

Sources of the problems involved when attempting to calculate or predict gestalt appearance responses and images are depicted in Figure 2–2. This outlines the makeup of a product or scene and its perception.

The *structure* of the scene may be described in terms of molecules arranged in particular geometries in space. The structure related to its illumination and environment provides the *stimulus.* The stimulus is modified by retinal and neural characteristics into the *appearance response,* which can be defined by critical analysis. Appearance response is converted, via the viewer's temperamental factors, into consumer images, quality judgments, and *preferences.* Although the structure and the stimulus may be physically handled and "measured," the later stages in the above sequence cannot. There is some understanding of the neural and temperamental factors associated with appear-

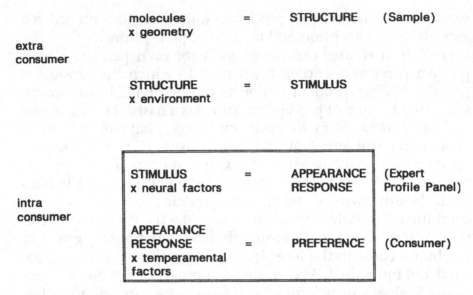

molecules   =   STRUCTURE   (Sample)
x geometry

extra
consumer

STRUCTURE   =   STIMULUS
x environment

intra
consumer

STIMULUS   =   APPEARANCE   (Expert
x neural factors   RESPONSE   Profile Panel)

APPEARANCE
RESPONSE   =   PREFERENCE   (Consumer)
x temperamental
factors

Figure 2–2 The Sample and Its Appraisal (following Drake 1968, Hutchings and Lillford 1988, with permission).

ance and texture, for which the original version of Figure 2–2 was suggested. Normally, however, not enough is known about them to allow safe predictions of preference to be made from a physical specification of structure.

Total Appearance was initially posited to consist of three groups of attributes: optical properties, physical form, and mode of presentation (Hutchings 1977), and attributes composing these groups were seen as combining in different ways to result in three groups of images: those produced when the product is on the shelf, while it is in preparation, and when it is on the table. This schema survived in an industrial food situation for some time. However, in world of larger and more diverse markets, it has been essential to extend the treatment to include a fuller appreciation of the product and the human factors involved in its perception.

Influencing the progress toward the present concept was the independent search initiated by the Inter Society Color Council Project Committee 33. A "list of factors believed to influence the human response to colour" was sought (Granville 1986). These factors were formulated into *Total Appearance* theory, a logical theory of the derivation of perceived appearance images (Hutchings 1988). According to this theory such images are constructed from the lighting of the scene, physically definable object properties, the perceiver's visual and other receptor mechanisms, the perceiver's inherited and learned responses, and the perceiver's immediate environmental factors.

This chapter discusses *Total Appearance* theory in terms of specific applications to areas within the food industry. However, the theory is relevant to all industries and to all human behavior patterns for which appearance images are important, applying to scenes that range from a piece of carrot to a piece of architecture and from a symphony concert to a ballet dance (Hutchings 1995). The total appearance theory concerns the analysis or derivation of a personal opinion of quality. The analysis can be divided into the properties of the scene and its design and the characteristics and response of the viewer.

FACTORS AFFECTING TOTAL APPEARANCE

The *Total Appearance* of a scene consists of the visual images within us. These images are controlled by viewer-dependent variables and scene-dependent variables.

Viewer-dependent variables are the viewer's individual visual characteristics, upbringing and preferences, and immediate environment. *Scene-dependent variables* are the physics of the constituent materials and their temporal properties, combined with the way these are put together, and the scene illumination providing light and shade to define the volume and texture of the scene. The model considers the buildup of appearance images.

Product Images

Appearance images are dynamic and depend on the product situation. They can be broadly classified under three headings. These are the *on-the-shelf* images contributing to the buying decision, the *in-preparation* images, which contribute to the decision to continue preparing the product, and the *on-the-table* images, which determine whether we eat the product. Properties change during preparation, and throughout this process experiences are compared with expectations. Images formed are a significant part of any repeat buying decision.

The *on-the-table* appearance image consists of a mixture of individual appearance attributes that must fulfill physical and emotional expectations. What is on the plate should be stimulating; should look sufficient to cope with the appetite, large or small, of the moment; and should live up to expectations, whether derived from advertising claims, prior experience, or a personal mental image of the product.

Within each of these product situations, it is postulated that there are two basic types of appearance image, immediate images and considered images.

The *immediate* image is the gestalt or impact image comprising the initial recognition of the object or scene plus an initial judgment of quality—for example, "I like the look of this meal." *Considered* images are of a more thoughtful kind—for example, sensory, emotional, and intellectual images. These descriptors are not rigidly categorical but are useful in prompting questions to be asked of a particular scene.

In an eating situation, a *sensory* image includes an assessment of visually perceived sensory properties—for example, "This yellow dessert will taste of lemon." It is a response mainly to the food and the decor but not to the event (i.e., the reason for the meal). An *emotional* image may have positive or negative connotations and includes the reasons for the meal—for example, "I always eat chocolate-colored blanc mange to celebrate my birthday" or "What a waste of money!"—emotion by physical association or recall. An *intellectual* image might raise questions of the food, decor, the company, or the occasion—for example, "What is the relationship between the colors and proportions of individual vegetables on my plate and the extent of the visual impact?" or "I wonder how this was made." Responses to these images range from eating with relish to abandoning the meal.

Cues from the food gained before consumption give rise to an expectation of quality. Discrepancy between expectation and performance is generally either minimized or maximized by the consumer. On the one hand, according to the cognitive dissonance or assimilation theory, the evaluation will be changed to make expectation and performance more consonant. On the other, the disparity will be magnified, and the product will be evaluated to be less favorable than if there had been no prior expectations for it (Deliza and MacFie 1996). Images are conditioned for the occasion because expectations differ according to the event. A dish highly appropriate for a picnic may be inappropriate for dinner. That is, quality is not an intrinsic property of a food; rather, it is coupled with fitness for the customer's use in a given context. Confusion is caused when a cue gives rise to the "wrong" experience. An example is the red apple Top Red, which has a flavor typical of a green variety. When the apple was presented whole, panelists gave it a low preference; when it was presented without the skin, the preference increased (Daillant-Spinnler et al. 1996). This halo effect is discussed in Chapter 5. Expectations influence perceived experiences and these in turn influence future expectations (Issanchou 1996).

From time to time, issues arise affecting the relationship between the customer and the retailer. These include concerns of the environment, animal welfare, energy or pesticide usage, value for money, food labeling, customer health, genetic modification of foods, and fat and sugar contents. Any of these aspects of appearance can dominate the gestalt judgment of degree of acceptability.

Viewer-Dependent Variables

Viewer-dependent variables are shown in Exhibit 2–1. They consist of the viewer's receptor mechanisms, the viewer's inherited and learned responses to specific events, and the viewer's immediate environment.

Receptor Mechanisms

Illumination and object properties interact to provide the stimulus for the receptor mechanisms. There are four parallel sys-

Exhibit 2–1 Total Appearance—Viewer-Dependent Variables

RECEPTOR MECHANISMS
Inherited and acquired sensory characteristics consisting of

- Color vision: adaptation, after-image, constancy, discrimination, metamerism
- Aging effects: cataract, glare, light intensity need, yellowing
- Other senses: hearing, smell, taste, touch

INHERITED AND LEARNED RESPONSES TO SPECIFIC EVENTS

- Culture
- Memory
- Preference
- Fashion
- Physiological effects
- Psychological effects

IMMEDIATE ENVIRONMENT

- Geographical factors: climate, landscape, seasonal change
- Social factors: crowding, personal space, degree of awareness
- Medical factors: survival and need, state of well-being, protection

tems concerned with different attributes of vision—one for motion, one for color, and two for form (Zeki 1992). Receptor response is governed by the inherited and acquired visual characteristics of the viewer, and perceptions of color and appearance are unique to the individual. They change with color vision ability, the state of visual adaptation, afterimage, color constancy, discrimination, and metamerism characteristics. These properties are affected by viewer age, but other aspects of aging affect response. These include cataract, response to glare, intensity need, and yellowing of the macular pigment. Information received about the object from the other senses, hearing, smell, taste, and touch, also affect appearance judgments. The weight given to each individual sense when making judgments is different for different people. Visual receptor mechanisms are discussed further in Chapter 4.

Inherited and Learned Responses

Information from the receptor mechanisms is processed, and any judgments made will be modified according to the perceiver's specific upbringing and experience, which include inherited and learned responses to specific subjects and activities. Governing these are such considerations as culture, tradition, fashion, memory, preference, and prejudice. These in turn control our feelings for what we need and our feelings for the aesthetic attributes of a scene or situation. Survival and need include factors such as communication for identification and safety, the food we should eat, and the protection and shelter we require.

Our behavioral pattern in response to a stimulus is organized in the following sequence: (1) perception of the stimulus; (2) check against internal representation of adequacy; (3) behavior. Our internal representation of adequacy determines whether we accept the stimulus as genuine by comparing the perception against our experience. If the result is satisfactory, we are then in a condition for the next perception to take place. Our individual internal representation of adequacy is patterned by those aspects that are innate, residual emotions from infancy, information learned through upbringing, and cultural conditioning. Eating patterns are part of group behavior, and some foods may be seen to have a higher social status than others (Watson 1981).

Information is gleaned from many sources. These are of two types: internal (those generated from the memory) and external (those for which we actively search). Memory sources are actively or passively acquired. Active acquisitions are the memorized results of previous external searches and those that have been absorbed through prior personal experience. Passive acquisition arises in two ways: through interrupts and through low-involvement learning. The former occur when we are confronted with an unexpected stimulus that causes us to stop, notice, and remember. Low-involvement learning is acquisition of information without its being sought, as in the case of information learned from viewing television commercials, whether by active and cognitive questioning of the advertisement or, if we really are not interested, by passive absorption (Beales et al. 1981). The effectiveness of passive acquisition by children from television food advertisements is well established (e.g., Jeffrey et al. 1982).

The human information-processing system has been modeled by Barker (1982). The main elements of our response to environmental events may be divided into sensory systems, the permanent and working memories, the central processor, and the response system. The permanent memory is built up from direct stimulation from the sensory receptors and from the short-term working memory. It also has the power to build up dictionaries of appearances and other sensory properties. This enables us to recognize good and bad quality. Our judgments of good and bad are individual. Unless our quality judgments have been deliberately trained—say, for participation in an expert panel—they will be purely the result of our own personal past experiences and prejudices.

Many of the inconsistencies and low repeatabilities that occur in liking as well as in panel data can have their basis in cultural anthropology. A cultural anthropology framework for food selection has been proposed. Use of a particular food is determined by three groups of factors. Extrinsic determinants are culture, society, and economy. Intrinsic determinants include heredity, sex, age, and activity. Perception is determined by knowledge, belief, convenience, price, prestige, familiarity, risk assessment, naturalness, taste, tolerance, and satiety (Krondl and Lau 1982). All contribute to appearance in the wider sense.

Extrinsic determinants in the form of behavioral patterns differ throughout the world and take many forms. However, all cuisines are governed by a communally determined and accepted set of rules. Eating itself acts as a social bond (Rozin 1982). It is also a form of communication; boys do not give girlfriends chocolates because they think they are hungry (King 1979). Food color in cultural anthropology and oral tradition is discussed in Chapter 1.

Intrinsic determinants in terms of food selection are strong markers of age, ethnic group, and class. For example, young children prefer strong bright colors, and this extends to their choices of sweets and desserts. By adolescence, color preference has changed to pastel shades (Lyman 1989). In this respect, James (1979) has discussed the place played by *kett* in the life of a child. In the adult world, this word refers to a despised and inedible substance; in the world of the child, it refers to a revered sweet. Manufacturers communicate with children by deliberately seeking novelty and variety, appealing to vivid imaginations and love of the sensational. One method of doing this is by using saturated red, blue, lime green, black, yellow, and blue colors. Teenage girls pay more attention to dietetics and snack less than boys. Before puberty, teenagers reject many foods they previously liked. Afterwards, they begin to appreciate "adult" foods. These changes are due to social and cognitive influences (Nu et al. 1996). Older people are more willing to taste unusual foods (Otis 1984). Preferences for soup and vegetables increase with age but preferences for beverages, cereals, desserts, and fruits decrease (Kahn 1981).

The perception group of factors governing food use includes decisions made during shopping. We make decisions guided by our personality type. Jung classified individuals according to four scaled criteria: extravert to introvert, sensing to intuitive, thinking to feeling, and judging to perceiving. The Myers-Briggs personality scale is a widely used instrument based on these categories (Myers and McCauley 1985). There are "adult" responses to products in terms of health and other adult-orientated messages, and "childlike" responses to products in terms of messages of sensory reinforcement and fun (Moskowitz 1993). Factors affecting purchase choice are our perceptions of taste, nutrition, variety in preparation, compatibility with other foods, usefulness as snacks, availability, cost, attainability within the food budget, level of expectation, and quickness and ease of preparation. At the eating stage, foods may be disliked because of quality, size of serving, food temperature, monotony, overfamiliarity, menu cycles, and standardization of preparation and presentation (Kahn 1981).

Preference for individual colors is said to be linked to personality type. However, lack of stimulus and surround color control can render conclusions from studies questionable (Pickford

1972). Preferences are subject to culture and fashion. For meaningless shapes, preference is relatively independent of color (Ekehammar et al. 1987), and where there is an arbitrary association of object and color, color choice is often irrelevant (Smith 1962). Association of colors with objects means that a liking for color may be related to a liking for the object. Also, color preference for clothing and decor is heavily dictated by fashion and, though consequently relevant to color used in the food environment, is irrelevant to the color of food itself. Food color is vital to our well-being; our visual characteristics probably evolved to ensure the successful selection of good-quality food. Human beings have made use of color vision in such a way that change of color use from time to time can be accommodated. This freedom is not permitted for food. Hence, we can redecorate the store in different colors from time to time, but we cannot change the color of staple food items without customer rejection. For objects, color is stored as an appendage to the object (Davidoff 1991).

Although extrinsic and intrinsic determinants and perception govern food selection, it is possible for new products to gain acceptance. The diffusion theory considers the four elements affecting acceptance of a new product to be the innovation itself, the communication concerning the innovation, the diffusion within the social system, and time. Time is important in three respects. The first involves the time from when the individual gains initial knowledge of the innovation through to its adoption or rejection. The second involves the relative time that the individual takes to adopt the innovation compared with others in the group. The third concerns the rate of adoption within the group (Rogers and Shoemaker 1971; Kleyngeld 1974). A model based on the concept of cognitive dissonance has been used to study acceptance of novel foods. This concept recognizes that experiences conforming to prior expectancies result in greater satisfaction. If the characteristics of a new food are similar to those of the food being replaced, it is more likely to be accepted. Expectancies of the sensory qualities of a new food and what it should taste like depend to a great extent on information supplied about the food. This information can be provided by the product name, its package, its shape, the method of serving, and the means provided for its consumption (Cardello et al. 1985). Reduced acceptability leading to the rejection of a product influences interaction between the eater's expectations and the perception of the food. Both expectations and perception may be innate, temporal, contextual, or acquired. Methods that partially control the resulting market segmentation lie in advertising, which controls expectations, and manipulation of the product (Land 1988). Educational materials can be successful in changing and widening our food selection (Lynch et al. 1986).

Color Symbolism and Meaning

Color is undoubtedly a perception; it exists in the mind. However, reflected color looks so much a part of the surface that there is a strong case for it to be regarded as such. After all, to change the color of a wall we apply paint, and that affects the surface. The philosopher's approach is to call color a virtual property of a material. Colors are located in space on the surface. They are virtual in that although located in space they are not there. "The surfaces do not have the colors I or anyone else perceives them as having. The colors are virtual colors" (Maund 1995).

This property of not being real results in colors' being given any "meaning" we want. Thus, red colors can be "warm" and blue colors "cold," or colors can be ascribed the property of altering emotional states (e.g., Wilson 1966). Colors can make us feel more or less "aggressive" (Schauss 1985). They can "accelerate" or "decelerate" the passage of time (Smets 1969) or increase "grip strength" (O'Connell et al. 1985). However, there are only a few well-documented effects of light and color on behavior, those concerning the treatment of jaundice and the production of bouts of epilepsy (Kaiser 1984). Poor experimental design has led to conclusions that color alone affects heart rate, that wall colors affect perceived temperature, and that time passes faster in "red" light. Such associations of emotions with colors may be innate or learned. For example, color memory can be based on internal color space, and such associations may reside here also.

The associations may arise through association with particular objects, with red being linked with, for example, fire, blood, and danger. However, there is evidence that mood tends not to be linked when hues are presented as isolated patches and that emotional connotations come from a color's lightness and saturation (Wright and Rainwater 1962). Studies using Japanese subjects produced the conclusions that positive affects were more likely to be associated with lightness (perhaps associated with daylight) and saturation (perhaps associated with ripe fruit) than with specific object color (Kunishima and Yanase 1985). Neither red nor blue stimuli induced any significant changes in blood pressure or heart rate of 16 young men (Yglesias et al. 1993). In conclusion, if subjects respond to color in the environment, the changes are subjective and psychological; they are within us and are not objective physiological measures (Kunishima and Yanase 1985).

Color in the environment affects our moods, feelings, and efficiency. Different degrees of performance in routine office tasks can be expected. For example, subjects made significantly more proofreading errors in a white office than in a red or blue office of the same design. Females indicated more depression, confusion, and anger in offices of low-saturation colors, whereas males indicated increased depression, confusion, and anger in highly saturated environments. Subjects revealed that they would be the least likely to work in the orange or purple offices (Kwallek et al. 1996).

There are four types of color symbolism: associational symbolism, acculturated symbolism, symbolism of the familiar, and archetypal symbolism. These are relevant to many areas of life, including foods. *Associational symbolism* concerns personal experience that arises from upbringing and education. Color forms a major part of associational symbolism in the marketplace. White is associated with dairy products, and softer brown and golden tints are associated with expensive luxury foods. In some stores, color is used to identify specific locations. On a cold day, we feel warm at the sight of a hot meal, but the same material can change in association according to current environmental conditions. The perception of a meal changes from being welcome to being not needed when an un-welcome visitor arrives. *Acculturated symbolism* concerns cultural influences. For example, roast beef and Yorkshire pudding are the essence of Englishness. At Jewish New Year celebrations, green foods represent a new beginning; round things such as onions, ring-shaped pastries, and meatballs signify continuity and the hope the year will be rounded; gold items such as sliced carrots, pumpkin, squash, and honey represent prosperity; and chickpeas and beans represent abundance. The dark and black of eggplant, olives, or chocolate is nowhere to be seen (Roden 1997). Food itself is symbolic and can be accepted as such without regard to its nutritional value. Some religious foundations specify rigid food prescriptions or taboos. *Symbolism of the familiar* concerns the routine of everyday life. Everyday foods of correct appearance eaten at normal mealtimes represent security. An incorrectly or unnaturally colored food arouses suspicion and is unsettling. *Archetypal symbolism* is the symbolism of psychologists. It is that which may lie in the subconscious part of our personality. Red and yellow colors are called "warm," blue and green "cold."

Colors have symbolic significance, but the picture is not straightforward. Green is the color of fresh growth and vegetables, so green is a welcome color that symbolizes "good." But green is also the color of putrid meat, so it symbolizes "bad." Such associations are made worldwide, certainly within Hindu, Buddhist, and Moslem cultures. Similar arguments can be used for all major colors—they can all be positive or negative symbols. It seems that colors are too few and too important to be symbolic of just one thing or emotion. So they are symbolic within strictly defined boundaries that are set by context, by our upbringing and education, or by usage or are created within us by strength of advertising (Hutchings 1997).

Immediate Environmental Factors

Immediate factors of the environment affecting *Total Appearance* include geography, season, climate, and landscape character. Our physical and social situation comprises crowding, our personal space, the company we are in, how much money we have, our general medical state, and our appetite. While shopping, we are aware of the way foods are packaged and arranged in displays, the choices available, and what others are buying. We are aware of store colors, cleanliness, and how easy it is to get around. Each perception involves memories, concepts, and attitudes (Bartley 1958), and these in turn are affected by the way we feel at the time and affect the way we feel at the time. The cognitive set of an individual—that is, the "I feel great" or "I feel tired" condition—has been found to affect acceptance of chocolate bars (Siegel and Risvik 1987). Many such features in the environment, independent of the food we are eating, influence our opinion of the food and success of the occasion. These are "halo effects" and are discussed in Chapter 5.

To a large extent, the immediate environment controls where we shop. Retailers and suppliers of services can be discussed on a scale ranging from "need" to "want." Efficiency of performance is the only quality we need from the supplier of our everyday foods. At the other end of the scale, "want" may be governed solely by the sort of appearance that can be satisfied by a carton of caviar in a Harrods' labeled bag (Smith 1979).

Greater prosperity allows individuals, if so inclined, to live to eat. A reduced need for food requires the appearance of what we are eating to stimulate the appetite. To some, eating is regarded merely as a maintenance function. Complete meals requiring the minimum of preparation reduce interruptions to work and leisure activities. In these situations, color and appearance must still tempt. Both serious and casual eaters prefer food images to be positive. Illness has psychological repercussions affecting attitudes, often taking the form of pickiness and finickiness. Older people tend to be more health conscious (Smails 1996), and their olfactory and taste sensitivities decline with age (Cowart 1989). Although vision sensitivity also declines, older members of the population are more sensitive to color cues and less sensitive to changes in flavor concentration (Philipsen et al. 1995). Since they also perceive colors as having less chromatic content than do those who are younger, there is a case to be made for meals of greater coloredness to be served to older people in hospitals. Meals for the sick should be attractive, appealing, and appetizing, even to the extent of sacrificing nutrition (Pumpian-Mindlin 1954). Ways of increasing the acceptability of hospital food include reducing the frequency of similar foods on the menu, providing choices among the entrees and varying food combinations, providing menus with more description of the items, and providing attractive trays (Kahn 1981). Most or all of these suggestions have visual appearance aspects.

Scene-Dependent Variables

Scene-dependent variables are shown in Exhibit 2–2. They consist of (1) the design tools—that is, scene materials and lighting—and (2) the design.

The Design Tools

Materials and lighting are design tools. There are three types of physically definable properties of materials: optical properties,

Exhibit 2–2 Total Appearance—Scene-Dependent Variables

The designer designs for

- Communication for identification, safety, symbolism
- Aesthetic reasons in different forms, from one-dimensional writing, through two-dimensional pictures, to three-dimensional architecture, to four-dimensional performing arts.

Material physics of the scene and scene elements

- Material properties
  - Optical properties: spectral, reflectance, transmission, goniophotometric
  - Physical form: shape, size, surface texture
  - Temporal aspects: movement, gesture, rhythm
- Lighting of the scene
  - Illumination type: primary, secondary, tertiary
  - Spectral and intensity properties
  - Directions and distributions
  - Color-rendering properties

physical form, and temporal properties. Optical properties include light distribution occurring over the surface and within the depth of the material and reflectance, transmission, spectral, and goniophotometric factors. These properties can be used in specifications of visually perceived object color, translucency, transmission, and gloss. Physical form includes parameters of size, shape, pattern, surface texture, and those strength or viscosity properties that become visible by virtue of the temporal aspects of the object. Temporal properties include attributes that change with time, such as the wobbliness of jelly under an applied force or the performance as the waiter prepares a flambé at the table. Object properties are discussed in Chapters 8 and 9.

Light and shade define the volumes and textures of the scene in view. Lighting properties, such as spectral intensity, directional distribution, and color-rendering ability, also affect the perceived color, gloss, and translucency of objects within the scene. Under tungsten lamps and all fluorescent lighting designed for the home, the eye can become accommodated. That is, white or near-white objects will appear white. Although butchers normally take care to present their raw beef under red-biased illumination, some meat cases are illuminated with very red light. This defeats the purpose of the exercise, as the eye cannot become accommodated: whites become pink, and the meat can appear unnaturally red and almost fluorescent. Lighting can be used to great effect in the food industry, although no lamp is yet available that can attractively present all foods. Although cranberry juice appears almost black under normal fluorescent light containing a high blue component, it appears much redder under low–blue component "warmer" fluorescent light. When illuminated with back lighting using a "warm" fluorescent tube, it appears startlingly red (Francis 1995). Lighting is further considered in Chapter 3.

The Design

Design involves the creation of a whole from materials building blocks and lighting. A design has two functions. It makes a product or scene communicate, indicate, or symbolize, and it makes the product or scene look good. Color can be a primary part of an object—for example, the red of the traffic light or the white of milk. Color can also be a secondary property, such as that incorporated in a barely noticed room decoration. However, for most foods, the color is a primary property that tells us about the eating quality.

So, food color identifies, it advertises that a fruit is ripe and directs us to harvest, it informs us that roast meat is ready to eat, it arouses our expectation and anticipation, and it motivates us to eat. If the meat is green, the color commands us not to eat it. Color of fruit and vegetables is a symbol that reinforces our belief that we are eating a healthy meal that will perhaps initiate a cure for our current condition. We eat a piece of traditional white and blue wedding cake so it will bring good luck to the happy couple.

Color and design are used for aesthetic reasons: that is, they contribute to conspicuousness, and they decorate, please, or placate. It is helpful to consider aesthetic judgments in terms of performance or dimensional complexity. The written word has a single dimension, a painting has two dimensions, a package, a live landscape, or an architectural scene has three dimensions, and an active artistic performance has four dimensions. Food products by themselves or with their packaging are mainly three dimensional, and with a fourth dimension revealed during food service, perhaps in the restaurant.

TOTAL APPEARANCE AND THE GESTALT

Gestalt is the view of the whole. Although the world is made up of many small elements, these play a minor part in our perception in everyday life. Although we can see the individual bricks of a building, it is the whole structure that creates the impact and tells us whether it is appropriate for, and fits into, our environment. So it is with a plate of food: the whole means more than each constituent. Even abstract shapes, such as lines drawn on a sheet of paper, have meaning in the gestalt sense. According to Lyman (1979), people associate smooth, rounded shapes with feelings of calm, amusement, friendliness, being needed, and self-confidence and angular shapes with anger, resentment, frustration, jealousy, and worry. The translation of this to emotions evoked by food shapes is problematical. Perhaps smooth, rounded foods evoke calmer feelings, and angular foods carry unpleasant meanings. However, smooth, rounded foods are usually molded like jellies and are undistinguished in flavor and texture. A scoop of mashed potatoes can be less appealing than the normal shapes of the less manipulated product. Irregularly shaped foods are more common and natural (Lyman 1989).

Illusions and impressions of shape and size fit under the heading of gestalt. Food portions appear larger if they are served on a small plate. Apparent sizes of disliked or well-liked foods can be manipulated in this way. Objects of value and those evoking pleasant or unpleasant feelings are estimated to be larger than objects arousing neutral feelings. Chocolate bars are estimated to be larger than pieces of wood of the same dimensions. Children have estimated the weight of candy-filled jars to be greater than jars filled with the same weight of sand and sawdust. The main dish of a Western dinner is the entree, so it is the largest physically, and it has a dominating effect in determining the overall meal acceptance (Turner and Collison 1988). It is considered wrong for the portion of custard or cream to be too large for the slice of pie served for dessert. The apparent volume ratio between meat and vegetables should give the meat the proper emphasis. This applies also to garnishes, which should complement the meat or fish they accompany. In contrast to Western custom, portions of equal size are often served in Chinese meals. "The size of any serving should be appropriate to the setting, the physical characteristics of the food, the relation of the food to the other foods being served, and the meanings and significance one wishes the serving to express" (Lyman 1989).

In 1912, gestalt psychologists challenged the prevailing views of structuralists, who held that a square was just the experience of a particular set of points stimulating the retina. Gestalt theorists maintain that the parts of the square interact with one another and produce a perceived whole, which is distinct from the sum of the parts (Rock and Palmer 1990). This seems to accord with the general experience of life, whether it is walking around the town, attending a concert, or eating a plate of food.

Stored knowledge occurs in three forms, *isa, hasa,* and *associative* (Davidoff 1991). *Isa* concerns object function; an *isa* network for meat includes connections, for example, for mutton and chicken. *Hasa* information concerns sensory properties. For example, lemon and grapefruit are yellow and have an acid taste. *Associative* knowledge is that gained through experience with the entry-level representations. In the process of recognition, the most critical stage is activation of the entry-level representation. After categorization at this level, naming can occur, and the *isa* and *hasa* networks can be activated.

Although color is an intrinsic guide to quality, it is not usually needed for object recognition. For example, most types of fruit can be clearly recognized from a black and white photograph as easily as from a colored picture. Exceptions are structurally similar objects—for example, competing packs of yellow spreads. Also, degraded images of meat (e.g., of diced beef and pork) require the presence of color for separation. The primary task of the visual system is to categorize boundaried surfaces at the pictorial register as objects. This has precedence over categorization of surface color (Davidoff 1991). Overuse of color is not helpful in, for example, competing displays of small numbers of highly colored irregularly patterned small packs. Larger displays are different in that colors of aligned packs, for example, create their own repeated impact patterns.

However, we often need to analyze a scene or a product, and this requires a structured and disciplined approach. Certainly the chef, food technologist, manufacturer, and store designer need to look critically at a product, whether it is on a plate, in the kitchen, on a table in a restaurant, or in the store. Understanding of raw materials and products is as primary a requirement of a successful food production and marketing business as it is of a construction company.

The gestalt images and the on-the-shelf, in-preparation, and on-the-table images are bases for consumer and market research. The *Information Transfer Process* model and total appearance approach provide means of analyzing the results and improving the product. They provide a logical framework that enables answers to be found to such questions as "Why do you like this product?"

DETERMINING PERSONAL RESPONSES TO THE ENVIRONMENT BASED ON THE DERIVED IMAGES

The *Total Appearance* approach discussed in this chapter defines the origins and derivation of the individual's images (i.e., attitudes) of the object and scene out in front. It provides a mechanism by which images, and hence attitudes to images, can be studied. This can be looked at further, first in terms of images formed by specific colors and elements in the scene and second in terms of subsequent behavior.

We can aim to predict and understand social behavior in terms of reasoned action derived from the individual's attitudes and subjective norms (i.e., derived from scene images) (Ajzen and Fishbein 1991). Behavior arises from the individual's attitude toward performing the behavior as well as the individual's belief of what others think about his or her performing the behavior. Social pressures—that is, the subjective norms—contribute toward our *good* or *bad* behavior. For example, the attitude toward the intention to go out to restock the larder depends on the person's own attitude toward buying, say, fresh produce or highly processed foods and on current social attitudes that fresh produce is a healthier option and that others in the supermarket will witness the purchase of the unhealthy alternative. The model is developed on the basis of information gained from interview and questionnaire. This approach has been discussed in terms of, for example, the use of table salt and low-fat milk (Shepherd 1988) and adolescent food choice (Dennison and Shepherd 1995).

Methods of assessing the impact of a particular space include observation of behavior and structured interviews and questionnaires. Postoccupancy techniques are used to study the impact of color and design of hospital rooms (Preiser et al. 1988; Burton 1989).

THE INFORMATION TRANSFER PROCESS AS A SCIENCE

Consideration of the complex, interacting, and interdisciplinary nature of the individual factors composing *Total Appearance* leads to the conclusion that study of the *Information Transfer Process* should be regarded as a science in its own right. The approach is applicable to the food industry and all businesses involved in developing, producing, and marketing products. It can also be extended and applied to the study of industries that have a human interface, such as packaging, design, and construction.

Formalization enables a logical view to be taken of parameters influencing a product's appearance, whatever its situation. It can thus increase technologists' awareness of potential and actual appearance problems relevant to their product. The *Information Transfer Process* makes it clear that food quality is not an inherent property of a food. A product must confer sufficient benefit to the customer before a customer image of "high quality" can be recorded.

Appearance science can also be applied to the development of a rational means of studying the factors that influence product appearance on the shelf. The boundary between the disciplines of technology and marketing is sometimes found difficult to deal with, define, and reconcile. Formalization of appearance science provides a framework against which this may be achieved.

How things look tells us about the world. The *Information Transfer Process* model clarifies the task of the product and scene developer in terms of *material properties* and *basic perceptions* and quantifies the success of the product in terms of the customer's *derived perceptions* and *expectations*.

REFERENCES

Ajzen I., Fishbein M. (1980). *Understanding attitudes and predicting social behavior.* Englewood Cliffs, NJ: Prentice Hall.

Barker LM. (1982). Building memories for foods. In *The psychobiology of human food selection,* ed. LM Barker, 85–99, Westport, Conn: Avi.

Bartley SH. (1958). *Principles of perception.* New York: Harper & Row.

Beales H, Mazis MB, Salop SC, Staelin R. (1981). Consumer search and published policy. *J Consumer Res* 8 (June):11–22.

Burton CM. (1989). Developing methodologies for post-occupancy evaluation of color and light for interior environments: analysis of an international survey. In *Proceedings of the Sixth Congress of the International Colour Association,* vol. 2, 184–186. Buenos Aires: Grupo Argentino de Color.

Cardello AV. (1994). Consumer expectations and their role in food acceptance. In *Measurement of food preferences,* ed. HJ MacFie, DMH Thomson, 253–297. Glasgow: Blackie.

Cardello AV, Maller O, Masor HB, Dubose C, Edelman B. (1985). Role of consumer expectancies in the acceptance of novel foods. *J Food Sci* 50: 1707–1714, 1718.

Cowart BJ. (1989). Relationship between taste and smell across the adult life span. In *Nutrition and the chemical senses in aging: recent advances and current research needs,* ed. C Murphy, WS Cain, DM Hegsted, 39–55. New York: New York Academy of Sciences.

Daillant-Spinnler B, MacFie HJH, Beyts PK, Hedderley D. (1996). Relationships between perceived sensory properties and major preference directions of 12 varieties of apples from the Southern Hemisphere. *Food Qual Pref* 7: 113–126.

Davidoff J. (1991). *Cognition through color.* Cambridge, Mass: MIT Press.

Deliza R, MacFie HJH. (1996). The generation of sensory expectation by external cues and its effect on sensory perception and hedonic ratings: a review. *J Sensory Stud* 11:103–128.

Dennison CM, Shepherd R. (1995). Adolescent food choice: an application of the theory of planned behaviour. *J Hum Nutr Diet* 8:9–23.

Drake B. (1968). The biorheological process of mastication. In *Rheology and texture of foodstuffs,* 29–38. London: Soc Chem Ind.

Ekehammar B, Zuber I, Nilsson I. (1987). *A method for studying color-form reaction to visual stimuli.* Report no. 661. Stockholm: University of Stockholm, Psychology Department.

Francis FJ. (1995). Quality as influenced by color. *Food Qual Pref* 6:149–155.

Granville WC. (1986). List of factors believed to influence the human perception response to color. *Inter Soc Color Council Newslett* 303 (Sept–Oct):8. Note.

Hutchings J. (1977). The importance of the visual appearance of foods to the food processor and consumer. *J Food Qual* 1:267–278.

Hutchings J. (1988). Factors affecting total appearance. *Inter Soc Color Council Newslett* 314 (July–Aug):12–13. Note.

Hutchings J. (1995). The continuity of colour, design, art and science: part 1, the philosophy of the total appearance concept and image measurement, and part 2, application of the total appearance concept to image creation. *Color Res Appl* 20:296–312.

Hutchings J. (1997). Folklore and symbolism of green. *Folklore* 108:55–63.

Hutchings JB, Lillford PJ. (1988). The perception of food texture: the philosophy of the breakdown path. *J Texture Stud* 19:103–115.

Issanchou S. (1996). Consumer expectations and perceptions of meat and meat product quality. *Meat Sci* 43:S5–S19.

James A. (1979). Confections, concoctions and conceptions. *J Anthropol Soc Oxford* 10:83–95.

Jeffrey DB, McLellarn RW, Fox DT. (1982). The development of children's eating habits: the role of television commercials. *Health Educ Q* 9:174–189.

Kahn MA. (1981). Evaluation of food selection patterns and preferences. *CRC Crit Rev Food Sci Nutr* 15:129–153.

Kaiser PK. (1984). Physiological response to color: a critical review. *Color Res Appl* 9:29–36.

King S. (1979). Presentation and the choice of food. In *Nutrition and life styles,* ed. M Turner, 67–78. London: Applied Science.

Kleyngeld HP. (1974). *Adoption of new food products.* Tilburg, Netherlands: Tilburg University Press.

Kramer A, Szczesniak AS. (1973). *Texture measurement of foods.* Boston: D Reidel.

Krondl M, Lau D. (1982). Social determinants in human food selection. In *The psychobiology of human food selection,* ed. LM Barker, 139–155. Westport, Conn: Avi.

Kunishima M, Yanase T. (1985). Visual effects of wall colors in living rooms. *Ergonomics* 28:869–882.

Kwallek N, Lewis CM, Lin-Hsiao JWD, Woodson H. (1996). Effects of nine monochromatic office interior colors on clerical tasks and worker mood. *Color Res Appl* 21:448–458.

Land DG. (1988). Negative influences on acceptability and their control. In *Food acceptability,* ed. DMH Thomson, 475–483. London: Elsevier.

Lyman B. (1979). Representation of complex emotional and abstract meanings by simple forms. *Percept Motor Skills* 49:839–842.

Lyman B. (1989). *A psychology of food: more than a matter of taste.* New York: Van Nostrand Reinhold.

Lynch NM, Kastner CL, Kropf DH. (1986). Consumer acceptance of vacuum packaged ground beef as influenced by product color and educational materials. *J Food Sci* 51:253–255, 272.

Maund B. (1995). *Colours: their nature and representation.* Cambridge, UK: Cambridge University Press.

Moskowitz HR. (1993). Consumer-designed concepts for cereals. *Cereal Foods World* 38:811–816.

Myers IB, McCauley MH. (1985). *A guide to the development and use of the Myers-Briggs Type Indicator.* New York: Consulting Psychologists Press.

Nu CT, MacLeod P, Barthelemy J. (1996). Effects of age and gender on adolescents' food habits and preferences. *Food Qual Pref* 7:251–262.

Otis LP. (1984). Factors affecting the willingness to taste unusual foods. *Psychol Rep* 54:739–745.

O'Connell BJ, Harper RS, McAndrew FT. (1985). Grip strength as a function of exposure to red or green visual stimulation. *Percept Motor Skills* 61:1157–1158.

Philipsen DH, Clydesdale FM, Griffin RW, Stern P. (1995). Consumer age affects response to sensory characteristics of a cherry flavored beverage. *J Food Sci* 60:364–368.

Pickford RW. (1972). *Psychology and visual aesthetics.* London: Hutchinson.

Preiser WFE, Rabinowitz H, White E. (1988). *Post-occupancy evaluation.* New York: Van Nostrand Reinhold.

Pumpian-Mindlin E. (1954). The meanings of food. *J Am Diet Assoc* 30: 576–580.

Rock I, Palmer S. (1990). The legacy of gestalt psychology. *Sci Am* 263(6):48–61.

Roden C. (1997). Happy Rosh Hashanah. *Independent,* 27 Sept, 50.

Rogers EM, Shoemaker FF. (1971). *Communication of innovations.* 2nd ed. New York: Macmillan.

Rozin E. (1982). The structure of cuisine. In *The psychobiology of human food selection,* ed. LM Barker, 189–203. Westport, Conn: Avi.

Schauss AG. (1985). The physiological effect of color on the suppression of human aggression: research on Baker-Miller pink. *Int J Biosoc Res* 7:55–64.

Shepherd R. (1988). Consumer attitudes and food acceptance. In *Food acceptability,* ed. DMH Thomson, 253–266. London: Elsevier Applied Science.

Siegel SF, Risvik E. (1987). Cognitive set and food acceptance. *J Food Sci* 52:825–826.

Smails S. (1996). Heart of the matter. *Supermarketing* 25 (Oct):34–35.

Smets G. (1969). Time expression of red and blue. *Percept Motor Skills* 29:511–514.

Smith RG. (1979). Color as a marketing tool. *Color Res Appl* 4(2):78–82.

Smith SL. (1962). Color coding and visual search. *J Exp Psychol: Gen* 109:377–392.

Turner M, Collison R. (1988). Consumer acceptance of meals and meal components. *Food Qual Pref* 1:21–24.

Watson RHJ. (1981). The importance of colour in food psychology. In *Natural colour for food and other uses,* ed. JN Counsell, 27–40. London: Applied Science.

Wilson GD. (1966). Arousal properties of red versus green. *Percept Motor Skills* 23:947–949.

Wright B, Rainwater L. (1962). The meaning of color. *J Gen Psychol* 67:89–99.

Yglesias M, Stewart KT, Gaddy KT, Zivin JR, Doghramji G, Thornton W, Brainard GC. (1993). Does color influence blood pressure and heart rate? *Proceedings of the Seventh International Colour Association Congress,* 123–124. Budapest: Technical University of Budapest.

Zeki S. (1992). The visual image in mind and brain. *Sci Am* 267(3):43–50.

CHAPTER 3

Light and Its Interaction with Food Materials

LIGHT AND LIGHTING

Light is that part of the electromagnetic spectrum perceived by the eye. We see light in the form of a direct source, such as a lamp, but otherwise we can see it only after it has been scattered or reflected by a substance with which it has come into contact. Light used for general illumination is white, but there are many types of white light, each having different spectral power outputs. The most common light emitter is a hot body such as the sun. The spectral distribution of sunlight varies considerably during the day because the sun varies in angle above the horizon. Its light passes through different depths of atmosphere and different concentrations of atmospheric pollutants at different times of the day. Direct sunlight has a reasonably balanced spectrum. The color of the sky away from the sun is dominated by scattered blue light. The sun low in the sky looks red. This is light from which the high-energy short wavelengths have been removed by scattering. As pollution levels increase, an increased amount of shorter wavelength light is preferentially scattered until the sun appears red. The same phenomena account for the appearance of a glass of dilute skimmed milk, which has a blue reflected color but is red brown when viewed by transmission. In this case, the scattering elements are casein micelles in the milk (Dunkerley et al. 1993).

Black-Body Radiators

The black-body or Planckian radiator is a fundamental laboratory light source having a controllable spectral output. It consists of an empty enclosure surrounded by a jacket through
which fluid at a constant temperature can be circulated. The inside of the sphere has a nonselective absorption coating that
permits only nonselective absorption and scattering processes
to occur inside the enclosure. Radiation is emitted through a
hole in the sphere wall. Any radiation entering the sphere
rapidly becomes absorbed and is not re-emitted. Radiation emitted by the black body depends only on the temperature. As the
temperature is increased, more blue light relative to red is emitted. The color of the radiation changes from redder at a temperature in the region of 2,000K, to whiter at 4,000 or 5,000K, to
bluer at 8,000 to 10,000K, where K is the temperature in Kelvin.
Hence, the color of the radiation can be described in terms of
color temperature. This black-body color locus is shown plotted
on the 1931 chromaticity diagram (see Chapter 7) of the Commission Internationale de l'Eclairage (CIE) in Figure 3–1.

Standard Illuminants

The color of an object depends upon the light illuminating it.
Therefore, for color measurement and assessment, the spectral
distribution of the illumination must be standardized. The CIE
has defined a number of standard illuminants. The three most
commonly reported in the food literature are illuminants A, C,
and D_{65}. Illuminant A is provided by a gas-filled tungsten lamp
operating at a color temperature of 2,856K. Illuminant C, designed to represent daylight from an overcast sky, is tungsten
light filtered to give a color temperature of 6,774K. Source D_{65} is
based on measurements of total daylight of sun plus sky and has
a correlated color temperature of 6,500K. The spectral distributions of these illuminants are shown in Figure 3–2.

The positions of the three illuminants are shown on the CIE
chromaticity diagram in Figure 3–1. Illuminant A is yellower,
while C and D_{65} are bluer. Whichever of these illuminants is

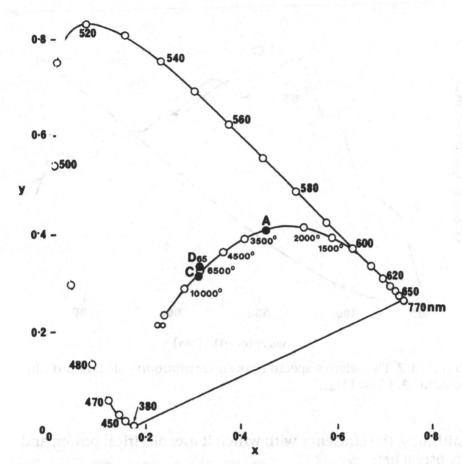

Figure 3–1 CIE (1931) chromaticity diagram showing the Planckian locus and the chromaticities of standard illuminants A, C, and D$_{65}$. The color temperature is in K (Judd and Wyszecki 1975, copyright, reprinted with permission of John Wiley and Sons Inc).

used in a measurement or assessment, its location in the chromaticity diagram defines the reference white point.

Light Source Quality Specification

Light source quality can be discussed in terms of color temperature (discussed above), the ability of the lamp to render colors

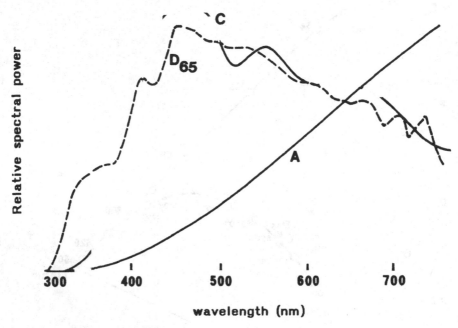

Figure 3–2 The relative spectral power distributions of standard illuminants A, C, and D_{65}.

faithfully, the efficiency with which it uses electrical power, and its useful life.

Lamps render colors to different degrees of faithfulness. Color rendering is the effect of an illuminant on the color appearance of objects, in conscious or subconscious comparison with their color appearance under a reference illuminant. Although colors look approximately the same under different phases of daylight and under tungsten lighting, they can look very different when viewed under other lamps. An approximate measure of color rendering is available. This is based on the principle of assessing the magnitude of the chromaticity shift produced when eight specific Munsell samples are irradiated in turn by a reference and the test lamp. The *Color Rendering Index* (R_a) of the test lamp is determined from the average of the chromaticity shifts. It is scaled so that an exact reproduction of the performance of the reference illuminant yields an R_a value of 100. A 60-watt tungsten lamp is normally used as the reference. A high R_a does not guarantee a high degree of color rendering of a particular material; it can only be taken as a guide when evaluating a lamp for practical lighting tasks. This applies especially to food materials such as meat, greens, and fruit because their reflectance spectra are less broad band in form than those of the colors used in the determination of R_a.

Chromaticities of white light fluorescent lamps fall near the black-body locus shown in Figure 3–1. Such sources can be given a *correlated color temperature* (cct) by taking the nearest black-body chromaticity match. Such isotemperature lines are shown on Figure 3–3, which also indicates the band within which the chromaticities of white fluorescent lamps fall. The chromaticities of two specialist lamps, Rosetta and Grolux, are also shown.

The rendering of an object's color depends on the spectral qualities of light source, object, and visual mechanism of the viewer. If any one of these factors changes, the perceived color changes. Light sources having good color-rendering properties also have better *visual clarity*. Using such lamps, lower levels of illumination are required for the performance of visual tasks (Aston and Bellchambers 1969).

These measures are attempts to describe the effectiveness of practical illumination systems. *Efficacy* is a valid measure of performance, but cct and R_a are merely guides for use in specific situations.

Lamp efficiency, or *luminous efficacy*, is determined from the amount of visible radiation emitted per unit of electrical energy input—that is, lumens per watt. *Lumen maintenance* is an indication of the decrease in light output with lamp life.

Color temperature is a valid basis for comparing two lights only if the spectral distributions are similar. For example, two tungsten lights may be compared. Two fluorescent tubes may be compared if they contain the same phosphors and approximately the same admixture of mercury emission lines. Comparison of tungsten and fluorescent lamps using cct has little meaning. In these circumstances, color temperature is useful

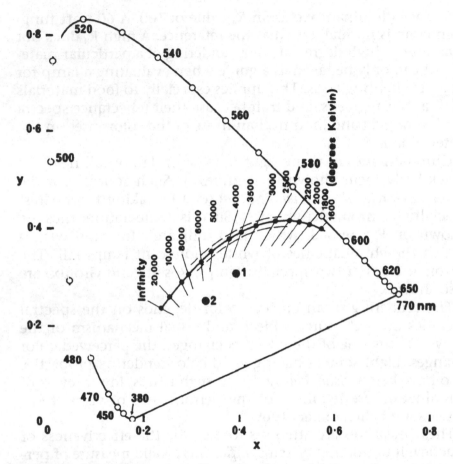

Figure 3–3 Isotemperature lines for the evaluation of a correlated color temperature for non-Planckian radiators. The band encloses the chromaticities of white fluorescent lamps. Also indicated are the chromaticities of two specialist lamps: 1, Rosetta; 2, Grolux.

only for comparing the color of the lamps on a yellow-blue scale. The R_a value is valid only for the Munsell samples tested. Even then, the color difference measurements used in the calculation of R_a are not directional. Hence, the R_a is no firm measure of the effectiveness of a lamp for any particular viewing situation.

The need for a *color preference index* has been proposed. A warm white lamp with a high index would indicate that objects illuminated would appear "agreeable." Such a lamp might be called a *flatter* or *glamor* lamp. It would be used, for example, in the social areas of hotels, restaurants, and some areas of public buildings. The human face was found to be the most sensitive subject used in determination of the index. Lamps that render the face agreeably will also render most food well. Tomatoes are redder and green vegetables greener, while butter changes little. However, carrots look too red. The use of flatter light in sales areas is ethically doubtful (Einhorn 1976).

Practical Light Sources

Practical light sources are of three types: incandescent, discharge, and fluorescent. Incandescent tungsten lamp radiation arises solely from heat. Spectral output is close to that of a blackbody radiator and has a spectral form similar to source A in Figure 3–2. Tungsten filament lamps used for general lighting operate within a color temperature range of 2,500 to 3,000K. Their life is restricted because the metal evaporates and is deposited inside the glass envelope. The tungsten halogen lamp has a silica case, which permits a higher running temperature, and halogen gas, which prevents formation of tungsten deposits on the envelope. Halogen lamps are more efficient and have a longer life. Tungsten lamps are useful in retail display in that they have good color-rendering properties, are flattering to human skin tones, and are easily controlled. However, only 5 percent of the emitted energy is in the form of light, and the lamps have relatively short lives.

Discharge lamps depend on bombardment of the atoms of a vapor with electrons from an electric current. Narrow-wavelength bands of radiation are emitted, so colors illuminated by them tend to be severely distorted. There are four types of discharge lamp: low-pressure sodium, high-pressure sodium, mercury, and metal halide. Compared with incandescent lamps, they run cooler, give a greater light output, and are more efficient. The low-pressure sodium lamp is one of the most efficient light sources, but because it renders color very badly, it is normally used only for orange-yellow street lighting. The output of high-pressure sodium lamps is broader, and its color rendering is sufficiently good for sports halls and general external lighting in shopping areas. High-pressure mercury-based lamps common in industrial interiors and warehouses emit a bluish light having poor color rendering. This is improved when the inside of the lamp case is coated with a red-emitting phosphor. This contributes light of wavelength greater than 600 nm to the output (Figure 3–4).

Addition of metal halides to the mercury considerably improves color rendering. Metal halide lamps have a higher efficiency and a greater lumen maintenance and can have good color-rendering properties. They are a suitable substitute for daylight and are useful when used with low natural daylight levels, although some can have a cold appearance at night.

The high-pressure xenon arc lamp has a spectral output close to average daylight; hence its use for floodlighting, in film projectors, and in the laboratory. Its spectrum is near to that of D_{65}, shown in Figure 3–2.

Fluorescent lamps are gas discharge tubes that have a coating of fluorescent phosphor on the inside of the glass. Ultraviolet radiation absorbed by the phosphor is re-emitted as visible light. The many phosphors developed have led to a number of tubes emitting white light of varying approximations to daylight. They can be obtained in a range of sizes, have a longer life, and are three to four times more efficient than tungsten halogen lamps. Standard fluorescent tubes incorporate conventional halophosphate phosphors. The first triphosphor fluorescent or prime color lamps used narrow-band phosphors emitting discrete spectral lines in addition to the halophosphate phosphor. Modern three-band tubes, however, have dispensed with the halophosphate phosphor (Figure 3–5). These offer relatively high light outputs, lower energy costs, and good color-rendering properties; hence their wide use in retail stores. Their lumen maintenance is also better, retaining 95 percent of their initial lumen output after 8,000 hours of burning, as against 70 to 75 percent for standard fluorescent tubes. However, some colors

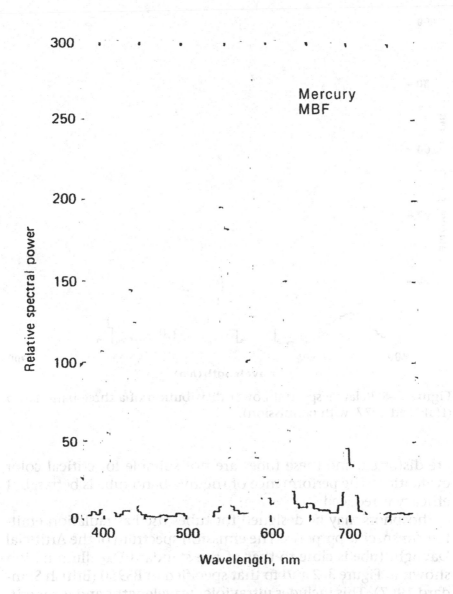

Figure 3–4 Relative spectral power distribution for a high-pressure mercury lamp that has a red-emitting phosphor coated on the inside of the envelope. *Source:* Reprinted with permission from R.W.G. Hunt, *Measuring Colour*, 3rd ed., p. 79, © 1998, Fountain Press.

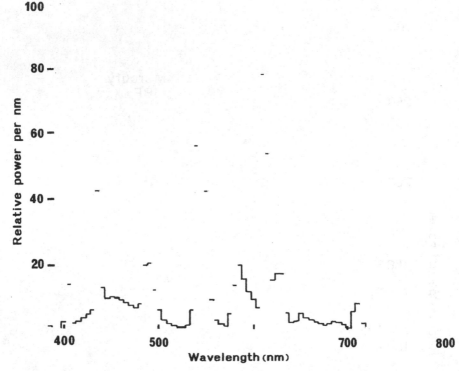

Figure 3–5 Relative spectral power distribution of a three-band lamp (Halstead 1977, with permission).

are distorted, and these tubes are not suitable for critical color evaluation. The performance of the five-band tube is better, but efficacy is reduced.

Phosphors may be designed for tubes used as radiation emitters for specific purposes. The emission spectrum of the Artificial Daylight tube is close to that of the standard D_{65} illumination shown in Figure 3–2 and to that specified in BS950 (British Standard 1967). This includes ultraviolet wavelengths and is for critical color-viewing work on samples, especially those that fluoresce. Northlight gives a similar full–visible spectrum output, excluding ultraviolet wavelengths. Fluorescent tubes are sold specifically for illumination of food displays, especially for meat and bread. The red-enhanced spectral output of Gourmet, one of these tubes, is compared with the output of a Daylight tube in Figure 3–6.

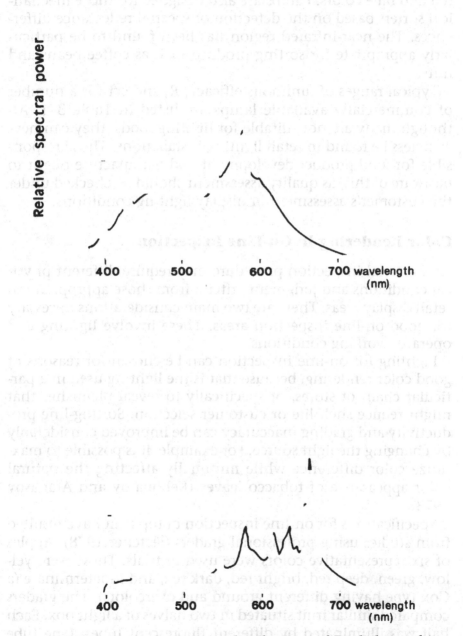

Figure 3–6 Relative spectral power distributions of Daylight, a white lamp, and Gourmet, a lamp with increased red output designed for food displays.

Grolux is designed for florist shops and aquaria for enhancing red and blue colors. Lamps are also designed for those mechanical sorters based on the detection of spectral reflectance differences. The near-infrared region has been found to be particularly appropriate for sorting produce such as coffee beans and nuts.

Typical ranges of luminous efficacy, R_a and cct for a number of commercially available lamps are listed in Table 3–1. Although many are not suitable for lighting foods, they can nevertheless be found in retail lighting installations. Those responsible for food product development and manufacture ought to be aware of this, as quality assessment should be checked under the customer's assessment or display lighting conditions.

Color Rendering in On-Line Inspection

Commercial inspection procedure may require different physical conditions and judgment criteria from those appropriate to retail display areas. There are two main considerations necessary for good on-line inspection areas. These involve lighting and operator working conditions.

Lighting for on-line inspection can be chosen for reasons of good color rendering, because that is the lighting used in a particular chain of stores, or specifically to reveal blemishes that might reduce shelf life or customer selection. Sorting-line productivity and grading inaccuracy can be improved considerably by changing the light source. For example, it is possible to maximize color difference while minimally affecting the natural color appearance of tobacco leaves (Kehlibarov and Atanasov 1973).

Specifications for on-line inspection of top fruit have resulted from studies using professional graders (Fletcher 1978). Apples of six representative colors were used in trials. These were yellow, green, deep red, bright red, dark red, and indeterminate (a Cox type having different ground and overcolors). The graders compared similar fruit situated in two halves of a light box. Each half was illuminated by different fluorescent tubes. One tube was Kolor-rite, which was used as the control. The base of each

Table 3–1 Performance Figures of Some Commercial Lamps (taken from manufacturers' literature)

	Correlated Color Temperature (cct) K	Color-Rendering Index (R_a)	Luminous Efficacy (lm/W)
Tungsten (240 V, 40–100 W)	2,650–2,750	100	10–13
Tungsten halogen	3,000	100	21
Mercury (color corrected)	3,800	45	20
Metal halide	3,000–6,000	65–93	70–100
Sodium—high pressure	2,000	25	100
Sodium—low pressure	*	*	150
Xenon	5,290	93	25
Fluorescent			
Artificial Daylight[a]	6,500	92	45
Northlight[b]	6,500	93	55
Cool White	4,200	58	81
Natural Deluxe	3,700	92	46
Warm White	2,950	51	83
Warm White Deluxe	2,700	62	48
White	3,450	54	83
Natural	4,050	73	72
Daylight	6,500	76	67
Three-band	4,000	85	96
Five-band	4,000	95	65
Kolor-rite	4,000	89	51
Grolux[c]	*	*	23
Food retail display lamps[c,d]	3,300–3,800	75–92	37–49

* Data that ought not be calculated because the spectral distribution is too dissimilar to that from a black-body radiator (see Figure 3–6).

[a] A full-spectrum output (including UV) that meets BS950 for critical viewing work of samples, particularly if they fluoresce.

[b] A full–visible spectrum output (excluding UV) that meets BS950 for critical viewing work.

[c] Used in florist shops and aquaria for enhancing red and blue colors.

[d] Designed to enhance and differentiate colors of meat and other foods. Spectral distributions of some of these lamps are too dissimilar from that of a black-body radiator; hence, cct and R_a ought not be calculated.

half of the box consisted of rollers, which could be white or black. The inside of the box was a neutral reflecting surface. Each fruit was judged separately under each pair of lamps. Panelists were asked to select the lamp under which they preferred to view that apple. Also, they were asked to choose the tube capable of showing all top fruit varieties and their blemishes in an acceptable manner for grading purposes.

A number of conclusions relevant to produce grading were made. Kolor-rite or Colour 38 fluorescent lamps were recommended. Tubes lower in the order of merit were Natural, Daylight, Warm White, and White (Fletcher 1978). The Kolor-rite tube has been recommended in Britain by the Department of Health and Social Security for good color rendering of skin tones and for use in health buildings (Medical Research Council 1965). Ambient daylight should be excluded from the grading area. A minimum light intensity of 500 lux is suggested for surfaces used for grading. For other inspection areas, 300 lux is sufficient. Where no natural light is available, an intensity of 50 lux for the general building area is suggested. Visual distraction should be minimized (Fletcher 1978). For example, alternate pools of light and dark should be avoided in all areas (Meneer 1984). The following are recommendations made by the Ministry of Agriculture, Fisheries, and Food in Britain:

- All operators working on quality inspection should be checked for color deficiency (see Chapter 4).
- Color cards or similar aids should be used as references.
- Light fittings and tubes should be cleaned regularly at the start of each session.
- Tubes should be replaced after a maximum of 7,500 hours (approximately three years) of service.
- Quality inspection tables should reflect less light than the fruit.
- The superstructure of the quality inspection tables should be a neutral color (Fletcher 1978).

Uniformity in the use of lighting installations for grading throughout the top fruit industry was advocated. Also, as degree of ripeness/freshness and defects are the most important aspects of top fruit in the market, it was suggested that suitable lighting conditions for assessment should be provided in retail outlets (Fletcher 1978).

ABSORPTION AND SCATTERING OF LIGHT

A number of physical phenomena may occur when light impinges on a material (Figure 3–7). The interactions involved are

- *From the surface*—specular and diffuse reflection, and refraction into the body of the material
- *Within the material*—internal diffusion (scattering) and absorption
- *Through the material*—regular and diffuse transmission

Absorption

During transmission of light through a medium, the amount of light absorbed depends on the wavelength, the properties of the

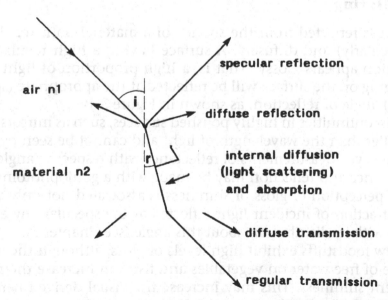

Figure 3–7 The interaction of light with a material containing light-absorbing and scattering elements.

medium, and the distance traveled. For nonscattering low-concentration dye solutions of a monomolecular form, this is achieved through the Beer-Lambert law:

$$A = -\log_{10}(T_i) = \log_{10}(I_1/I_2) = ecp$$

where T_i is the transmittance, I_1 the light intensity entering the sample, I_2 the intensity leaving the sample, e the molar extinction coefficient, c the molar concentration of absorbing dye, and p the path length or sample thickness. A is the absorbance, which is additive.

This equation relates to monochromatic radiation. Dyes present in aggregate concentrations greater than the monomolecular form tend to have a lower absorbance than if present as separate molecules (Bridgeman 1987). The application of this law to transparent food materials should not be taken for granted. For example, some red wines obey it only within certain ranges of dilution. Chemical reactions and pH changes occurring on dilution are reasons for nonlinearity (Negueruela et al. 1988).

Scattering

Light is reflected from the surface of a material both regularly (specularly) and diffusely. A surface having a high regular reflection appears glossy. That is, a high proportion of light impinging on the surface will be reflected at the appropriate (specular) angle of reflection, as shown in Figure 3–8.

Discontinuities in highly polished surfaces, such as mirrors, are smaller than the wavelength of light and cannot be seen by the naked eye. Measurements of reflectance with respect to angles of incidence and reflection may be made with a goniophotometer. The perception of gloss or shininess is associated not only with the fraction of incident light reflected at the specular angle but also with its distribution about this angle (see Chapter 7).

Few foodstuffs exhibit high levels of gloss, although the presence of free water on vegetables and fish can increase their apparent shininess. This may increase the visual derived perception of *freshness*. Cheeses and yellow spreads vary considerably and characteristically in their gloss characteristics. In these

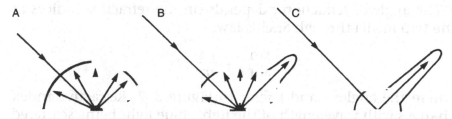

Figure 3-8 The polar distribution of reflected light. A, the diffuse reflectance of a matte finish; **B**, the diffuse and specular reflectance from a semimatte finish; **C**, a high-gloss finish from which most of the incident light is specularly reflected.

products, the extent and type of gloss largely depend upon the method of manufacture. This controls fat configuration and distribution.

The amount of light specularly reflected depends upon the difference between the refractive indices of the mediums involved. Fresnel's equation for reflection at the angle of normal incidence is

$$R = \left(\frac{n_2 - n_1}{n_2 + n_1}\right)^2$$

where n_1 and n_2 are the refractive indices for the two media involved.

As the difference in refractive index increases, the amount of light regularly reflected increases. The refractive index of air is low compared with that of solids and liquids encountered in everyday life; hence, the potential for shiny surfaces is great. Fresnel reflection occurs from all surfaces, but only those that are smooth appear shiny. Scattering takes place at every boundary at which there is a change of refractive index. This reflection can involve from 4 percent to 90 percent of the incident radiation. Our love for shiny surfaces plays a large part in the success of the costume jewelry industry. Parts of the food industry have acknowledged this; hence, wax coatings are used to increase the attractiveness of apples by making them shinier. The wax fills small pits in the apple surface with a polishable material of similar refractive index.

The angle of refraction r depends on the refractive indices of the two media through Snell's law:

$$\frac{\sin i}{\sin r} = \frac{n_2}{n_1}$$

where the angles i and r refer to Figure 3–7. Refractive index changes with wavelength of the light, blue light being scattered more than red.

The extent to which light is transmitted through gases or liquids depends upon the size of discontinuities present in the transmitting medium, compared to wavelength of the light. Rayleigh scattering occurs when particles much smaller than the wavelength are present in the medium. For an unpolarized incident wave, the scattering intensity I of light wavelength λ, scattered by a Rayleigh scatterer at an angle θ to normal incident radiation, is given by

$$I = \frac{8\pi^4 r^6}{d^2\lambda^4} \left(\frac{n^2 - 1}{n^2 + 2}\right)^2 (1 + \cos 2\theta)$$

where r is the particle radius, n the refractive index ratio, and d the distance to the detector.

Hence, for a constant particle size of Rayleigh scatterer, scattering is greater at lower wavelengths. There is also a particle size effect, smaller spheres scattering blue wavelengths preferentially. For those of slightly larger radius (approximately 1 μm), the scattered light is yellower (Kerker et al. 1966). The Mie theory must be used to calculate the degree of scattering when particle sizes are greater than approximately 0.1 μm. At these particle sizes, scattering is a form of diffraction, with most of the light being scattered forward in a narrow cone. Red wavelengths are spread over a wider cone than blue wavelengths; hence, the continuing beam is bluer.

Understanding of light/particle interaction in highly dilute solutions has been furthered by dynamic light-scattering spectroscopy. However, this technique cannot be used directly with foods, as these are normally concentrated solutions or solids. As dilution usually affects structural properties, application to foods is limited (Dalgleish and Hallett 1995).

An indication of the complexity of the interaction between particle size and scattering may be gauged from the work of Birkett (1985). Scattering coefficients were determined for nonabsorbing, light-scattering model systems of paraffin-in-water emulsions and dispersions of glass ballotini in a thickened aqueous phase. Layers 2 mm deep in a glass cell were measured using an abridged-reflectance spectrophotometer. Corrections were made for errors arising from external diffuse and total internal reflectance at the air/glass cover slip interface. The surface weighted mean diameter of the particles was used in subsequent calculations. Rayleigh and Mie theory values of the scattering coefficients were computed for a range of particle diameters, using values for the refractive index ratio of 1.10, dispersed-phase volume of 0.50, and wavelength of 630 nm. The results, with the experimental determinations, can be seen in Figure 3–9.

The behavior of two types of scattering coefficient is shown. One has been calculated using the Rayleigh and Mie equations (S_{MIE}), while the other (S_{KM}) is intended to be consistent with results calculated using the Kubelka-Munk theory. Considering the former, scattering from smaller particles (to a diameter of approximately 0.4 μm) is accounted for by the Rayleigh equation. Thereafter, the Mie and many-flux theories are appropriate (Billmeyer and Richards 1973; Bayvel and Jones 1981). The diameter at which maximum scattering occurs (approx. 1 μm) is dependent on the absorbing and refractive index properties of the pigment. Using the Mie theory, Brockes (1964) calculated that the diameter relative to wavelength can vary between 0.2 and 1.0 for inorganic and organic paint systems typically used in the paint industry. The complexity of scattering is caused by the changing dependency on particle radius and light wavelength, as indicated in Figure 3–9. At the largest particle sizes, the laws of geometric optics apply, and scattering is independent of wavelength but dependent on r^2. When a number of particles are present, and as their diameter increases, the appearance of suspensions changes. They first increase in opacity from being transparent through different types of haze and varying degrees of opacity, eventually becoming transparent again.

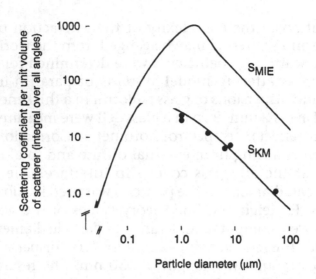

Theory Required	Rayleigh[a]	Mie and Many-Flux Theories [a,b,d]		Optical Limit
Scattering per particle: depends on particle radius:	r^6	$r^{4(c)}$	$r^{2(c)}$	r^2
and depends on wavelength:	$\lambda^{-4(c)}$	$\lambda^{-2(c)}$	λ^0	λ^0

[a] Billmeyer and Richards (1973)
[b] Brockes (1964)
[c] Francis and Clydesdale (1975)
[d] Bayvel and Jones (1981)

Figure 3–9 The effect of particle size on the scattering of light of wavelength 630nm (Birkett 1985, with permission).

The true scattering (S_{MIE}) is not directly comparable with the experimental scattering coefficient derived from the Kubelka-Munk model (Figure 3–9) because the latter assumes isotropic scattering. The Mie function must be modified by an angular term to yield a scattering coefficient (S_{KM}) equivalent to experimentally derived values. The relationship between them is

$$S_{KM} = .75(1 - \langle \cos \theta \rangle) S_{MIE}$$

where $\langle\cos \theta\rangle$ is the intensity weighted average of $\cos \theta$. As can be seen from Figure 3–9, a reasonable agreement was found between the experimental and theoretical values. In addition, a linear relationship was found between S_{KM} and visual magnitude estimates of *Opacity* (Birkett 1985).

The peak of the S_{KM} curve occurs when the particle diameter is approximately one quarter of the wavelength of light ($\lambda/4$). Pigment costs can be reduced by taking advantage of this during product formulation. For example, in a granulated instant drink formulation, it may be possible to include dispersible colorant particles of diameter $\lambda/4$ and having a refractive index different from that of the substrate in which they are embedded (Francis and Clydesdale 1975).

Scattering changes with angle of view. This is because perceived scattering is the sum of horizontally and vertically plane-polarized components and light interference. For example, Figure 3–10 is a vector diagram showing the radial distribution of scattering for particle sizes less than the wavelength of light.

The broken and dotted lines indicate the vertically and horizontally plane-polarized components of the scattered light. Their sum is shown as a continuous line. The vertically polar-

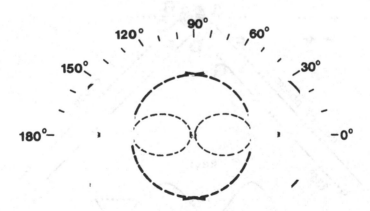

Figure 3–10 Vector diagram indicating the radial distribution of scattered light from a small particle about the scattering center (Thorne 1963 cited in Francis and Clydesdale 1975).

ized component has a uniform intensity around the scattering center independent of direction. The horizontally polarized component has the form of $\cos^2 \theta$. The resulting total scatter is the sum of these two components.

Interference occurs when the particle size is of the same order as the light wavelength. This effect is shown in Figure 3–11.

Light waves are shown as alternating continuous and dotted lines, each denoting half-wavelength pulses. Scattering is shown taking place from A and B, two points on a particle of appreciably wavelength size. Scattering in two directions, 45° and 135°, is considered. Light scattered at 45° contains two crests, ADE and, passing through the particle, ABC. These rays are in phase at C and E and thus undergo constructive interference.

The light scattered at 135° contains two crests, AF and ABD. These rays are out of phase at F and G: there is destructive interference, and no scattering occurs. Hence, interference produces asymmetrical scattering profiles (predicted by the Mie theory), such as those shown in Figure 3–12.

The profiles X and Y are for different particle diameters. The extent of the asymmetry can be quantified by the ratio (Z) of the light scattered at two angles 45° and 135°. Thus $Z = 1$ for the

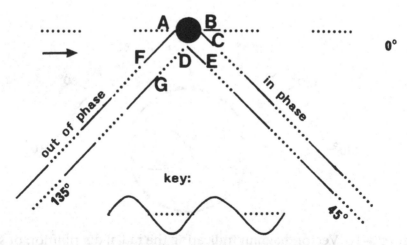

Figure 3–11 Interference of a ray of scattered light from a large particle (Thorne 1963 cited in Francis and Clydesdale 1975).

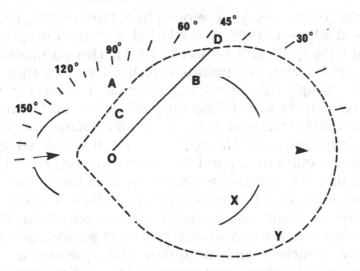

Figure 3–12 Nonsymmetrical light scatter diagram for two solutions containing particles of different sizes (Thorne 1963 cited in Francis and Clydesdale 1975).

profile shown in Figure 3–10. For the larger particles of Figure 3–12, $Z = 1.7$ and 5.0. Such patterns of light scattering are of great importance in the assessment and measurement of foods such as beer and tea, for which haze can be a significant problem. They are also the cause of inconsistent nephelometer readings when the solutions are viewed at different angles, as can be seen in Figure 3–12.

Kubelka-Munk Theory for Multiple Scattering

The ability to handle, specify, and predict the light scattering and absorption properties of materials is vital to solving many practical problems arising within many industries. These problems range from calculating how many layers of paint are needed to cover an existing contrasting color already on a surface to how to select colorants and predict the concentrations required to match an existing color. Food industry problems include specifying translucency and determining the interaction between product translucency and color. All these problems can be approached using turbid-medium theory. The most useful application of the theory, developed by Schuster (1905, cited in Atkins and Billmeyer 1968), is the Kubelka-Munk concept of the colorant layer (Judd and Wyszecki 1975). This postulates two light fluxes traveling in opposite directions across a thin layer within a strongly light-scattering material. The layer is thick compared with the size of pigment particles but thin compared with the sample thickness. It has an infinite lateral dimension. As each light flux passes through the layer, it is affected by the absorption coefficient (K) and the scattering coefficient (S) of the material. A number of assumptions have been made. The treatment is relevant to monochromatic radiation, and it does not allow for light lost through Fresnel reflection at the medium/air boundary. Also, the pigments are assumed to be randomly orientated, and the differential equations are concerned only with diffuse light traveling in two directions.

The upward flux (j) is

- decreased by absorption: $= -Kj\,dx$
- decreased by scattering: $= -Sj\,dx$
- increased by backscatter from the downward-proceeding flux: $= +Si\,dx$

Hence, the total change in the upward flux is

$$dj = -(K + S)j\,dx + Si\,dx$$

Similarly the downward-proceeding flux (i) is changed by

$$-di = -(K + S)i\,dx + Sj\,dx$$

A number of solutions of these differential equations, for a layer of thickness X, have been discussed by Judd and Wyszecki (1975). Those most quoted in the food literature concern the following relationships between K, S, and reflectance properties.

$$\frac{K}{S} = \frac{(1 - R_\infty)^2}{2R_\infty}$$

where R_∞ is the reflectance of an infinitely thick layer of the material;

$$a = \frac{1}{2}\left(R + \frac{R_0 - R + R_g}{R_0 R_g}\right) = \frac{S + K}{S}$$

$$R_\infty = a - b$$

where R is the reflectance of a colorant layer backed by a known reflectance R_g, R_0 is the reflectance of the layer with an ideal black background, and $R_g = 0$;

$$T_i{}^2 = (a - R_0)^2 - b^2$$

where T_i is the internal transmittance of the layer; and

$$SX = \frac{1}{b} \text{arctgh}\left(\frac{1 - aR_0}{bR_0}\right)$$

where $b = (a^2 - 1)^{1/2}$ and arctgh is an inverse hyperbolic function such that

$$\text{if } v = \text{ctgh } u, \text{ then } u = \text{arctgh } v$$

and

$$\text{ctgh } u = (\cosh u)/(\sinh u)$$

Tables are widely available for calculating such hyperbolic functions (e.g., Judd and Wyszecki 1975).

Two measurements of a thin layer of the sample, over backgrounds of white and black, enable K and S to be calculated. These tend to behave linearly with dye and scatterer concentration, and they can be used additively to predict paint formulations. This may be carried out by using K/S as follows:

$$(K/S)_{\text{mixture}} = a(K/S)_A + b(K/S)_B$$
$$+ c(K/S)_C + W(K/S)_{\text{BASE}}$$

where a, b, c, W are, respectively, the concentrations of colorants A, B, C, and the base material. Alternatively, K and S may be treated independently in the same way as the ratio (Billmeyer and Abrams 1973).

The Kubelka-Munk theory does not take into account the reflectance from the front surface of the sample or the diffuse light incident upon this surface from the inside. The latter effect is illustrated in Figure 3–13.

Part of the light approaching the surface from a particle P within the sample is scattered back into the sample. If the ray approaches the surface at an angle greater than the critical angle, it will be completely scattered back (solid lines). The scales of these reflections are shown in Figure 3–14.

The internal and external reflectances for completely diffused incident flux are shown as functions of the refractive index ratio n_2/n_1. Reflectance for perpendicular incidence is also included. As is evident, light loss can be substantial. Approximately 50 percent of the light attempting to emerge from a boundary of refractive index ratio of 1.33 (air and water) is backscattered. A correction for the reflectance data has been developed (Saunderson 1942).

$$R_c = \frac{R_m - k_e}{(1 - k_e)(1 - k_i) + k_i R_m - k_e k_i}$$

where R_c is the corrected reflectance, R_m the measured reflectance, k_e the fraction of incident light reflected externally, and k_i the fraction lost internally.

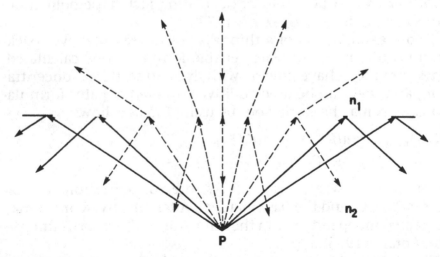

Figure 3–13 The bending of rays by refraction away from the perpendicular as they enter a less dense medium n_1. If the rays are incident at the boundary at an angle greater than the critical angle, they will suffer total internal reflection.

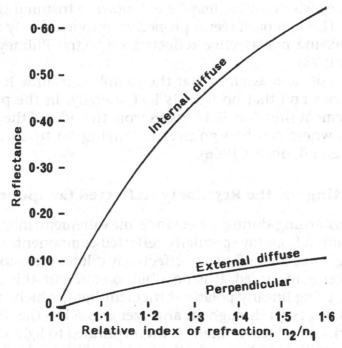

Figure 3–14 Internal and external reflectances of completely diffused incident flux as functions of the refractive index ratio n_2/n_1. Reflectance for perpendicular incidence is also shown (Judd and Wyszecki 1975, copyright, reprinted with permission of John Wiley and Sons, Inc).

The value of k_e is found from the Fresnel equation above; the maximum value is 0.04. For diffuse irradiation, k_e can be found from tables (Campbell and Billmeyer 1971). As shown in Figure 3–14, k_i can be considerable, and for a typical paint film of $n = 1.5$, it has a value of 0.6 for a perfectly diffuse internal flux. Saunderson (1942) recommended the use of 0.4 for plastics (Billmeyer and Abrams 1973). Values for $k_e = 0.04$ and $k_i = 0.6$ have been used for plastics (Brockes 1964).

The Kubelka-Munk treatment was developed for monochromatic radiation because K particularly is sensitive to wavelength. However, Mackinney et al. (1966), Huang et al. (1970), and Gullett et al. (1972) have indicated that the treatment, when used with caution, may be extended to tristimulus values X, Y, Z. This approach seems properly appropriate only for materials having nonselective reflectance spectra (Billmeyer and Abrams 1973).

The model also assumes that the sample is infinite in lateral dimensions and that no light is lost laterally. In the practical measurement situation, light losses from the sides of the sample and elsewhere can be significant, ranging up to 25 percent (Atkins and Billmeyer 1966).

Correcting for the Regularly Reflected Component

Problems arising during reflectance measurement include the contribution from the specularly reflected component. This imparts a gloss to the surface and effectively dilutes the color of the object being measured. It is possible to eliminate this component by using linearly polarized incident light, which, after reflection, is passed through an analyzer crossed to the direction of polarization. Measurement is then confined to light depolarized by multiple scattering in the body of the sample. For fresh meat, the reflectance spectrum so obtained is still not identical to the transmission spectrum of oxymyoglobin. This is probably due to wavelength-dependent scattering (Pfau 1966).

An extension of the principle is used in the clinical examination of skin condition. It is possible to highlight preferentially the light reflected from the dermis and to suppress light reflected from the surface of the stratum corneum, and vice versa. Hence, by using polarized illumination, it is possible to produce measurements and photographs specific to various clinical conditions by analyzing with respect to polarization angle (Philp et al. 1988).

REFERENCES

Aston SM, Bellchambers HE. (1969). Illumination, colour rendering and visual clarity. *Lighting Res Technol* 1:259–265.

Atkins JT, Billmeyer FW Jr. (1966). Edge-loss errors in reflectance and transmittance measurement of translucent materials. *Mater Res Stand* 6:564–569.

Atkins JT, Billmeyer FW Jr. (1968). On the interaction of light with matter. *Color Eng* 6(3), 40–47.

Bayvel LP, Jones AR. (1981). *Electromagnetic scattering and its applications*. London: Applied Science.

Billmeyer FW Jr, Abrams RL. (1973). Predicting reflectance and paint color of paint films by Kubelka-Munk analysis. *J Paint Technol* 45:23–38.

Billmeyer FW Jr, Richards LW. (1973). Scattering and absorption of radiation by lighting materials. *J Color Appearance* 11(2):4–15.

Birkett RJ. (1985). The appearance of concentrated colloidal dispersions. In *Proceedings of the Fifth Congress of the International Colour Association*, Monte Carlo.

Bridgeman I. (1987). The nature of light and its interaction with matter. In *Colour physics for industry*, ed. R McDonald, 1–34. Bradford, UK: Society of Dyers and Colourists.

British Standard. (1967). *British Standard specification for Artificial Daylight for the Assessment of Colour, Part 1, Illuminant for Colour Matching and Colour Appraisal*. BS950, part 1.

Brockes A. (1964). The relationship between the colour strength and particle size of coloured pigments according to Mie's theory. *Optic (Stuttgart)* 21:550–565.

Campbell ED, Billmeyer FW Jr. (1971). Fresnel reflection coefficients for diffuse and collimated light. *J Color Appearance* 1(2):39–41.

Dalgleish DG, Hallett FR. (1995). Dynamic light scattering: applications to food systems. *Food Res Int* 28:181–193.

Dunkerley JA, Ganguli NC, Zadow JG. (1993). Reversible changes in the colour of skim milk on heating suggest reversible heat-induced changes in casein micelle size. *Aust J Dairy Technol* 48:66–70.

Einhorn HD. (1976). Colour preference index. *In CIE compte rendu, 18th session*, ed. Commission Internationale l'Eclairage 36;297–304. London.

Fletcher TOT. (1978). *Top fruit grading and blemish detection in artificial light*. R & D Project no. 6FR 77. London: Ministry of Agriculture, Fisheries, and Food.

Francis FJ, Clydesdale FM. (1975). *Food colorimetry: theory and applications*. Westport, Conn: Avi.

Gullett EA, Francis FJ, Clydesdale FC. (1972). Colorimetry of foods: orange juice. *J Food Sci* 37:389–392.

Huang I-L, Francis FJ, Clydesdale FC. (1970). Colorimetry of foods: color measurement of squash using the Kubelka-Munk concept. *J Food Sci* 35:315–317.

Judd DB, Wyszecki G. (1975). *Color in business, science and industry*. 3rd ed. New York: John Wiley.

Kehlibarov T, Atanasov B. (1973). On the choice of illumination for increasing the contrast of colours in the grading of tobacco leaves. In *Proceedings of the Second Congress of the International Colour Association*, 544–545. London: Adam Hilger.

Kerker M, Farone WA, Jacobsen RT. (1966). Colour effects in the scattering of white light by micron and submicron spheres. *J Opt Soc Am* 56:1248–1255.

Mackinney G, Little AC, Brinner L. (1966). Visual appearance of foods. *Food Technol* 20:1300–1306.

Medical Research Council. (1965). *Spectral requirements of light sources for clinical purposes*. Mem. no. 43. London: Medical Research Council.

Meneer RR. (1984). *Lighting of horticultural packing houses*. Leaflet 846. London: Ministry of Agriculture, Fisheries, and Food.

Negueruela AI, Echavarri JF, Ruiz MJ. (1988). Contribution to the study of wine's color: colorimetry of dilutions of red wines. *Optica Pura Aplicada* 21:63–71.

Pfau A. (1966). Regulare und diffuse Reflexion beim Fleisch. *Naturwissenschaften* 53:553.

Philp J, Carter NJ, Lenn CP. (1988). Improved optical discrimination of skin with polarized light. *J Soc Cosmet Chem* 39:121–132.

Saunderson JL. (1942). Calculation of the color of pigmented plastics. *J Opt Soc Am* 32:727–732.

CHAPTER 4

Vision

Vision is useful because sight helps to keep us from bumping into things. Most objects and events within the natural world that are dangerous reflect or emit radiation within the wavelength range to which our eyes are sensitive. This sensitivity applies to only a small fraction of the electromagnetic radiation that is around us. The full spectrum spans wavelengths from 10^{-5} nm for gamma radiation through X-rays and ultraviolet to the visible range, approximately 300 to 700 nm, and increases through infrared, shortwave radio, radar and microwaves, television at 10^{10} nm, broadcast radio at 10^{11} nm, and long radio waves at wavelengths greater than 10^{16} nm.

Color vision evolved to help us find food as it presented itself or grew against the color of its natural background. It helps us evaluate the quality of what we select to eat. Vision detects objects and lights in the scene and provides information on color, form, movement, alignment, and position.

THE EYE

The cornea and lens form an image that falls upon the retina, a light-sensitive area covering a large portion of the inside of the eye (Figure 4–1). Before reaching the retina, light must travel through the vitreous humor and the preretinal structures. The latter contain the nerve cells and their interconnections, which transmit signals stimulated by the image to the optic nerve. Accuracy of focus is governed by the volume of the aqueous hu-

91

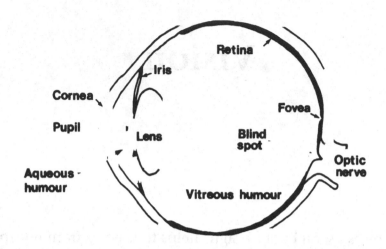

Figure 4–1 Horizontal cross-section of the human eye.

mor, which controls the bulge of the cornea. The ability to change focus is governed by the circumferential muscle around the lens. The iris regulates the amount of light falling onto the retina through control of the pupil size. The vitreous humor provides pressure to stabilize eye shape, keeps the light path to the retina clear, and prevents retinal detachment.

The retina acts as a transducer between light entering the eye and the processes of light and color perception taking place in the visual cortex of the brain. The light-detecting elements of the retina are the morphologically distinct 120 million rods and 7 million cones. Approximately 1 million ganglion cells carry information from the retina to the optic nerve. At the fovea the convergence is one to one, while at the periphery it is several hundred to one. This degree of convergence determines spatial resolution and sensitivity. Hence, at the faveola, a spot 2 mm in diameter in the center of the fovea containing only cones, there is low light sensitivity but maximum resolution to 1 minute of arc. The number of rods increases to a maximum approximately 20° from the fovea. Here, there is high light sensitivity and low spatial resolution. This area is colored yellow probably as a protection against ultraviolet light. This is the angle used for improved sight in the dark, when we use averted vision. Outside this area, color discrimination is zero, and the remainder of the retina is probably used solely for the detection of movement.

There are three types of cone, each having a characteristic distribution of wavelengths over which it responds to incoming radiation. Rods function under low levels of illumination; this is *scotopic* vision. Under these conditions, cones do not operate, and there is no color discrimination. Cones operate at higher levels of illumination; this is *photopic* vision. *Mesopic* vision occurs when the illumination intensity is between that required for photopic and scotopic vision. In this range, there is a combination of cone and rod response and a gradual shift from one type of response to the other.

Purkinje observed in 1825 that under daylight conditions the eye is most sensitive to yellow green but that at night maximum sensitivity shifts to blue green and that as light levels decrease, blues tend to appear progressively brighter and reds darker relative to one another (Hunt 1998). The overall spectral luminosity functions of the rod and cone responses shown in Figure 4–2 illustrate these observations. Such curves are obtained by asking

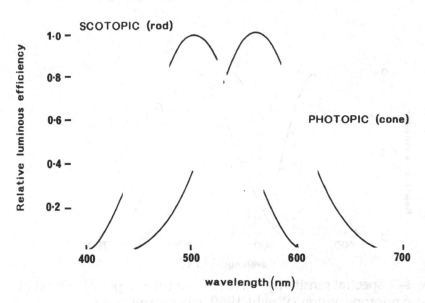

Figure 4–2 Spectral sensitivity curves of cone and rod visual mechanisms (Wright 1969, with permission).

observers to adjust the relative strengths of light at different wavelengths for quality of perceptual intensity, using low- or high-intensity stimuli.

The spectral sensitivity of each type of cone is shown in Figure 4–3. These curves are obtained using color-matching experiments, but spectral sensitivities can be confirmed from direct observations on individual cones using a microspectrophotometer. The three cone types have peak sensitivities in the blue, green, and yellow-green parts of the spectrum. They are called respectively, the blue (β), green (γ), and red (ρ) receptors. The different sensitivities to wavelength provide the foundation for color vision. Light of different wavelengths impinging on the retina induce responses in the β, γ, and ρ cones of magnitudes dictated by the spectral sensitivities.

There are six specific areas, V1 to V6, within the visual cortex responsible for the perception of different aspects of appearance. V1 responds to orientation and real and imaginary boundaries and has some wavelength response. It detects overlapping features in the scene. Cells in V2 are sensitive to color, motion, orientation, and stereoscopic features. V3 is sensitive to form and depth. V4 is sensitive to color and is the site responsible for

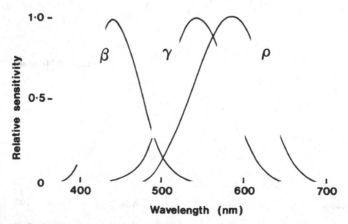

Figure 4–3 Spectral sensitivity curves of the three types of cone that compose photopic vision (Wright 1969, with permission).

the maintenance of color constancy. V5 analyzes motion, and V6 is responsible for analyzing the absolute position of an object in space (Lee 1997). Information from the retina is processed in two broad systems, one concerned with identification of the object and the other concerned with relative spatial position (Tovée 1996).

Visual recognition occurs when a retinal image of an object matches a representation stored in the memory. The ability to recognize objects depends on visual neurons' being able to form networks that can transform a pattern of points of different luminous intensity into a three-dimensional representation. This enables the object to be recognized from any viewing angle. Neuron cells responsible for specific recognition tasks have a columnar organization. Adjacent cells usually respond to very similar feature configurations. These simple shapes form a "visual alphabet" from which a representation of more complex shapes can be constructed. Some neurons, called face cells, may form a neural substrate for face processing. Such cells respond only to faces, whether in life, as plastic models, or on the screen.

Moving objects can be tracked using pursuit eye movement. Considerable neuron interconnections are required to follow objects that are continuously displaced from one point to another. Two other types of eye movement occur. Continual small movements are needed to destabilize the image and prevent the retina adapting to a continuous stimulus, and larger short movements permit the eye to scan the visual environment. Three distances appear to be relevant within this depth perception. These concern the personal space occupied by our body, peripersonal space within reach, and extrapersonal space beyond (Tovée 1996).

Eye function and vision have been reviewed by Krauskopf (1998), Hunt (1998), Lee (1997), Tovée (1996), and Baylor (1995).

COLOR PERCEPTION

There is a highly complex series of connections between the detector and the fibers making up the optic nerve that take the resulting signals to the visual cortex. This link can be depicted using a simplified hypothetical framework (Hunt 1998). There appear to be not four but three types of nerve fiber. One carries achromatic or brightness information, and the other two carry color information—the opponent color mechanisms. This model is shown in Figure 4–4.

Color information from the retina is obtained from a series of signal differences:

$$\rho - \gamma = C_1$$
$$\gamma - \beta = C_2$$
$$\beta - \rho = C_3$$

These are transmitted as two signals, C_1 and $(C_2 - C_3)$.

The achromatic signal under photopic conditions comprises all three ρ, γ, and β responses, weighted to compensate for the differing number of each type of cone (40:20:1) in the retina. That is:

$$2\rho + \gamma + \beta/20$$

Hence, the total achromatic signal can be described as

$$A = 2\rho + \gamma + \beta/20 + S$$

where S defines the rod response. When $\rho = \gamma = \beta$, $C_1 = C_2 = C_3 = 0$. The value of A indicates the level of brightness and grayness (degree of whiteness or blackness).

BASIC COLOR ATTRIBUTES

Hunt (1998) defined three color attributes.

1. *Brightness*—the attribute of a visual sensation according to which an area appears to exhibit more or less light (adjectives *bright/dim*)
2. *Hue*—the attribute of a visual sensation according to which an area appears to be similar to one, or proportions of two, of the perceived colors red, yellow, green, and blue
3. *Colorfulness*—the attribute of a visual sensation according to which an area appears to exhibit more or less of its hue

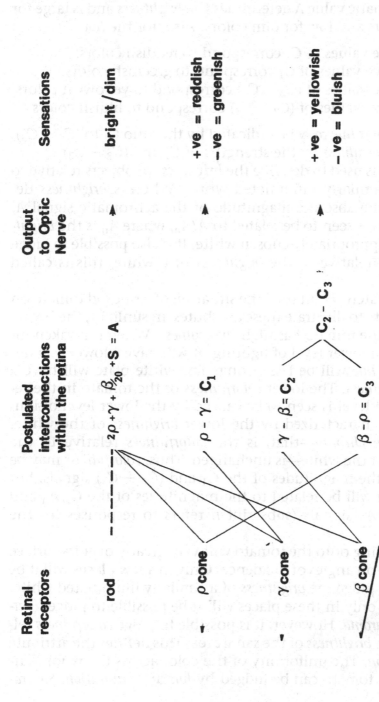

Figure 4-4 A postulated representation of the interconnections between receptors and the subsequent signals on which color sensations depend (Hunt 1991, reproduced from *Measuring Color*, with permission from Ellis Horwood, Ltd., Chichester).

The achromatic value A determines the *brightness* and is large for bright colors and low for dim colors. Also, for the *hue:*

- Positive values of C_1 correspond to reddish colors.
- Negative values of C_1 correspond to greenish colors.
- Positive values of $(C_2 - C_3)$ correspond to yellowish colors.
- Negative values of $(C_2 - C_3)$ correspond to bluish colors.

The particular *hue* may be indicated by the ratio C_1 to $(C_2 - C_3)$, and the *colorfulness* by the strengths of C_1 and $(C_2 - C_3)$.

Lightness is used to describe the *brightness* of objects relative to that of a similarly illuminated white. Whereas *brightness* depends on the absolute magnitude of the achromatic signal A, the *lightness* is seen to be related to A/A_n, where A_n is the *brightness* of an appropriately chosen white. It is also possible to judge *colorfulness* relative to the *brightness* of a white; this is called *chroma.*

Appropriately, Hunt used the situation of a ripe red tomato on a white plate to illustrate these attributes. In sunlight, the strong red *hue* of the tomato has high *colorfulness.* When it is taken indoors into a lower level of lighting, it will have a lower *colorfulness,* as its *hue* will be less strong. The white plate will have a lower *brightness.* The lower *colorfulness* of the tomato in the reduced light level is seen to be caused by the lower level of illumination, characterized by the lower *brightness* of the white. Hence, the *chroma*—that, is the *colorfulness* relative to the *brightness* of the white—is unchanged. Thus, *colorfulness* may be related to the magnitudes of the C_1 and $(C_2 - C_3)$ signals, but the *chroma* will be related to the magnitudes of the C_1/A_n and $(C_2 - C_3)/A_n$ signals (subscript n refers to responses for the white).

Light falling onto the tomato will vary greatly over its surface depending on angles of incidence. Only in a few places will it be possible to judge the *brightness* of a similarly illuminated white, and hence only in these places will it be possible to judge *lightness* and *chroma.* However, it is possible to judge *colorfulness* relative to the *brightness* of the same area. This defines the attribute of *saturation.* The uniformity of the color across the whole surface of the tomato can be judged by *hue* and *saturation. Satura-*

tion may be related to the magnitudes of the C_1/A and $(C_2 - C_3)/A$ signals.

Hunt (1998) defined these three relative subjective terms as follows:

1. *Lightness:* the *brightness* of an area judged relative to the *brightness* of a similarly illuminated area that appears to be white or highly transmitting (adjectives: *light* and *dark*)
2. *Chroma:* the *colorfulness* of an area judged in proportion to the *brightness* of a similarly illuminated area that appears to be white or highly transmitting (adjectives: *strong* and *weak*)
3. *Saturation:* the *colorfulness* of an area judged in proportion to its *brightness*

The six subjectively judged attributes may be used to describe our perception of our environment, whether in the home, the supermarket, or the kitchen. Hunt's scheme may be applied to the assessment of complex scenes.

ADAPTATION AND CONSTANCY

Lightness and color constancy contribute to a more stable perceived world. Light intensity visible to human beings ranges from that of the sun at noon (10^{10}) to the weakest 10^{-6} candelas/m^2. However, the intensity ratio within the optic nerve is approximately 100:1, the operating range of the visual system changing with ambient light level. This is *light adaptation*. It allows *light constancy* to be possible. That is, a surface will appear equally light, relative to the surround, over a range of illumination.

Cones respond to different wavelengths of light, but the relationship between wavelength and the color we perceive is not simple. Wavelengths reflected from an object vary according to the quality of light falling onto it. Yet whatever the light, from dawn to dusk, coal still appears black and leaves green. When a green object is brought indoors from natural daylight to an area of tungsten lighting, it reflects yellow-green light. However, this change does not appear to apply to the object. The color change is recognized as belonging to the light, and we perceive the object color relative to this. That is, the object remains almost as green as it did in daylight. As we become adapted to the new illumination, light reflected from the object is perceived to have almost the same color as when daylight was reflected from it. This phenomenon is *object-color constancy* (Judd and Wyszecki 1975). However, color constancy of, for example, spectrally selective materials is not absolute. Reddish lighting on fresh beef is sufficient to conceal the first visual signs of the brown metmyoglobin, characteristic of older meat, that would be revealed under bluer cooler lighting.

Color constancy is achieved through a comparison of wavelengths reflected by nearby objects with their overall brightness. White tends to appear white, whatever the light shining on it, as long as our eyes are adapted. In a scene containing a series of achromatic colors (white or gray), the brightest will be seen as white, while the others will appear to be gray. If this white is removed, the next brightest gray will come to be called white. That is, our perceptual system notes the highest brightness achromatic in the field of view as white and then relates all other colors to it. In the absence of grays, other colors may act as reference whites also. This phenomenon occurs for a wide range of illuminants, certainly all common daylight-type fluorescent and tungsten lamps. As the perceived white is a reference for all other colors, these tend to be perceived as constant under different illuminations. Grayness is perceived as a relative reduction in the proportion of light reflected. That is, there must be a reference to the total amount of light present. Mistakes in perception can occur if there is no clue to the total amount of light present. Brocklebank (1967) has used the example of a gray wall in a room being the lightest surface in view. This wall may be mistaken for white and the level of illumination judged lower than it actually is. When someone walks in front of the wall, either his or her complexion will appear unnaturally vivid, or the wall will suddenly change from white to gray.

Adaptation taking place on a change of illumination is not complete. The color shift on adaptation is a mixture of an objective (colorimetric) shift and a subjective (adaptive) color shift. The changed spectral composition of the light source causes a change in the spectral composition of the light leaving the object, and the result is a colorimetric shift. The adaptive color shift is caused by *chromatic adaptation* (Judd and Wyszecki 1975).

There is another type of constancy. When we see shadows on a plain plate, we do not immediately think that different colored paints have been used in the finish. We subconsciously assume that it is uniformly colored. That is, we make an immediate adjustment to allow for the change in illumination falling on the different areas. Similarly, parts of the surface of a green apple in a bowl of fruit may be in shadow or may be altered by reflections from adjacent differently colored fruit. However, we perceive the apple to be uniformly green. A bruise on the apple will probably be seen because we can recognize that the damaged surface is not the same color or shape that would result from a shadow or reflection from other items in the bowl. When we look at our red car in the drive, we do not examine the color in detail. If other factors, such as size, shape, and design, are appropriate, it must be the same color as it was before. That is, for this type of constancy, our memory and other appearance properties govern the color perception. This might be termed *total appearance constancy*.

Problems caused by chromatic adaptation can occur when serving food. When outside in the sunlight, our plate will appear white and the tomato red. When we bring them indoors and look at them under tungsten light, the plate will still be white and the tomato red, but neither will appear to be so bright. Food prepared in the bright light of the kitchen may be presented in a redder, less bright ambience of the restaurant. Hence, the customer's perception may be totally different from that intended by the chef. This occurs with restaurant decor, particularly with dimly lighted decor. If the decor possesses the wrong balance of brightness or color, the derived perception of it may be dirty and therefore unhygienic.

Part of the physiology of color constancy is based on the Young-Helmholtz theory of color vision. This postulates the presence of three types of cone, sensitive to shortwave, middle-wave, and long-wave parts of the visible spectrum. When the eye is exposed to the reddish light emitted by a tungsten lamp, the red receptors and to a lesser extent the green receptors are partially desensitized. The blue receptors are hardly affected, and adaptation to reddish light results in a relative gain in sensitivity by these receptors. Then, when the reddish illumination is changed for one with a bluer bias, the response from the blue receptors is proportionally greater (von Helmholtz 1866). In accepting this approach, it has been postulated that adaptation can be accounted for simply by reducing the three cone sensitivities by constant factors (von Kries 1905). The correction necessary is actually more complex (Halstead 1977). Methods for calculating corrections for color adaptation and constancy are described by Hunt (1998).

In survival terms, adaptation has significance for human beings in maintaining the stability of some perceptions in circumstances of an ever-changing external environment. Much research has been carried out into the mechanisms of adaptation in theoretical terms and in terms of such industrial implications as color rendering of light sources. This area has been reviewed by Hunt (1998) and Hill (1987).

CONTRAST PHENOMENA

Lightness contrast is well known; it is the reason we can see stars at night but not in the daytime. Color appearance is also affected by contrast with surround. Faced with a plate of food, our eyes constantly move across the scene, focusing on different patches of color that come into view. Our state of adaptation, chromatic sensitivity, and color perception change as outlined above. *Afterimage* effects also contribute to our perception. For example, if before the meal arrives, our eyes have fixated on an orange tablecloth, food on the plate will be tinged with complementary blue. Such complementary color effects can be uncomfortable when, for example, sorting green peas on a conveyor belt induces perception of red spots.

The color of an object in a complex scene depends, through various contrast mechanisms, on other colors that also happen to be in the scene. Unlike afterimage effects, these contrast phenomena require no fixation for their manifestation.

Successive contrast occurs when the eye is first exposed to one color and then immediately to a second. As described for afterimage, on exposure to the first color, say red, the ρ cones will tend to become desensitized. When the second color is substituted, the signal from the ρ cones will tend to be low in comparison to the γ cones' signal. The resulting $(\rho - \gamma) = C_1$ signal will be more negative, and the object will therefore appear to be too green. So a yellow area can appear green if the eyes are first adapted to red for a short while.

Simultaneous contrast occurs when we look at colors placed side by side. One color assumes some of the hue of the complementary of the adjacent color. For example, a gray circle on a green background will appear pinkish. When a carrot and a potato are viewed together on a plate, the potato has conferred upon it the blue that is complementary to the carrot's orange. When complementary colors are placed together, both are exaggerated. The orange of our carrot is intensified when served on a blue plate. In the window of the butcher's shop, the presence of green herbs around a piece of raw beef enhances the redness of the meat. Sometimes the effects of adjacent complementaries are uncomfortable in their boldness.

Other types of contrast are of interest to designers. *Border contrast* occurs when changes in the border zone between two colors tend to increase the perceived difference between the colors. *Contour contrast* is effective in enhancing apparent brightness or volume. For example, if, within a uniformly colored surface, the circumference of a circle is marked by a small area of shading, the surrounded circle will appear to be a lighter color. *Vibration,* an intense shimmering effect, can be induced along contours, particularly when complementary hues of approximately the same lightness are involved. In *optical mixing,* pointillist artists exploit the fact that, from a distance, the eye cannot perceive the edges of small shapes. These tend to blend into each other, the result being seen as a mixture. An illusion of *transparency* occurs when there are three adjacent colors if the middle one is a mixture of the other two (Bloomer 1976).

These and other complex visual effects appear to be caused through the mechanisms of simultaneous and successive contrast and an interaction of excitatory and inhibitory mechanisms in adjacent cones. Contrast phenomena arise as a result of the different functions within the visual cortex of color selectivity, contrast selectivity, temporal properties, and spatial resolution. These result in the different visual functions of color, depth, movement, form, and orientation perception (Davidoff 1991).

COLOR MEMORY

Color and appearance memory are used in making decisions about food quality. For food, we judge freshness, ripeness, naturalness, extent of cooking, bruising, and spoilage. Of the environment, we judge perhaps cleanness and safety. These judgments are governed by our upbringing, culture, and how we are feeling at the time, as described in Chapter 2. Food color and appearance are compared with mental images formed from past experience. If something about a food or environment does not correspond with our memory of wholesomeness and hygiene, the product may be left on the shelf or the plate.

The presence of a mental image is a significant factor in how quickly we recognize an object when it appears in front of us (Finke 1986). If the real object is, say, differently aligned from the image, our recognition speed is decreased. Possibly similar mechanisms exist by which we are disappointed if the quality of the object does not match the quality pictured by our mental image. Finke also postulated that once an image is formed, "It can begin to function in some respects like the object itself, bringing about the activation of certain types of neural mechanisms at lower levels of the visual system." Whatever defects exist in our original observation of an object, they carry through to the mental image of the object. Hence, "Mental images may come to acquire visual characteristics and may serve to modify perception itself." This has profound cross-cultural implications that are considered further in Chapter 2.

Humans do not memorize colors well. A tie purchased to match a shirt in the wardrobe may not be a good match on the return home. In testing such an assertion, Bartleson (1960) determined the "memory colors" of 10 naturally occurring objects, such as flesh, red brick, green grass, and blue sky. Subjects selected their memory colors from an array of 931 Munsell color chips. Each memory color tended to be more characteristic of the dominant chromatic attribute of the object in question. That is, grass was greener, and bricks were redder. Also in most cases, saturation and lightness increased in memory.

Individuals can be trained to memorize color for a particular application. For example, experienced daffodil breeders could correctly describe the color of a particular bloom in terms of the Royal Horticultural Society Colour Chart code without reference to the actual charts. However, these individuals found great difficulty in giving other yellow flowers their correct codings (Battersby 1978). From an evolutionary point of view, these are abstract situations. Human beings were never meant to perform them successfully. We have, however, developed the ability to recognize particular stages in the natural ripening and spoilage processes of foods. Color memory is used for tasks vital to our well-being. Even this ability may not be tested very often, since in most cases, contrast colors are present during our visual ripeness and spoilage judgments. Examples include bruising of fruit or the appearance of a slight green tinge to meat. Hence, it is likely that it is the aptitude of the visual system in judging contrasts that accounts for many of our food quality decisions. Judgments of product quality are frequently made from a visual image. Hence, when disciplined judgments of flavor and texture must be made, it is essential that they be made in the absence of cues supplied by the product's visual properties (see the discussion of the halo effect in Chapter 5).

Color memory in the abstract may be formally assessed using the Burnham-Clark-Munsell test. The test uses 85 Farnsworth-Munsell 100 Hue Test discs. Discs are of different hues but similar chroma and value. Alternate discs from the set are laid out in a circle. The subject is shown duplicates of 20 of the discs, one at a time, for a standard period. After a given interval, one disc from the circle is selected as identical to the test disc. The accuracy of the selection is used to quantify color memory ability (Burnham and Clark 1954). Color memory is best among those between the ages of 25 and 39 (Lakowski and de Beck 1973).

COLOR HARMONY

Color harmony occurs when two or more colors seen in neighboring areas produce a pleasing effect. It is a purely personal reaction to a perception. Harmony is used and exploited by designers in their bid to communicate and create an aesthetically pleasing scene (see Chapter 2). Appreciation of the designer's efforts depends on the viewer's learning and appreciation of fashion, on whether the viewer has become tired of, or has begun to appreciate, certain color combinations, and on the shapes, layout, and relative sizes of colored areas. Harmony is also affected by the absolute angular size of areas. For example, a Roman mosaic multiplied 10 times would produce a garish effect. We also back away from large paintings to obtain a better view. Undesirable contrasts can occur through chromatic adaptation, and area effects are discussed above under "Adaptation and Constancy." Manuals containing examples of suggested color harmony for use in various design situations have been published (e.g., Kobayashi 1984; Hihara et al. 1989; Shibukawa and Takahashi 1990).

Personal skill tests are available for assessing some elements of observer potential for working with color. The Japanese Color Aptitude Test determines color discrimination ability using a series of triangle tests, as well as knowledge and skills of color scaling in terms of the three Munsell dimensions, Hue, Value, and Chroma (Billmeyer and Kawakami 1995). A wider ranging color sense test can be used to assess a subject's depth of interest and understanding in the use of color in color harmony and design. As well as including standard color vision, discrimination, memory, and identification tests, it requires subjects to select color combinations suitable for given house designs and for a range of products (Minato 1989).

IMAGE COLOR QUALITY

Increasing use of photographs and video images to represent food has called for consideration of such picture attributes as *quality* and *naturalness*. As already noted under "Color Memory," observers often prefer scene elements of a picture to be different from the original. For example, skies remembered and preferred are bluer, skin is more saturated, and sunsets are redder (Hunt 1982).

Quality, colorfulness, and *naturalness* judgments were made of a series of four colored images. The images were of a young woman, fruit, a terrace scene with white furniture and a colored umbrella, and a town hall with concrete and some greenery. A series of each image was produced by changing each pixel chroma. In every series, there was one image judged as having maximum *quality* that also had a chroma greater than the average chroma of the original—in agreement with Hunt. The difference between the original image and the changed image was most marked in an image of low average chroma.

Colorfulness in this experiment was defined as *chromaticness,* that attribute of visual sensation according to which the perceived color of an area appears to be more or less chromatic. *Colorfulness* increased with increase in chroma for all four scenes and was the main perceptual attribute determining *quality*. *Colorfulness* depended on the distance of image colors from the neutral gray and the distance between individual image colors. This was modeled using a linear equation containing the average chroma within the picture and the standard deviation of the distribution of the chroma values of all individual colors. The high multiple correlation coefficient of 0.977 indicated some validity for the model.

There is a strong link between image *quality* and *naturalness,* but they are not identical. In images of natural scenes, especially the woman, the relationship was strong. It was proposed that individuals sometimes use different criteria when assessing these attributes (Fedorovskaya et al. 1997).

COLOR VISION DEFICIENCY

Approximately one man in 13 and one woman in 250 perceive colors in a markedly different way from the remainder of the population. An example concerns the farmer's son who had trouble picking cherries "because the blamed things match the leaves in colour" (Chapanis 1951). Color vision deficiency in relation to viewing foods has not been well researched, but if visual impressions are important, color-defective individuals may substitute other parameters for color in their food appreciation. Individuals with poor color perception learn the socially approved color names for many objects (Amerine et al. 1965).

The other 92 percent of the population do not perceive colors in exactly the same way. For instance, the wavelength at which the peak in the photopic curve occurs (Figure 4–2) varies by 20 nm. People, as well as foods derived from natural sources, belong to biological populations in which large variations occur. This is one of the facts of life in the study of food color and appearance.

Color-deficient vision can be inherited or acquired as a result of retinal or optic nerve pathology. Deficiency can arise from illness or ingestion of neurotoxic agents, although the most common acquired form is from cataract and glaucoma in old age. As we grow older, our color perception and color sensitivity decline. Ability to discriminate yellows decreases as yellowing in the eye lens increases between the ages of 60 and 90 (Yoshida 1997). Discrimination in 80-year-olds is worse in the blue-green region than for red-yellow shades (Wijk et al. 1997). The change can be extensive (Granville 1991). Subjects in their 60s and 70s perceive colored surfaces as having less chromatic content than do those under 30. This is not caused by physical changes within the eye but is possibly caused by changes in neural mechanisms responsible for processing hue (Schefrin and Werner 1993).

Defects arise when the response functions of the cones, or very occasionally the rods, are different in some way. The most common causes relate to a deficiency (5 percent; deuteranomaly) or an absence (1 percent; deuteranopia) of the γ cones or to a deficiency (1 percent; protanomaly) or an absence (1 percent; protanopia) of the ρ cones. Defect or absence of the β cones is unusual. The presence of these conditions can be detected by using Ishihara charts (Kanehara Shuppan, Tokyo), which consist of a series of plates made up of spots of different colors. The spots may wrongly be described as being the same color; hence, the charts are known as pseudoisochromatic tests. These can be used to detect the presence of a color vision deficiency but cannot necessarily give information as to the type.

Color deficiency is a significant handicap only in certain technical situations. Individuals with one of these conditions will not be able to perform exacting color tasks, such as mixing colors to produce a generally acceptable color match. These skills are required, for example, in the textile dyeing, ceramics, and paint industries. Problems can also occur in the inks, photographic, printing, and electronics industries and among air traffic controllers, color television technicians, and art restorers. Recently occurring problems in the food industry include picking tomatoes too early while they are still unripe and producing green sweets instead of red ones.

A list of jobs, careers, and industries that have different degrees of vulnerability to abnormal color vision has been compiled by Fletcher and Voke (1985). Included are the following areas of interest to the food and agriculture industries. There are three categories of defect:

1. "Where defective colour vision is a handicap and in which important consequences might result from errors of judgement." This includes buyers (e.g., of fruit, cocoa, and tobacco), food chemists, market gardeners (e.g., of fruit), meat inspectors, tobacco graders, and cotton graders.
2. "Where good colour vision is desirable but in which defective colour vision would not necessarily cause a handicap." This includes bar staff, farmers, bakers, fish mongers, botanists, foresters, brewers, butchers, grocers, buyers (e.g., of fruit, cocoa, and tobacco), horticulturalists, waiters, and cooks or chefs.

3. "Jobs and careers requiring perfect colour vision." These include navigators and fishermen.

There has been a feeling—for example, in the armed services—that only dichromats (those who have lost the response from one cone type) need be excluded from occupations such as pilot or driver. That is, only those with protanopia and deuteranopia should be excluded from signals-dependent occupations (Fletcher and Voke 1985).

In summary, those within the food industry who select and grade raw materials and products on the basis of color ought to be chosen with care. For a potential grader in the industry, a screening with the Ishihara charts, although not perfect, will probably be sufficient to detect unsuitable candidates. The probability that a person with perfect color vision will pass the test—that is, make no more than three plate errors—is 0.99. The probability that a person who has defective color vision will fail the test (have more than three mistakes) is 0.82. These charts must be used according to the strict instructions supplied and under the recommended lighting and viewing conditions. Further investigations using more complex color vision–testing techniques must be carried out by someone with training and experience in their application and interpretation.

REFERENCES

Amerine MM, Pangborn RM, Roessler EB. (1965). *Principles of sensory evaluation of food.* New York: Academic Press.

Bartleson CJ. (1960). Memory colors of familiar objects. *J Opt Soc Am* 50:73–77.

Battersby BKA. (1978). Accuracy in colour memory: a study of colour assessment in flowers. In *Proceedings of the Second Congress of the International Colour Association*, 409–412. London: Hilger.

Baylor D. (1995). Colour mechanisms of the eye. In *Colour: art and science*, ed. T Lamb, J Bourriau. Cambridge, UK: Cambridge University Press.

Billmeyer FW Jr, Kawakami G. (1995). New edition of Japanese color aptitude test. *Inter Soc Color Council News* 355:5–7.

Bloomer CM. (1976). *Principles of visual perception.* New York: van Nostrand.

Brocklebank R. (1967). Perception of colour. *Building Mater* 27(7):51–52, 55.

Burnham RW, Clark JR. (1954). A color memory test. *J Opt Soc Am* 44:658–659.

Chapanis A. (1951). Color blindness. *Sci Am* 203 (March):48–53.

Davidoff J. (1991). *Cognition through color.* Cambridge, Mass: MIT Press.

Fedorovskaya EA, de Ridder H, Blommaert FJJ. (1997). Chroma variations and perceived quality of color images of natural scenes. *Colour Res Appl* 22:96–110.

Finke RA. (1986). Mental imagery and the visual system. *Sci Am* 254 (March):76–83.

Fletcher R, Voke J. (1985). *Defective colour vision.* Bristol: Hilger.

Granville WC. (1991). Colors do look different after a lens implant! *Leonardo* 24:351–354.

Halstead MB. (1977). Colour rendering: past, present and future. In *Proceedings of the Third Congress of the International Colour Association,* 97–127. Bristol, UK: Hilger.

Hihara M, Kodama A, Matsui H. (1989). Interior color coordination system. In *Proceedings of the Sixth Congress of the International Colour Association,* vol. 2, 46–49. Buenos Aires: Grupo Argentino del Color.

Hill AR. (1987). How we see colour. In *Colour physics for industry,* ed. R McDonald, 211–282. Bradford, UK: Society of Dyers and Colourists.

Hunt RWG. (1982). Chromatic adaptation in image reproduction. *Col Res Appl* 7:46–49.

Hunt RWG. (1998). *Measuring colour.* 3rd ed. Kingston-upon-Thames, UK: Fountain Press.

Judd DB, Wyszecki G. (1975). *Color in business, science and industry.* 3rd ed. New York: John Wiley.

Kobayashi S, ed. (1984). *Colour image coordination book.* Tokyo: Nippon Color and Design Research Institute.

Krauskopf J. (1998). Colour vision. In *Color for science, art and technology,* ed. K Nassau. Amsterdam: Elsevier.

Lakowski R, de Beck J. (1973). Age effect in the Burnham-Clark-Munsell color memory test. In *Proceedings of the Second Congress of the International Colour Association,* 406–408. London: Hilger.

Lee J. (1997). Before your very eyes. *New Sci* 153(2073).

Minato S. (1989). The CPC test on colour sense. In *Proceedings of the Sixth Congress of the International Colour Association,* vol. 2, 513–515. Buenos Aires: Grupo Argentino del Color.

Schefrin BE, Werner JS. (1993). Age-related changes in the color appearance of broadband surfaces. *Col Res Appl* 18:380–389.

Shibukawa I, Takahashi Y. (1990). *Designer's guide to color 4.* London: Angus and Robertson.

Tovée MJ. (1996). *An introduction to the visual system.* Cambridge, UK: Cambridge University Press.

von Helmholtz H. (1866). Die Lehre von den Gesichtsempfindungen. In W Nagel, *Handbuch der Physiologische Optik,* vol. 2. Leipzig: Voss.

von Kries J. (1905). Die Gesichtsempfindungen. In W Nagel, *Handbuch der Physiologie des Menschen*, vol. 2, 109–282. Braunschweig: Vieweg.

Wijk H, Berg S, Sivik L, Steen B. (1997). Aspects of colour perception in an elderly Swedish population. In *Proceedings of the Eighth Congress of the International Colour Association*, vol. 1, 191–194. Tokyo: Colour Science Association of Japan.

Wright WD. (1969). *The measurement of color*. 4th ed. London: Hilger and Watts.

Yoshida CA. (1997). Sense of the elderly in colour discrimination. In *Proceedings of the Eighth Congress of the International Colour Association*, vol. 1, 94–99. Tokyo: Colour Science Association of Japan.

CHAPTER 5

Sensory Assessment of Appearance Methodology

There are many reasons for sensory testing. In-house applications range from the development of new products, product matching and improvement, cost reduction, and process or raw material change to quality control, testing of storage stability, and product grading. The other major area of sensory testing seeks consumer opinion and judgments of product acceptability and preference. Product appearance is involved in all sensory testing through two enormous effects: first, the intrinsic importance of appearance in the total mix of product properties, and second, the ability of appearance to influence judgments of texture and particularly flavor. Product appearance ought to be concealed when objective judgments of flavor and texture are required.

Research, development, and quality control assessments are made by panels numbering normally from 10 to 20 individuals. These may be drawn from local members of staff or from a pool of external assessors. The latter work part time, and payment is made on an hourly basis. Such groups may meet for one or two sessions each week, with each session lasting approximately three hours. For the majority of these assessors, attendance at the panels is their only paid employment, and motivation can be very high. This, the lack of time constraints, and the absence of any direct involvement in the development of the product usually lead to a thorough, cost-effective, and efficient assessment process (Foster and Kilcast 1988). The number used for expert profiling is approximately 6 and for group discussion is 10 but for an in-home consumer test may need to be 400 (Nelson and Smith 1977).

Sensory evaluation can be objective or subjective. Howgate (1977) has used the term *objective* to refer to a sensory or instrumental procedure in which biases are minimized. The term *subjective* refers to a sensory test in which biases are not minimized. Biases are always present in all sensory testing, but they can be minimized by careful design of the scoring system and by rigorous training for assessor objectivity. Consumer testing, however, seeks to determine the actual likes (or biases) of potential customers.

Problems of establishing the objectivity of judges differ with industry. In daily life, the appearance of a paint on a wall is mainly a subliminal experience. The color it is painted is normally incidental to the business carried on in the room. With food color, the situation is different. We have evolved to judge visually whatever we may be about to eat. There is an interaction between the color and the decision involving whether we eat. It is all too easy to carry this visual judgment into the expert panel situation. In training for objectivity, such instincts must be eliminated. Therefore, asking objectivity-orientated panelists to answer subjective, hedonic questions in the same panel is not recommended (American Meat Science Association [AMSA] 1991).

Within the food industry, vision is the most commonly used means of assessing attributes of appearance. In the paint industry, however, color measurement is firmly established because color can be isolated as a single visual and instrumental property. Full confidence can be placed in the results obtained, even for the high precision and accuracy needed for matching paints. On the other hand, the complex makeup of the appearance properties of foods, their heterogeneous nature, and the short season or the batch nature of manufacture often make sophisticated color measurement and matching techniques unnecessary, unworkable, and uneconomic. Problems arising from agreeing on and defining individual appearance attributes and from the labile nature of foods can normally be solved more easily by discussion. When assessors agree, scales can be devised and samples scored in a disciplined way.

Also, in the food industry, unlike most other colorant-using industries, the number of colorants permitted is strictly limited. Many manufacturers wish to reduce their colorants to a minimum or to market their product with a "no artificial colorants added" label. In these cases, the scope of the producer for matching to a preconceived color depends solely on the skill with which ingredients can be selected and manipulated. If the processor is using synthetic colors in artificial foods such as sweets, adherence to the recipe and process is needed to obtain a consistent product. However, changes of ingredients, including changes in noncolorants, can result in an undesirable change of product appearance. Batches of new materials should be monitored and compared with physical or sensory standards, with the aim of keeping products consistent.

At the heart of this difference between the food and other color-using industries lie the concepts of perceptibility and acceptability of a color difference. In paint, textile, and ceramic industries, inspectors deal with close tolerances. The just-noticeable or just-perceptible difference between two samples is close to the just-acceptable difference. For a resprayed vehicle wing, if a difference can be seen between the new and adjacent old paintwork, it is probably large enough to be unacceptable. The situation with natural foods is different. Large differences in color can exist among slices of roast beef, and all are probably acceptable. That is, the definition of a small difference in color may be meaningless because of the large variation in acceptability. Appearance control is still necessary because appearances acceptable in products made at home may not be acceptable in products to be purchased. For example, homemade mushroom soup is brown, but the purchased version must have a creamier color. Appearance control is necessary also because mass marketing methods have trained customers to expect a high degree of product uniformity. This places additional constraints on the producer.

There are some manufactured products for which consumers seem to have a special sensitivity to color. Examples of these are branded tomato ketchup marketed in glass bottles and brewed tea containing milk. In such cases, it is essential that raw material selection, categorizing, and blending techniques be rigorously established and that manufacturing and marketing techniques be carefully defined and controlled. Visual assessments are normally used for this.

The tea industry provides two examples of skilled disciplined judges working alone. Tea buyers are responsible for judging raw tea's quality, suitability for inclusion in the company stocks, value, and subsequent purchase. On a busy day, a buyer may have to judge as many as 1,500 samples. The tea blender has the task of blending available raw teas into a product consistent throughout the year in flavor, strength, appearance, and overall quality. He or she may have to judge 50 or 60 blends per day (Theobald 1977). Other lone judges are those production line operatives by whose decision processing is changed, continued, or stopped. Milling and baking are examples of traditional craft industries requiring sensory judgments to be made during processing.

Color acceptability can vary across an industry. A study of three experienced meat buyers from different supermarket chains indicated that there were significant differences in what each considered acceptable. Using a 15-point hedonic scale to score 317 carcasses, only moderate agreement was obtained between the judges (mean r^2 approx. 0.4) (Truscott et al. 1984). Although a difference in policy between the supermarket chains may not have been surprising, a better agreement among the judges might have been expected.

A common factor across all industries to which color and appearance are important is that formal sensory assessments of appearance must be carefully planned, controlled and executed. The objectives of the assessment and the purpose of the work must be understood and agreed upon. The running of sensory panels can be expensive in time and labor; therefore, the questions asked of the project and the techniques employed by those in control of the testing must be relevant to agreed-upon objectives. Adequate selection and training of panelists are essential for the effective and efficient performance of the panel.

PANEL SELECTION, SCREENING, AND TRAINING

Selection and training of appropriately qualified and motivated persons for participation in analytical panels are essential to effective performance. The 8 percent of males and 0.5 percent of females possessing color vision deficiencies are unsuitable for participation in panels involving color judgments. Potential panelists for such judgments should be screened using a color vision test such as the Ishihara charts (see Chapter 4).

Also appropriate for those who may be expected to detect and quantify a wide selection of appearance attributes is a test that quantifies speed of response. The Bodmann test is a search task consisting of numbers randomly scattered over a sheet of paper. Subjects are asked to find the positions of specific numbers, and the time taken is an indication of their search speed. The Land Halt Ring and Maze Tests involve the subject in finding gates in circles or mazes. These are recommended for screening those who, for example, sort fruit on high-speed grading lines (Fletcher 1980). A battery of tests have been assembled for defining inspection performance in the engineering industry. These include acuity, inspection, sorting, and memory tests (Gallwey 1982).

Much has been published concerning the training of members of sensory panels, particularly those concerned with quantifying flavor, taste, and aroma (e.g., Dethmers et al. 1981). It has been found necessary to give potential panelists tests for

- *sensitivity,* the ability to recognize basic tastes
- *differences detection,* the ability to detect in a reproducible manner specific variations of the test product
- *differences measurement,* the ability to measure differences reproducibly using appropriate scales for the product under investigation

Where just-noticeable color differences are being determined, the Farnsworth-Munsell 100 Hue Test is beneficial in categorizing panelists passing the Ishihara test. Sensitivity to food color differences can be tested by ranking browning, using solutions of caramel and heat-treated milk (Jellinek 1985). Sensitivity to other appearance factors, such as translucency and gloss, may be determined by discussion and trial. This involves using a series of samples, preferably food, possessing a range of the attributes in question.

Distribution and uniformity of color, translucency, and gloss can be quantified according to the specific problem. Proportions of a discontinuity within an area of color (e.g., proportion of green on the surface of a tomato) may be expressed as a percentage of the surface area or by using a simple rating scale.

Each product application has specific requirements involving the possible development of suitable definitions, scales, and anchor points. There are many areas of the food industry in which such expert experienced judgments are required. An example is the Campden (1988) specification for the internal appearance of corned beef. "The internal colour of the meat shall be a reddish colour typical of the country of origin, with a fine, even distribution of fat. Small air voids may be present but they should be reasonably evenly distributed with no large voids. Overall appearance to show fairly regularly sized discrete pieces of beef muscle." Use of such grade definitions implies a wide experience with the product, as many aspects of this statement require decisions to be made about what is *typical, fine, even, small, large,* and *fairly.* Also, whether one or more than one of these criteria are needed to downgrade the sample is implied through the grader's experience. Such grading procedures are satisfactory and can work well, but discussion, agreement on attributes composing the grade, adequate training, and retraining are vital to their success and uniform application within the particular industry.

Factors Affecting Panel Performance

Many personal and environmental factors upset panelist performance. Care must be taken in panel design to counterbalance or eliminate all possible extraneous variables, including practice and fatigue (O'Mahony 1979). Individuals behave differently, but score variability can be reduced by normalizing scores. This can be achieved using the hypothesis that subjects behave in scaling like instruments working with different sensitivities of range and scales of response and different zero settings. The score of each panelist is recalculated by normalizing sensitivity of each subject to the averaged sensitivity of the panel (Myers 1990; Weiss and Zenz 1989).

Panelists not performing as they normally do can sometimes be eliminated. For example, wine judge performance, in terms of agreement, reliability, discrimination, and stability, varies with time. During any series of panel assessments, performance can be monitored. When judges become unreliable and nondiscriminating, their results can be omitted from analysis (Brien et al. 1987). Monitoring can take the form of a calculation of the Euclidean distance of scores from the mean for each subject. Judges not performing in the same way as the rest of the panel can be identified from the larger distances (Anthoney et al. 1984). A control chart technique for monitoring each panelist's performance may be preferred to an analysis of variance method (Gatchalian et al. 1991).

How we feel at the time of the assessment may affect the judgments made. In a series of acceptance tests, two groups of panelists completed a questionnaire about their current physical state. For the "positive" group, all statements in the questionnaire were positive (e.g., "I feel great"). For the "negative" group, statements were negative (e.g., "I feel tired"). Acceptance ratings of dairy bars were significantly higher for the positive group (Siegel and Risvik 1987). Hunger and monetary reward also affect performance. Rewarded subjects and hungry subjects rated samples of breaded fish higher than did either the unrewarded or the prefed subjects (Bell 1993).

Many tasting installations incorporate a computer keyboard for recording judgments. Although many people are interested in and enjoy interacting with computers, fears have been expressed over possible increases in anxiety and feelings of dehumanization. However easy the panelist's task, cost in financial and human terms may become a problem (Armstrong et al. 1997).

Rapid alternate tastings of samples prior to difference testing can improve performance in triangle tests (O'Mahony et al.

1988). This work involved only tasting, but such an effect might be demonstrated for visual tasks.

In many reports in the food research/development/quality literature, there is an implicit assumption that, for the product reported, all members of the population have identical preferences. Attempts to justify this assumption are few. Booth (1987) has warned of the dangers of premature aggregation of consumer judgments, stating that "maybe very few individuals have the average preference and only a small minority of citizens show a conjunction of the commonest attitudes or habits." The discussion of total appearance in Chapter 2 includes examples of the almost inevitable linking of preference to race, environment, and background. Changing lifestyles affect food choice. Two groups of people illustrate the influence of social trends on health perception: those whose attitudes can be termed *return to nature* and *acceptance of disorder*. The majority in the former group believe that special measures are necessary for health care and that only products that state nutritional values should be purchased. They believe that too many artificial ingredients are added to food and are concerned about eating sugar, carbohydrates, animal fat, white bread, and salt (Lowe 1979). Such preconceptions affect panel scores (Lundgren 1981).

There can be problems with scale use: scales used successfully in Europe and North America cannot be transferred directly to an urban African population without pretesting. Unfamiliar concepts include equal-interval and continuous rating scales, graphic and visual scales, and the giving of marks in numerical scales. Graphic scales themselves can generate confusion because the meaning of the line is not clear. Responses can lie beyond scale boundaries. Line scales may be used as 3-point scales. Five-point "Smiley" scales can be unsuccessful, as the urban African pays more attention to the eyes of a smiling face than the mouth. Hence, as the eyes are all the same, this scale is not a basis for discrimination, and 61 percent of subjects could not place the faces in the correct ordinal sequence. Difficulties are also encountered with numerical scales. Common in South Africa is the concept that the number "1" means "the best." When ratings on a scale of 1 to 10 are requested, the number 1 may be given to the most preferred samples (Morris and van der Reis 1980).

The Whorfian hypothesis states that the language of a culture determines how members of that culture think (Whorf 1956). That is, it will be difficult to discriminate between two stimuli if they cannot be individually described or labeled. Errors of appearance discrimination were illustrated during an experiment in communication between two Liberian rice farmers (Cole and Scribner 1974). The two farmers sat on either side of a screen, and both had sets of 10 very differently shaped wooden sticks. The experimenter, sitting by one of the farmers (the transmitter), selected a stick, and the farmer attempted to describe it to his neighbor (the receiver) so that it could be set to one side. The poor performance of the receiver was reversed when a college-educated man from the same tribe acted as the transmitter. This indicated that the problem was not sensory or linguistic but cognitive. Regarding pure color terms, Berlin and Kay (1969) found that although different languages encode within their vocabularies different numbers of basic color categories, a total universal inventory of exactly 11 basic color categories exists. This does not mean that different peoples have different abilities to discriminate or see color. It indicates that communication about color may be difficult.

Whatever the requirements and questions, those in charge of a panel need to make certain that panel members are thoroughly conversant and practiced with definitions, descriptions, and scales to be used before the assessments proper start. Those in charge must also be aware, as far as possible, of factors that affect individuals and groups and may influence direction and magnitude of scores.

The Halo Effect

Color and appearance are powerful indicators of object quality. This applies particularly to food. Human beings have different sensitivities to flavor, and it is relatively easy to confuse tasters by giving them inappropriately colored foods (Moir 1936). Colored raspberry-flavored fruit jellies were given to 18 panelists.

When the jellies were red, 11 panelists identified the flavor as a red fruit—raspberry (8 tasters), red currant (2), and strawberry (1). When the jellies were colored yellow or green or blue, 4, 5, and 6 panelists, respectively, identified the flavor as a red fruit (Scheide 1976). Atypical colors of fruit-flavored beverages induce incorrect flavor responses that are characteristically associated with the typical color (DuBose et al. 1980). Unusual color/flavor combinations reduced identification of raspberry-flavored drink more than that of orange flavor (Stillman 1993). In study that involved the tasting of six identical versions of the same confectionery product covered in wrappers of identical design but different colors, significant differences were found between the responses of men and women. For example, women rated the product wrapped in yellow twice as favorably as men, while men had a greater preference than women for the orange- and red-wrapped products (Scott 1976). Men have been shown to have a more marked preference than women for particular apples under particular illuminations (MB Halstead, personal communication on results of CIE British Panel TC 3.2, 1992). Food folklore has it that diners eating in the dark can be made physiologically sick by switching on the lights, thus revealing that they have been eating inappropriately colored food (Kostyla and Clydesdale 1978). In special circumstances, the appearance response can be overcome, and subjects will happily eat incorrectly colored foods. The subjects of one experiment were university students (presumably hungry), the food looked normal (apart from its color), and a trust existed that although the food was an incorrect color, it was safe to eat (Watson 1981).

The existence of the halo effect results in some foods' having to be tailored for a particular market. For example, the French prefer their rosé wines to be an onionskin color, the British prefer a light pink, and the Germans and Swiss prefer a deeper pink. An example of the halo effect in a slightly different form is to be found in the United States, where use of the descriptor *blush* instead of *rosé* increased sales. Milk substitute feeds for veal calves yield a similar example. French farmers believe that a feed that is slightly yellow contains color additives and is therefore undesirable. British farmers, however, tend to believe that such feeds contain more butterfat and are therefore more beneficial for their animals. Lozano (1989) has drawn attention to the work of Noel and Goutedonega (1985), who investigated colorant mixtures for vegetable protein frankfurters. Two populations of panelists were found. One group preferred pale colors because saturated colors seemed unnatural. The other group preferred saturated colors because pale colors suggested a high fat content. Examples of such divisions within the population include United Kingdom preferences for the color of tomato soup. There are two distinct groups of people, those who prefer orange red and those who prefer a blood or cherry-red color. Further evidence for the existence of multimodal populations is given in Chapter 8.

Prior to 1914, the yellow color of beef carcass fat was derived from the animal's feed. Gradual replacement of natural grass feeding led to the development of white fat, which is now preferred. This consumer preference might be reversed if the yellowness of adipose tissue came to be regarded as an indication that the animal had been reared in a traditional manner (Swatland 1988).

In some accounts in the literature, experiments are described in which all sensory attributes are judged at the same time, under the same conditions. Scores may be given for the flavor and texture attributes of a product that also varies in color. Sometimes high correlations between attributes are reported without comment of the fact that panelists may be influenced by appearance. That is, a halo effect from the color may affect scores for other properties.

Types of Halo Effect

Halo effects arise from a number of sources during sensory testing. Panel scores of in-mouth attributes are influenced by sample appearance, the environment, and the panel organizer's attitude. Panelists may be influenced by the way in which instructions are given: clues may be given to the answers expected. A halo from the physical environment arises when branded food packages or other foods are seen in the vicinity of the test area. Influences arise from the wider environment, perhaps a plush hotel or a run-down community hall.

During optimization of product flavor or texture, low illumination levels or colored lighting is used in the tasting area. Although the actual color of the product may be completely lost, conclusions may still be drawn about the sample color from the intensity of light reflected. In Britain, dark green peas are preferred, lighter samples being regarded as faded. Sensory evaluation of in-mouth texture properties of low-fat milks is greatly enhanced when the product can be viewed under normal illumination conditions (Phillips et al. 1995). These phenomena are discussed further in the section dealing with panel organization. The possibility of all such halo effects must always be taken into consideration. Variations in appearance must be eliminated while judgments of flavor and texture are made. This can then be followed with greater confidence by an investigation of interactions that might occur between appearance and other sensory properties.

In the panel situation, not everyone is affected by the look of the product being tasted. There are two groups of subjects, the *field independent* and the *field dependent*. Members of the field-independent group attend to their taste and smell perceptions when classifying flavor, without regard to what may be an inconsistent visual stimulus. The field-dependent subjects make more mistakes when trying to identify flavors in the absence of visual cues to their origin (Moskowitz 1983). There may be an age dependency, as older subjects were more strongly influenced by an off color in orange juice (Tepper 1993) and as older subjects (over 60 years) were more sensitive to visual cues and less sensitive to changes in flavor concentration of a cherry-flavored drink (Philipsen et al. 1995). During sensory work on port, all panelists were field dependent. Flavor and aroma judgments of all expert and nonexpert assessors appeared to be influenced by appearance (Williams et al. 1984).

Harvest time affects the sensory qualities of pecans. A later harvest yields darker, tougher kernels possessing greater off flavor. Subjects evaluating these nuts indicate a decreased preference for their color, appearance, texture, and flavor; a lower overall acceptability; and a consequent lower intention to buy. Pecans are, therefore, a good example of a food material for which it is essential to eliminate all visual cues when making detailed studies of flavor and texture (Resurreccion and Heaton 1987). Similarly, hedonic scores for red apples having the flavor of green apples increased when the flavor of the peeled apples was assessed (Daillant-Spinnler et al. 1996).

The point of view of the judge who works alone has been put forward by Presswood (1977). He commented upon the impression gained by an itinerant butter and cheese grader when he first enters a store room. "A grader has an impression on entering a store; and whilst grading is a much more precise estimation than an impression, the finish and appearance (of a cheese or box of butter) will undoubtedly make a grader aware of a need to look closely if he is unfavorably impressed." The grader should be left to get on with his work without interruption. Due to the itinerant nature of the graders' work in the United Kingdom, "Generally he [the grader] finds himself overlooked by someone with a vested interest in the result of the grading, be it proprietor or maker or merchant." This Presswood deplored.

Size constancy effects arising from a knowledge of sample size have been considered. Does it make any difference whether the panelist sees the sample in general sensory testing? Some compensation may be achieved when making sensory judgments on samples of different sizes. This has been shown to occur in olfaction. An increase in volume of odorant flowing across the olfactory receptor surface of a passive subject produces an increase in perceived intensity. If, however, the subject is aware of the higher odorant volume, a reduction in sniffing force will result in the same intensity being recorded (Teghtsoonian et al. 1978; Teghtsoonian and Teghtsoonian 1982).

This may also occur when samples of different size are presented to a subject for sensory texture analysis. It is argued that if samples are presented to the taster in reasonably sized pieces, all of them should have the same texture. That is, texture scores ought to be independent of sample size (Peleg 1983). This has been examined in a study of the effects of cue condition and sample volume on *hardness* and *chewiness* determinations (Cardello and Segars 1989). Five variables were manipulated:

1. Samples were presented sequentially or simultaneously, to control for specific knowledge of size difference.
2. Samples were presented randomly or in ascending order of size, to give more or less clue as to size differences.
3. Panelists were blindfolded or allowed to see the samples, to control for visual size cues.
4. Samples were or were not handled prior to placement in the mouth, to control for tactile size cues.
5. Panelists were given specific operational definitions of the attributes or were allowed to use idiosyncratic definitions. The latter provide less of a clue to size difference than the use of definitions, which draw attention to how the sample is manipulated in the mouth.

It was found that the greater the number of cues concerning actual differences, the smaller the difference in judged *hardness* and *chewiness*, but the authors drew two major conclusions:

1. Perceptual size constancy—that is, compensation for visually or tactilely detected differences in size—did not occur.
2. Control of cue condition and sample size must be maintained during sensory testing (Cardello and Segars 1989).

Additional complications to the nonexistence of perceptual size constancy can affect product appearance testing. For example, the British have been subjected to the persistent advertising claim that "smaller peas are sweeter." An unwary approach to the paneling of peas might result in erroneous findings founded upon a knowledge of this claim.

A product brand has physical, cultural, and emotional associations. In itself, the brand has an impact on the perception of sensory quality of the product. When assessing a product on a blind basis, consumers may or may not be found to like the brand they commonly use. A different result may be obtained when the assessor has knowledge of the brand involved. This phenomenon occurs with appearance factors. In an investigation concerning instant coffee, users of the Maxwell House brand tasted their own product and Folgers coffee with and without knowledge of the brand. Panelists scored for *like appearance* and *darkness*. In judgments of Maxwell House coffee, the *like appearance* score increased from 48 when users were judging blind to 70 when users were judging with knowledge of the brand. When users were judging Folgers, the scores increased from 60 (blind) to 66. In this instance, brand loyalty seemed to play a significant part. Smaller effects were observed with the *darkness* judgments (Moskowitz 1985).

Branding is used to position products in the mind. In a blind test, beer drinkers were unable to discern differences between five brands of beer. However, when the same beers were reevaluated with their brand names visible, brand-loyal users assigned significantly higher ratings to their preferred brands (Allison and Uhl 1964). Brand-loyal subjects have an enhanced perception of product properties for that brand. When the branding is obvious, the belief raises expectations, and consequently higher scores are obtained (Sheen and Drayton 1988).

The traditional research and development function of the manufacturer may concentrate too much on the product alone. Likewise, the marketing end of the business may focus too much on product concepts and the lifestyle and attitudes of potential customers. Little account is taken of the extreme influence of brand (Martin 1990).

Preconditioning influences eaters' beliefs. When a familiar object is viewed in a scene, the color perception of it tends to be changed in the direction of the color previously perceived to belong to the object (Adams 1923). This preconditioning halo effect extends to packaging. Knowledge of the product's fat content affects consumer responses to the product. Identical samples with different labels were perceived as the same by one third and as different by two thirds of the subjects (Lundgren 1981). Subjects given information about the flavor of a particular solution change their flavor ratings of the solutions. When consumers expected a very sweet solution but a less sweet one was presented, they rated it sweeter than when they had rated it blind. This tended not to occur with trained panelists (Deliza

and McFie 1996). Preferences of young children are ea
changed by identification of the product with hero figures. T
can learn to like tastes that initially they did not like, especi
if the product is sweet (Berger 1979).

It cannot be assumed that subjects will be honest in rep
given to questions. In a study comparing reported dietary
takes and measured expenditure, it was found that the self
porting of implausibly low energy intakes is common am
overweight and normal-weight subjects (Mela 1997).

A number of factors are relevant to these phenomena.
have different sensitivities to flavor, and each of our senses
different sensitivities in particular situations. In arriving ¿
judgment in the panel situation, we normally use a combi
tion of our senses, although in everyday life we may rely mai
on sight. As a species, humans have survived by adaptatior
the environment. We accept a challenge, and in what ough
be the pleasant or remunerative atmosphere of the taste or n
ket research panel, we like to please. If we are asked to make,
a flavor judgment using only the taste sense, it may be diffi
not to use other cues when scoring. Sensory hierarchy is
volved. If one sense is not sensitive enough, we will try to i
one that is. When handling human beings, we work with
adapting biological world, not with the rigid mathemat
structures of the physical world.

Psychophysical Relationships Between Color and Otl Senses

The existence of synesthetic or cross-modality effects has b
known for some time. Musicians such as Rimsky-Korsakof, ¡
abin, and Messiaen have associated particular colors with
tain musical keys (Scholes 1965; Griffiths 1978). New
wanted to develop an analogy between colors and the mus
scale, so he included indigo and orange with the five princ
colors in the spectrum (McLaren 1985). To the nineteenth-c
tury poet Arthur Rimbaud, the vowel sounds were endo
with colors (Marks 1975). There are distinct color asso
tions with odors, and there is a good agreement among

ropean women about what colors go well with particular fragrances (Jellinek 1988). For objects of equal size and shape, one that is red appears heavier than one that is yellow (Pinkerton and Humphrey 1974), and a red or black medicine capsule is seen to be stronger and more potent than one that is white (Sallis and Buchalew 1984).

Color (red, green, yellow) can significantly influence taste threshold concentrations. In a test using water solutions, subjects did not associate sour flavor with yellow and green. Red was not associated with a bitter taste, and yellow and green bitter solutions were detected at significantly higher concentrations than the control. None of the colors affected salt thresholds. Subjects did not associate yellow color with a sweet taste, sweetness was detected in the green solution at a significantly lower concentration than in the colorless control, and red had no effect (Maga 1974). When red was added to sucrose solutions, sweetness in darker colored solutions was rated 2 to 10 percent higher than in the lighter reference, even though the actual sucrose concentration was 1 percent less (Johnson and Clydesdale 1982).

Psychophysical relationships have also been found between color and flavor. Kostyla and Clydesdale (1978), in reviewing early work, found a certain amount of disagreement between investigators and advocated suitable methodology for further research. Using this methodology, Kostyla (1978) found many instances in which color affected the flavor of fruit-flavored beverages:

- Addition of red increased perceived sweetness by 5 to 10 percent.
- Addition of blue to cherry or to strawberry decreased tartness and fruit flavor by up to 20 percent.
- Addition of blue to raspberry decreased tartness by 1 percent.
- Addition of yellow to cherry and strawberry could decrease sweetness by 2 percent and fruit flavor by 3 to 5 percent.
- Addition of green to these beverages did not produce consistent changes in perceived flavor sensations.

The color of oxidized white wine was found preferable to that of unoxidized wine, although the flavor of the latter was preferred. Hence, a delicate balance between desirable color and flavor may be required. It was also found that tasters could specify the quality of red wine from its color but that the wine color could not be specified from its flavor and aroma. Provided the flavor did not depart too much from that expected from its color, the color appeared to determine the quality (Timberlake 1982). Williams et al. (1984) found that the assessment of aroma and flavor was influenced by the ability to see the port wines under assessment. Darker lagers of identical flavor were perceived to be more alelike in taste, although untrained individuals were found to be more open to influence than trained tasters (Butcher 1988).

Work on model systems showed that although color tended to confuse salt perception, the effect was not significant (Gifford and Clydesdale 1986). Increase of suitable coloring in chicken-flavored broths neither altered salt perception nor affected flavor preference (Gifford et al. 1987). It is perhaps not surprising that color/salt effects are few. Many foods of different colors are salty: pretzels are brown, olives are green and black, and popcorn is white (Maga 1974).

The effect of color on aroma judgments has been demonstrated using a range of foods including cheese, margarine, bacon, orange drink, and gelatin. With the exception of bacon, appropriately colored foods were perceived by sighted panelists to have a stronger and better quality aroma. The same trends were observed for flavor, but these were not statistically significant. Judgments of texture were not affected by color addition. The state of the panelists' hunger was also tested, but no extra effect due to color was observed (Christensen 1983). In later experiments involving processed cheese and grape-flavored jelly, Christensen (1985) failed to confirm that color alters the perceived aroma or flavor intensities. Visual cues can override olfactory cues in identification of type and intensity of fruit odors (Blackwell 1995).

These effects arise through a variety of causes. The musical and literary associations are individual and may have arisen through impressions gained in childhood. Each person perceiving the colors has a different pattern of association. The purely color strength phenomena, such as exist for medicine potency and apparent weight, may have their origin in contrast-of-extension effects. For example, a saturated yellow appears to be balanced in weight by only one quarter of the area of its complementary violet.

The literature is fairly extensive, and disagreements are evident (Clydesdale 1993). One reason is that an implied product association may exist for some panelists in experiments designed to be of a purely model nature. However, color influences other sensory characteristics, and hence food discrimination, through learned associations. The major mechanism prevailing within color/taste and color/aroma interactions is one of association with specific product and specific product type. An increase (within limits) in color associated with a product in many cases reinforces the flavor (cherriness) or flavor characteristic (sweetness) of the product. Characteristic color/taste/flavor associations of specific fruits arose with evolution of color vision skills. We appear to have a learned response to color and sweetness because orange or red fruits are normally riper, more edible, and sweeter than others (Kostyla and Clydesdale 1978). Experiments on specific products have revealed significant relationships between product quality and levels of color and flavor. Examples are pear nectar (Pangborn 1960), cherry and orange beverages and cake (Dubose et al. 1980), orange juice (Fellers et al. 1986; Fellers and Barron 1987), fruit punch beverages (Clydesdale et al. 1992), strawberry drink (Johnson et al. 1983), cherry drink (Johnson et al. 1982; Philipsen et al. 1995), lemon and lime drink (Roth et al. 1988), dessert gels (Imram 1998), and potato chips (Maga 1973).

Responses depend on age of subject. College-age students produced statistically nonsignificant effects of color on flavor, tartness, and aftertaste intensity ratings of a cherry-flavored soft drink containing artificial sweeteners (Holcomb et al. 1995). This conclusion was later confirmed in a study involving young adults. However, for panelists over the age of 60, perceived flavor intensity of a cherry drink increased with color intensity.

Sweetness perceptions of neither group were affected by color (Philipsen et al. 1995). Children aged 5 to 10 years presented with colored fruit drinks did not show the expected effect of darker colors' raising sweetness judgments. The 11- to 14-year-old group showed a trend in the opposite direction from adults (lighter red color was judged as sweeter). The concept of sweetness may be more strongly associated with sweet aroma characteristics at an early age than with specific colors (Lavin and Lawless 1998).

Color can modify texture perception. For example, butter of a stronger yellow color is perceived as being easier to spread. The educated consumer may be aware of effects of natural variations in fat composition with feeding and season of production and their relationship with color and texture (Rohm et al. 1997).

The role of color in food selection is difficult to quantify. Clydesdale (1993) has observed that where a cause-effect relationship is not quantified there is a denial of such a relationship. In a world of increased product development of nontraditional products, perhaps formulated to meet the special dietary needs of an aging population, the need for an understanding of the role of color is growing.

Expert Tasters as Predictors of Consumer Response

Although it may be unwise to ask trained panelists to give hedonic judgments while they are performing analytically, they may still be asked to give such judgments as a separate exercise. It is relevant to enquire whether judges trained to observe a food will give the same response as consumers.

Moskowitz (1985) asked consumers and experts from the industry to judge the appearance attributes of chocolates. Using a 0- to 100-point anchored scale, they judged appearance scoring for darkness, shininess, size (area), and thickness. Values of r^2 were used to compare responses (Table 5–1). These were encouragingly high except for the last attribute.

From studies on red wine, it was concluded that it was dangerous to use a highly trained panel to predict the response of an inexperienced group. Different preferences were found for

Table 5-1 Relationships Between Expert and Consumer Opinions of the Properties of Chocolates (Moskowitz 1985)

Attribute	r^2
Darkness	0.90
Shininess	0.96
Size	0.97
Thickness	0.70

color (Ough and Amerine 1970). Similar conclusions were reached regarding the use of the refined palate of expert beer tasters, who were more acutely affected by color (Butcher 1988). The pattern of overall preferences for tomato soups (Shepherd et al. 1988) and egg yolks (Shepherd and Griffiths 1987) differed between trained and untrained panelists. In daily life, our responses arise from experience within everyday life, imaginative application of that experience, and specific training. These results lead to the conclusions that reasonable agreement may possibly be found when judges with specific training and consumers are judging specifically defined appearance attributes, but although everyone may be a consumer at heart, those who have been specifically trained to look critically at food appearance have different preferences from those who have not.

PHYSICAL REQUIREMENTS FOR FOOD APPEARANCE ASSESSMENT

There are four types of conditions under which formal appearance assessments are made. Two of these are normally located where other types of sensory assessment are performed: that is, in a booth or a training/discussion room. The third location is a larger area in which supermarket lighting or package designs can be tested in mock or actual retail layouts. The fourth occurs wherever market researchers happen to find themselves, perhaps in the street or in a hotel.

Detailed guidelines for the layout of general sensory evaluation laboratories have been published by the American Society for Testing and Materials (ASTM). Design features particularly important for appearance assessment include the recommendation that "all evaluation areas, such as panel booths and training/testing rooms, should provide a comfortable, neutral, nondistracting environment. Color tones in the white, cream or grey category are the preferred choices" (Eggert and Zook 1986). Kitchen and preparation areas are sometimes colored, but this is not advisable.

There should be no other materials present in the examination area when samples of one color are being evaluated, and there should be no distracting or brightly colored objects in the vicinity. Bright or highly colored clothing should be covered by a smock of neutral gray color. No one should wear colored lenses.

Computer terminals are installed in many individual booth areas. The ASTM-recommended booth bench size—30 inches (76 cm) wide and 15 inches deep—may not be sufficient to house the terminal as well as the sample. The terminal may also introduce color and contrast into the booth via the finish and screen colors. These could supply distractions during an appearance assessment. Contrast introduced may be particularly intrusive during translucency assessments or color assessments of translucent samples. Therefore, the color of the terminal and its organization into the booth should be considered with care. Translucency becomes evident when illumination is from behind or from in front of the sample. In both cases, perception is caused by the presence of contrast from the lamp or from the background. This also affects the perceived color of the translucent object. Hence, as few contrasts as possible normally ought to be present in the viewing area. Introduction of contrast into the viewing area should be deliberate so that the viewing situation is under control.

It is necessary to control and standardize sample handling procedures. Acceptance appraisals should be conducted under conditions that simulate those under which consumers make their selections. For meat, this includes retail cases, lighting, case and meat temperature, defrost cycling, overwrap, and packaging (AMSA 1991). Foods such as hot milked tea brewed using tea bags change appearance quickly and can present a problem for assessment. A room is a better location than a booth for such assessments. A suitable procedure involves use of a table around which there is space to walk. Labeled cups, each containing a tea bag, are placed around the periphery of the table. Next to the cups are aliquots of milk and a teaspoon. At each corner of the table is placed a kettle for boiling water for the brewing. Panelists then take their places standing around the table. At appropriate times, the panel organizer orders the boiling of the water, the filling of the cups with boiling water, the standard stirring of the contents, the removal of the bag, the addition of the milk, and the repeat stirring. Again at the proper time, panel members are asked to file around the table, scoring the appearance attributes of each sample on paper as they walk. Many temperature-labile samples can thus be looked at by many people in a short time.

Although informal assessments take place on the production line, conditions for viewing the sample and matching to a memorized or formal color standard are rarely ideal. The environment tends to be insufficiently consistent for reliable and repeatable assessments to be made. If possible, production line assessments should be transferred to the laboratory, where it is easier to provide the standard physical conditions necessary. Where this is not possible, good consistent viewing conditions are necessary on line (see Chapter 3).

Market researchers often find themselves in conditions unsuitable for the making of reliable appearance assessments. A specification of the environment of the test, including types of illumination and color of the test surrounds, should be noted in the final report. This will make it easier for the occurrence of such metameric problems discussed in Chapter 8 to be minimized. If the consumer's views of product color and appearance are important, standard lighting and viewing conditions should be provided.

Lighting for Appearance Assessment

The light by which we view food materials is critical to our assessment of color and appearance (see Chapter 3). Many types of lighting installation operating over a wide range of intensities are to be found in development, manufacturing, retail, preparation, and dining areas. Appropriate light sources and levels are essential for obtaining reproducible, reliable assessment results. Lighting of panel booths is of concern in two ways. It may be used to provide illumination for the quality assessment, and it may be required to hide sample color.

Choice of lighting installation for color assessment depends upon the products and needs of the concern and on the illumination used by customers in their assessment rooms, display areas, dining rooms, or stores. Constant, extended-area, controllable incandescent and/or fluorescent light sources are required to provide even illumination free from shadows. It should be sufficiently directional to allow perception of a textured surface. Various types of diffusing screen for fitting beneath the light source are available. Translucent plastic diffusing screens and open plastic grids can be used, but checks are required from time to time so that any change in their color can be detected. If this occurs, screens must be replaced. Extraneous light sources should be excluded from the booth area. For color assessment, specular reflection of the source (i.e., where the angles of incidence and reflection are approximately the same) should be minimized. This can be achieved by ensuring that the light source is vertically above and that the viewing takes place at approximately 45° to the sample. These angles may be reversed. Lighting installations should be ventilated, especially when tungsten illumination is provided.

Dimmable systems capable of operating between approximately 750 lux, the level recommended for small offices (van Ooyen et al. 1987), and 1,200 lux are recommended (Eggert and Zook 1986). An intensity of 807 to 1,614 lux is recommended for evaluation of fresh meat (AMSA 1991). The British Standard BS950, part 1, is concerned with defining the spectral distribution of a light source of correlated color temperature 6,500K, which can be used to replace daylight for visual appraisal and color-matching tasks (British Standard 1967). No practical source exactly meets this standard. However, the Artificial Daylight tube, which includes an ultraviolet component, and Northlight, which does not contain such a component, are currently regarded to be the closest approximations available. These are widely used in industries in which color assessment is crucial. Recommended American lamps for meat (fresh or cured in oxygen-permeable or impermeable films) in display studies are SPX-30 (a high-efficiency lamp), Color-Gard 32, Deluxe Cool White, Deluxe Warm White, Soft White, Natural White, or Natural. Among those not recommended are the American Cool White, Daylight, or Standard Grolux (AMSA 1991). Some metal halide lamps with good color rendering properties are suitable substitutes for daylight.

Specific viewing recommendations for using the Munsell color atlas are similar to those applicable to other atlases and physical standards. Matching must be done by individuals with normal color vision, using standard viewing and lighting conditions. There should be no interference from extraneous light sources or from reflections from the ceiling or other objects in the room. To eliminate gloss effects, a 0°/45° (or vice versa) relationship must be maintained between the illumination, the sample, and the viewer. The sample should be placed on a neutral middle-gray to white background and assessed against one chip at a time, a gray mask being placed over other chips on the atlas page. Estimates of Munsell Value, Chroma, and Hue are made in that order (ASTM 1994). Lighting and viewing conditions to be used when making these comparisons have been recommended. Illumination within the viewing booth should consist of an extended-area source located above the specimens, the illuminance being uniform over the viewing area, within ±20 percent. For general evaluation of intermediate-lightness colors, the intensity should be between 810 and 1,880 lux (ASTM 1994). The spectral quality of the light should approximate that of daylight under a moderately overcast sky at a color temperature of 7,500K ± 200K. With the light source directly over the sample, the product observation is made at an angle of 45° and at a distance of 12 inches or more from the product (U.S. Department of Agriculture [USDA] 1978). Specific instructions for inspection and for the use and care of Macbeth lighting units with the disc colorimeter have been issued by the USDA (1962, 1975).

The assessment of gloss in sensory booths can be difficult because of the area, lighting, and angular viewing needed for its perception, and alternative arrangements may be needed. A specially designed lamp is used for visual evaluation of gloss differences. This is a desk fluorescent lamp that has a black reflecting surface behind the tube and a wire mesh grid in front (ASTM 1985). Such lighting allows the observer to assess any of the several types of gloss occurring in foods (see Chapter 7).

Very low levels of colored lighting are often used when the experimenter does not want panelists to see the sample color. This tactic often fails in its objective. For example, neither dark red, blue, green, mauve, nor sodium lighting was successful at masking color differences of orange juices, so judges wore blindfolds (Barbary et al. 1993). A different method is to make sure the sample is in total shadow. This may be accomplished by illuminating the opaque sample container from beneath. Even so, light from extraneous sources can increase ambient light levels to an unacceptable extent. Half-filled, dark-colored glasses are suitable for fluids.

If doors are included at the back of each booth, consideration must be given to illumination quality and level in the sample feed area. If these are inconsistent with conditions in the booth, samples should be introduced through a light baffle.

Lighting units should be regularly checked. Lighting levels and color temperature should be monitored and lamp replacements made as recommended by the manufacturer. Filters and light units need regular checking and cleaning every three months when the units are cool, again as specified by the manufacturer. A maintenance log should be completed.

Adaptation is another area of potential concern for panel organizers (see Chapter 4). The lighting of the briefing room can be selected to ensure that each panelist is adapted to the illumination of the booth. When lighting is changed during the experiment, time must normally be allowed for the subject to become adapted. Adaptation is of practical concern in the supermarket also. Illumination quality and intensity change when shoppers enter. Also, some concerns use different types of lighting to highlight different parts of the store. The effects produced can range from lightening the chore of shopping to causing downright confusion and possibly annoyance.

TYPES OF SENSORY TEST

Discrimination and Descriptive Testing

There are three broad types of sensory test:

1. *Discrimination:* Are product differences perceived at appropriate levels of significance?
2. *Descriptive:* What are the perceived sensory characteristics, and what are their relative intensities?
3. *Affective:* How well is the product liked? (Stone 1988)

Discrimination or difference tests indicate whether a difference can be seen between two samples. The samples can be presented in a pair: in threes, two of which are the same; or in a duo-trio test. These tests can determine to a statistical level of significance not only whether a difference can be seen but also the individual's level of sensitivity to the stimulus difference being offered. The absolute threshold occurring in a paired-comparison test of a graduated series of samples is normally taken as the point in the series at which 75 percent of a judge's responses are correct. Tests are described in detail by, for example, Lawless and Heymann (1998) and Meilgaard et al. (1987).

Although well represented in the literature of other color-using industries, very little mention of threshold tests is made in the food color literature. The reason for this lies in the wide variation normally found within each food sample. Bacon samples, for example, presented as fresh and faded pairs, are clearly separated by panelists, but less discrimination is achieved when samples are presented one at a time (MacDougall and Moncrieff 1988). Knowledge of color difference thresholds may be valuable when dealing with uniformly colored foods—for example, eggs, in which pinkness can be caused by incorrect feeding; tea; enzymically browned apple products; and milk browned by heating. Just-noticeable or just-detectable color differences can be determined for a particular region of food color space using model foods. Alternatively, it may be possible to add dyes and/or light-scattering material to the base food.

Descriptive tests are commonly used in product control and development. They are widely used to specify appearance characteristics of materials in a qualitative or quantitative manner. They are flexible enough to be used to describe appearance properties from the raw material, through processing, to the finished-product stages. They are of two types: category scaling and ratio scaling.

Category scaling is the most popular method of sensorially assessing foods. The Sensory Evaluation Division of the Institute of Food Technologists described the process:

> Coded samples are presented simultaneously or sequentially in a balanced order which differs among the individual panel members. A single-sample product evaluation is seldom employed; most perceptual judgements are relative. Category scales consisting of a series of word phrases (adverbial or adjective modifiers) structured in ascending or descending order of intensity are used to measure the specific attribute (e.g. sweetness, off flavor, etc.). An alternate scaling procedure is an unstructured vertical or horizontal line with verbal anchors at each end to describe or limit the attribute. For analysis purposes, successive digits are later assigned to each point represented on the scale, usually at the end representing zero intensity. This follows the convention of having higher numbers represent a greater magnitude or more of a given quality. A statistical analysis (e.g. analysis of variance) of the mean intensity scores for each sample is used to determine significant differences among the mean scores for the sample represented. (Dethmers et al. 1981)

Ratio scaling or magnitude estimation is also a powerful sensory tool (e.g., Moskowitz 1977; Dethmers et al. 1981). It is used to estimate the relationship between physical intensity and sensory magnitude and to obtain comparative ratings of specific attributes of two or more products. This method allows each subject to use a wide range of positive numbers (not zero) without restriction. The ratios of the numbers must reflect ratios of sensory intensity. After sufficient practice, this method is simple to apply to a wide range of sensory experiences. Samples are presented successively in a balanced order. If a reference is used, it must be presented first, but it may also be reintroduced later. The numerical rating given to the first sample presented is the choice of the subject. Ratings given to the succeeding samples should be in proportion to the rating assigned to the first. The numbers the subject uses do not influence the scale; only the ratios between the numbers convey information. The numbers given to the samples are normalized, typically by calculation of the geometric mean. The general form of the relationship between physical stimulus (P) and the sensory intensity (S) is

$$S = kP^n$$

This is a relationship yielding the straight line

$$\log S = k + n(\log P)$$

This form has been found to be valid for a very wide range of stimuli, including loudness, pressure, and the four basic tastes of sweetness, sourness, saltiness, and bitterness.

Some modalities exhibit a deviation from this power law. This sometimes occurs when the level of stimulus approaches threshold value. This occurs with brightness. The relationship can then be modified to an equation similar to

$$S = k(P - P_0)^n$$

where P_0 is related to the absolute threshold (Galanter and Messick 1961). It is unlikely that these simple relationships will apply to all psychophysical functions.

Ratio scaling normally requires the subject to generate numbers reflecting his or her perception of the ratio between the intensity levels of a particular stimulus. The cross-modality method involves the direct generation of an intensity of a second continuum reflecting this perception. For example, the subject might be given a hand grip with the instruction to squeeze it in proportion to the perceived ratio between the experimental stimuli. The placing of markers along the edge of an infinite bench is another method (Stevens 1975). Although this is normally regarded as a method for specialist panels only, Williams et al. (1986) have demonstrated that ratio scaling can be used successfully by visitors at a wine fair.

Other established analyses are (Dethmers et al. 1981):

- *flavor profile analysis,* in which the panelist assigns intensity to separate aroma and flavor attributes
- *texture profile analysis,* in which a systematic approach is applied to the assessment of mechanical, geometrical, fat, and moisture characteristics
- *appearance profile analysis (APA),* in which a systematic approach is applied to all aspects of appearance, including basic perceptions and derived perceptions (to be discussed further in Chapter 7)
- *quantitative descriptive analysis,* in which panelists characterize sample properties in order of their appearance. It is used to determine significant differences among products. Results are presented in terms of appearance, flavor, and texture attributes.

The profile analyses are absolute in that they seek to catalogue all relevant properties exhibited by each sample studied. In contrast, Quantitative Descriptive Analysis of appearance is comparative.

Affective (Preference and Acceptance) Testing

Affective or hedonic testing involves judgments made on scales of liking, pleasure, desirability, satisfaction, and preference. A hedonic judgment of liking or disliking appearance is a summation of the subject's response to an undefined combination of all aspects of total appearance (see Chapter 2). Each judgment is the sole opinion of the particular person making it. It is not, as in the above profile tests, a scoring by trained assessors on a previously discussed and agreed-upon scale.

A ranking for *liking* may be used and preference inferred from the relative positions of the samples. Structured 5- or 9-point scales that have each point described by a modified like or dislike description are often used. Words may be substituted by facial caricatures that display different degrees of pleasure (a "Smiley" scale). Alternatively, the scale can be completely unstructured, with **like** at one end and **dislike** at the other. Any of these scales can be converted to numerical scores and statistical analysis applied to determine the difference in degree of liking. Very few actual foods will receive a rating of less than 4 on a 9-point scale ranging from **like extremely** to **dislike extremely**. If something were disliked by everyone, it would not become a food at all (Harper 1977). In the laboratory, consumption can be better than acceptance as a predictor of acceptability (Meiselman et al. 1988). Liking is sometimes implied in scales rather than being explicity stated. For example, scales for *fatness* and *overall attractiveness* of raw pork muscle included such terms as **much too fatty** and **moderately attractive** (Rhodes 1970).

The Food Action Rating Scale (Dethmers et al. 1981) is a general 7- or 9-point attitude scale quantifying the panelist's overall opinion of the product in terms of such phrases as "I would eat (or buy) this product often" or "I would never eat (or buy) this product." This type of scale can be readily applied to food appearance.

The hedonic test is simple in concept. The taster may only have to say whether he or she does or does not like the product. This can make the final interpretation of the result more difficult because different populations, ethnic groups, sexes, and age groups may have totally different opinions and reproducibility of hedonic responses can thus be poor. In work on fruit juices, there was little difference in judgments of *intensity* over a six-week period, but *degree of liking* varied considerably (Trant and Pangborn 1983).

Stimuli from a wide range of attributes can be rated, but little attention is normally paid to determining their relative importance (Moskowitz and Jacobs 1986). *Annoyance* scales lead to the discovery of those product attributes to which the panelist attends and that influence acceptance. Those attributes that fail to annoy are those that have no dramatic importance. In a study on snack chips, panelists scored on a scale of 0 (**no annoyance**) to 100 (**extreme annoyance**). Appearance factors studied were *small/large, thin/thick,* and *light/dark.* There were four levels for each defect: **much too much, too much, too little,** and **much too little.**

- Chips that were much too small (mean score = 65) caused more annoyance than those that were much too large (25).
- Chips that were much too thick (57) caused more annoyance than those that were much too thin (31).
- Chips that were much too dark (74) caused more annoyance than those that were much too light (25) (Moskowitz 1985).

Other routes to consumer opinion involve the use of *relative-to-ideal* and *just right* affective scales. These are used to measure the pleasantness or desirability of the intensity of a specific attribute in order to determine the optimum level of an ingredient in a product. Personal tastes can be assessed more easily by these methods than by using conventional hedonic scales. Both have been used to determine the optimum level of sweetener in products (McBride 1985; Vickers 1988). *Relative-to-ideal* scale techniques have also been used to determine whether yolk color of range eggs was closer to consumers' ideal than yolk color of litter and cage eggs (Shepherd and Griffiths 1987). Relationships between specific sensory attributes (and instrumental measurements) and acceptability are discussed in Chapter 8.

STATISTICS AND MATHEMATICS IN APPEARANCE SENSORY TESTING

The statistics discipline is used in the food industry, as in other areas of natural and social sciences, for experimental design and analysis of results. The use in experimental design is to eliminate subject and observer bias and to design an efficient, cost-effective, and manpower-effective program that can adequately cover the variables the organizer wishes to include. The statistics for the control and analysis of day-to-day sensory food quality assessment tasks have been detailed by, for example, Lawless and Heymann (1998) and Meilgaard et al. (1987).

Analytical processes may be required

- to search for population segmentation
- to indicate areas of possible data redundancy, perhaps simply using a correlation matrix
- to summarize the data produced by the experiment
- to indicate the success of the experiment in terms of significance and predictiveness, indicating possible cause-and-effect relationships

These manipulations involve reasonably straightforward methodologies. Methods of examining rates of change of appearance attributes with time are discussed in Chapter 8. More sophisticated mathematical techniques are also used in sensory work. These have two aims. The first is to determine the number and hopefully the identity of the sensory dimensions needed to describe the class of food under investigation. The second aim is to identify those scales that discriminate between the samples and so to reduce the volume of data used in the final analysis. This can be achieved using conventional multidimensional analyses or neural networks (see Chapter 8).

Multidimensional scaling (MDS) and individual differences scaling (INDSCAL) are used to develop spatial representations or maps of psychological or other stimuli. For example, a map of a country can be constructed from a knowledge of distances between cities. Such a two-dimensional problem can be solved using a piece of paper, a ruler, and a pair of compasses. The orientation of the map in terms of north, south, east, and west may be incorrect, but this can be overcome simply by rotation. The main use in research, however, is in mapping psychological responses for which the number of dimensions of the map is unknown. An application concerns determining the degree of similarity between 12 countries as perceived by 18 individuals. The two-dimensional solution of the MDS analysis corresponded to *communist/noncommunist* and *underdeveloped/developed* axes. The ability of INDSCAL to use data from each individual revealed that subjects could be divided into three populations. A questionnaire revealed that the subjects could be divided into *hawks, moderates,* and *doves.* There was a systematic relation between the relative importance of the dimensions and each subject's political orientation. The political alignment dimension was more important for hawks than doves, while economic development was more important for the doves. This powerful technique can be used to map each country according to its geographical location and its appearance according to the attitudes of the person making the judgment (Wish and Carroll 1973).

This type of analysis has been applied successfully to color perception (e.g., Indow and Uchizono 1960; Indow and Kanazawa 1960; Burnam et al. 1970; Helm 1964, cited in Wish and Carroll 1973). Wish and Carroll (1973) have used the Indow and Kanazawa (1960) experimental data derived from color vision–normal subjects who judged the closeness of pairs of colors. Samples ($N = 24$) varying in Hue, Value, and Chroma were included. INDSCAL was used to reconstruct the stimulus spaces on the basis of this information alone. The first two dimensions of stimulus space are shown in Figure 5–1.

Dimensions 1 and 2 reveal an approximate hue circle with red-green and blue-yellow dimensions. Concentric "circles" in this plot correspond to different Munsell chromas. Dimension 3 reflects Munsell value. Different subject spaces are produced by those with color vision anomalies (Helm 1964, cited in Wish and Carroll 1973).

Multidimensional analyses have been carried out on flavor alone (e.g., Lyon 1988) and texture alone (e.g., Howard 1976). No analysis of total appearance alone using these techniques has been published. Principal components analysis (PCA) is a method whereby related observed (sensory) variables are described linearly in terms of new uncorrelated variables (PCs). Thus, some sensory attributes contribute more than others to each PC. The technique extracts first the PC accounting for

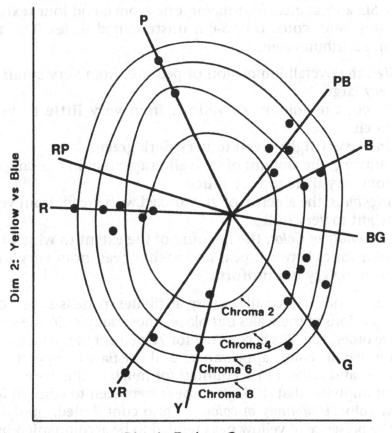

Figure 5-1 Dimensions 1 and 2 of the INDSCAL group stimulus color space, showing Indow and Kanazawa (1960) data on similarities among colors differing in Hue, Value, and Chroma. The equi-Chroma and Hue radii were drawn by hand to clarify interpretation. R, red; G, green; B, blue; Y, yellow; P, purple (Wish and Carroll 1973, copyright ASTM, reprinted with permission).

most of the variance. Subsequent extractions produce other PCs that explain next-most variances and are orthogonal to each other (Tabachnick and Fidell 1983).

Data were obtained from a consumer panel in which 15 descriptive sensory attributes were scored for peas (Sanford et al.

1988). Six appearance, four flavor, one aroma, and four texture attributes were scored on 15-cm unstructured scales. The appearance attributes were

- *Size,* the overall impression of pea size, from **very small** to **very large**
- *Yellow,* the amount of yellow, from **very little** to **very much**
- *Green,* **very light green** to **very dark green**
- *Shriveling,* the amount of shriveling apparent in the sample, from **very little** to **very much**
- *Brightness,* the absence of gray mixed with green, from **very bright** to **very dull**
- *Uniformity of Color,* the measure of the extent to which the color varies between peas and within peas, from **very uniform** to **very nonuniform**

It was found that for all sensory attributes, panelists used different portions of the scales but placed most samples in the same relative order. Five PCs accounted for 72 percent of the variance: texture, flavor, color, appearance, and off flavor respectively. *Green* (−) and *yellow* (+) accounted for much of the color component implying that dark green peas were seen to contain less yellow color. *Uniformity of color* (+) also contributed, implying that the presence of yellow peas tended to be accompanied by a greater visual nonuniformity. *Shriveling* (+) and *brightness* (+) accounted for much of the appearance component, with *uniformity of color* and *size* also contributing—dull, highly shriveled peas tended to be small and nonuniform in color. Size (+) also contributed in a lesser way to the texture component: size increased with growth, maturation, and toughness. Color and appearance attributes tended not to contribute to the flavor and off-flavor components (Sanford et al. 1988).

Derived canonical variates have been used to summarize the sensory characteristics of samples of fresh roast pork, cooked to various internal end-point temperatures. Canonical variate analysis on nine sensory attributes resulted in the derivation of the variates shown in Figure 5–2. The arrows denote increasing values of each sensory attribute projected onto the two canoni-

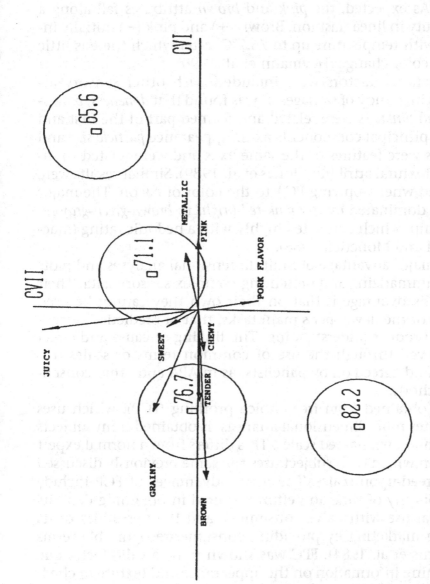

Figure 5–2 Projection of sensory data on pork roasts, canonical variates I and II. Attribute loadings and factor scores for four internal end-point temperatures are shown. Ellipses indicate the 95-percent confidence interval for the specific temperature (Heyman et al 1990, with permission).

cal variates CV1 and CV2. The ellipses denote the 95-percent confidence limits for samples of four specific end-point temperatures. As expected, the *pink* and *brown* attributes fall along a continuity in linear fashion. Brown (+) and pink (−) initially increase with temperature up to 76.7°C, after which there is little further color change (Heymann et al. 1990).

Appearance factors were included with other sensory attributes in a study of sausages. It was found that *doneness, brownness,* and *pinkness* were related and formed part of the first and second principal components axes. Appearance *particle size* and *evenness* were features of the same axes and were related to in-mouth textural attributes (Jones et al. 1989). Similar results were obtained when applying PCA to the color of bacon. The major PC was dominated by the *pink-red-bright* to *brown-green-gray* relationship, which correlated highly with a hedonic rating (MacDougall and Moncrieff 1988).

The major advantage of multidimensional analyses and plots is for summarizing and presenting complex sensory data. Their major disadvantage is that on their own they cannot be used for one of the developer's main tasks, that of prediction, essential for product understanding. The linking of cause and effect is achieved through the use of common terms or scales discussed and agreed on by panelists, as in APA and other consensus methods.

Data obtained from free choice profiling (FCP), which uses Procrustes multidimensional analysis, is obtained from subjects using their own named scales. This differs from a normal expert panel, in which each subject uses the same previously discussed and agreed-upon scales. The main advantages of FCP include the economy of time and effort required in obtaining data, its potential use with naive consumers, and the possibility of its helping marketing by providing consumer-recognizable terms (Williams et al. 1984). FPC was shown to be a valid technique for gaining information on the appearance and texture of cheddar cheeses from a group of untrained consumers. The first dimension related to appearance, color being of particular importance. Dimension 2 related to style and origin of the cheese and dimension 3 to texture (Jack et al. 1993).

However, the statistical analysis is more complex, interpretation is sometimes confounded, analysis of variance of individual attributes and individual subjects is impossible, and spider's web or star diagrams (almost essential for product development) are difficult to derive. A spin-off from the Procrustes method is that words commonly used by naive subjects are found, but these can be obtained by simpler means (Thomson 1989).

REFERENCES

Adams GK. (1923). An experimental study of memory colour and related phenomena. *Am J Psychol* 34:359–407.

Allison RI, Uhl K. (1964). Influence of beer brand identification on taste perception. *J Marketing Res* 1:36–39.

American Meat Science Association Committee. (1991). *Guidelines for meat color evaluation*. Chicago: National Live Stock and Meat Board.

American Society for Testing and Materials. (1985). *Colour and appearance measurement: standard method for visual evaluation of gloss differences between surfaces of similar appearance*. D 4449–85. Philadelphia: ASTM.

American Society for Testing and Materials. (1994). *Colour and appearance measurement*. Philadelphia: ASTM.

Anthoney K, Ennis D, Cook P. (1984). Lemon juice color evaluation: sensory and instrumental studies. *J Food Sci* 49:1435–1437.

Armstrong G, McIlveen H, McDowell D, Blair I. (1997). Sensory analysis and assessor motivation: can computers make a difference. *Food Qual Pref* 8:1–7.

Barbary O, Nonaka R, Delwiche J, Chan J, O'Mahony M. (1993). Focussed difference testing for the assessment of differences between orange juices made from orange concentrate. *J Sensory Stud* 8:43–67.

Bell R. (1993). Some unresolved issues of control in consumer tests: the effects of expected monetary reward and hunger. *J Sensory Stud* 8:329–340.

Berger AA. (1979). Reflections on the American diet. *J Am Culture* 2:433–435.

Berlin B, Kay P. (1969). *Basic color terms*. Berkeley: University of California Press.

Blackwell L. (1995). Visual cues and their effect on odour assessment. *Nutr Food Sci*, no. 5, 24–28.

Booth DA. (1987). Objective measurement of determinants of food acceptance: sensory, physiological and psychosocial. In *Food acceptance and nutrition*. ed. J Solms, DA Booth, RM Pangborn, O Raunhardt, 1–27. New York: Academic Press.

Brien CJ, May P, Mayo O. (1987). Analysis of judge performance in wine-quality evaluations. *J Food Sci* 52:1273–1279.

British Standard. (1967). *Standard specification for artificial daylight for the assessment of colour. Part I. Illuminant for colour matching and colour appraisal.* BS950, part 1.

Burnam RW, Onley JW, Witzel RF. (1970). Exploratory investigation of perceptual color scaling. *J Opt Soc Am* 60:1410–1414.

Butcher KN. (1988). Flavour: in the mind or matter? In *Proceedings of the Second Aviemore Conference on Malting, Brewing, and Distilling,* ed. I Cambell, FG Priest, 371–374. London: Institute of Brewing.

Campden Food and Drink Association. (1988). *Campden imported canned food specification: canned corned beef.* (November). Chipping Campden, UK: Campden Food and Drink Association.

Cardello AV, Segars RR. (1989). Effects of sample size and prior mastication on texture judgements. *J Sensory Stud* 4:1–18.

Christensen CM. (1983). Effects of color on aroma, flavor and texture judgements of foods. *J Food Sci* 48:787–790.

Christensen CM. (1985). Effect of color on judgements of food aroma and flavor intensity in young and elderly adults. *Perception* 14:755–762.

Clydesdale FM. (1993). Color as a factor in food choice. *Crit Rev Food Sci Nutr* 33(1):83–101.

Clydesdale FM, Gover R, Philpson D, Fugardi C. (1992). The effect of color on thirst quenching, sweetness, acceptability and flavor intensity in fruit punch flavored beverages. *J Food Qual* 15:19–38.

Cole M, Scribner S. (1974). *Culture and thought: a psychological introduction.* New York: John Wiley.

Daillant-Spinnler B, MacFie HJH, Beyts PK, Hedderley D. (1996). Relationships between perceived sensory properties and major preference directions of 12 varieties of apples from the Southern Hemisphere. *Food Qual Pref* 7:113–126.

Deliza R, MacFie HJH. (1996). Information affects consumer assessment of sweet and bitter solutions. *J Food Sci* 61:1080–1084.

Dethmers AE, et al. (1981). Sensory evaluation guide for testing food and beverage products. *Food Technol* 35:50–59.

DuBose CN, Cardello AV, Maller O. (1980). Effects of colorants and flavorants on identification, perceived flavor intensity and hedonic quality of fruit-flavored beverages and cake. *J Food Sci* 45:1393–1399, 1415.

Eggert J, Zook K, eds. (1986). *Physical requirement guidelines for sensory evaluation laboratories.* ASTM PCN 04-913000-36. Philadelphia: American Society for Testing and Materials.

Fellers PJ, Barron RW. (1987). A commercial method for recovery of natural pigment granules from citrus juices for color enhancement purposes. *J Food Sci* 52:994–995, 1005.

Fellers PJ, de Jager G, Poole MJ. (1986). Quality of retail Florida-packed frozen concentracted orange juice as determined by consumers and physical and chemical analyses. *J Food Sci* 51:1187–1190.

Fletcher TOT. (1980). *Lighting of horticultural packing houses.* London: Ministry of Agriculture and Fisheries.

Foster T, Kilcast D. (1988). Sensory panels: the Leatherhead advantage. *Food Manuf* 63 (November):45–46.

Galanter EH, Messick S. (1961). The relationship between category and magnitude scales. *Psychophys Rev* 68:363–372.

Gallwey TJ. (1982). Selection tests for visual inspection on a multiple fault type task. *Ergonomics* 25:1077–1092.

Gatchalian MM, de Leon SY, Yano T. (1991). Control chart technique: a feasible approach to measurement of panellist performance in product profile development. *J Sensory Stud* 6:236–254.

Gifford SR, Clydesdale FM. (1986). The psychophysical relationship between colour and sodium chloride concentrations in model systems. *J Food Prot* 49:977–982.

Gifford SR, Clydesdale FM, Damon RA Jr. (1987). The psychophysical relationships between color and salt concentrations in chicken flavored broths. *J Sensory Stud* 2:137–147.

Griffiths P. (1978). Catalogue de couleurs. *Musical Times* 119:1035–1037.

Harper R. (1977). Our senses and how we use them. *Proceedings of the Sensory Quality Control Conference,* ed. HW Symons, JJ Wren, 4–18. Aston, UK: University of Aston.

Heymann H, Hedrick HB, Karrasch MA, Eggeman MK, Ellersieck MR. (1990). Sensory and chemical characteristics of fresh pork roasts cooked to different endpoint temperatures. *J Food Sci* 55:613–617.

Holcomb LM. Clydesdale FM, Griffin RW. (1995). Effects of color and sweeteners on the sensory characteristics of soft drinks. *J Food Qual* 18:425–442.

Howard A. (1976). Psychometric scaling of sensory texture attributes of meat. *J Texture Stud* 7:95–107.

Howgate P. (1977). Measurement of fish freshness by an objective sensory method. *Proceedings of the Sensory Quality Control Conference,* ed. HW Symons, JJ Wren, 41–48. Aston, UK: University of Aston.

Imram N. (1998). Sensory perception of colour and appearance attributes in a dessert gel formulation. *Proceedings of the International Conference on Culinary Arts and Sciences,* ed. JSA Edwards, D Lee-Ross, 421–430. Bournemouth: UK Worshipful Company of Cooks Centre for Culinary Research at Bournemouth University.

Indow T, Kanazawa K. (1960). Multidimensional mapping of colours varying in hue, value and chroma. *J Exp Psychol* 59:330–336.

Indow T, Uchizono T. (1960). Multidimensional mapping of colours varying in hue and chroma. *J Exp Psychol* 59:321–329.

Jack FR, Piggott JR, Paterson A. (1993). Discrimination of texture and appearance in cheddar cheese using consumer free-choice profiling. *J Sensory Stud* 8:167–176.

Jellinek G. (1985). *Sensory evaluation of food.* Chichester, UK: Ellis Horwood.

Jellinek JS. (1988). Perfumes and colors: a European consumer attitude study. *Dragaco Rep* 1:14–29.

Johnson JL, Dzendolet E, Damon E, Sawyer M, Clydesdale FM. (1982). Psychophysical relationships between perceived sweetness and color in cherry flavored beverages. *J Food Prot* 45:601–606.

Johnson JL, Dzendolet E, Clydesdale FM. (1983). Psychophysical relationships between perceived sweetness and redness in strawberry flavored drinks. *J Food Prot* 46:21–25, 28.

Johnson J, Clydesdale FM. (1982). Perceived sweetness and redness in colored sucrose solutions. *J Food Sci* 47:747–752.

Jones RC, Dransfield E, Crosland AR, Francombe MA. (1989). Sensory characteristics of British sausages: relationships with composition and mechanical properties. *J Sci Food Agric* 48:63–85.

Kostyla AS. (1978). *The psychophysical relationships between color and flavor.* PhD thesis, University of Massachusetts.

Kostyla AS, Clydesdale FM. (1978). The psychophysical relationships between color and flavor. *CRC Crit Rev Food Sci Nutr* 10:303–319.

Lavin JG, Lawless HT. (1998). Effects of color and odor on judgements of sweetness among children and adults. *Food Qual Pref* 9:283–289.

Lawless HT, Heymann H. (1998). *Sensory evaluation of food.* New York: Chapman & Hall.

Lowe M. (1979). Influence of changing lifestyles on food choice. In *Nutrition and lifestyles,* ed. Turner, 806. London: Applied Science.

Lozano RD. (1989). Colour in foods. *Proceedings of the Sixth Congress of the International Colour Association,* vol. 1, 115–139. Buenos Aires: Grupo Argentino del Color.

Lundgren B. (1981). Effect of nutritional information on consumer responses. In *Criteria of Food Acceptance Symposium proceedings,* ed. J Solms, RL Hall, 27–33. Zurich: Forster.

Lyon BG. (1988). Descriptive profile analysis of cooked, stored and reheated chicken patties. *J Food Sci* 53:1086–1090.

MacDougall DB, Moncrieff CB. (1988). Influence of flattering and tri-band illumination on preferred redness-pinkness of bacon. In *Food acceptability,* ed. DMH Thomson. London: Elsevier.

Maga JA. (1973). Influence of freshness and color on potato chip sensory preferences. *J Food Sci* 38:1251–1252.

Maga JA. (1974). Influence of color on taste thresholds. *Chem Senses Flavor* 1:115–119.

Marks LE. (1975). Synesthesia. *Psychol Today* 9:48–52.

Martin D. (1990). The impact of branding and marketing on perception of sensory qualities. *Food Sci Technol Today* 4(1):44–49.

McBride RL. (1985). Stimulus range influences intensity and hedonic ratings of flavour. *Appetite* 6:125–131.

McLaren K. (1985). Newton's indigo. *Color Res Appl* 10:225–229.

Meilgaard M, Civille GV, Carr BT. (1987). *Sensory evaluation tech*, vols. 1 & 2. Boca Raton, Fl: CRC Press.

Meiselman HL, Hirsch ES, Popper RD. (1988). Sensory, hedonic and situational factors in food acceptance and consumption. In *Food acceptability*, ed. DH Thomson, 77–87. London: Elsevier.

Mela D. (1997). Honest but invalid: what subjects say about their dietary intake. *Inst Food Res News* 3(2):1.

Moir HC. (1936). Some observations on the appreciation of flavour in foodstuffs. *Chem Ind* 55:145–148.

Morris N, van der Reis AP. (1980). The transferability of rating scale techniques to an African population. *Technol Transfer Res* 11:417–436.

Moskowitz HR. (1977). Magnitude estimation: notes on what, how, when and why to use it. *J Food Qual* 3:195–227.

Moskowitz HR. (1983). *Product testing and sensory evaluation of foods*. Westport, Conn: Food and Nutrition Press.

Moskowitz HR. (1985). *New directions for product testing and sensory analysis of foods*. Westport, Conn: Food and Nutrition Press.

Moskowitz HR, Jacobs BE. (1986). The relative importance of sensory attributes for food acceptance. *Acta Alimentaria* 15:29–38.

Myers RH. (1990). *Classical and modern regression with applications*. 2nd ed. Boston: PWS Kent.

Nelson G, Smith DL. (1977). Sensory evaluation in the confectionery industry. *Proceedings of the Sensory Quality Control Conference*, ed. HW Symons, JJ Wren, 85–87. Aston, UK: University of Aston.

Noel P, Goutedonega R. (1985). Saucisses thermiques de type francfort a base de proteines vegetales: stabilité et coloration de l'emulsion. *Proceedings of the 31st European Meeting on Meat Research Work*, Albena, Bulgaria, 786–789.

O'Mahony M. (1979). Psychophysical aspects of sensory analysis of dairy products: a critique. *J Dairy Sci* 62:1954–1962.

O'Mahony M, Thieme U, Goldstein LR. (1988). The warm up effect as a means of increasing the discriminability of sensory difference tests. *J Food Sci* 53:1848–1850.

Ough CS, Amerine MA. (1970). Effect of subjects' sex, experience and training on their red wine color-preference patterns. *Percept Motor Skills* 30:395–398.

Pangborn RM. (1960). Influence of color in the discrimination of sweetness. *Am J Psychol* 73:229–238.

Peleg M. (1983). The semantics of rheology and texture. *Food Technol* 37(11):54–61.

Philipsen DH, Clydesdale FM, Griffin RW, Stern P. (1995). Consumer age affects response to sensory characteristics of a cherry flavored beverage. *J Food Sci* 60:364–368.

Phillips LG, McGiff ML, Barbano DM, Lawless HT. (1995). The influence of nonfat dairy milk on the sensory properties, viscosity and color of lowfat milks. *J Dairy Sci* 78:2113–2118.

Pinkerton E, Humphrey NK. (1974). The apparent heaviness of colours. *Nature* 250:164–165.

Presswood JB. (1977). Butter and cheese grading. *Proceedings of the Sensory Quality Control Conference*, ed. HW Symons, JJ Wren, 54–59. Aston, UK: University of Aston.

Resurreccion AVA, Heaton EK. (1987). Sensory and objective measures of quality of early harvested and traditionally harvested pecans. *J Food Sci* 52:1038–1058.

Rhodes DN. (1970). Meat quality: the influence of fatness of pigs on the eating quality of pork. *J Sci Food Agric* 21:572–575.

Rohm H, Strobl M, Jaros D. (1997). Butter colour affects sensory perception of spreadability. *Z Lebensm Unters Forsch A* 205:108–110.

Roth HA, Radle LJ, Gifford SR, Clydesdale FM. (1988). Psychophysical relationships between perceived sweetness and color in lemon and lime flavored drinks. *J Food Sci* 53:1116–1119, 1162.

Sallis RE, Buchalew LW. (1984). Relation of capsule color and perceived potency. *Percept Motor Skills* 58:897–898.

Sanford KA, Gullett EA, Roth VJ. (1988). Optimization of the sensory properties of frozen peas by principal components analysis and multiple regression. *Can Inst Food Sci Technol J* 21:174–181.

Scheide J. (1976). Flavour and medium: mutual effects and relationship. *Ice Cream Frozen Confect* 228–230.

Scholes PA. (1965). *The Oxford companion to music*. 9th ed. Oxford, UK: Oxford University Press.

Scott R. (1976). *The female consumer*. London: Associated Business Programmes.

Sheen MR, Drayton JL. (1988). Influence of brand label on sensory perception. In *Food acceptability*, ed. DH Thomson, 89–99. London: Elsevier.

Shepherd R, Griffiths NM. (1987). Preferences for eggs produced under different systems assessed by consumer and laboratory panels. *Lebensm Wiss Technol* 20:128–132.

Shepherd R, Griffiths NM, Smith K. (1988). The relationship between consumer preferences and trained panel responses. *J Sensory Stud* 3:19–35.

Siegel SF, Risvik E. (1987). Cognitive set and food acceptance. *J Food Sci* 52:825–826.

Stevens SS. (1975). *Psychophysics: introduction to its perceptual, neural and social prospects.* New York: John Wiley.

Stillman JA. (1993). Color influences flavor identification in fruit-flavored beverages. *J Food Sci* 58:810–812.

Stone H. (1988). Using sensory resources to identify successful products. In *Food acceptability,* ed. DMH Thomson, 283–297. London: Elsevier.

Swatland HJ. (1988). Carotene reflectance and the yellowness of bovine adipose tissue measured with a portable fibre-optic spectrophotometer. *J Sci Food Agric* 46:195–200.

Tabachnick BG, Fidell LS. (1983). *Using multivariate statistics.* New York: Harper & Row.

Teghtsoonian R, Teghtsoonian M. (1982). Perceived effect in sniffing. *Percept Psychophys* 31:324–329.

Teghtsoonian R, Teghtsoonian M, Berglund B, Berglund U. (1978). Invariance of odor strength with sniff vigor. *J Exp Psychol: Hum Percept Perform* 4:144–149.

Tepper BJ. (1993). Effects of slight color variation on consumer acceptance of orange juice. *J Sensory Stud* 8:145–154.

Theobald DJ. (1977). Sensory evaluation in tea buying and blending. *Proceedings of the Sensory Quality Control Conference,* ed. HW Symons, JJ Wren, 69–72, 76–77. Aston, UK: University of Aston.

Thomson D. (1989). Recent advances in sensory and affective methods. *Food Sci Technol Today* 3:83–88.

Timberlake CF. (1982). Colours in beverages. *Food Flavourings Ingredients Processing Packaging* 4(7):14–17.

Trant AS, Pangborn RM. (1983). Discrimination, intensity and hedonic responses to color, aroma, viscosity and sweetness of beverages. *Lebensm Wiss Technol* 16:147–152.

Truscott TG, Hudson JE, Anderson SK. (1984). Differences between observers in assessment of meat colour. *Proc Austr Soc Animal Prod* 15:762.

U.S. Department of Agriculture. (1962). *Facts about colour.* USDA File Code 131-A-105. Washington, DC: USDA.

U.S. Department of Agriculture. (1975). *Use and maintenance of Macbeth lights.* USDA File Code 131-A-31. Washington, DC: USDA.

U.S. Department of Agriculture. (1978). *US standards for grades of tomato juice.* 43FR19814. Washington, DC: USDA.

van Ooyen MHF, van der Weijgert JAC, Begemann SHA. (1987). Preferred luminances in offices. *J Illum Eng Soc* 16:152–156.

Vickers Z. (1988). Sensory specific satiety in lemonade using a just right scale for sweetness. *J Sensory Stud* 3:1–8.

Watson RHJ. (1981). The importance of colour in food psychology. In *Natural colour for food and other uses*, ed. JN Counsell, 27–40. London: Applied Science.

Weiss J, Zenz H. (1989). Reduction of panel variances by a simple two-step normalisation procedure for graphic line scale. *Acta Alimentaria* 32:313–323.

Whorf BL. (1956). *Language, thought and reality*. New York: John Wiley.

Williams AA, Langren SP, Timberlake CF, Bakker J. (1984). Effect of colour on the assessment of ports. *J Food Technol* 19:659–671.

Williams AA, Baines CR, Finnie MS. (1986). Optimization of colour in commercial port blends. *J Food Technol* 21:451–461.

Wish M, Carroll JD. (1973). Concepts and applications of multidimensional scaling. In *Sensory evaluation of appearance of materials*, ed. RS Hunter, PN Martin, no. STP545, 91–108. Philadelphia: American Society for Testing and Materials.

CHAPTER 6

Appearance Scales

Economic and commercial stability of growing and processing industries require specific, easily understood, and easily interpreted statements of quality. Assessment methods and standards include specific instructions for sample preparation, examination, and measurement. At one level, there are words and phrases used by the layperson; at another level, there are the terms and words known only to those in the specific business. Diagrams are used where words are cumbersome or insufficient.

In general, there are two types of appearance sensory descriptor and scale: memory and comparative. Memory descriptors and scales occur in two forms. One relies on the assessor's absolute appearance memory, not related to the food. An example is the memory for **pink** or **yellowy green** called for in the assessment of tomatoes. The second depends on the memory of appearance as applied to the food in question. An example of this is the visual judgment of **just ripe**. Comparative descriptors and scales depend on specific physical comparators. The scales consist of, for example, transparent glasses, plastic sheets, and photographs.

This chapter discusses the memory descriptors and scales as well as physical comparative scales used to describe *basic* and *derived perceptions*.

BASIC PERCEPTIONS: COLOR AND APPEARANCE MEMORY DESCRIPTORS AND SCALES

Basic perceptions of a product can be described in terms of the elements *visual structure, surface texture, translucency, gloss,* and *color.* The *derived perceptions* of *visual expectation* are discussed initially in terms of *visual identification* and *visually assessed texture, flavor,* and *satisfaction.* Each of these may be described in terms of memory scales and some in terms of comparative scales. The following memory scales can be found in the literature.

Visual Structure

A description of overall structure depends upon the product. Bacon, for example, has critical dimensional attributes relevant to different end uses by different consumers. A slice has a *size* and *shape* and can be subdivided into areas of *fat, lean, rind,* and *cartilage.* The sizes, shapes, distributions, and area (or apparent volume) ratios of each component can be specified geometrically. Fresh, stale, and cooked bacon have characteristic shapes. Slices of processed meat may comprise two muscles, each a different color. The area of each can be assessed as a percentage of the total sample area.

Restructured beef steaks distort during cooking. Such a surface may be defined in terms of *macro distortion,* the degree to which the overall is uneven and warped, and *micro distortion,* the degree to which small sections of the cooked surfaces look uneven or rough. Both types of distortion increase with the particle size of the meat used in the manufacture (Berry and Civille 1986). *Muscle cut resemblance* scales have been used to assess storage effects on restructured beef and turkey steaks (Marriott et al. 1987).

Product Definition

Where produce is marketed in different forms, identification and definition of the forms may be needed. For example, detailed descriptions of green beans and waxed beans are used for canning. These include definitions of style, uniformity, and size

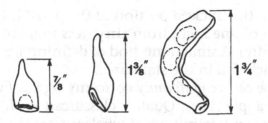

Figure 6–1 Methods of measuring length units of green and wax beans (USDA 1972).

(U.S. Department of Agriculture [USDA] 1972). Methods for size measurement are shown in Figure 6–1.

Whole style consists of "whole pods, which, after the removal of either or both ends, are less than $2^3/_4$ inches in length, or transversely cut pods not less than $2^3/_4$ inches in length and, except for *vertical* pack or *asparagus* style, are not arranged in any definite position in the container." *Whole vertical pack* style consists of *whole* beans packed parallel to the sides of the container. *Whole asparagus* style consists of *whole* beans cut at both ends, of substantially equal lengths, and packed parallel to the sides of the container.

Cut or *cuts* style consists of pods that are cut transversely into pieces less than $2^3/_4$ inches but not less than $^3/_4$ inches long and may contain shorter end pieces that result from cutting. *Short cut* or *short cuts* style consists of pieces of pod, of which no fewer than 75 percent are less than $^3/_4$ inches long and no more than 1 percent are more than $1^1/_4$ inches long. *Mixed* or *mixture* style means a mixture of two or more of the styles *whole, sliced lengthwise, cuts,* or *short cuts.*

Thicknesses of *round-type* and *flat-type* beans may also be defined. *Round-type* beans are defined as having a width not greater than $1^1/_2$ times the thickness, while *flat-type* beans have a width greater than $1^1/_2$ times the thickness. Each round-type bean unit may then be given a *size (thickness)* grade from **tiny** (1) through **small, medium, medium large, large,** and **extra large** (6). The size of the cut bean "is determined by measuring the thickness at the shorter diameter of the bean transversely to the long axis at the thickest portion of the pod." It is defined in terms of 64ths of one inch, from **tiny** (less than $14\frac{1}{2}$) to **extra large** (27 or more). A similar method of defining size of flat-type units is also included in the standard.

Another type of *size uniformity* concerns *degree of wholeness* of foods such as apple slices. Quality classifications for *uniformity of slice thickness* and definition of *wholeness* are shown in Table 6–1 (USDA 1957).

These measurements describing and quantifying visually assessed structure are necessarily written in precise terms. Only then is it possible to achieve and maintain consumer confidence and respect for industrywide standards.

Surface Texture

Each geometrical element of a *visual structure* possesses *surface texture* individual to the product and material. Material descriptors include *smooth, porous, fibrous, striated, hairy, rough, textured, connected, compacted, cellular, crystalline, homogeneous, watery,* and *powdery* (Abend 1973).

Consumer definitions of angel cake *surface texture* were *fluffy, smooth on top, fine texture, dense,* and *grainy* (Moskowitz 1985).

Table 6–1 Apple Slices—Uniformity of Size Classifications (USDA 1957)

Classification	Of the Drained Weight:
A	$\geq$ 90% consists of whole or practically whole* pieces of length $\geq$ 1¼ inches, or $\geq$ 90% consists of units uniform in thickness that does vary by $>$¼ inches
C	$\geq$75% consists of whole or practically whole pieces of length $\geq$1¼ inches
Substandard	A product failing to meet classification C. This is a limiting rule; i.e., such a product cannot be graded above substandard, regardless of the total score for the product.

*A *practically whole* slice may have been cut or broken, but at least ¾ of the apparent original slice remains.

Visually perceived attributes describing *surface texture* of gels were *grainy, wrinkles, bubbles, pinholes, firm,* and *bouncy* (response to shaking) (Clark 1990). Consumer terms for white fish flesh included *layered stratum, pocked surface,* and *porous* (Sawyer et al. 1988). *Gaping* in fish is caused by individual myotomes or muscle flakes separating in rigor mortis. This is an unsatisfactory attribute that continues to be apparent after cooking (Kent 1985). Ice cream has the properties of *crystallinity, consistency,* and *graininess* (Dolan et al. 1985). During storage, there is a decrease in consistency of *visual structure* and *surface texture* as *shrinkage, patchy iciness, yellowness,* and *surface skin* increase (Spencer 1977).

Appearance of bread slices is judged in terms of *crumb fineness, evenness of distribution of cell size, crumb coarseness,* and *crust darkness* (Swanson and Penfield 1988). *Size* and *surface texture* properties are critical to the definition of peanut maturity (see Table 7–1). Apart from size and shape, whole fruit and vegetable produce can be described, for example, in terms of *roughness* or presence of *surface ridges,* which on apples run vertically down from the stalk (Daillant-Spinnler et al. 1996). *Core-to-layer ratio* and the presence of *fibers* and *layers* are important to quality palm hearts (Lawless et al. 1993). Defective coffee beans can be *immature, black, waxy, foxy,* or *whitish* (Raggi and Barbiroli 1993). Appearance attributes of cooked rice include *looseness,* with anchor points from **cooked sweet rice** for least loose to **raw rice** for loosest (Yau and Huang 1996). Canned salmon curd can be evaluated by visual scoring (and weighing). A 7-point visual scale ranged from **little or no curd** to **maximum curd coverage** of the surface (Wekell and Teeny 1988).

Graininess of cooked sausage emulsion may be assessed on the scale **coarse grain** to **medium grain** to **fine grain** to **smooth**. *Evenness* may be scored from **uneven** to **even** (Haq et al. 1972). The structural character and internal appearance of cooked sausages may be described in terms of *particle size* (**fine** to **coarse**) and *evenness* (**uneven** to **even**) (Jones et al. 1989). Patterning of the lean areas of meat also may be defined because they are constructed with fibers, some of which are large enough to be visible. These arguments can be extended to the presence of cartilage. Meat has characteristic structures including fiber size and alignment, as well as the pores that appear after cooking of carbon dioxide–packed packed beef (Bruce et al. 1996).

Geometrical properties of particulate material have been classified by Brandt et al. (1963) (Tables 6–2 and 6–3). The *visual structure* of tea from the packet consists of the elements leaf, stalk, and dust, each having a size and shape distribution. The *surface texture* concerns the state of these elements. Terms used in the tea trade to describe the physical condition of the elements include

- *Blistered*—leaf is swollen and hollow inside
- *Bold*—pieces of leaf are big and might be cut smaller
- *Clean*—leaf is free from fiber and dust
- *Crepy*—leaf is crimped in appearance
- *Even*—leaf is true to grade and consisting of pieces of roughly equal size
- *Flaky*—leaf is in flakes, as opposed to being well twisted
- *Stalk*—red stalk is present
- *Uneven*—tea is composed of uneven irregular pieces (Rietz 1961)

Table 6–2 Properties of Particulate Material Related to Particle Size and Shape (based on Brandt et al. 1963, with permission)

Property	Example
Powdery	Confectioners' sugar
Chalky	Raw potato
Grainy	Cooked cream of wheat
Gritty	Pears
Coarse	Cooked oatmeal
Lumpy	Cottage cheese
Beady	Cooked tapioca

Table 6–3 Properties of Particulate Material Related to Particle Shape and Orientation (based on Brandt et al. 1963, with permission)

Particle Shape	Examples
Flaky	Boiled haddock
Fibrous	Breast of chicken, base of asparagus shoot
Pulpy	Orange sections
Cellular	Raw apples, cake
Aerated	Whipped cream, milk shake
Puffy	Puffed rice, cream puffs
Crystalline	Granulated sugar

Produce Sorting and Attribute Hierarchy Using Visual Structure and Surface Texture

Tomatoes received at the factory after mechanical harvesting may arrive in various conditions. An example of a formalized appearance analysis for a specified tomato variety (Leonard et al. 1977) can be presented according to an attribute hierarchy in which qualities are divided into levels. Each level has a distinct criterion in the evaluation process based on successive elimination of lower grades (Exhibit 6–1).

Laying out a grading scale using a hierarchical approach helps to reduce the potential ambiguities sometimes arising when a purely verbal scale is used. It also provides a first stage in the design of a mechanical sorting system.

Definitions of Structural Defects

Definitions of structural and appearance defects occur in many specifications. Examples are given for selected raw and processed foods. Whole or dressed fish can be downgraded for the following visually perceived structural defects (USDA 1986b):

- *Abnormality* includes an appearance of dryness, chalkiness, granularity, or fibrosity.
- *Defects* concern the consistency of the flesh, eyes, gills, and skin; the odor; and excessive blood or drip.

Exhibit 6–1 An Example of an Attribute Hierarchy for Tomatoes (data from Leonard et al. 1977)

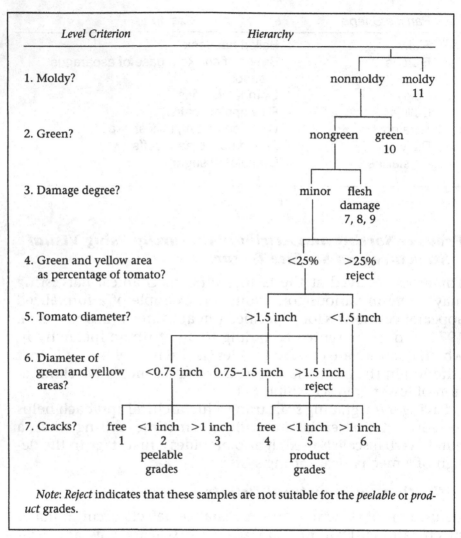

Level Criterion   *Hierarchy*

1. Moldy?   nonmoldy   moldy
                                                      11

2. Green?   nongreen   green
                                                      10

3. Damage degree?   minor   flesh
                                                  damage
                                                  7, 8, 9

4. Green and yellow area   <25%   >25%
   as percentage of tomato?   reject

5. Tomato diameter?   >1.5 inch   <1.5 inch

6. Diameter of
   green and yellow   <0.75 inch   0.75–1.5 inch   >1.5 inch
   areas?   reject

7. Cracks?   free <1 inch >1 inch   free <1 inch >1 inch
                   1   2   3   4   5   6
                        peelable   product
                         grades   grades

Note: Reject indicates that these samples are not suitable for the *peelable* or *product* grades.

- *Dehydration* is the loss of moisture from exposed flesh occurring during frozen storage or the degree of dullness and shrinkage of skin-on fish.
- *Surface defects* refers to occurrence of attached or loose scales; accumulation of blood spots; occurrence or absence

of fins or pieces of fin; presence of dark or light inner layers of skin for skinless fillets or, for semiskinned fish, the presence of dark outside layers; the presence of damaged portions of fish muscles (bruising); and damage to the protective coating, such as voids in the ice glaze or breaks in the skin.
- *Cutting and trimming defects* refers to body cavity cuts; improper heading; evisceration defects, such as inadequate cleaning; improper washing; and belly burn, an enzymic effect causing the flesh to have a burned or discolored appearance.

Specific terms are used to describe the external condition of a fruit or vegetable (Mohsenin 1968b):
- *Abrasion*—abrasive injury, which may vary in severity from separation of the periderm or skin to the removal of part or all of the pericyclic cortex
- *Bruising*—damage caused by external forces, causing physical change in texture and/or eventual chemical alteration of color, flavor, or texture; bruising does not break the skin
- *Distortion*—a change in the shape of intact fruit or vegetable, one that is not characteristic of the variety
- *Crack*—a cleavage without complete separation of the parts
- *Cut*—a penetration or division by a sharp edge
- *Puncture*—a small hole or wound on the surface, made by a pointed object or stem
- *Shatter cracks*—one or more tortuous cracks radiating from a point of impact
- *Skin break*—fracture of the periderm or a break limited to the skin
- *Skinning*—a separation of periderm from the plant part by scraping or rubbing
- *Feathering*—like skinning, but the separated periderm is still attached to the unseparated periderm
- *Split*—division into parts
- *Stem end tearing*—skin break caused by separation of stem from the fruit
- *Swell cracking*—cracking due to uptake of water by osmosis

Defects in whole fruit and vegetable produce can be described, for example, in terms of *browning, wilting,* and *sliminess.* Defective broccoli can have low *turgor pressure, yellow florets,* and *black or brown decay spots* (Paradis et al. 1996). Plant products suffer from storage deterioration, *surface pitting, water rot,* general loss of *structural integrity,* and poor *color development* (Jackman et al. 1988), which occurs, for example, with apples (Mohammed and Wickham 1997). Peach cultivars differ in their susceptibility to impact bruising, which can be characterized by *fiber tears, voids,* and areas of *brown discoloration* (Vergano et al. 1995). Damage to pimiento peppers can be classified as *growth cracks, mechanical, shriveled, bacterial, insect, fungal,* and *sunscald* (Shewfelt et al. 1985). Pineapple blemishes include *water blister, yeasty rot, black heart,* and *brown spot* (Bartolome et al. 1996). Fruit and vegetable deterioration also includes *mold growth, shriveling, bruising,* and *internal decay* (Hotchkiss and Banco 1992).

Examples of structure defect specifications used for processed foods occur in standards for various apple products and beans. The standards for canned apple (USDA 1957), apple butter (USDA 1955), apple juice (USDA 1982a), and applesauce (USDA 1982b) provide examples of how different products using the same raw material give rise to different critical criteria for defects.

In canned apples, *absence of defects* refers to the degree of freedom from harmless extraneous matter, such as whole or portions of leaf stem, cores, and seeds, damaged or seriously damaged slices, and carpel tissue. A *damaged unit* (apple slice) can be defined in terms of aggregate areas of green, red, and bruising, pathological or insect injury, or other detraction from high quality. Terms such as *practically free* and *fairly free* are used to describe the extent of various defects in terms of the ratio of weight per total product weight.

For apple butter, the problem of *uniformity* does not assume an individual significance and is only included among the *defects.* However, the preparation of the sample for its assessment is specific: "This factor is evaluated by observing a layer of the product on a smooth white surface. Such a layer is prepared by drawing a scraper, with an indentation $^3/_{32}$ inches high by 7 inches long for clearance, rapidly through the product in two horizontal planes so as to form an approximate square." Defects in a classification A "do not more than slightly affect the appearance or edibility of the product." Defects for applesauce and apple juices center on the numbers of seed, peel, and discolored apple particles in the sauce and the presence of amorphous and nonamorphous sediment or residue in the juice.

In green and wax beans, *absence of defects* refers to the freedom from blemishes (scars and pathological, insect, or other injury), unstemmed units, detached stems, leaves and extraneous vegetable matter, split units, loose seeds and pieces of seed, small pieces of pod (less than ½ inch long), ragged-cut units (so that cutting "seriously affects the appearance of the pod"), and damage by mechanical injury. Each classification is defined in terms of the maximum allowable percentages of drained weights of blemished units in each can or by count of each defective unit in the can. Defects also include the *degree of sloughing* of the epidermis (USDA 1972).

Many appearance defects occur in poultry. Standardized inspection procedures include examination of the unopened and opened pack, as well as examination of the carcass for general quality and processing damage in terms of *bruises, burns, blisters, irregular plucking,* and *cutting* (Forsyth 1998). After raw fish reaches specified size and conformity, the general appearance *freshness* is judged using specific criteria.

Translucency

Translucency is the property by which light penetrates and disperses into and/or through a material. Both light-scattering and light-absorbing entities within the material contribute. It is a critical property of many foods and drinks. Sweets are formulated to have different degrees of translucency. Nonuniformity of translucency can be used to vary the appearance of layered desserts—transparent and opaque layers are, perhaps, alternated. Normally, vanilla ice creams are perceived as opaque and sorbet as less so. Depending on the material, translucency can be scaled in terms of *transparency* for nominally clear foods such as beer or *opacity* or *translucency* for materials that are less light transmitting. Care should be taken when using the term *translucency* for scaling. An increase in translucency may mean an increase in transparency to some panelists while meaning an increase in opacity to others. To overcome this, panelists can be asked to score on a scale from **opaque** to **translucent** (MacDougall and Moncrieff 1988). Mungbean starch noodles have been scored for *extent of visibility* (Galvez and Resurreccion 1990). The phenomenon of translucency is amplified in Chapter 3, instrumental specification in Chapter 8, measurement in Chapter 9, and light scattering as a driving force in food appearance in Chapter 10.

The *clearness* of liquor from canned beans and canned wax beans has been defined (USDA 1972). **Practically clear** means that a liquor may possess a slight tint and no more than a trace of suspended or sediment material. **Reasonably clear** means that the liquor may be cloudy and contain a small quantity of sediment. **Fairly good** implies that the liquor may be dull in color but not off color, may be cloudy, or may possess a noticeable accumulation of sediment. **Substandard** liquor is definitely off color, is excessively cloudy, or has a seriously objectionable quantity of sediment.

Other foods for which translucency can be scaled include gravy, which can vary widely in *translucency* from **clear** (transparent) to **opaque**; milk (Tuorila 1986); and fish cornea and the outer water slime on fish. Fresh fish fillets have a translucent property. On cooking, protein precipitates, and the flesh becomes more opaque. Whole and dressed fish flesh is abnormal if it appears either jellied—that is, having a "gelatinous, glossy, translucent appearance"—or milky—having a "milky-white, excessively mushy, pasty, or fluidized appearance" (USDA 1986b).

Gloss

Gloss is the property by which a material appears shiny or lustrous. There are six types of gloss: specular gloss, sheen, contrast gloss or luster, absence-of-bloom gloss, distinctness-of-image gloss, and surface uniformity gloss (Hunter and Harold 1987). Although he gave no specific examples, Hunter (1974) implied that three of these—specular, contrast, and distinctness-of-image gloss—are relevant to the appearance of foods.

Gloss has relevance to particle size composition and molding of chocolate bars, with high gloss normally being associated with high quality (Musser 1973). A 5-point scale (1, **heavy bloom**; 2, **considerable bloom**; 3, **slight bloom**; 4, **fair, some dulling, but no bloom**; 5, **perfect glossy appearance**) can be used for its assessment (Talbot 1995). Mungbean noodles differ in the amount of *surface shine* (Galvez and Resurreccion 1990). *Glossiness, crystallinity,* and *graininess* are three undefined ice cream attributes used by judges (Dolan et al. 1985). The *glossiness* is probably caused by specular or directional reflectance. *Crystallinity* is probably an expression of the uniformity of the specular component across the surface. *Graininess* may be the degree of uniformity of the diffuse reflectance over the surface of the ice cream. *Glossiness* of rice, defined as the amount of shine on the surface of the kernel, is an important attribute for some consumers. Reference anchors are **raw rice** for the lowest gloss and **oily cooked rice** for the highest (Yau and Huang 1996).

Color

The construction of visual memory color scales is often reasonably simple when changes in the product are constrained by a relatively straightforward pigment system. In any one naturally based product, although the same pigment system is normally present, seasonal differences result in variations in the concentration balance of pigments present. However, such seasonal variations rarely cause difficulties in the comparison of sensory judgments. The situation may be different when more complex pigment systems are present, when color additives are used, when different processes are used, when different varieties of fruit or vegetables are being compared, or when agreement is sought with instrumental color measurement.

Color scales are often required to describe the changing colors occurring during ripening, processing, or aging. These colors are transitional and change progressively. The color that was dominant becomes the background color as maturity develops. There are many examples of memory descriptors and scales for color in the food literature. In this chapter, these are discussed in terms of scale types: five types of single scales—those using color names, those based on product lightness and/or darkness, hybrid scales, linear and additive scales, and single multidimensional scales—as well as multiple scales. Examples of each type are given, followed by examples of *color uniformity*.

Single Scales Using Specific Color Names

Sensory scales describing meat color have been devised, for example, by the American Meat Science Association (AMSA). They recommend scales for categorizing *oxygenated, deoxygenated,* or *discolored* beef, lamb, and pork; scales for different types of raw ground meat; scales for assessing *color stability* of vacuum and non–vacuum-packed meat on display; and scales for lean meat *browning*, fat *color*, surface and internal cooked meat *color*, and meat *discolorations* (AMSA 1991). Scales are also available for veal (USDA 1980).

Scales for raw beef color are used to study consumer behavior. For example, bloomed beef steaks were visually examined by a trained panel, wrapped, and randomly displayed in three stores for consumer evaluation (Jeremiah et al. 1972). The scale and the percentages for consumer decisions that samples were **desirable** or **undesirable** are shown in Table 6–4.

A similar but 7-point scale has been used to grade raw beef. On this scale, **pale red** exudative meat is scored 2.0; normal, soft, exudative meat, 3.3; meat of normal color, firmness, and exudation, 3.7 (between **light cherry red** at 3.0 and **bright cherry red** at 4.0); and dark, firm, dry meat, 6.4 (between **dark red** at 6.0 and **very dark** at 7.0) (Hunt and Hedrick 1977).

The *color* and *doneness* of roast semitendinosus muscle have been assessed using the two scales shown in Table 6–5. Single muscles were subdivided into three roasts, each weighing 500 to 700 g. Two sets of three roasts were assigned target internal temperatures of 60, 68, and 75°C and were cooked in ovens set at 121 and 177°C. Cooking was replicated four times. The two cen-

Table 6–4 A Standard Scale for Beef Muscle Color, with Consumer Acceptance Decisions (Jeremiah et al. 1972)

Color Score	Trained Panel Description	Consumer Decision	
		Desirable (%)	Undesirable (%)
1	Very pale pink		
2	Pale pink	56.0	26.0
3	Pink	82.8	6.1
4	Slightly pale red	76.5	8.9
5	Cherry red	65.8	15.9
6	Slightly dark red	65.5	22.5
7	Moderately dark red	13.3	71.0
8	Dark red		
9	Very dark red		

ter slices were dissected from the roast. The edges were removed, and each slice was allowed to bloom for 10 minutes.

The cooking conditions produced a very wide range of colors. For example, of the panelists, 85 percent scored the 60°C sample **medium red** or **very red**, and 71 percent scored the 75°C sample **gray brown** or **brown**. There was a considerable varia-

Table 6–5 Color and Doneness Scales for Roast Beef (Lyon et al. 1986, with permission)

Color	Doneness
Very red	Rare
Medium red	Medium rare
Pink	Medium
Slightly pink	Medium-well done
Pinkish gray	Well done
Gray brown	
Brown	

tion in the *doneness* score applied to the middle colors. For example, the **pinkish-gray** samples produced responses ranging from **medium rare** to **well done**. Specifically, they were described by 14 percent as **medium rare**, by 30 percent as **medium**, by 54 percent as **medium-well done**, and by 2 percent as **well done**. Hence, *color* is a more well-defined attribute than *doneness* (Lyon et al. 1986). The latter is a hedonic attribute that depends on personal taste: in the restaurant, a steak may be **medium-well done** to one diner but **rare** to another.

Colors expected of fresh beef—that it should be pink or red—can be carried through to processed meat products. Consumers prefer meat patties to be the cherry red color of ground beef rather than the pale or grayish-pink color normally associated with ground pork. In Britain, colorant is added to many brands of pork sausages to make them the traditional pink (Parizek et al. 1981). In extending this work, it was found that 20-percent mature beef (lean) extended with up to 60 percent mature pork and 20 percent beef fat created the necessary cherry red color and high raw appearance scores (Miller et al. 1986). The internal cooked color of ground beef patties depended on pH. Gray patties resulted from meat in the normal muscle pH range (5.3–5.7). Muscle with a pH of 6.2 produced a slightly red color, but reddest cooked color was most intense inside patties with the highest pH and the highest total pigment concentration. Bull meat has a higher pH and a greater level of total pigments. Hence, specifications for raw material used in fresh ground beef should include pH (Mendenhall 1989). For restructured steaks, it is necessary to assess both internal and external *colors* (Miller et al. 1988). A *discoloration* scale is also useful when assessing effects of storage on restructured beef and turkey steaks (Marriott et al. 1987).

For cured meats, a simple color scale is normally sufficient. For example, both green and matured Parma ham can be evaluated on a **very pale** to **pale** to **normal** to **dark** to **very dark** scale (Chizzolini et al. 1994).

Single-color scales are useful for fruit. An example concerning the maturing of peanut mesocarps is included in Table 7–1 (Williams and Drexler 1981). A color maturation index for ripening olive fruits consists of the stages **intense green**, **green**,

yellowing green, small reddish spots, and **turning color** (Minguez-Mosquera et al. 1989).

The definition of tomato maturity has been described solely in terms of color. According to the Australian Fruit and Vegetable Grading and Packing Regulations:

Mature means that the tomato has reached its maximum growth, and has reached or passed the stage when:

a. its skin has changed from a dull green to a bright green colour, and
b. the contents of the seed cavity have changed from a light green to a deep amber or to a deep amber tinged with pink. (quoted in Adams 1977)

A wholly color word–descriptive scale, such as **mature green, green orange, orange green, orange, red,** can be used to describe tomatoes (Dixon and Hobson 1984). The USDA color maturity descriptions for red-fleshed tomatoes are from **green** to **breakers** to **turning** to **pink** to **light red** to **red**. They also have a **mixed-color** grade (USDA 1977a). With this scale, the effects of packaging and storage temperature on the development of color maturity have been investigated (Yang et al. 1987). To delay ripening during storage, tomatoes must be seal packaged prior to the **turning** and **pink** stages when stored at 21°C and 12°C respectively. No delay in ripening was observed at either temperature when the tomatoes were packaged beyond the **turning** stage.

Changes in fruit color are typical of type. Limes change from **yellow** to **yellow green** to **green** to **dark green** (Sierra et al. 1993). Bananas undergo color and color uniformity changes from **green** to **light green/greenish yellow** to **yellowish green** to **yellow with green tips** to **fully yellow** to **yellow with brown spots**. These color changes, typical at 20°C, do not occur at higher ambient temperatures, and as the fruit remains green, the ripening stage is judged by texture (Thomas and Janave 1992). Pepino, a fruit of the eggplant-potato-tomato family, undergoes complex color changes. The unripe fruit is initially fully **green**; then **purple stripes** appear on the green; then the green fades, and the background becomes first **pale cream**, then **yellow**, then **orange**, and finally **deep golden orange** (O'Donoghue et al. 1997). Browning is a major problem associated with preparation and storage of fruit pulp. For example, custard apple pulp can be evaluated using the scale **blackish brown** to **dark brown** to **light brown** to **light brown or pink** to **creamy pink** to **creamy** (Gamage et al. 1997).

Cucumber, broccoli, and spinach can be scored on a line scale having the anchor points **green, yellow green**, and **yellow**. The anchor **red** was added for the scoring of tomatoes (Gnanasekharan et al. 1992). Yellow corn can be graded from **very pale yellow** through **bright clean yellow** to **reddish yellow**, while the scale for white corn is from **very pale dull white** to **pale dull white** to **bright clean white** to **cream-colored white** to **off-colored white** (Floyd et al. 1995). Color and color uniformity categories of fermented cocoa beans are from **slaty** to **fully purple** to **three-quarters purple** to **half brown** to **three-quarters brown** to **fully brown** (Ilangantileke et al. 1991).

Arctic char is a salmonid fish that can be scaled from **chestnut red brown** to **dark amber** to **medium amber** to **pale amber–gray** to **yellow gray** to **cream gray** (Swatland et al. 1997). Cooked Atlantic salmon is judged on a 5-point scale. High-quality fish (5) is a "bright, fresh and uniform salmon red; clear distinction between white and red muscle; shiny clear lipid beads in the surrounding meat juice"; score 4 is "slightly greyish and weaker, but uniform salmon red; slight pink lipid beads"; score 3 is "greyish, yellowish and not uniform color; pink lipid beads"; score 2 is "yellowish and uneven color"; and score 1 is "clear discolored" (Sigholt et al. 1997). Surface color of squid is described in terms of paling using a scale from **reddish brown** to **pale white**. This change is caused by the oxidation/reduction pathway of the ommochrome pigment responsible for squid integumental coloration (Hincks and Stanley 1985).

Single Scales Based on Product Lightness and/or Darkness

It is convenient to judge achromatic or near-achromatic—that is, hueless—products using a lightness or darkness scale. For example, words used by consumers when describing the color of 18 species of cooked Atlantic fish are listed with frequency of use in logical groups in Exhibit 6–2. A total of 16 color terms, including both *color* and *color uniformity* terms, were included.

The *color* terms may be divided into four groups, as shown. The first group, which made up the bulk of responses, contains the achromatic colors **white**, **gray**, and **neutral** (the sample population contained a number of white fish). The second group contains chromatic words having a firm color reference. Group 3 contains the most common color-modifying words **light** and **dark**, and group 4 contains words related to color uniformity. On the basis of this, in subsequent studies the authors defined all cooked fish on a scale of *darkness* (Sawyer et al. 1988). Most fish is nonwhite, increasing in darkness by being either **gray** or **brown**; examples include two species of cooked mackerel (Hong et al. 1996; Silva and Nunes 1994).

A similar color scale has been used for Pacific fish. A total of 387 housewife consumers were given the options **dark**, **inter-**

Exhibit 6–2 Color-Related Terms Generated by Consumers To Describe Cooked Atlantic Fish Divided into Groups of Related Color Attributes (data from Sawyer et al. 1988)

Color Group 1		Color Group 3	
White	61	Dark	20
Off white	3	Light	5
Gray	11	Pale (?)	1
Gray black	1		
Neutral	2		
Color Group 2		**Color Group 4**	
Brown	10	Mottled	1
Light brown	2	Uniform	2
Tan	2		
Yellowish	1		
Ivory	1		
Off gray	1		

mediate, and **light** when scoring color. The overwhelming majority scored skipjack as **dark** and flying fish, plaice, and channel rockfish as **light**. Common horse mackerel caused the most confusion, with 16 percent scoring **dark**, 75 percent **intermediate**, and 9 percent **light** (Hatae et al. 1988). These three descriptors are also used to describe tuna. **White** is reserved for the species albacore (*Thunnus germo*) (Food and Drug Administration 1987). Canned tuna can be scored on a 10-point scale. A score of 10 was given to **very light, almost white** samples and 1 to **dark, very red, or very brown** fish (Little 1969).

The color of fish minces is affected by species, handling prior to recovery of the mince, and the presence of blood (Young and Whittle 1985). The overall appearance of mince recovered by various treatments is affected by different constituents of the whole fish (Table 6–6). Grayness arises from melanin in the skin, black from skin or membrane, and red or red brown from blood or dark muscle. Presence of minor black blemishes is a prime de-

Table 6–6 Factors Affecting the Appearance of Fish Minces (Young and Whittle 1985, with permission, Crown copyright reserved)

Material Used	Factors	Dark Muscle	Black Belly Lining	Skin	Spinal Cord	Swim Bladder	Kidney
Fillets	Skin off	+	+				
	Skin on	+	+	+			
	Untrimmed	+	+	+			
Trimmings	V cuts		+	+			
	J cuts		+	+			
Cutlet		+	+	+			
Dorsal split	− Bone	+	+	+			
	+ Bone	+	+	+	+		
Frame	Spines						
	Posterior				+		
	Anterior				+	+	+
	Whole				+	+	+
Headed and gutted fish		+	+	+	+	+	+

fect in the United Kingdom. Overall grayness, red-brown discoloration, and white blemishes from the swim bladder or spinal cord also lower quality. The National Marine Fisheries Service (1975) proposed the use of the color descriptors **white**, **light**, and **dark** in their interim grade standard for frozen minced fish blocks.

Meat products can sometimes be judged on this type of scale. The color of fried breaded broiler thighs has been scored by an untrained sensory panel using a 14-point category scale ranging from **extremely light** to **extremely dark**. The rate of browning decreased above 177°C. This was probably caused by depletion of ingredients in the breading material participating in the nonenzymic browning (Lane and Jones 1987).

White dairy products, such as milk, can be scored on whiteness scales (Rankin and Brewer 1998). A scale for rice whiteness contains as anchor points **dry baked rice** for least white and **white paper** for the whitest (Yau and Huang 1996). Darkness scales have also been useful in the description of orange juice (Moskowitz 1985) and an ice cream product (HR Moskowitz, personal communication, 1990).

Single Hybrid Scales

A hybrid scale is one in which attributes other than color are included in a single scale. This becomes necessary when, for example, developing color during ripening does not follow consumer-important properties. For example, whole tomatoes are often picked when green and are stored in the presence of ethylene gas. This reddens color without optimizing texture. It has been recommended that tomatoes be cooled rapidly to hold color development and that green fruit be ripened under controlled temperature conditions (Adams 1977). A number of color scales have been developed for red varieties, the first appearing in 1926 (Goose 1978). Scales are used in specific situations to control harvesting, marketing, and processing. Different methods have been used to describe the ripening stages of the fresh fruit.

A hybrid scale for tomatoes includes three types of assessment (Table 6–7). This hybrid scale illustrates a difficulty when work-

Table 6–7 Maturity Scale for Whole Tomatoes (Hutchings et al. 1969)

Maturity Stage	Description
1	8–12 days
2	24–26 days
3	34–36 days
4	44–46 days full green
5	Just coloring
6	Yellow
7	Orange
8	Just ripe
9	Ripe
10	Overripe

Note: Day 0 is the day of petal drop.

ing with biological systems. A sensitive enough scale based on one type of measurement or assessment can be produced for the whole maturity scale. For example, the **number of days from petal drop** may be adequate to describe the early stages of maturity. Subsequent to maturity stage 4, however, tomatoes ripen at different rates, and color names can be used. Redness at the **ripe** stage changes little, so the scale basis was again altered to a judgment of the extent of ripeness. Gloss seems to be a major cause of appearance change during the **ripe** to **overripe** stages (Hutchings et al. 1969).

The Campden scale for canned whole peeled tomatoes (Chipping Campden 1988), shown in Table 6–8, is also a hybrid scale with terms incorporating uniformity (**venation** and **nonuniform**) and hedonic attributes (**unattractive**). No description of the early stages of maturity is required for commercial grading; quality depends solely on appearance.

Linear and Additive Scales

Very few sensory color scales are designed to be linear and additive, but the scale for brewed tea with milk is one example. Two major pigment groups, theaflavins and thearubigins, are involved in the color of brewed, black leaf tea. Both are chemically complex and difficult to analyze, but the balance between the two plays a major role in tea color, especially when it contains milk (Smith and White 1965). Expert tea taster vocabulary includes **colory brown** for teas coming from Assam, **golden** for teas from Sri Lanka, and **rosy pink** for Dooars teas. **Colory** teas tend to have higher theaflavin concentrations, while **gray** teas result from a higher proportion of thearubigins. Different brewing conditions tend to favor the extraction of the two groups. Brewing in hard water decreases the theaflavin extraction and the *brightness*. Boiling the water for a few minutes before brewing increases the *brightness* but spoils the flavor. Theaflavins have a lower solubility rate, so a short infusion time favors extraction of the thearubigins. Homogenization of the milk increases the number of fat globules. Hence, its whitening power is increased, and the *brightness* of the infusion increases. The use of heat-treated milk causes pinkish brown colors to develop and produces an increase in *brightness*. Increasing the concentration of the milk, which itself can be scaled from **yellowish** to **bluish**, increases the whiteness and reduces the variation in color of tea infusions (Tuorila 1986).

Table 6–8 The Campden Scale for Grades of Canned Whole Peeled Tomatoes (Chipping Campden 1988, with permission)

Description	Grade
Bright, uniform deep red in color. May be slightly pale red or orange red, with distinct, but not unattractive, pale venation near the surface of the fruit.	Campden Grade A
Rather yellow/yellow orange in color. Rather dull, with pronounced venation near the surface of the fruit. Rather nonuniform.	Campden Grade B
Uneven in color, with some fruits bleached by the sun (sunburn). Rather dull.	Campden Grade C
Uneven, patchy, shades of yellow, orange, or green. Dull and unattractive.	Ungraded

For commercial assessment, samples are brewed in a rigorously controlled manner; Theobald (1977) described the routine:

> In our company we weigh 6.4 grams into a standard pot, pour on freshly boiling water from a large kettle (each pot holds 300 ml), [and] allow it to infuse for six minutes. We then strain the liquor into a standard bowl and tip the wet leaves (or the infusion as we call it) on to the lid. Before straining the liquor into the bowl we will have added milk (12.5 ml) to the bowl first. Tea must be added to the milk, and not vice versa, so as to avoid scalding the milk; dropping a small volume of cold milk into a very hot bowl of liquid would decompose the milk and would impart a taint.

Expert tasters use two scales for assessing value and blending quality of tea batches. *Brightness* and *color* are each scored on the 6-point scales (Theobald 1977; Fox et al. 1978) shown in Table 6–9. These characteristics are estimated to one decimal place. Maximum variation within a retail blend is usually ±0.3 in *brightness* and ±0.2 in *color*. The scales are so constructed that they behave in a linear and additive manner. That is, it is a fundamental requirement that two teas scored 2 and 6 for *color*, when mixed in equal quantities, result in the same visual ap-

Table 6–9 Scales Used for Milked Teas (data from Theobald 1977)

Color Description	Score
Weak yellow or grayish	1
Golden yellow	2
Yellowish brown	3
Brown	4
Brownish red	5
Pink/reddish	6

Note: Brightness scores range from *very dull* (1) to *very bright* (6).

pearance as a tea scored 4. As a check on the anchor points for these scales, master standard teas for each scale are always available for reference. These standards are regularly checked and replenished to minimize drift.

Single Multidimensional Scales

Color may change in different hue directions during processing. For green vegetables, the chemical pathways and pigment combinations available on growing, storage, and processing are so numerous that it is possible for a range of colors to develop. Commercial scales range from **bright green** to **brown** or some other **off shade** (e.g., for cooked snap beans, Shewfelt et al. 1986). Such approaches are common in the United Kingdom (e.g., for Brussels sprouts, Chipping Campden 1969), and the United States (e.g., for sauerkraut, USDA 1963b, and for potatoes, USDA 1986a). Hence, for green vegetables, different colors appear on the same sensory *color* scale grade. One grade for peas may be described as **pale** or **yellow** or **khaki**. This situation is shown schematically in Figure 6–2.

Fresh green vegetables starting with a color grade 1 **bright green** can change, in order of decreasing perceived lightness, to the green shades of yellow, pale, khaki, or black. A small change in any of these directions, or combination of directions, results in the color classification grade 2, and greater changes result in classification in the lower grades 3 or 4. This scale is sufficient for a sensory quality specification, but it cannot be related easily to physical color standards or instrumental color measurements (Lozano 1989).

A less complex case of a multidimensional color change occurs with cooked Atlantic fish. Most fish is nonwhite, increasing in darkness by being either gray or brown (see Exhibit 6–2). This situation was simplified by using a single scale for *darkness* (Sawyer et al. 1988).

Multiple Scales for Color

A color can be described in terms of its constituent hues. For example, an orange color can be scaled in terms of its *redness* and *yellowness*. (A multicolored surface containing discrete patches

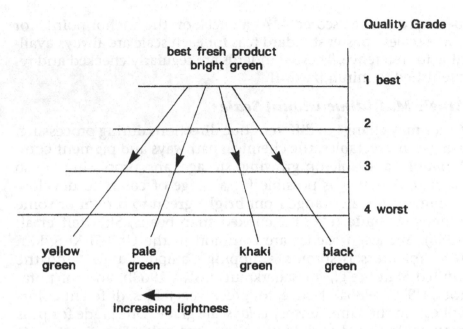

Figure 6–2 Industrial color grading of green vegetable products.

of red and yellow requires, on the other hand, a *color uniformity* description.)

Color changes occurring on the exterior of Brussels sprouts can be described in terms of *green, yellow, brown, depth of color,* and *brightness*. Insides can be described in terms of *yellow, brown,* and *pink* (Lyon et al. 1988). Professional tea tasters scale brewed milked tea in terms of *color* and *brightness* (see Table 6–9). A comparison of different varieties of apple requires the use of a number of color scales. Twelve varieties from the Southern Hemisphere were assessed for background *yellow* and *green,* streak *redness,* and near-stalk *brown* (Daillant-Spinnler et al. 1996). A group of nine port assessors used scales of *red, brown, yellow/golden, blue, purple,* and *gray,* as well as *clarity, intensity,* and *depth of color at edge.* All contribute significantly to quality (Bakker and Arnold 1993). *Browning* may be estimated in the presence of the normal color of the product, as in black currant syrup (Skrede et al. 1983) and in spinach (Loeff 1974).

Tomato pastes of different concentrations have been described in terms of the amount of perceived *red, orange, yellow, brown*, *purple*, and *gray* (Figure 6–3). These color elements changed with concentration, the largest occurring for *red* and *orange*. Panelist use of hue descriptors was linear with the logarithm of the concentration. Individuals found it impossible to assign values to the achromatic color *gray* (MacDougall 1987).

Diluted black currant drinks can be described in terms of *red*, *brown*, *purple*, *yellow*, *richness*, *natural*, and *transparency*. Principal component analysis of the results is shown in Figure 6–4.

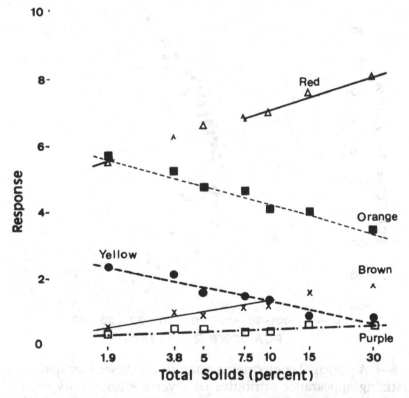

Figure 6–3 Relationship between panel mean response to the color profile descriptors and the logarithm of the total soluble solids content of a series of tomato paste suspensions. *Source:* Reprinted with permission from D.B. MacDougall, Optical Measurements and Visual Assessment of Translucent Foods, in *Physical Properties of Foods 2*, R. Jowett, F. Esher, M. Kent, B. McKenna, and M. Roques, eds., © 1987.

Principal component 1 accounted for 81.5 percent of the variance. *Richness, purple, red,* and *natural* were related positively with principal component ($r > -0.82$), which was related negatively with *transparency* ($r = 0.94$). Principal component 2 (11.4 percent) was difficult to interpret but was loosely correlated with the contrast between *red* ($r = -0.54$) and *purple* ($r = 0.41$) (Brennan et al. 1997).

Scales for *pinkness* and *brownness* are used to describe colors of meat products. Small, say 5- or 6-point, scales may be suitable for natural-colored products like fresh meat, ham, and bacon.

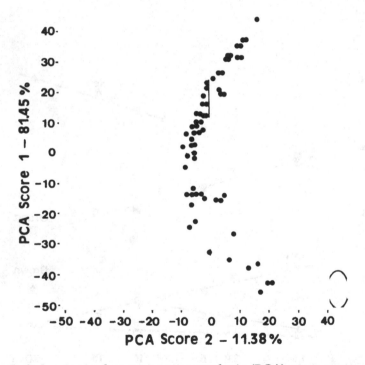

Figure 6–4 A principal components analysis (PCA) sensory space map summarizing appearance attributes of diverse *Ribes* genotypes. PC1 contrasts translucency with richness/purple character; PC2 contrasts red and purple character. *Source:* Reprinted from *Food Research International*, Vol. 30, Brennan, Hunter, and Muir, Genotypic Effects on Sensory Quality of Blackcurrant Juice Using Descriptive Sensory Profiling, pp. 382–390, Copyright 1997, with permission from Elsevier Science.

For products such as luncheon meat, which contain artificial colorants, a 10-point scale is needed to describe *pinkness*. For cured meat having low *pinkness* levels, a third scale, *grayness*, should be included (Lyle 1977). *Brownness* and *pinkness* have also been used in the sensory assessment of the outside and inside of experimental and commercial brands of British sausages. Both attributes were scored on unstructured 100-mm scales, marked **nil** to **extreme** at the ends (Jones et al. 1989).

Bacon has been described in terms of nine hue or hue-related terms and three achromatic terms. These were *red, pink, orange, yellow, cream, brown, green, blue, purple, white, gray,* and *black*. Panel members were also asked to judge on scales of *darkness* (**dark** to **pale**) and *brightness* (**dull** to **bright**). Relationships emerged within two principal components: (1) *red, pink,* and *brightness* and (2) *brown, green,* and *gray* (MacDougall and Moncrieff 1988).

Processing of crabcakes results in different colors and glosses. These can be assessed using *gray/blue, yellow/green,* and *sheen* scales. Products cryogenically frozen changed in color less than those receiving other preservation treatments (Henry et al. 1995).

Color Uniformity

Nonuniformity of color—that is, the presence of more than one color—can arise from deliberate nonhomogenized mixtures of food components—for example, fat and meat inclusions in bologna—or through the patchy change of pigment or pigment form. Examples of this include the gradual appearance of brown metmyoglobin in the red oxymyoglobin matrix of fresh beef, and changes in apple skin appearance as chlorophyll fades and carotenoids and anthocyanins are revealed or synthesized. In all these products, there are surfaces of mixed colors.

Marbling is a *color uniformity* attribute of fresh meat. The USDA specification for marbling in carcass beef employs seven degrees: **practically devoid, traces, slight, small, modest, moderate,** and **slightly abundant** (USDA 1997). This scale can be extended at both ends using **devoid, moderately abundant,** and **abundant** (Cross et al. 1975). These authors also used scales for *marbling distribution* (**even** to **uneven**) and *marbling texture* (**fine** to **medium** and **coarse**). Visual assessments of marbling can be imprecise, and panelists may be unable to discern thoroughly between marbling fleck size and abundance (Gerrard et al. 1996). This is important, as it is recommended that each marbling fleck be considered equal, regardless of size (USDA 1997).

Color uniformity is also a feature of bacon. Across the fat, there may be a gradation of white or yellow. Associated with each area of lean, there is normally a range of colors produced by either diffuse reflectance or interference. *Cured color uniformity* is included among the visual traits of concern to the consumer of bacon rounds. Scale ends were anchored as **nonuniform** and **uniform** (Stites et al. 1989).

The extent of two-toning of lean meat can be visually estimated from **uniform** to **slightly two-toning** to **small amount** to **moderately** to **extremely two-toning**. Similarly, surface discoloration can be estimated as a percentage of the total surface area (AMSA 1991). Two muscles, each a different color, may be present in slices of processed meat. Estimates of the area occupied by each color as a percentage of the total slice area can be coupled with a color description. Thus, two scales of *pinkness* and *brownness* make up a statement of *color uniformity*. For example, 80 percent of a slice may have a *pinkness* score of 7 and a *brownness* score of 4, and 20 percent may have a *pinkness* score of 3 and a *brownness* score of 3 (Lyle 1977).

Color uniformity is used in commercial specifications. For example, the Queensland, Australia, Fruit and Vegetable Industry Board recommended the following scale for tomatoes shipped in bulk (Adams 1977).

- **Green** refers to tomatoes that are completely green in color. The shade of green color may vary from light to dark.
- **Backward color** is used when all fruit in the package are showing a definite break in color from green to tarnish yellow, pink, or red, but not on more than 30 percent of the surface.
- **Semicolor** is used when all fruit in the package have more than 30 percent, but not more than 60 percent, of the surface showing a pink or red color.

- **Forward color** is appropriate when on all fruit in the package have more than 60 percent of the surface showing pinkish red or red. The fruit must not be a full red color.
- **Red ripe** is used when the full surface of the fruit has developed a red color.

Canadian grades for green-seeded dry peas are based on the proportion bleached. A bleached pea is one that has more than one eighth of the cotyledon surface bleached to a distinct yellowish color in marked contrast to its natural color. Grades 1, 2, and 3 contain a maximum of 2, 3, and 5 percent of the peas bleached, respectively (Canadian Grain Commission, quoted in Mazza and Oomah 1994). This type of specification has been used more precisely in a definition of apple maturity. As apples mature, their seeds brown, and a seed color index has been devised to describe its extent in terms of *color uniformity*. It assigns points 1 to 6 for 10, 25, 50, 75, 90, and 100 percent of the surface browned (Dhanaraj et al. 1986).

Nonspecific color specifications are also used for color uniformity. An example is the USDA standard for canned green and waxed beans (USDA 1972):

- **Practically uniform bright typical color** is a color that is typical, bearing in mind the variety, with not more than 5 percent by count of units that vary markedly from this typical color.
- **Reasonably uniform typical color** is defined similarly to the above grade, except that no more than 10 percent should vary markedly.
- **Fairly uniform typical color** beans have no more than 15 percent that vary markedly.
- **Substandard** beans are definitely off color or fail to meet the **fairly uniform typical color** grade.

Color uniformity of banana, pepino, and apple has already been discussed above. Catchall terms may be used for the uniformity of appearance. For example, for snap beans, *uniformity of appearance* may be assessed on the memory scale **not uniform** to **uniform** (Shewfelt et al. 1986). Color uniformity terms were included in consumer descriptions of cooked Atlantic fish (see Exhibit 6–2).

DERIVED PERCEPTIONS

The above *basic perceptions* contribute to the *derived perceptions* of *visual expectation*—that is, *visual identification* and *visually assessed texture, flavor,* and *satisfaction*. These attributes are individual to the particular food product, but, like *basic perceptions,* they can be scaled. *Derived perceptions* are the basis of the customer's assessment of quality, expectations, and decisions of whether to buy or to eat.

In the shop, where we are unable to eat the product, appearance may be the only guide to expectation. A purpose of the label on fresh produce is to aid *visual identification.* Pictures aid identification by reinforcing words naming the product within. Sample tasting is provided where stores wish to add a flavor experience to a new product, reinforce an identification of a readily identified product, or change a flavor identification associated with a particular appearance.

Visually assessed texture may be static or dynamic. In a static situation, it is that sum of individual properties that makes the viewer think that a product is, say, *crisp* (as in biscuits, lettuce, or bacon), *chunky* (chocolate), or *tough* (meat). Dynamic visual texture properties become apparent when the product is moved or distorted. Examples of this include *strength* (of a jelly from its wobbliness), *toughness* (of meat from its visually perceived behavior when prodded with a fork), or *sliminess* (as of yogurt from the way it pours).

The overall appearance of the product contributes to an impression of a specific flavor or flavor type. Although color is important to this process, it may not be dominant. For example, a product may be yellow, but we may know from its presentation that it ought to taste of vanilla, not lemon or mustard. *Visually assessed flavor* attributes include the four basic tastes, the visually perceived flavor, and flavor highlights evoked by sight of the product. Flavor and flavor highlights might be, for example, *lemon, beef, musty, stale, spicy,* or *juicy.* The visual flavor of gravy might be *starchy, meaty,* or *watery. Tasty looking* has been used to describe the apparent flavor of white fish (Sawyer et al. 1988).

Each cooked bacon slice gives an impression of what it will taste like when it is being eaten. If it looks well cooked and is curled, it may appear as if it will be crisp in the mouth (*visually assessed texture*), as well as giving a guide to the taste (*visually assessed flavor*). Depending on whether we like well-cooked bacon, it will also contribute in a positive or negative way to our appetite and expectation of, for example, the degree of potential satisfaction (*visually assessed satisfaction*). This is also determined by the quantity or proportion of the bacon on the plate.

In different ways and in different combinations, these components lead to an overall *visual expectation*, which in turn contributes to the drive to purchase or to eat.

COMPARATIVE SCALES OF APPEARANCE

Color Atlases

Although the eye is good at discriminating between colors, our capacity for remembering them is comparatively poor. This may pose less of a problem for other industries, in which paint or textile standards, although not permanent, can be stored under controlled conditions and used to refresh the memory. For most foods, this is not possible: last season's fruit will not be available for comparison with the current crop. As far as color is concerned, a method of overcoming these difficulties is to match the sample to a color chip obtained from a color order system. Alternatively, some form of a color reference plaque or tile may be used.

Advantages to the food industry of using such comparative standards include

- *low cost*—only a few color chips covering the relevant range need be purchased
- *stability and consistency*—although chips may become easily contaminated, commercial color atlas chips are reproduced to precise and accurate tolerances
- *portability*—communication between manufacturer and purchaser can be eased

Disadvantages include

- the necessity for a standard viewing and lighting system
- the necessity to check that the observer has normal color vision and is consistent at the task of matching colors
- the comparatively large steps between adjacent chips in established color order systems (often only an approximate match is possible)
- the saturated colors missing from atlases (this is not a problem for foods)
- the difficulty of matching intense, dark colors
- the difficulty of imagining that a piece of uniformly colored card or plastic looks like a particular food

Many color atlases have been produced (Hale 1986). The most well known are the Hungarian Coloroid System, the German DIN System, the Inter Society Color Council/National Bureau of Standards System, the Munsell System, and the Natural Colour System (NCS). Associated with these systems are atlases consisting of books of chips covering a wide color range organized in some logical system of color space. A few atlases have been developed with natural materials in mind. These include the Syme Atlas of 1814, developed for naturalists; the Ridgeway nomenclature of 1886 for ornithologists; the French atlas of 1905 for horticulture; and the currently available Horticultural Colour Charts, produced by the Royal Horticultural Society of England (RHS). Most work on food has been limited to the RHS charts, the Munsell System, and the NCS. Further comment is restricted to these and to a possible future system for the food and allied industries. Using a color atlas to match a particular color is reasonably straightforward. The separation of surface color into its subcomponents is difficult even for color experts, and patient training is necessary. Specific recommendations made for using the Munsell atlas (see Chapter 5) are also relevant in principle to the use of other atlases or plaques.

Royal Horticultural Society Charts

Two examples of the use of the 1966 RHS Colour Charts have been recorded (Arthey 1975). The scale sequence developed for raw peas is 150D (very pale yellow green) to 145C, 145A, 144B, 143C, 143A, 141B, 135B, 135A, and 133A (dark blue green). This was developed over three seasons and was used to aid growers and processors in selecting the best cultivars for marketing. Although both pale and dark cultivars are used for canning, those lighter than 144B are suitable only for canning. Only darker peas are acceptable for the fresh market, freezing, and dehydration.

The color sequence for raw carrots is pale (yellow orange) 9C, 17B, 24A, 28B, and dark (deep orange) 30C. Colors of the core and flesh are assessed separately, as is the freedom from shoulder and internal discoloration. Only carrots with intense coloring are suitable for canning (Arthey 1975).

An appearance scale for picking raspberries includes specific definitions within each grade (see Table 6–10). This includes a verbal color and appearance description with reference to RHS charts (O'Donoghue and Martin 1988).

Table 6–10 Picking Standards for Raspberry Fruit cv Ohau Early (O'Donoghue and Martin 1988, reproduced by courtesy of the *Journal of Horticultural Science*)

Grade	Color	Appearance
1	42A	Pale green calyx, flattened druplets, hard. Orange-red.
2	46B	Calyx tips turned upwards. Flattened, hard druplets. Bright scarlet red.
3	46A	Calyx tips shriveling. Druplets expanding and softening. Deep red.
4	187B	Calyx brown and withered. Druplets swollen and spongy. Deep crimson purple with dull surface.
5	±187B	Calyx and stem shriveled. Torus discolored. Very soft to touch. Speckled appearance.

Note: Color specifications refer to the Royal Horticultural Society charts.

The Munsell System and Atlas

The Munsell System is the color order system most widely quoted in food industry literature. Munsell color dimensions are *Hue* (H), *Value* (V), and *Chroma* (C). Figure 6–5A is a sketch of the color solid. The *Hue* circle consists of 10 major hues, each divided into 10, visually equally spaced steps (Figure 6–5B). The central achromatic *Value* (lightness) axis consists of 10 visually equal steps, extending from *ideal black* (0) to *ideal white* (10). The distance from this axis indicates an increase in *Chroma*—that is, an increase in hue content and departure from gray. The *Chroma* is zero at the achromatic axis and increases in visually equal steps to /10, /12, /14, or greater for particularly saturated colors (see Figure 6–5B). The Munsell atlas consists of pages of colored chips, with a separate page for each hue. Chips are arranged so the vertical axis of the page represents an increase in V, the horizontal axis an increase in C. That is, each page is a vertical section through the model, taking one hue at a time. The Munsell description of a yellow-red color of Hue 3YR, Value 5/, and Chroma /6 is 3YR 5/6. Interpolation between whole units is possible, and single decimal places may be used where appropriate.

Selection of a series of color chips to represent the color gamut of a particular natural material can be a lengthy task. This is because large variations in color occur with cultivar, season of growth, and poststorage treatment. For example, over a four-year period, Munsell color dimensions for a pecan crop from 21 clones at 11 locations and four storage methods were monitored. Hue ranged from 10 Red to 2.5 Yellow, Value from 2.5 to 8.0, and Chroma from 1.0 to 8.0. A simplified color rating system with 6 color classes was eventually developed for general use by the industry (Thompson et al. 1996).

A disadvantage of color atlases is the comparatively large visual steps between individual chips. The problem may be overcome using a spinning disc. In this technique, two or more colors in the form of interleaved discs are mixed by rapid spinning. Within the color saturation limits of the discs, a large range of colors can be produced. In his study of the storage stability of

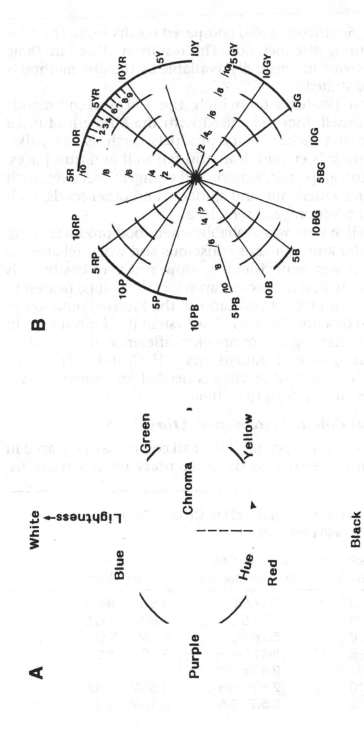

Figure 6–5 The Munsell Colour System. (A) Dimensions of the Munsell Colour System solid. (B) Organization of Hue and Chroma in the Munsell System solid.

canned peas, Southcott (1968) compared results using the atlas and the spinning disc methods. The results are listed in Table 6–11. The increase in sensitivity available by the disc method is amply demonstrated.

Agricultural products for which the USDA recommends matching Munsell discs to be used with the Macbeth-Munsell Disk Colorimeter include dairy products, such as egg yolks, cheese, and milk; beef, both lean and fat; fruit and fruit juices, such as apricot, grapefruit, lemon, and orange; vegetables, such as beets, beans, sauerkraut, and spinach; and other foods, such as chocolate, macaroni, and semolina.

The Munsell System was available when food producers were becoming color and standards conscious and when reliable instrumentation was nonexistent or slow and very costly. It is used in the sense that producers can select the chip(s) nearest to the color of their product and can use the Munsell notation of the chip(s) to describe the color. The system itself is not used in the sense of specifying a color or color difference in terms of visually estimating Munsell dimensions of H, C, and V. This is because much training and practice is needed for consistent estimates of these dimensions to be made.

The Natural Colour System and Atlas (NCS)

The NCS uses Hering's postulate that all colors may be placed in a system with reference to six elementary color sensations.

Table 6–11 The Storage Stability of the Color of Canned Peas (Southcott 1968, with permission)

Storage Time (Months)	Munsell Chip (Nearest Match)	Munsell Disc
0	5GY 5/8	4.5GY 4.8/8.5
3	5GY 5/6	4.6GY 5.1/6.6
6	5GY 5/6	4.75GY 5.1/5.2
8	5GY 5/6 to 2.5GY 7/10	3.8GY 5.2/5.8
10	2.5GY 5/6	3.5GY 5.3/5.8
12	2.5GY 5/6	3.1GY 5.2/5.8

These are whiteness (V), blackness (S), yellowness (Y), redness (R), blueness (B), and greenness (G). Any one color may be specified in terms of the percentages of two chromatic and two achromatic attributes. Thus, a particular color may contain V10S30Y30R30. This is specified as a hue (30 to 30 = 50 percent, i.e., 50 percent yellow with 50 percent red) of Y50R, a chromatic content (= 30Y + 30R) of 60 percent, and an achromatic content of S30 (white makes up the 100 percent and need not be specified). Hence, the color is 3060-Y50R. The NCS dimension system is simpler than the Munsell in concept. Color atlases are valuable in documenting specific areas of color in produce. For example, green tomatoes lie in the range 1060-G60Y to 2060-G60Y, orange from 0060-Y70R to 0090-Y70R, and red from 0080-Y80R to 1080-Y80R (Hetherington and MacDougall 1992). Product-specific grading charts can be built up from such ranges.

An Atlas for the Foods and Allied Industries

Existing color atlases are not easily applied to the problems of food specification. However, it ought to be possible to construct an atlas that is simpler and quicker to use by those less experienced in visually specifying colors. Such an atlas would consist of sequences of chips that would follow the color changes occurring in specific food materials. Color sequences would be required, for example, to follow the chlorophyll-to-carotenoid changes occurring on the ripening of fruits such as tomato, orange, or banana; categorize the greens occurring in different fresh green vegetables; follow the chlorophyll pigment family changes occurring on processing; and follow the purple-to-red-to-brown changes of globin pigments occurring in fresh meat.

More than one series may be required for each sequence. This would enable species, varietal, and seasonal differences to be encompassed. The range of colors would be smaller than in the conventional atlas, but the color distance between chips could be smaller. For maximum usefulness, such colors would be robust enough for use in the field. The colors ideally would be nonmetameric—that is, would match with the food material under all conditions of lighting. This might be impossible to fulfill in certain cases—for example, fresh beef—because of the lack of suitable inorganic pigments. In these cases, it would still be necessary to use a standardized lighting and viewing system.

This design principle would overcome another aspect of the complexity of using a conventional atlas. The color change routes of natural pigment systems run skew to the orthogonal layout of existing atlases. In the proposed atlas, the color routes of atlas and pigment system would be the same. Hence, such an atlas would be easier and less daunting to use. It would require less training and practice than conventional atlases and would probably yield more reproducible results.

Appearance Scales and Anchors Produced for Specific Foods

Plaques of card or plastic, liquids, and photographs are used to specify color quality of specific items of agricultural produce. Requirements for them are that they color-match the food material under specified viewing and lighting conditions, are manufactured to a close tolerance, have a high standard of reproducibility batch to batch, and, ideally, are rugged, easily cleaned, cheap, and permanent. It is difficult to visualize pieces of colored card or plastic as real food materials. This can be overcome to a certain extent by using textured surfaces; some plastic standards are made this way. An alternative may be to be use photographs, but these are difficult to produce to the high degree of color matching necessary, and they are very light sensitive. All standards should be protected and stored in a cool, dry, dark location. Packing material should not chemically attack or scratch the standard (Detroit Color Council 1994).

Photographs are widely used for visual grading in the meat industry (Speight 1979). Beef, sheep, and pig carcass photographic classifications have been developed by the USDA Agricultural Marketing Service and the UK Meat and Livestock Commission. The Japanese pork and beef color standards consist of plastic discs of meatlike appearance (Nakai et al. 1975). Guides for beef color have been published by Kansas State University, the National Live Stock and Meat Board (United States), and Iowa State University. Standards for beef color and discoloration are made by Agriculture Canada. Guides for beef fat color are produced by the National Institute of Animal Husbandry (Japan), and for ground beef patty cooked color by Kansas State University. The National Live Stock and Meat Board has produced photographic standards for the internal degree of doneness of cooked beef. Lamb color standards are made by Kansas State University and the American Lamb Council of Chicago. As well as the pork color discs from Japan, pork standards are produced by Agriculture Canada, the National Pork Producers Council (United States), the University of Wisconsin, and Iowa State University. Marbling guides are published by the National Cattlemen's Beef Association (United States) and Iowa State University for pork. Photographic color scales have also been produced for minced fish blocks (King and Ryan 1977). The Roche Yolk Colour Fan consists of 15 standards ranging in color from pale lemon yellow to deep orange yellow and is widely used in the assessment of egg yolk color (Vuilleumier 1969). The fan colors do not match those of broiler skin and shank; this leads to a lack of repeatability among judges (Waldroup and Johnson 1974).

A color classification chart for Granny Smith apples has also been developed (Jungman et al. 1984). This has enabled acceptance standards for the Argentine apple export market to be established. The visual evaluation of stripe density of apples can be assisted by a photographic scale (Cliff et al. 1996). In peach maturity, ripening, and storage studies, Lyon et al. (1993) used the six standard color chips developed for ground color assessment (Delewiche and Baumgardner 1985). Bananas can be graded according to eight standard skin color plates produced by the United Fruit Sales Corporation (1964). Color standards for canned tomatoes (USDA 1964), tomato paste (USDA 1977b), and canned juice (USDA 1985) are specified in terms of a defined physical scale. Other USDA plastic color scales include those for apple butter, canned mushrooms, canned clingstone and freestone peaches, frozen peas, and peanut butter. Single-color quality guides include those for canned and frozen lima beans, frozen red tart cherries, canned pimientos, canned pumpkin/squash, canned sauerkraut, and canned tomatoes (USDA 1989).

The color grading of processed orange juice is aided by a series of six standard plastic tubes. These standard tubes are placed on a gray background in a rack so that tubes of orange juice can be placed between them for direct comparison (Figure 6–6).

The plastic tubes are designated OJ1 to OJ6. Tubes OJ2 to OJ6 are assigned, respectively, the visual scores 40, 39, 37, 36, and 34. The relationship between these scores and the Hunter Instrumental Citrus Red and Citrus Yellow scales is shown in Figure 9–17 (USDA 1963a).

There are four Munsell color grades for walnuts specified by the USDA (Ravai 1992). Color photographs have been found

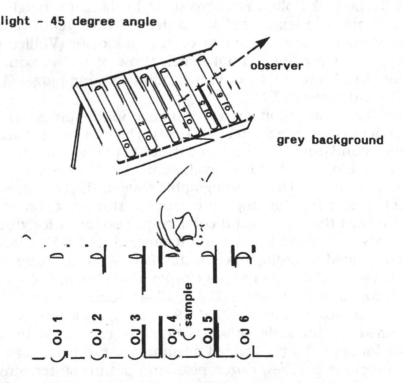

Figure 6–6 Method of use of the USDA orange juice color standards (USDA 1963b).

useful for grading coffee beans (Barbiroli 1967) and cocoa beans (Barbiroli et al. 1969). Color cards have been made to help producers assess occurrence of green tinge in sultanas ("Green-Tinge" 1964) and cucumbers (Schouten et al. 1997).

Phosphate buffer mixed with bromothymol blue in various proportions is used to classify virgin olive oils (Gutierrez and Gutierrez Rosales 1986). Color plates have also long been used to assess the color of sherry and port wines. As early as 1885, potassium permanganate and potassium dichromate solutions were being used as wine standards; the methodology is detailed by Heredia and Guzmán-Chozas (1993). Visual judgments of water color are made with reference to the solutions originally developed by Hazen in 1892, as well as to calibrated colored glass discs (Hongve and Åkesson 1996).

Scales for frozen french-fried potatoes are made by Munsell. The color intensity of potato chips (crisps) can be assessed on a 5-point scale using color standards produced by the Potato Chip/Snack Food Association (United States) (Hawrysh et al. 1996). The color produced by potatoes when fried can be predicted. Filter paper discs are impregnated with the juice from potato tubers and then fried in oil. The color produced closely resembles that of actual chips made from the same tuber. A color chart was developed to complement the method. This technique has been found suitable for use in potato-breeding programs for which routine production of chips becomes unmanageable (Chubey and Walkof 1971). It can also be used as a model system for color development during frying (Roe and Faulks 1991).

Glass standards for maple syrup and honey have been developed for the USDA. These are in the form of glass inserts for portable viewing box comparators. An optical cell is filled with the fluid, and its color is compared in transmission with the standard filters. Glass inserts are available for the maple syrup grades of light, medium, and dark amber and for the honey grades of water and extrawhite, white, extralight amber, light amber, and amber (USDA 1989). A color scale of 45 plastic plates ranging from very light amber to very dark amber (MacAdam 1943) can be used to monitor maple syrup color (Underwood et al. 1974). Scales have also been developed for the specification of sugarcane syrups and molasses (USDA 1959). Color standards have been recommended for the classification of refiners' syrup, the liquid product obtained from refining cane or beet sugar. These involve the use of acid solutions of copper, cobalt, and iron chlorides (USDA 1967). Lovibond glasses are used to define the color of many products, including beer, butter, milk, and oils.

Photographs are useful for depicting various forms of visual texture, such as fish fillet gaping (Kent 1985), chocolate crystallization (Tscheuschner and Markov 1986), and peanut pod maturity (Williams and Drexler 1981). Photocopying can be used to record cake shape and crumb structure (e.g., Miller and Trimbo 1965). Surface form or roughness is a characteristic feature of many products, such as restructured meats, yellow spreads, and coated fish dishes. Oblique-angle photography can be used to create for each product a roughness atlas that can be incorporated into a routine visual assessment procedure. The technique of distortion morphing can be used to construct altered images from original pictures. So, for example, by using pictures, the holes in a slice of bread can be made larger or smaller. A series of surface textures can be produced for use as standards (WN Hale, personal communication, 1995).

Standard solutions are available for classifying turbidity in beer (Francis and Clydesdale 1975). The degree of cloudiness in sugarcane molasses can be standardized using fluid suspensions (USDA 1959). The possibility of making standard references for surface mottling has been demonstrated for pharmaceuticals (Armstrong and March 1974).

Shape can be defined by digital analysis. However, the traditional method for shape definition is the charted standard, consisting of a number of named longitudinal and cross-section shapes. Charted standards for apples, peaches, and potatoes are shown in Figure 6–7 (Mohsenin 1965). The method is simple, but, as with all visual comparison methods, precautions must be taken to ensure that personal prejudice is reduced to a minimum. Only then can reasonable reproducibility among observers be ensured (Mohsenin 1968a).

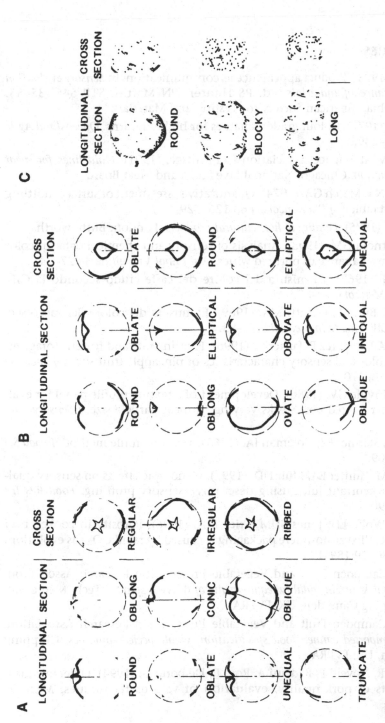

Figure 6–7 Examples of charted standards for describing shapes of fruits and vegetables. (A) apples, (B) peaches, (C) potatoes (Mohsenin 1968a, with permission).

REFERENCES

Abend CJ. (1973). Product appearance as communication. In *Sensory evaluation of appearance of materials*, ed. RS Hunter, PN Martin, STP 545, 35–53. Philadelphia: American Society for Testing and Materials.

Adams RH. (1977). Colour grade tomatoes for higher returns. *Queensland Agric J* 103:337–339.

American Meat Science Association Committee. (1991). *Guidelines for meat color evaluation*. Chicago: National Live Stock and Meat Board.

Armstrong NA, March GA. (1974). Quantitative assessment of surface mottling of colored tablets. *J Pharmacol Sci* 63:126–129.

Arthey VD. (1975). *Quality of horticultural products*. London: Butterworths.

Bakker J, Arnold GM. (1993). Analysis of sensory and chemical data for color evaluation of a range of port red wines. *Am J Enol Viticulture* 44:27–34.

Barbiroli G. (1967). La misura del colore del caffe crudo secondo la CIE. *Quaderni Merceol* 6:79–88.

Barbiroli G, Biino L, Garutti MA. (1969). La misura del colore del cacao secondo la CIE. *Quaderni Merceol* 8:71–83.

Bartolome AP, Ruperez P, Fuster C. (1996). Freezing rate and frozen storage effects on color and sensory characteristics of pineapple fruit slices. *J Food Sci* 61:154–156, 170.

Berry BW, Civille GV. (1986). Development of a texture profile panel for evaluating restructured beef steaks varying in meat particle size. *J Sensory Stud* 1:15–26.

Brandt MA, Skinner EZ, Coleman JA. (1963). Texture profile method. *J Food Sci* 28:404–409.

Brennan RM, Hunter EA, Muir DD. (1997). Genotypic effects on sensory quality of blackcurrant juice using descriptive sensory profiling. *Food Res Int* 30:381–390.

Bruce HL, Wolfe LH, Jones SDM, Price MA. (1996). Porosity in cooked beef from controlled atmosphere packaging is caused by rapid CO_2 gas evolution. *Food Res Int* 29:189–193.

Chipping Campden Fruit and Vegetable Preservation Research Association. (1969). *Quick freezing quality requirements of Brussels sprouts*. Tech. Memo. no. 91. Chipping Campden, UK: FVPRA.

Chipping Campden Fruit and Vegetable Preservation Research Association. (1988). *Imported canned food specification: whole peeled tomatoes*. Chipping Campden, UK: FVPRA.

Chizzolini R, Novelli E, Badiani A, Rosa P, Delbono G. (1994). Objective measurements of pork quality, evaluation of various techniques. *Meat Sci* 36:49–75.

Chubey B, Walkof C. (1971). A color chart for estimating potato chip color. *Can Inst Food Technol J* 4:136.

Clark RC. (1990). Flavour and texture factors in model gel systems. *Food Technol Int Eur* 271–277.

Cliff MA, Dever MC, MacDonald RA, Flemming WW. (1996). Development of a photographic scale for the visual evaluation of apple stripe density. *J Food Qual* 19:31–40.

Cross HR, Abraham HC, Knapp EM. (1975). Variations in the amount, distribution and texture of intramuscular fat within muscles of the beef carcass. *J Animal Sci* 41:1618–1626.

Daillant-Spinnler B, MacFie HJH, Beyts PK, Hedderley D. (1996). Relationships between perceived sensory properties and major preference directions of 12 varieties of apples from the Southern Hemisphere. *Food Qual Pref* 7:113–126.

Delewiche MJ, Baumgardner RA. (1985). Ground color as a peach maturity index. *J Am Soc Hort Sci* 110:53–57.

Detroit Color Council. (1994). *Procedure for visual evaluation of interior and exterior automotive trim.* Bull. no. 3. Detroit, Mich: Detroit Color Council.

Dhanaraj S, Krishnaprakash MS, Arvindaprasad B, Ananthakrishna SM, Krishnaprasad CA, Narasimham P. (1986). Effect of orchard elevation on maturity and quality of apples. *J Food Qual* 9:129–142.

Dixon TJ, Hobson GE. (1984). A general method for the instrumental assessment of the colour of tomato fruit during ripening. *J Sci Food Agric* 35:1277–1281.

Dolan KD, Singh P, Wells JH. (1985). Evaluation of time-temperature related quality changes in ice cream during storage. *J Food Processing Preservation* 9:253–271.

Floyd CD, Rooney LW, Bockholt AJ. (1995). Measuring desirable and undesirable color in white and yellow food corn. *Cereal Chem* 72:488–490.

Food and Drug Administration. (1987). Regulations for canned tuna. 21 CFR § 1.

Forsyth A. (1998). The role of national competitive testing as a force for the maintenance and improvement of poultry meat quality for the hotel and catering industry. In *Proceedings of the International Conference on Culinary Arts and Sciences*, ed. JSA Edwards, D Lee-Ross, 220–224. Bournemouth, UK: Worshipful Company of Cooks Centre for Culinary Research at Bournemouth University.

Fox B, Hilditch JF, Theobald DJ, White GW. (1978). Measurement of the colour of milked tea liquors. In *Food colour and appearance: proceedings of a Colour Group (GB) symposium*, 36–40. University of Surrey.

Francis FJ, Clydesdale FM. (1975). *Food colorimetry theory and applications.* Westport, Conn: Avi.

Galvez FCF, Resurreccion AVA. (1990). Comparison of three descriptive scaling methods for the sensory evaluation of noodles. *J Sensory Stud* 5:251–263.

Gamage TV, Yuen CMC, Wills RBH. (1997). Minimal processing of custard apple (*Annona atemova*) pulp. *J Food Processing Preservation* 21:289–301.

Gerrard DE, Gao X, Tan J. (1996). Beef marbling and color score determination by image processing. *J Food Sci* 61:145–148.

Gnanasekharan V, Shewfelt RL, Chinan MS. (1992). Detection of color changes in green vegetables. *J Food Sci* 57:149–154.

Goose PG. (1978). Colour in tomato products and its value as an index of quality. In *Food colour and appearance: proceedings of a Colour Group (GB) symposium*, 41–45. University of Surrey.

Green-tinge in sultanas. (1964). *Food Technol Austr* 16:701.

Gutierrez GR, Gutierrez Rosales F. (1986). Quick method for defining and classifying the colour of virgin olive oils. *Grasas y Acietes* 37:282–284.

Hale WN. (1986). *AIC annotated bibliography on color order systems*. Rep. of Committee E-12. Philadelphia: American Society for Testing and Materials.

Haq A, Webb NB, Whitfield JK, Morrison GS. (1972). Development of a prototype sausage emulsion preparation system. *J Food Sci* 37:480–484.

Hatae K, Yoshimatsu F, Matsumoto JJ. (1988). An integrated quantitative correlation of textural profiles of fish. *J Food Sci* 53:679–683.

Hawrysh ZJ, Erin MK, Kim S, Hardin RT. (1996). Quality and stability of potato chips fried in canola, partially dehydrogenated canola, soybean and cottonseed oils. *J Food Qual* 19:107–120.

Henry LK, Boyd LC, Green DP. (1995). The effects of cryoprotectants on the sensory properties of frozen blue crab (*Callinectes sapidus*) meat. *J Sci Food Agric* 69:21–26.

Heredia FJ, Guzmán-Chozas M. (1993). The color of wine: a historical perspective. Part I, spectral evaluations. Part II, trichromatic methods. *J Food Qual* 16:429–449.

Hetherington MJ, MacDougall DB. (1992). Optical properties and appearance characteristics of tomato fruit (*Lycopersocon esculentum*). *J Sci Food Agric* 59:537–543.

Hincks MJ, Stanley DW. (1985). Colour measurement of the squid *Illex illecebrosus* and its relationship to quality and chromatophore ultrastructure. *Can Inst Sci Food Sci Technol J* 18:233–241.

Hong LC, Leblanc EL, Hawrysh ZL, Hardin RT. (1996). Quality of Atlantic mackerel (*Scomber scombrus* L) fillets during modified atmosphere storage. *J Food Sci* 61:646–651.

Hongve D, Åkesson G. (1996). Spectrophotometric determination of water colour in Hazen units. *Water Res* 30:2771–2775.

Hotchkiss JH, Banco MJ. (1992). Influence of new packaging technologies on the growth of microorganisms in produce. *J Food Protection* 55:815–820.

Hunt MC, Hedrick HB. (1977). Chemical, physical and sensory characteristics of bovine muscle from four quality groups. *J Food Sci* 42:716–720.

Hunter RS. (1974). Objective methods for appearance evaluation. In *Objective Methods for Food Evaluation Symposium Proceedings*, 215–219. Washington, DC: National Academy of Sciences.

Hunter RS, Harold RW. (1987). *The measurement of appearance.* New York: John Wiley.

Hutchings JB, Wood FW, Young R. (1969). An objective colour method for the determination of tomato maturity. *J Food Technol* 4:45–49.

Ilangantileke SG, Wahyudi T, Bailon MG. (1991). Assessment methodology to predict quality of cocoa beans for export. *J Food Qual* 14:481–496.

Jackman RL, Yada RY, Marangoni A, Parkin KL, Stanley DW. (1988). Chilling injury: a review of the quality aspects. *J Food Qual* 11:253–278.

Jeremiah LE, Carpenter ZL, Smith GC. (1972). Beef color as related to consumer acceptance and palatability. *J Food Sci* 37:476–479.

Jones R, Dransfield E, Crosland AR, Francombe MA. (1989). Sensory characteristics of British sausages: relationships with composition and mechanical properties. *J Sci Food Agric* 48:63–85.

Jungman D, Lozano RD, de Bellora M. (1984). Measurement of color on foods: some experiences at INTI, Buenos Aires. In *Review and evaluation of appearance*, ed. JJ Rennilson, WN Hale Jr., ASTM STP 914, 62–68. Philadelphia: American Society for Testing and Materials.

Kent M. (1985). Water in fish: its effect on quality. In *Properties of water in foods in relation to quality and stability*, ed. D Simatos, JL Multon, 259–276. Dordrecht, the Netherlands: Martinus Nijhoff.

King FJ, Ryan JJ. (1977). Development of a color measuring system for minced fish blocks. *Marine Fisheries Rev* 39(2):18–23.

Lane RH, Jones SW. (1987). Influence of fry color on quality of coated broiler thighs. *J Food Qual* 10:239–244.

Lawless H, Torres V, Figueroa E. (1993). Sensory evaluation of hearts of palm. *J Food Sci* 58:134–137.

Leonard S, Marsh GL, Tombropoulos D, Buhlert JE, Heil JR. (1977). Evaluation of tomato condition in bin loads of processing tomatoes harvested at different levels of ripeness. *J Food Processing Preservation* 1:55–68.

Little AC. (1969). Reflectance characteristics of canned tuna. *Food Technol* 23:1301–1308.

Loeff HW. (1974). Instrumental colour measurement of processed spinach and other leaf vegetables. *Confructa* 19:120–130.

Lozano RD. (1989). Colour in foods. In *Proceedings of the Sixth Congress of the International Colour Association*, vol. 1, Buenos Aires 115–139.

Lyle J. (1977). Visual assessment of meat products. In *Proceedings of the Sensory Quality Control Conference*, ed., HW Symons, JJ Wren, 49–53. Aston, UK: University of Aston.

Lyon BG, Greene BE, Davis CE. (1986). Color, doneness and soluble protein characteristics of dry roasted beef semitendinosus. *J Food Sci* 51:24–27.

Lyon BG, Robertson JA, Meredith FI. (1993). Sensory descriptive analysis of cv. Cresthaven peaches: maturity, ripening and storage effects. *J Food Sci* 58:177–181.

Lyon DH, McEwan JA, Taylor JM, Reynolds MA. (1988). Sensory quality of frozen Brussels sprouts in a time-temperature study. *Food Qual Pref* 1:37–41.

MacAdam DL. (1943). Specification of small chromaticity differences. *J Opt Soc Am* 33:18–22.

MacDougall DB. (1987). Optical measurements and visual assessment of translucent foods. In *Physical properties of foods 2*, ed. R Jowett, F Esher, M Kent, B McKenna, M Roques, 277–330. London: Elsevier Applied Science.

MacDougall DB, Moncrieff CB. (1988). Influence of flattering and tri-band illumination on preferred redness-pinkness of bacon. In *Food acceptability*, ed. DMH Thomson, 443–458. London: Elsevier.

Marriott NG, Phelps SK, Graham PP. (1987). Restructured beef and turkey steaks. *J Food Qual* 10:245–254.

Mazza G, Oomah BD. (1994). Color evaluation and chlorophyll content in dry peas. *J Food Qual* 17:381–392.

Mendenhall VT. (1989). Effect of pH and total pigment concentration on the internal color of cooked ground beef patties. *J Food Sci* 54:1–2.

Miller BS, Trimbo HB. (1965). Gelatinisation of starch and white layer cake quality. *Food Technol* 19:640–648.

Miller MF, Davis GW, Seideman SC, Ramsey CB, Rolan TL. (1988). Effects of papain, ficin and spleen enzymes on textural, visual, cooking and sensory properties of beef bullock restructured steaks. *J Food Qual* 11:321–330.

Miller MF, Davis GW, Williams AC, Ramsey CB, Galyean RD. (1986). Effect of fat source and color of lean on acceptability of beef/pork patties. *J Food Sci* 51:832–833.

Minguez-Mosquera MI, Garrido-Fernandez J, Gandul-Rojas B. (1989). Pigment changes in olives during fermentation and brine storage. *J Agric Food Chem* 37:8–11.

Mohammed M, Wickham LD. (1997). Occurrence of chilling injury in golden apple (*Spondias dulcis*, Sonn.) fruits. *J Food Qual* 20:91–104.

Mohsenin NN, ed. (1965). *Terms, definitions and measurements related to mechanical harvesting of selected fruits and vegetables*. Progress Rep 257. Pennsylvania Agricultural Experiment Station.

Mohsenin NN. (1968a). *Physical properties of plant and animal materials*, vol. 1, part 1. Pittsburgh: Pennsylvania State University.

Mohsenin NN. (1968b). *Physical properties of plant and animal materials*, vol. 1, part 2. Pittsburgh: Pennsylvania State University.

Moskowitz HR. (1985). *New directions for product testing and sensory analysis of foods*. Westport, Conn: Food and Nutrition Press.

Musser JC. (1973). Gloss on chocolate and confectionery coatings. In *Proceedings of the 27th PMCA Production conference*, 46–50.

Nakai H, Saito F, Ikeda T, Ando S, Komatsu A. (1975). Standard models of pork colour. *Bull Nat Inst Animal Ind, Chiba, Jpn* 29:69.

National Marine Fisheries Service proposed interim grade standards, frozen minced fish blocks. (1975). *Fed Reg* 40(50):11729–11731.

O'Donoghue EM, Martin W. (1988). Juice loss as a simple method for measuring integrity changes in berry fruit. *J Hort Soc* 63:217–220.

O'Donoghue EM, Somerfield CD, de Vré LA, Heyes JA. (1997). Developmental and ripening-related effects on the cell wall of pepino (*Solanum muricatum*) fruit. *J Sci Food Agric* 73:455–463.

Paradis C, Castaigne F, Desrosiers T, Fortin J, Rodrigue N, Willemot C. (1996). Sensory, nutrient and chlorophyll changes in broccoli florets during controlled atmosphere storage. *J Food Qual* 19:303–316.

Parizek EA, Ramsey CB, Galyean RD, Tatum JD. (1981). Sensory properties and cooking losses of beef/pork hamburger patties. *J Food Sci* 46:860–867.

Raggi A, Barbiroli G. (1993). Colour uniformity in discontinuous foodstuffs to define tolerances and acceptability. In *Proceedings of the Seventh Congress of the International Colour Association*. 96–97. Budapest.

Rankin SA, Brewer JA. (1998). Color of nonfat fluid milk as affected by fermentation. *J Food Sci* 63:178–180.

Ravai M. (1992). Quality characteristics of California walnuts. *Cereal Foods World* 37:362–366.

Rietz CA. (1961). *A guide to the selection, combination and cooking of foods*. Westport, Conn: Avi.

Roe MA, Faulks RM. (1991). Color development in a model system during frying: role of individual amino acids and sugars. *J Food Sci* 56:1711–1713.

Sawyer FM, Cardello AV, Prell PA. (1988). Consumer evaluation of the sensory properties of fish. *J Food Sci* 53:12–18, 24.

Schouten RE, Otma EC, Kooten O van. Tijskens LMM. (1997). Keeping quality of cucumber fruits predicted by the biological age. *Postharvest Biol Tech* 12:175–181.

Shewfelt RL, Esensee V, Heaton EK. (1985). The effect of postharvest handling techniques on cannery yields of pimiento peppers. *J Food Processing Preservation* 9:43–53.

Shewfelt RL, Resurreccion AVA, Jordan JL, Hurst WC. (1986). Quality characteristics of fresh snap beans in different price categories. *J Food Qual* 9:77–88.

Sierra CC, Molina EB, Zaldivar CP, Flores LP, Garcia LP. (1993). Effect of harvesting season and postharvest treatments on storage life of Mexican limes. *J Food Qual* 16:339–354.

Sigholt T, Erikson U, Rustad T, Johanson S, Nordtvedt TS, Seland A. (1997). Handling stress and storage temperature affect meat quality of farmed-raised Atlantic salmon (*Salmo salar*). *J Food Sci* 62:898–905.

Silva HA, Nunes ML. (1994). Sensory and microbiological assessment of irradiated bluejack mackerel (*Trachurus picturatus*). *J Sci Food Agric* 66:175–180.

Skrede G, Naes T, Martens M. (1983). Visual color deterioration in blackcurrant syrup predicted by different instrumental variables. *J Food Sci* 48:1745–1749.

Smith RF, White GW. (1965). Measurement of colour in tea infusions. *J Sci Food Agric* 16:205–219.

Southcott A. (1968). The Munsell Colour System: a food technologist's experience. *Proceedings of a Colour Symposium* 1(5):7–10.

Speight BS. (1979). *Standardized photography of beef carcasses*. Bristol, UK: Meat Research Institute.

Spencer HW. (1977). Sensory control of ice cream quality. In *Proceedings of the Sensory Quality Control Conference*, ed. HW Symons, JJ Wren, 107–116. Aston, UK: University of Aston.

Stites CR, McKeith FK, Bechtel PJ, Novakofski J. (1989). Processing and sensory properties of round pork bacon. *J Food Sci* 54:214–215.

Swanson RB, Penfield MP. (1988). Barley flour level and salt level selection for a whole-grain bread formula. *J Food Sci* 53:896–901.

Swatland HJ, Haworth CR, Darkin F, Moccia RD. (1997). Fibre-optic spectrophotometry of raw, smoked and baked Arctic char (*Salvelinus alpinus*). *Food Res Int* 30:141–146.

Talbot G. (1995). Chocolate fat: the cause and the cure. *Int Food Ingredients* 1:40–45.

Theobald DJ. (1977). Sensory evaluation in tea buying and blending. In *Proceedings of the Sensory Quality Control Conference*, ed. HW Symons, JJ Wren, 69–72, 76–77. Aston, UK: University of Aston.

Thomas P, Janave MT. (1992). Effect of temperature on chlorophyllase activity, chlorophyll degradation and carotenoids of Cavendish bananas during ripening. *J Food Sci Technol* 27:57–63.

Thompson TE, Grauke LJ, Young EF Jr. (1996). Pecan kernel color: standards using the Munsell color notation system. *J Am Soc Hort Sci* 121:548–553.

Tscheuschner H-D, Markov E. (1986). Instrumental texture studies on chocolate: three processing conditioned factors influencing the texture. *J Texture Stud* 17:377–399.

Tuorila H. (1986). Sensory profiles of milks with varying fat contents. *Lebens Wiss Technol* 19:344–345.

Underwood JC, Kissinger JC, Bell RA, White JW Jr. (1974). Color stability of maple sirup in various retail containers. *J Food Sci* 39:857–858.

United Fruit Sales Corporation. (1964). *Banana ripening guide*. Boston: UFSC.

United States Department of Agriculture. (1955). U.S. standard for grades of apple butter. *Fed Reg* 20:9609.

United States Department of Agriculture. (1957). U.S. standard for grades of canned apples. *Fed Reg* 22:3535.

United States Department of Agriculture. (1959). U.S. standards for grades of sugarcane molasses. *Fed Reg* 24:8365.

United States Department of Agriculture. (1963a). *Scoring color of orange juice products with the USDA 1963 orange juice color standards.* Washington, DC: Agricultural Marketing Services.

United States Department of Agriculture. (1963b). Standards for canned sauerkraut. *Fed Reg* 28:2573.

United States Department of Agriculture. (1964). U.S. standards for grades of canned tomatoes. *Fed Reg* 29:7909.

United States Department of Agriculture. (1967). U.S. standards for grades of refiners' syrup. *Fed Reg* 32:8575.

United States Department of Agriculture. (1972). U.S. standard for grades of canned green beans and canned wax beans. *Fed Reg* 37:587.

United States Department of Agriculture. (1977a). U.S. standards for grades of fresh tomatoes. *Fed Reg* 42:32514.

United States Department of Agriculture. (1977a). U.S. standards for grades for grades of fresh tomatoes. *Fed Reg* 42:32514.

United States Department of Agriculture (1977b). U.S. standards for grades of tomato paste. *Fed Reg* 42:41843.

United States Department of Agriculture. (1980). U.S. standards for grades of veal and calf carcasses. 7 CFR §53.107–53.111.

United States Department of Agriculture. (1982a). U.S. standard for grades of apple juice. *Fed Reg* 47:5875.

United States Department of Agriculture. (1982b). U.S. standard for grades of canned apple sauce. *Fed Reg* 47:5877.

United States Department of Agriculture. (1985). U.S. standards for grades of tomato juice. *Fed Reg* 50:29635.

United States Department of Agriculture. (1986a). U.S. standards for grades of peeled potatoes. *Fed Reg* 51:21133.

United States Department of Agriculture. (1986b). U.S. standards for whole or dressed fish. *Fed Reg* 51:34990.

United States Department of Agriculture. (1989). *Visual aids approved for use in ascertaining grades of processed fruits and vegetables.* File Code 105-A-15. Washington, DC: USDA.

United States Department of Agriculture. (1997). *U.S. standards for grades of slaughter cattle.* January 31. Washington, DC: USDA.

Vergano PJ, Testin RF, Newall WC Jr, Trezza T. (1995). Damage loss cost curves for peach impact bruising. *J Food Qual* 18:265–278.

Vuilleumier JP. (1969). The Roche Yolk Colour Fan: an instrument for measuring yolk colour. *Poultry Sci* 48:767–779.

Waldroup PW, Johnson ZB. (1974). Lack of repeatability among persons using the Roche Colour Fan to assess the shank color of broilers. *Poultry Sci* 53:437–439.

Wekell JC, Teeny FM. (1988). Canned salmon curd reduced by use of phosphates. *J Food Sci* 53:1009–1013.

Williams EJ, Drexler JS. (1981). A non-destructive method for determining peanut pod maturity. *Peanut Sci* 8:134–140.

Yang CC, Brennan P, Chinnan MS, Shewfelt RL. (1987). Characterization of tomato ripening process as influenced by individual seal-packaging and temperature. *J Food Qual* 10:21–33.

Yau NJN, Huang JJ. (1996). Sensory analysis of cooked rice. *Food Qual Pref* 7:263–270.

Young KW, Whittle KJ. (1985). Colour measurement of fish minces using Hunter L,a,b values. *J Sci Food Agric* 36:383–392.

CHAPTER 7

Appearance Profile Analysis

Product success depends on a mix of visual attributes. As well as having a shape and portion size, it may have more than one color, gloss, translucency, or surface irregularity. *Appearance Profile Analysis (APA)* can be used to catalog logically such properties. The technique can be used to define and understand product appearance, as well as customer response to appearance. It provides the methodology for development of designer products and the scientific basis upon which product appearance development can be promoted.

Product gestalt may tell us that something is wrong with the appearance. Perhaps it does not look identical with a competing product, or overall judgments of quality are discovered to be different when the product is launched in another country, or it does not look as it did last time. Once such an overall impression has been acknowledged, future success of the product may depend on the skill and speed with which the cause of the undesirable impression can be identified. APA provides the analytical approach necessary to deal with this. Any physical scene and the personal images arising from it can be examined and analyzed in a disciplined manner using this approach. Chapter 2 explained the formation of *Total Appearance* images resulting from a physical scene and how it can be understood and quantified in terms of *basic perceptions* and *derived perceptions*. *Basic perceptions* are composed of *visual structure, surface texture, color, gloss, translucency,* and *temporal* properties. *Derived perceptions,* or *visual expectations,* include *visual identification* and *visually assessed flavor, texture,* and *satisfaction.*

In illustration, we can consider a custard dessert topped with cherries. Many appearance factors may be important to the selling success of such a product. There are the visual properties of the dish as a whole, including the container, the visual properties of each component of the dish, and the contrasts and relationships between each component. That is, the custard and cherries have complementary as well as individual attributes. These include the perceived volume of the whole dish and the perceived volume contrast of each component; the symmetry or randomness of position of the cherries; their number, size, wholeness, and defects; the perceived color and color uniformity of each component and their color contrast, the perceived and contrast translucency and gloss of each component, and the perceived texture and texture contrast of both custard and cherries. In such cases, an APA can reveal the properties in sufficient detail to make disciplined comparative judgments between two products or between product and concept.

An analysis of the *basic perceptions* of the product considers the geometrically defined *visual structure* of the illuminated whole product and each element of the product, as well as the visually perceived *surface texture, color* and *color uniformity, gloss* and *gloss uniformity, translucency* and *translucency uniformity,* and the *temporal* properties of each element.

Derived perceptions provide the basis of expectation derived solely from appearance. Within APA, they consist of disciplined judgments of *visually assessed flavor, texture,* and *satisfaction.* For example, the *derived perceptions* of the cherry custard product include *identification* of its place within the meal—that it is a dessert—as well as provoking some level of *satisfaction.* In addition:

- for the custard—*lumpiness, thickness, yellowness, vanillaness*
- for the cherries—*cherriness, redness, transparency, sourness*
- for the dessert as a whole—*tastiness, sweetness, fillingness* (is there enough/too much?), *appetizingness, wholesomeness*

Such *derived perceptions* can be assessed by subjects using connotative scales. For example, a *lumpiness* scale ranging from **very smooth** to **very lumpy** can be devised for the custard. The relationships between *basic* and *derived perceptions* differ in their complexities, but they follow the same general pattern for the customer in the store, cook in the kitchen, and diner in the restaurant. Such an approach ensures that all the key product appearance factors have been included in the analysis. To this end, and as an internal check, a catchall term such as *overall impression of appearance* may also be included (Skrede 1982).

In many cases, it is not necessary to make a formal statement of the complete analysis. A decision may be reached as to the most important appearance aspects for a particular product in a particular situation, and other attributes may be eliminated from further consideration. However, this attribute elimination ought to be made positively, on an understanding of the appearance science of the product, not by neglect. Use of *basic perception* sensory descriptors and scales is straightforward only when discussion, agreement, training, practice, and continuing retraining are conscientiously undertaken. *Derived perceptions* are obtained from consumers or potential consumers of the product.

EXAMPLES OF APPEARANCE PROFILE ANALYSIS

An APA consists of visually perceived properties seen to belong to the product and its interaction with the customer. The following are brief examples of scales needed for a number of APAs of various degrees of complexity. Some of the examples are complete, some partially complete as appropriate to the product and its situation. Some of the examples taken from the literature were not performed as APAs, but they illustrate some principles of the technique.

Lemon Products

Three sweet lemon products—lemon curd, lemon cheese, and lemon jelly—were presented in identical clear colorless glass jars. The properties isolated for scoring were

1. *Basic perceptions*
   - *Surface texture*, in the form of *graininess*
   - *Color*, in terms of *depth of yellowness*
   - *Translucency*, in the form of *opacity*

- *Gloss*
- The respective *uniformities* of all of the above features
2. *Derived perceptions*
   - *Visually assessed flavor*, in terms of *acidity*, *sweetness*, *sourness*, *saltiness*, and *lemonness*
   - *Visually assessed texture*, in terms of *firmness* and *bounciness*

Each attribute was scored on a 0-to-5 scale. Alternative ways of depicting attribute scores are shown in Figure 7–1.

All attributes contributed to the individual nature of the products, and most contributed to the differences between them. The exceptions were *sourness* and *saltiness*, which were scored zero for all samples. This analysis and presentation provide a solid basis for product description, product development, quality control, and research.

Dessert Gels

An intensive appearance study of a model mousse dessert gel product was undertaken (Imram 1998). During the training period, the panel generated a vocabulary of the following 26 descriptors. They are listed in terms of the classification described above:

1. *Visual Structure*: all samples were presented in standardized amounts and containers; hence, the *visual structure*—that is, size and shape factors—of the product was constant
2. *Color*:
   - *color identity*—that is, identification of the color using a Munsell atlas—and *color strength* (both found to be sensitive to added color)
   - *color uniformity* (not sensitive to added cream or color)
   - *lightness*, *naturalness*, and *vividness* (sensitive to added color)
3. *Translucency*:
   - *opacity* (sensitive to added cream)
   - *opacity uniformity* (not sensitive)

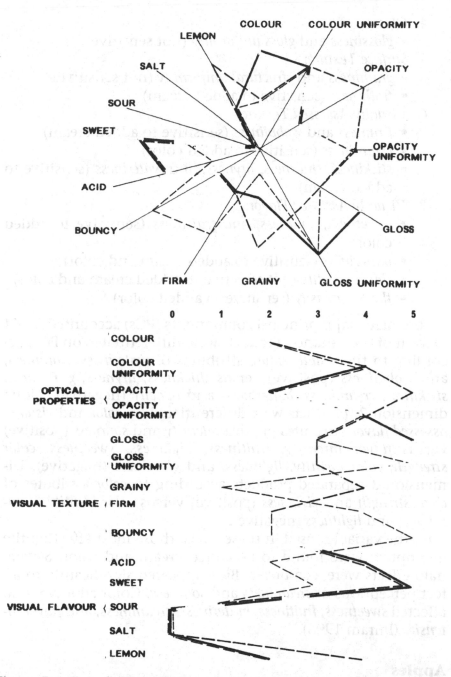

Figure 7-1 Two depictions of results from appearance profile analysis of three sweet lemon products: , lemon curd; − −, lemon cheese; — —, lemon jelly.

4. *Gloss*:
   - *glossiness* and *gloss uniformity* (not sensitive)
5. *Surface Texture*:
   - *graininess* and *structural uniformity* (not sensitive)
   - *frothiness* (sensitive to added cream)
6. *Visually Assessed Texture*:
   - *firmness* and *wobbliness* (sensitive to added cream)
   - *creaminess* (sensitive to added color)
   - *stickiness*, *thickness*, *dryness*, and *wateriness* (sensitive to added cream)
7. *Visually Assessed Flavor*:
   - *sweetness*, *sourness, and fruitiness* (sensitive to added color)
   - *butteriness* (sensitive to added cream and color)
   - *flavor identity* (not sensitive to added cream and color)
   - *flavor intensity* (sensitive to added color)

The three major principal components (PCs) accounted for 64 percent of the variance. Products were differentiated on PC 1 according to the *visual texture* attributes of *wateriness, frothiness,* and *wobbliness* (positive) versus *thickness, dryness, graininess, stickiness, creaminess, butteriness,* and *opacity* (negative). Along dimension 2, products were differentiated by *color* and *visually assessed flavor* attributes of *flavor identity* and *sourness* (positive) versus *flavor intensity, fruitiness, vividness, sweetness, color strength, color identity, lightness,* and *glossiness* (negative). Dimension 3 separated products according to *color* attributes of *color strength* and *vividness* (positive) versus *color identity, naturalness,* and *lightness* (negative).

Of the variables used in these trials, those most affecting the descriptors were found to be added cream and color. Several halo effects were also noted. Blue appeared significantly to affect perceptions of *creaminess* and *sourness*. Color additives also affected *sweetness, fruitiness, butteriness, creaminess,* and *flavor intensity* (Imram 1998).

Apples

After an apple is accepted for its *visual structure*, assessments are made of its elements in terms of *surface texture, color, translu-*

cency, and *gloss* attributes. Apple surface consists of the elements of the stalk, skin, and surface covering of dust and specks. For 12 varieties from the Southern Hemisphere, these were assessed on the following scales (Daillant-Spinnler et al. 1996):

1. Basic perceptions
   - *Skin color* and *color uniformity*
     - *yellow background*, the depth of yellow color in the background
     - *green background*, the depth of green color in the background
     - *red streaks*, amount of red streaks or coloration next to the stalk
     - *brown by stalk*, amount of brown coloration next to the stalk
   - *Skin gloss* and *gloss uniformity*
     - *shininess* of the apple surface, scored from **dull** to **shiny**
   - *Skin translucency* and *translucency uniformity*—no scales used
   - *Surface texture*
     - *ridges*, measure of how ridged the surface is; these run vertically from the stalk
     - *white dust*, amount of white dusty coating on the surface of the apple
     - *white specks*, amount of white specks on the surface of the apple
     - *dark specks*, amount of black or brown specks on the surface of the apple
2. Derived perceptions
   - *Visually assessed texture*, described as *looks hard*—that is, the appearance gives the impression that the apple will have a hard texture—scored from **soft** to **hard**

Lager

Important visually judged attributes of lager are *clarity, amber color, gray color, intensity of color, quantity of head, head density, head persistence, rising bubbles,* and *bubble size.* Eleven lagers were analyzed using these scales. A preference map, in the form of a principal components analysis, for the lagers is shown in

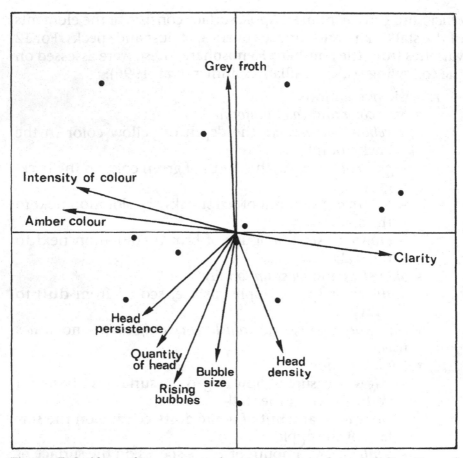

Figure 7–2 Ale preference map showing sensory data and direction of consumer preference. *Source:* Reprinted with permission from M. Spooner, Sensory Analysis in the Late 20th Century, *Food Manufacture*, Vol. 72, No. 4, pp. 34–36, © 1997, Miller Freeman UK.

Figure 7–2. The lagers, shown as individual points, can be described in terms of predominant qualities: for example, high gray froth, high clarity, large bubble size and head density, or high head persistence (Spooner 1997).

Milk

Important elements within the *basic perceptions* of milk and milk substitutes can be built up from the literature. It is necessary to make assessments at the edge as well as in the center of the sample.

- *Surface texture* includes *visual thickness*, which varies from **thin** (watery) to **thick** (creamy), and a *foamy* quality (Tuorila 1986). There is also *visual hang-up* (Phillips et al. 1995).
- *Color* properties are *center color* and *edge color*, both of which can be scored as *whiteness*, perhaps on a 14-point *whiteness* scale using **nonfat milk** and **whole milk** as anchors (Rankin and Brewer 1998).
- *Translucency* properties are scored as *opacity* (Phillips et al. 1995).

Hot Meat Gravy

This product can exist in a number of forms with and without fat. It can be colorless or off white or brown, and transparent or opaque. Samples of home-cooked and commercially available gravies were obtained from purchased reheat meat-in-gravy products. A market survey, carried out with a group of consumers, resulted in the identification of a range of discriminating attributes. When this product was eaten with a traditional meal of beef and vegetables, five groups of independent appearance properties were clearly recognized:

1. Basic perceptions
   - *Surface texture* revealed the presence of *fat* and *particulateness*. Particles clearly identified were fat globules, onion pieces, black pepper grinds, meat flakes, and flour.
   - *Color* varied between **very pale** and **very dark brown**; a **yellow-brown** hue was also identified in some samples.
   - *Gloss* revealed floating fat.
   - *Translucency* varied between **clear** and **opaque**.
   - *Uniformity of color, translucency*, and *gloss* were caused by local patches of floating fat and undissolved particles such as flour.
2. Derived perceptions
   - *Visually assessed flavor* may be scaled in terms of *starchiness, meatiness*, and *wateriness*.

- *Visually assessed texture*, in the form of *apparent viscosity*, ranged between **very thin** and **gelled**.

Mungbean Starch Noodles

The appearance of standard size and shaped mungbean starch noodles has been described (Galvez and Resurreccion 1990).

- *Color* is described in terms of hue name.
- *Translucency* is described in terms of *transparency*, the extent of visibility through the strands.
- *Glossiness* is scaled in terms of the amount of surface *shine*.
- *Uniformity* of *color, translucency* and *gloss* is expressed in terms of *speckledness*, that is, the presence of specks or particles within the strand.
- The *derived perception* of *visually assessed texture* of stickiness is described as the perceived degree of *adherence* of the noodle strands.

Peanuts

Peanut appearance has been specified in detail by Williams and Drexler (1981). Their scale was developed for the variety Florunner, the most widely grown cultivar in the United States. There are seven grades, the beginning and midpoint of each being anchored. The eighth attribute is concerned with the mesocarp physical structure and its response to pressure. Peanut pod maturity is described in terms of *visual structure* and *surface texture, color, color uniformity, visually assessed texture,* and *visual and tactile response to pressure.* The peanut pod maturity scale is listed in Table 7–1.

This approach has a broad application to the peanut industry. The speed and convenience of the judgments make this scale suitable for research purposes, as well as for the application of agronomic practices and harvesting.

Raw Fish

Natural changes occurring to a complex organism do not occur at the same rate. Hence, it is necessary to follow appearance

Table 7–1 Characteristics of Florunner Pods by Class Beginnings and Midpoints (Williams and Drexler 1984 with permission)

Stage	Mesocarp Color		Physical Size	Vein System	Surface Texture	Mesocarp Structure	Reticulations
	Dominant or Replacing	Being Replaced					
1.0	White		Initial development	Indistinct	Smooth	Soft, watery	Not present
1.5				Longitudinal apparent but indistinct	Smooth	Soft, watery	Not present
2.0	White			Longitudinal distinct	Smooth	Soft, watery	Not present
2.5			Maximum	Net venation begins on basal segment	Smooth	Soft, watery	Indistinct
3.0	Very pale yellow	White	Maximum	Net venation nearly complete or complete	Slightly rough	Somewhat soft, somewhat resilient	Becoming distinct
3.5				Net venation complete	Slightly rough	Resilient	Distinct
4.0	Deep yellow	Very pale yellow	Maximum	Net venation complete	Moderately rough	Somewhat rigid	Distinct
4.5				Net venation complete	Rough	Rigid	Very distinct
5.0	Orange to brownish orange	Deep yellow	Maximum	Net venation complete	Rough	Rigid	Very distinct
5.5						Very rigid	
6.0	Reddish brown to brown	Orange to brownish orange	Maximum	Net venation complete	Very rough	Very rigid	Very distinct
6.5							
7.0	Black	Reddish brown to brown	Maximum	Net venation complete	Very rough	Very rigid	Very distinct
7.5							

changes as a whole. An example can be taken from the visual scoring of fresh fish.

Once the *visual structure* is noted to be the correct size and conformation for the species, assessments are made of *surface texture*, *color*, *translucency*, and *gloss attributes* according to their placement on the fish. This is illustrated in Table 7–2. Hence, at different times, the condition of the eyes, skin, and surface slime are firm markers critical to the quality score (Davis 1988).

Cooked Fish Products

Visual structure and *surface texture* create a complex appearance situation confronting the fish technologist. For adult consumption, cooked whole fish should look natural and realistic. Properties such as size and shape of fillet, muscle structure, and skin-to-flesh ratio combine to convince potential eaters that they are looking at a natural product. However, the popularity of coated fish products has invited attempts to include comminuted material. This poses interesting psychophysical problems in determining how much of the alternative material can be included without its being detected by the consumer. In this situation, whole product assessment can be a detailed procedure and includes examination at every stage in the eating process.

Removal of the coating reveals the fish for an initial examination, and bending or breaking will reveal the structure. The proportion of *whole fillet flakes* and *compressed mince* may then be estimated. Surface examination reveals other consumer-important attributes such as the extent of *dye transfer* from the coating, as well as extents of *discoloration* and *opacity*. By using a fork to cut a piece from the sample, the *fragility* and *size of flakes* may be assessed. Sample behavior on further breakup will reveal whether the product has been constructed from *flakes*, *fibers*, *fiber bundles*, or *gels*. The presence of *bones* and *curd* may be assessed, as well as *color*, *translucency*, and *gloss* of inner flesh. At the same time, visual impressions may also be formed of *water release*, *stickiness*, and *sliminess*. Such an examination can reveal the market strategy of competitors, the technology they possess,

Table 7–2 Visually Assessed Criteria for Freshness of Whole Fresh Fish

	Score				
	5	4	3	2	0
Eyes	Perfectly fresh Pupil convex, black	Flat Gray	Slightly sunken Gray	Sunken White	Fully sunken
	Cornea translucent		Opalescent	Opaque	
Gills	Bright red*	Slight color loss	Some discoloration		Bleached or dark brown discoloration
Skin	Bright opalescent		Loss of bright opalescence		Bloom gone
	No bleaching		Some bleaching		Marked bleaching
Outer slime water	White or transparent	Opaque/milky	Thick knotted with some bacterial discoloration		
Bacterial slime			Thick knotted with some bacterial discolouration		Thick Yellow/brown
Head/ body					Shrunken

* Gill color is judged appropriate to the species. For example, *gill color* of salmon deteriorates from **bright red** to **pink** to **green brown** to **brown white** (Himelbloom et al. 1994).

Source: Data from H.K. Davis, Quality and Deterioration of Raw Fish, in *Fish and Fishery Products*, A. Ruiter, ed., pp. 215–242, © 1988, CAB International.

and the skill with which they use it. It can also reveal in a more detailed way whether the customer notices or objects to particular formulation tactics.

Pork

In the shop, the quality of prepacked fresh meat undergoes rigorous quality examination by the potential buyer. Product quality expected is largely inferred from appearance. Four *basic perception* appearance attributes of pork loin chops were found to be of importance to German consumers. These were the two *visual structure* attributes *share of fat* and *presence of liquid and meat juice and blood splashes; color;* and the *color uniformity* attribute of *fat marbling.* Major *derived perceptions* were the *visually assessed flavor* attributes of *taste, leanness,* and *juiciness,* the *visually as-*

sessed texture attribute of *tenderness,* and the *visually assessed satisfaction* attributes of *nutritional value, wholesomeness,* and *freshness* (Bredahl et al. 1998).

REFERENCES

Bredahl D, Grunert KG, Fertin C. (1998). Relating consumer perceptions of pork quality to physical product characteristics. *Food Qual Pref* 9:273–281.

Daillant-Spinnler B, MacFie HJH, Beyts PK, Hedderley D. (1996). Relationships between perceived sensory properties and major preference directions of 12 varieties of apples from the Southern Hemisphere. *Food Qual Pref* 7:113–126.

Davis HK. (1988). Quality and deterioration of raw fish. In *Fish and fishery products,* ed. A Reuter, 215–242. Oxford, UK: CAB International.

Galvez FCF, Resurreccion AVA. (1990). Comparison of three descriptive scaling methods for the sensory evaluation of noodles. *J Sensory Stud* 5:251–263.

Himelbloom BM, Crapo C, Brown EK, Babbitt J, Reppond K. (1994). Pink salmon (*Oncorhynchus gorbuscha*) quality during ice and chilled seawater storage. *J Food Qual* 17:197–210.

Imram N. (1998). Sensory perception of colour and appearance attributes in a dessert gel formulation. In *Proceedings of the International Conference on Culinary Arts and Sciences,* ed. JSA Edwards, D Lee-Ross, 421–430. Bournemouth, UK: Worshipful Company of Cooks Centre for Culinary Research at Bournemouth University.

Phillips LG, McGiff ML, Barbano DM, Lawless HT. (1995). The influence of nonfat dairy milk on the sensory properties, viscosity and color of lowfat milks. *J Dairy Sci* 78:2113–2118.

Rankin SA, Brewer JA. (1998). Color of nonfat fluid milk as affected by fermentation. *J Food Sci* 63:178–180.

Skrede G. (1982). Quality characterization of strawberries for industrial jam production. *J Sci Food Agric* 33:48–54.

Spooner M. (1997). Sensory analysis in the late 20th century. *Food Manuf* 72(4):34–36.

Tuorila H. (1986). Sensory profiles of milks with varying fat contents. *Lebens Wiss Technol* 19:344–345.

Williams EJ, Drexler JS. (1981). A non-destructive method for determining peanut pod maturity. *Peanut Sci* 8:134–140.

CHAPTER 8

Instrumental Specification

This chapter contains a description of the instrumental specification of color, translucency, and gloss. It forms the bridge between vision and the practical optical measurements used in food appearance specification. Emphasis has been placed on those aspects with wide applicability to the industry. The bulk is devoted to the understanding and specification of color. This is appropriate because all visually perceived surface properties have color and color contrast as a basis. For example, perception of translucency depends on a view of a contrast seen through the material or an awareness of light diffusion and color change within the body of the sample. Perception of gloss depends upon color intensity change as the eye or the light source is moved with respect to the object.

COLOR MEASUREMENT

Perceptions of objects occur inside the brain. Hence, as we have no direct access into visual mechanisms, we cannot measure color or any other sensory attribute. However, we can hope to obtain firm correlates between the perception and a measurement device. When an object is visually assessed, three physical factors must be present. There must be a source of light, the object, and a light receptor mechanism. Similarly, these factors are essential to the instrumental specification of color. Although the same principles of color measurement are valid for all materials, in this treatment the object will be assumed initially to be opaque.

It is possible to specify color additively by mixing colored lights and changing their relative intensities until a match with the sample is obtained (see Figure 8–1). The color of the object can be specified by the amounts of, say, red, green, and blue light needed to make a color match with the object.

This visual matching method is tedious to use routinely, but there are two other proven instrumental methods available. In the first, a direct imitation of the visual approach is attempted. This is illustrated in Figure 8–2. A light falls on an object at an angle of 45° to the normal, along which the reflected light is measured. In the instrument, the ρ, γ, and β cones of the eye are replaced with red, green, and blue filters and a photocell. If the combined responses of the filters and photocell relate to the cone responses, we can use the values obtained to specify the color of the object. This is the principle of the tristimulus colorimeter.

The second method involves measuring the spectral reflectance curve. This is the fundamental physics describing reflection characteristics. If spectral definitions are also available for the incident illumination and the receptor cones, it is possible to specify the color of the object in the same terms as the tris-

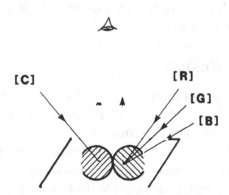

Figure 8–1 The principle of color measurement of a light [C] shining onto a screen. The intensities of three primary colored lights, [R], [G], and [B], shining onto an adjacent part of the screen are adjusted until a color match is obtained.

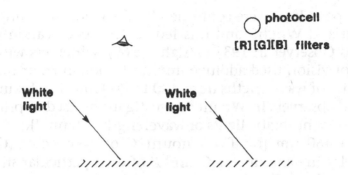

Figure 8–2 The eye assesses the color of an object using the ρ, γ, and β cones. An instrumental tristimulus measurement is made by replacing the cones with three filters, [R], [G], and [B], and measuring the light transmitted through each with a photocell.

timulus colorimeter. To further either of these approaches, it is essential to establish standard methods and procedures. These include defining suitable light sources and cone sensitivity functions. Black-body radiators are definable in terms of color temperature, and their spectral distributions act as a basis for defining standard illuminants (see Chapter 3). The Commission Internationale d'Eclairage (CIE) Standard Observer can be specified in terms of the average color-matching characteristics of the 92 percent of the population having normal color vision.

The foundation for the success of instrumental color specification lies in the postulates of Grassman (1853, quoted in Judd and Wyszecki 1975). The first states that any color can be matched by adding together suitable proportions of three primary colored lights. Here, *primary* means that none of these three lights can be matched by mixing together proportions of the other two. The second postulate states that only the visual effect of these lights is important. That is, their spectral composition is not important, only the fact that they are three primary colors—red [R], green [G], and blue [B]. A color can then be represented by the tristimulus values *R*, *G*, and *B*, the amounts of the three primaries required to make a match.

These postulates were confirmed by the color-matching work of Guild and Wright, and this led to the specification of the Standard Observer in 1931 (Wright 1969). Observers with normal color vision used additive mixing to obtain color matches for a series of wavelengths from 400 to 700 nm. To obtain these matches, observers in Wright's investigations used as primaries three monochromatic lights of wavelength 650 nm [R], 530 nm [G], and 460 nm [B]. The amount C of each color [C] was matched using amounts R, G, and B of each particular stimulus [R], [G], and [B]. That is,

$$C[C] = R[R] + G[G] + B[B]$$

For many problems, it is convenient to separate the color *quality*, represented by the proportions of R, G and B, from the total *intensity* of light, represented by the values of R, G, and B. This is done by dividing R, G, and B in turn by $(R + G + B)$ to give

$$1.0[C] = r[R] + g[G] + b[B]$$

where:

$$r = \frac{R}{R + G + B}, \quad g = \frac{G}{R + G + B}, \quad b = \frac{B}{R + G + B}$$

and r, g, and b are chromaticity coordinates. The results for Wright's color-matching experiments on the spectrum are shown in Figure 8–3. At many wavelengths, one of the coordinates is negative. Before a match can be obtained at these wavelengths, the spectral color has to be desaturated. That is, one of the three stimuli has to be added to the color being matched. For example, matches at wavelengths 450 to 530 nm yield a negative r coordinate:

$$1.0[C] + r[R] = g[G] + b[B]$$

From the above definitions,

$$r + g + b = 1$$

That is, if any two of r, g, and b are specified, the third can be calculated. Hence, the Figure 8–3 data can be shown in the form

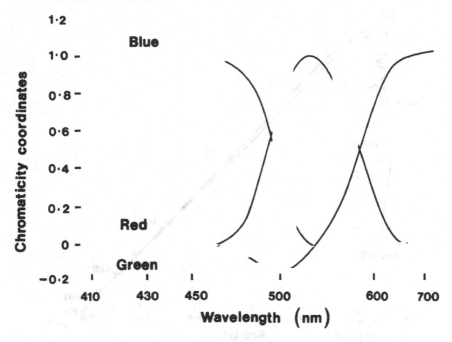

Figure 8–3 Chromaticity coordinates of the spectrum—the mean response obtained from 10 observers (Wright 1969, with permission).

of the two-dimensional chromaticity diagram Figure 8–4, in which *r* and *g* are plotted. The spectral colors for which the wavelengths are noted on the diagram are shown as a spectrum locus. The blue wavelength, approximately 460 nm near one end of the locus, is at the origin where *r* and *g* are zero and *b* = 1. The locus progresses to a wavelength of 530 nm, where *r* is zero and *g* = 1 (and *b* is therefore zero), and on to the red wavelength 650 nm, where *g* and *b* are zero and *r* = 1.

It can be seen from Figure 8–4 that the reason for the negative values of *r*, *g*, and *b* is that everywhere the spectrum locus is convex. Hence, no real primaries exist that will always yield positive values. Because of the inconvenience that it was felt these negative values would cause, the CIE decided that three unreal primaries, [X], [Y], and [Z], should be used. These were chosen so that the chromaticity coordinates *x*, *y*, and *z* would always be

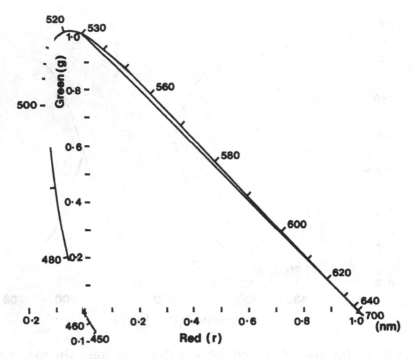

Figure 8-4 The data from Figure 8-3 plotted as a chromaticity chart. The matching stimuli used were 650, 530, and 460 nm (Wright 1969, with permission).

positive. To do this, the new primaries must all lie outside the spectrum locus, as shown in Figure 8–5. The Guild and Wright data were transformed accordingly and are shown relative to the real primaries [R], [G], and [B]. This figure shows the spectrum locus defined in terms of the stimuli 700, 546.1, and 435.8 nm and measured in units that are equal to one another for an equal-energy white E. Hence, the equal-energy white has $r = g = b = 0.333$. The line joining the two ends of the spectrum locus is the purple boundary. This marks the boundary of colors that are mixtures of high and low wavelengths and not part of the spectrum itself.

The new primaries were, in effect, deduced by lowering the baseline of Figure 8–3 so that all chromaticities would be positive. They were then converted to allow for the eye's sensitivity to different wavelengths. This was done by multiplying them, wavelength by wavelength, by the photopic luminous efficiency function (Figure 4–2). These manipulations result in the color-matching functions, $\bar{x}$, $\bar{y}$, and $\bar{z}$ shown in Figure 8–6. This figure is the foundation of the CIE system of color specification. The areas under the curves are equal, so equal proportions de-

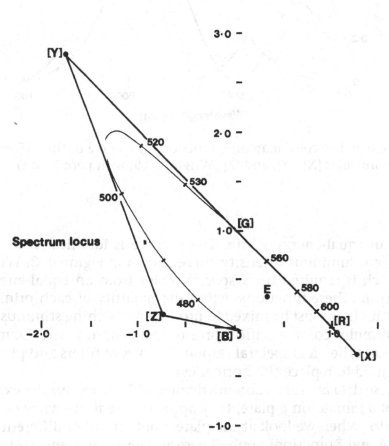

Figure 8–5 The spectrum locus determined using the matching stimuli 700, 546.1, and 435.8 nm. Also shown are the chromaticities of the real stimuli [R], [G], and [B] and the imaginary reference stimuli [X], [Y], and [Z] selected by the CIE in 1931. E is the equal-energy white (Wright 1969, with permission).

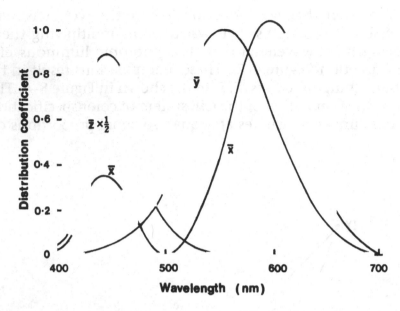

Figure 8–6 The color-matching functions $\bar{x}$, $\bar{y}$, and $\bar{z}$ of the CIE imaginary primaries [X], [Y], and [Z] (Wright 1969, with permission).

fine an equal-energy white. The $\bar{y}$ curve is identical with the photopic luminous intensity curve shown in Figure 4–2. When a match is required for a spectral color from an equal-energy spectrum, these functions define the quantity of each primary stimulus that must be mixed to produce a matching stimulus. In a tristimulus color specification situation, the $\bar{x}$, $\bar{y}$, and $\bar{z}$ curves represent the ideal spectral response curves of filters and photocells used to replace the normal eye.

These data are for a subtended angle of 2°. If we use the example of a tomato on a plate, they approximate to the view of the tomato. When we look at the plate, another rather different set of $\bar{x}$, $\bar{y}$, and $\bar{z}$ functions apply. These are the $\bar{x}_{10}$, $\bar{y}_{10}$, and $\bar{z}_{10}$ functions. This situation arises because of the different construction of the retina at larger subtended angles. The CIE 1931 (2°) and 1964 (10°) chromaticity charts are compared in Figure 8–7. The differences are significant, and the same measurement and calculation conditions as appropriate should be used in practical instrumental work.

The chromaticity coordinates of a number of foods and the Roche Egg Yolk Colour Fan are shown on the 2° CIE chromaticity diagram in Figure 8–8. These were determined with reference to Illuminant C. The chromaticity paths are shown for the ripening tomato and the Roche Yolk Colour Fan as it progresses from pale yellow to orange yellow. The cross denotes the *white point* defined by the coordinates of Illuminant C.

The foregoing treatment has led to the calculation of the chromaticity coefficients x and y. A complete definition of a color

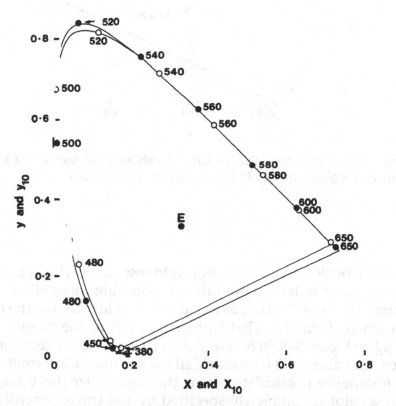

Figure 8–7 The CIE 1931 2° (●) and the 1964 10° (○) chromaticity charts (Wright 1969, with permission).

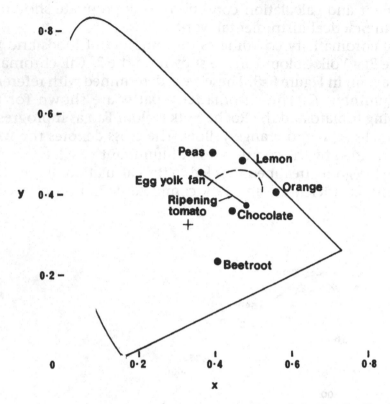

Figure 8–8 The chromaticities of some foods and the Roche Egg Yolk Colour Fan plotted on a 1931 CIE chromaticity diagram.

must also include an intensity or brightness factor. This is the *luminance*, and it is derived from the photopic luminous efficiency function (Figure 4–2). This, as stated above, is identical to the distribution coefficient $\bar{y}$. That is, the locations of the stimuli [X] and [Z] in Figure 8–5 were selected to lie on a line of zero luminance, the value of [Y] containing all the luminance information. The luminance is calculated from the area under the Y curve. Thus, a color is completely specified by the three dimensions (x, y, Y).

Tristimulus and Spectrophotometric Measurement

The tristimulus colorimeter, shown in principle in Figure 8–2, is the simplest instrumental method for specifying color. The light source (usually a tungsten lamp), filters, and photocell characteristics are so combined that direct evaluations of X, Y, and Z can be obtained.

However, the spectrophotometer may also be used for color specification. A spectral reflection or transmission curve, obtained from a spectrophotometer covering the visible wavelength range, can be converted into X, Y, Z data. The curve is integrated wavelength by wavelength, through the visible range, with the spectral emission of an illuminant (E) and the standard observer functions ($\bar{x}$, $\bar{y}$, $\bar{z}$). The areas under the resulting curves yield the values of X, Y, and Z. Mathematically, this process is described as

$$X = k \int_{380}^{710} RE\bar{x}\, d\lambda, \qquad Y = k \int_{380}^{710} RE\bar{y}\, d\lambda, \qquad Z = k \int_{380}^{710} RE\bar{z}\, d\lambda$$

where R is the reflectance obtained at set wavelength intervals over the visible range and k is a constant chosen so that $Y = 100$ for a perfect white.

The calculation process is illustrated in Figure 8–9. In commercial reflectance spectrophotometers, the wavelength intervals for these integrations normally range between 5 and 20 nm. The 20-nm interval has been found adequate for commercial dyehouse matches. The wavelength range used also varies, but all instruments cover at least 400 to 700 nm.

It will be realized that the X, Y, and Z values depend upon the measurement geometry, illuminant, and observer. If any of these are changed, the tristimulus values will also be changed. Therefore, a statement of these variables must be a part of the color specification. When results made on different occasions or at different laboratories are compared, care must be taken to ensure that all measurement and calculation details correspond.

Response and computation speeds are such that a spectral scan can be made as quickly as a set of tristimulus readings. An

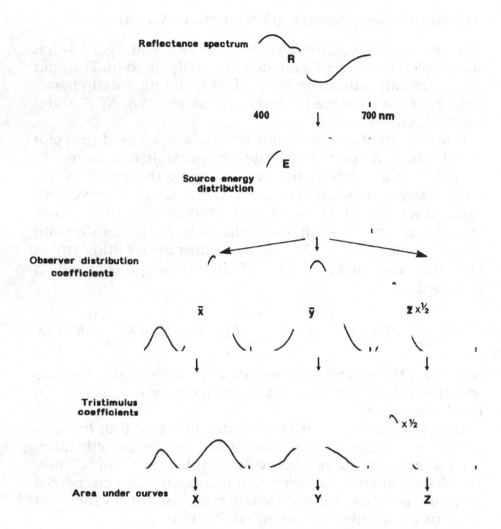

Figure 8–9 Derivation of tristimulus values X, Y, Z of an object from its reflectance spectrum. At each wavelength, the reflectance is multiplied by the relative energy of the light source and the color-matching functions $\bar{x}$, $\bar{y}$, $\bar{z}$ to yield three curves. The areas under these curves give the tristimulus values X, Y, Z.

advantage of a spectrophotometer is its ability to yield a spectral curve. In some instances, this can give useful information regarding, for example, possible chemical mechanisms or kinetics of a particular color change.

Instrumental Factors

Measurement Geometry

The appearance of an object depends on the angles of light incidence and viewing. Light reflected from objects viewed under large areas of illumination, such as the sky or fluorescent lighting, will always contain an element of the properties of the light source. That is, light is reflected without becoming selectively absorbed by surface pigments. This specular reflection from shiny surfaces adds white light to the surface, which thus looks less saturated. Glossy surfaces appear more saturated in directional than in diffuse illumination. With matte surfaces, all reflected light contains an element that has not interacted with the surface. Therefore, it will always appear desaturated. Unless glossy surfaces are illuminated very diffusely, matte surfaces are not normally as saturated as glossy surfaces. This influence of illumination and viewing angle on object appearance makes their standardization in any measuring and viewing system imperative.

The CIE recommends the four geometries shown schematically in Figure 8–10. All angles are in degrees, and, giving the illumination angle first, they are designated *45/0*, *0/45*, and, where integrating spheres are used, *diffuse/0* and *near 0/diffuse*. The 45/0 and 0/45 geometries exclude the spectral component of the reflected light from uniform flat surfaces. Measurements made using a near 0/diffuse geometry may include or exclude this component.

Standards

All measurements are made relative to a white standard, which the CIE nominates as a *perfect diffuser*. This is a material that diffusely reflects all incident light. Such a material does not exist,

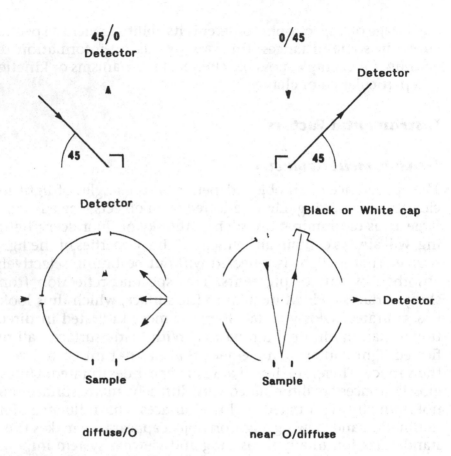

Figure 8–10 Standard illuminating and viewing conditions for reflectance measurements.

but working standards of known reflecting properties are available. During sample measurement and calculation, compensation is made for the nonperfect reflectance of the standard.

A colored substandard (or "hitching post" standard) is sometimes used to increase measurement reliability and sensitivity. A color near to that being measured is used for calibration of the instrument. The smaller the difference between the standard and the sample, the greater the measurement precision. The region of effectiveness around the standard is small enough to exclude the use of generic calibration tiles (i.e., red or green). The individual foodstuff should be targeted so that the range of applicability around the standard color is within 10 CIELAB units. As sample chroma increases, the precision produced by the hitching post method also increases. This technique has been found particularly useful when interlaboratory color quality specification procedures are being established (Kent et al. 1993).

Instrumental performance may be checked with highly stable standards. Transmission standards can be used to check the spectral response of transmission spectrophotometers. Colored opaque standards such as those produced by the National Physical Library and British Ceramic Research Association can be used to check the performance of reflectance spectrophotometers and colorimeters.

Dominant Wavelength and Purity Calculations

The luminance Y is an approximate correlate of perceived lightness, so any change in its value may be readily visualized. Visualization of the chromaticity (x, y) can be made easier through a conversion to dominant wavelength, an approximate correlate of perceived hue, and excitation purity, an approximate correlate of saturation. The methods of calculation are shown on the chromaticity diagram in Figure 8–11. A line drawn from the illuminant W, coordinates (x_w, y_w), through the measured color C, coordinates (x, y), is produced onto a point, coordinates (x_d, y_d), on the spectrum locus. This point defines the dominant wavelength (λ_d) of the color. The excitation purity (p_e) is defined as the ratio of the distances CW to $W\lambda_d$. That is:

$$p_e = \frac{x - x_w}{x_d - x_w} \quad \text{or} \quad \frac{y - y_w}{y_d - y_w}$$

These formulas are equivalent, but the accuracy of each will depend on the angle of the line $W\lambda_d$.

A purple color, which is a mixture of red and blue, has a complementary dominant wavelength (λ_c). The value is calculated by producing the line from the color C_1 back through the white point onto the spectrum locus at λ_c. Purity is measured using the intersection P, coordinates (x_p, y_p), with the purple boundary in the same way as above—that is:

$$p_e = \frac{x - x_w}{x_p - x_w} \quad \text{or} \quad \frac{y - y_w}{y_p - y_w}$$

The colorimetric purity (p_c) may also be calculated. This is determined from the ratio of the luminous flux of the wavelength λ_d that, when mixed with a flux of the white light, will match the color C. This concept is rarely used.

Hence, a color may be described as having coordinates (x, y, Y) or (λ_d, p_e, Y). Purity (p_e) falls in the range of 0 to 1, or it may

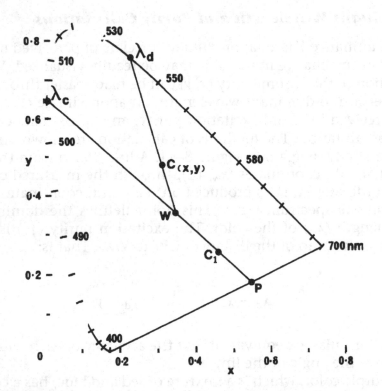

Figure 8–11 Derivation of dominant wavelength and excitation purity from the chromaticity coordinates.

be expressed as a percentage. Although λ_d and p_e are approximate nonlinear correlates of hue and saturation, the terms are never exchanged.

Discrimination of Color Attributes

The sensitivity with which the eye can distinguish between two adjacent colors differs according to the color. Just-noticeable differences (jnds) have been determined for wavelength, colorimetric purity, and luminance.

To determine wavelength discrimination, two halves of a 2° photometric field are illuminated with light of the same luminance and wavelength. Keeping the luminances equal, the wavelength of one half of the field is adjusted until a jnd is detected by the observer. Repeated measurements through the visible spectrum yield the wavelength discrimination plot shown in Figure 8–12. The three minima, where greatest discrimina-

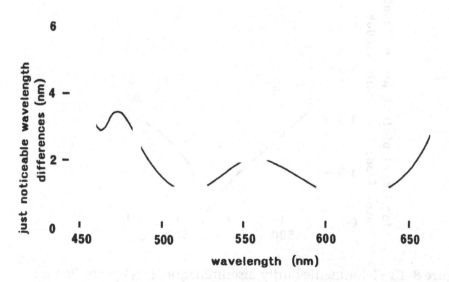

Figure 8–12 Typical wavelength discrimination for a normal observer using a 2° field (Wright 1969, with permission).

tion occur, are due in part to the spectral sensitivities of the three types of cone. Thresholds can vary widely with viewing conditions such as luminance level, field size, retinal eccentricity, and chromaticity of the surround. Maximum sensitivity is attained using an angular subtense greater than 2°, viewed foveally at a luminance in the photopic range, when the color and its surround are of similar luminance.

A bipartite field, in which a neutral light can be compared with a neutral light mixed with a monochromatic light, was used to obtain the spectral purity discrimination curve shown in Figure 8–13. Constant luminance was maintained during the judgments. This figure is plotted as the reciprocal of the jnd of the colorimetric purity (p_c). Thresholds are smallest and sensitivity is greatest at long and short wavelengths. That is, more steps can be perceived between white and deep red than between white and deep yellow. Purity discrimination is greatly

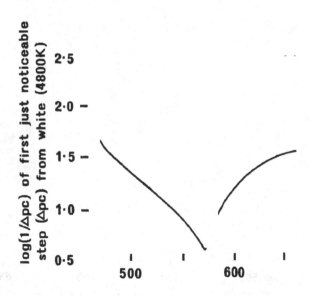

Figure 8–13 Colorimetric purity discrimination: $\log(1/p_c)$ of first just-noticeable step (pc) from white (Hill 1987, reprinted from Colour Physics for Industry, with permission from The Society of Dyers and Colourists).

dependent on stimulus wavelength and luminance, and this has profound implications for the setting of tolerances for color quality control.

Luminance discrimination has also been determined. The response of the cones is at its maximum at approximately 550 nm (Figure 4–2). Here, at the maximum stimulus, the threshold of luminance intensity is at its minimum. For many stimuli, including photopic luminance, the Weber-Fechner law holds. That is, the greater the stimulus, the more it must be increased for the eye to detect a change (Figure 8–14).

The Weber-Fechner ratio, $I_1/I_1 = I_2/I_2 = 0.01$, holds for a wide range of luminance (Hill 1987). Departure from this law occurs at low luminances, through interaction between rods and cones, and at high luminances, through saturation in receptor activity.

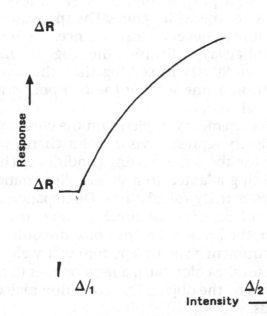

Figure 8–14 Relationship between response and stimulus intensity (Hill 1987, reprinted from Colour Physics for Industry, with permission from The Society of Dyers and Colourists).

As a rule of thumb, a difference between the color of two foods will be seen if the measured colors differ (1) in dominant wavelength by 1 or 2 nm, or (2) in excitation purity by 3 to 5 percent, or (3) in luminance by 1 or 2 percent. These limits are very approximate and vary for different colors. If any reliance is to be put on the degree of visual detection of measurement differences, they should be tested for the particular food. Such crude approximations are only possible for foods because of their inherent variability. These rule-of-thumb differences do not apply to more uniformly colored surfaces.

Properties of the Chromaticity Diagram

All colors can be located within the area enclosed in the x, y chromaticity diagram. The standard illuminant used in the measurement governs the position of the achromatic axis (the *white point*). Deep, spectrumlike colors have chromaticities near to the spectrum locus (or purple boundary). Paler, less chromatic colors fall nearer to the white point. The spectrum locus is always straight or convex, never concave. Hence, any mixture of spectral stimuli will always fall inside the diagram. Also, if two color stimuli are additively mixed together, the resulting chromaticity will fall on a line joining the two points representing the two original colors.

The behavior of measurements, depicted on the chromaticity diagram, will only closely represent visual color changes if instrument and viewer use the same viewing conditions. This includes the viewer's being adapted to a similar illumination to that used for the chromaticity calculations. Discrepancies may still be found when surfaces of nonuniform color are involved. Measurement of a mottled object having color discontinuities smaller than the instrument viewing aperture will yield some sort of average value for the color. Such a result relates to no actual property possessed by the object. This condition may easily arise with many foods.

Where possible, instrumental measurements should be accompanied by visual assessments of the same samples under the same conditions. This can be especially valuable for food materials that, during a trial, change markedly in optical properties other than color. Sometimes all normal precautions taken for color measurements still yield instrumental and panel results that are not in agreement. This can occur particularly with samples that change in translucency or gloss (see Chapters 9 and 10).

The chromaticity diagram is not perceptually uniform. That is, equal distances in different parts of the diagram do not yield equal visual impressions of color difference. This is illustrated in Figure 8–15. The length of each line represents subjectively equal color steps. These color steps were determined by varying

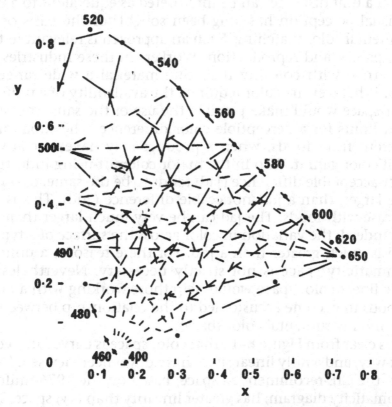

Figure 8–15 Subjectively equal color steps for a 2° observer plotted on the CIE 1931 chromaticity chart. The lines represent a color difference approximately three times greater than the just-noticeable difference (Wright 1969, with permission).

the mixture of two monochromatic radiations taken from opposite sides of the spectrum locus or a mixture of one monochromatic radiation with a purple color. The proportion of one was altered until a noticeable change was detected by comparison with a similar mixture seen in the other half of the field. Both stimuli were maintained at constant luminance.

Uniform Chromaticity Space and Uniform Color Scales

Ideally, the lines in Figure 8–15, representing identical differences in perceived color, should be the same length. A space in which a unit distance can be interpreted as equivalent to a unit of visual perception has long been sought for the basis of instrumental color matching. Such an approach is relevant to textiles, paints, and reproduction. Workers in these industries are concerned with coloring their one material a wide range of hues. Whatever the color required, the availability of a uniform color space would make possible the use of the same measurement limits for a perceptible color difference. The situation is different from foods, when a product may have a relatively small color gamut. Also in the major colorant-using industries, a just-acceptable difference (jad) tends to be the same, or only a little larger, than a just-noticeable difference (jnd). This is not the case with foods. The jad can be very much larger than the jnd. Indeed, the variation in color across the surface of a typical food is much greater than a jnd. For these reasons, a uniform chromaticity space is not strictly necessary. Nevertheless, a more linear color space would help those working with a range of foods to become accustomed to the relationship between visual and instrumental color spaces.

It is clear from Figure 8–15 that color space is curved in a complex way, and a fully linear space has not yet been devised. However, the CIE-recommended space, based on the 1976 uniform chromaticity diagram, has greater linearity than (xy) space. This has axes u' and v' (CIE 1986). The transformations are

$$u' = \frac{4X}{X + 15Y + 3Z} = \frac{4x}{-2x + 12y + 3}$$

$$v' = \frac{9Y}{X + 15Y + 3Z} = \frac{9y}{-2x + 12y + 3}$$

This space (Figure 8–16) contains a selection of lines equivalent to those in Figure 8–15. The increase in linearity achieved can be judged from the ratios of the lengths of the longest and shortest lines shown in Figures 8–15 and 8–16, 20 to 1 as against 4 to 1 for the transformed space. As in the (*xy*) diagram, additive mixtures of two components lie on a straight line.

The CIE recommended two ways of combining luminance and chromaticity information. These are the CIE 1976 (*L*u*v**),

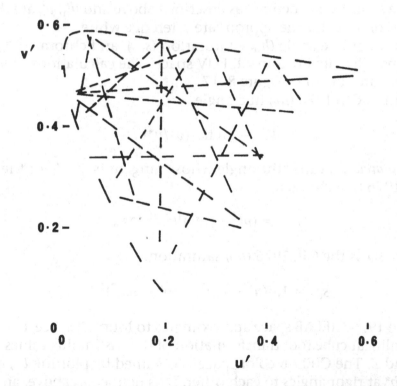

Figure 8–16 The more uniform CIE (*u'v'*) chromaticity diagram, with a selection of the lines from Figure 8–15 (Hunt 1991, reproduced from *Measuring Colour*, with permission from Ellis Horwood Ltd., Chichester).

or CIELUV, and the CIE ($L*a*b*$), or CIELAB, spaces. The CIELUV space was primarily intended for industries involved with color additive mixing, such as lighting and television.

The 1976 CIELUV space chromaticity diagram is a linear transform of (xy) space. The CIELUV color space is obtained by plotting $L*$, $u*$, and $v*$ at right angles to each other.

$$L* = 116(Y/Y_n)^{1/3} - 16 \quad \text{for} \quad Y/Y_n > 0.008856$$

$$L* = 903.3(Y/Y_n) \quad \text{for} \quad Y/Y_n \le 0.008856$$

$$u* = 13L*(u' - u'_n)$$

$$v* = 13L*(v' - v'_n)$$

where u' and v' are defined as described above and u'_n, v'_n are the values of u', v' for the appropriate reference white.

Values of hue angle (h_{uv}), saturation (s_{uv}), and chroma ($C*_{uv}$) may be calculated within CIELUV space. The calculations of h_{uv} and s_{uv} are shown in Figure 8–17.

For the CIE 1976 (uv) hue-angle,

$$h_{uv} = \arctan(v*/u*)$$

where *arctan* means "the angle whose tangent is. . . ." For the CIE 1976 (uv) chroma,

$$C*_{uv} = (u*^2 + v*^2)^{1/2} = L*s_{uv}$$

where s_{uv} is the CIE 1976 (uv) saturation:

$$s_{uv} = 13[(u' - u'_n)^2 + (v' - v'_n)^2]^{1/2}$$

The 1976 CIELAB space approximates to Munsell space, and is a nonlinear cube root transformation of the tristimulus values X, Y, and Z. The CIELAB color space is obtained by plotting $L*$, $a*$, and $b*$ at right angles to each other. $L*$ is defined as above, and

$$a* = 500[(X/X_n)^{1/3} - (Y/Y_n)^{1/3}]$$

$$b* = 200[(Y/Y_n)^{1/3} - (Z/Z_n)^{1/3}]$$

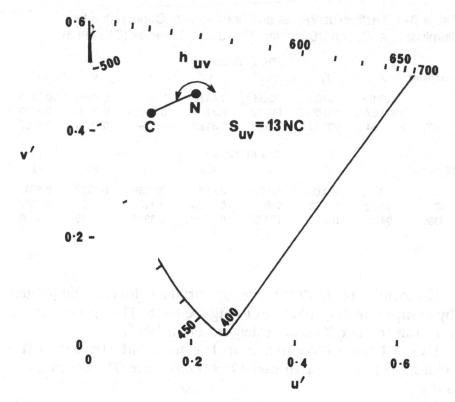

Figure 8–17 The derivation of h_{uv} and s_{uv}. N is the reference white and C is the color (Hunt 1991, reproduced from *Measuring Colour*, with permission from Ellis Horwood Ltd., Chichester).

where X_n, Y_n, and Z_n are values of X, Y, and Z for the appropriate reference white.

Values of hue angle and chroma can be similarly calculated for CIELAB space. As the ratios of the tristimulus values are used in the equations as cube roots, there is not a chromaticity diagram for CIELAB space; hence, there is no correlate of saturation.

The tristimulus values and chromaticity coordinates (x, y) and (u', v') for the illuminants A, C, and D_{65} and the 2° and 10° observers are shown in Table 8–1.

Table 8–1 Tristimulus Values and Chromaticity Coordinates for Illuminants A, C, and D65 for the 2° and 10° Observer (CIE 1986)

Illuminant	X_0	Y_0	Z_0	x	y	u'	v'
			The 2° Observer				
A	109.89	100.00	35.58	0.4476	0.4074	0.2560	0.5243
C	98.07	100.00	118.23	0.3101	0.3162	0.2009	0.4609
D65	95.04	100.00	108.89	0.3127	0.3290	0.1978	0.4683

Illuminant	X_{10}	Y_{10}	Z_{10}	x_{10}	y_{10}	u'_{10}	v'_{10}
			The 10° Observer				
A	111.15	100.00	35.20	0.4512	0.4059	0.2590	0.5242
C	97.28	100.00	116.14	0.3104	0.3191	0.2000	0.4626
D65	94.81	100.00	107.33	0.3138	0.3310	0.1979	0.4695

Uniformity of $(L*a*b*)$ space for surface colors can be judged by comparing ellipsoid sizes in Figure 8–18. The space is more uniform than (xyY) space (Melgosa et al. 1997).

Use of Hunter colorimeters in the food industry led to the widespread use of the Hunter 1958 (Lab) space. The coordinates are

$$L = 10Y^{1/2}$$

$$a = \frac{17.5(1.02X - Y)}{Y^{1/2}}$$

$$b = \frac{7.0(Y - 0.847Z)}{Y^{1/2}}$$

where X, Y, and Z are the 1931 CIE tristimulus values (percent).

Axes L, a, and b are mutually perpendicular, as shown in Figure 8–19. An increase in the negative value of a represents an increase in the green component of a color. An increase in a represents an increase in red; $-b$, an increase in blue; and $+b$, an increase in yellow. An increasing value of $L*$ represents an increase in whiteness or lightness. Neutral colors are close to ($a = b = 0$). The simplicity of this concept has often led to the description of a as *redness* and b as *yellowness*. This is incorrect, as

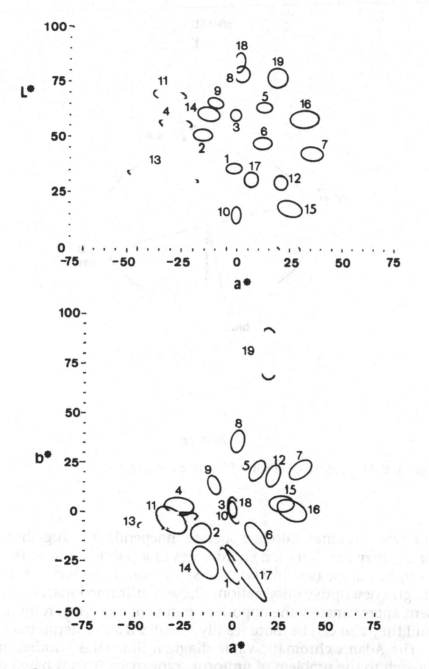

Figure 8–18 Sections of $L*a*b*$ CIELAB ellipsoids, with semiaxes enlarged threefold. *Source:* Reprinted with permission from Melgosa, Hita, Poza, Alman, and Berns, Suprathreshold Color-Difference Ellipsoids for Surface Colors, *Color Research Application*, Vol. 22, pp. 148–155, © 1997, John Wiley & Sons, Inc.

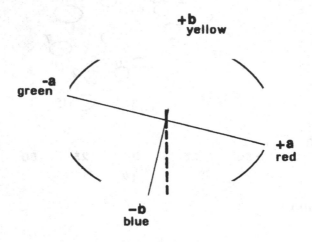

Figure 8–19 A representation of Hunter (*Lab*) space.

the two coordinates do not operate independently. Together as (*a*, *b*), they are Cartesian coordinates of a point in space. However, hue angle (arctan *b/a*), and saturation index [$(a^2 + b^2)^{0.5}$] do give descriptive information. These transformed spaces represent approximately the opponent nature of our visual response, and they also can be more readily visualized and interpreted.

The Adams chromatic value diagram illustrates another approach to the problem of uniform color space. This is based on the reasonable perceptual uniformity of the Munsell space and the relationship between the *Y* value of samples, which are equally spaced in lightness, and their corresponding Munsell values V (Adams 1942). The definition of Munsell value is (Priest et al. 1920)

$$V = Y^{1/2} \text{ (maximum value of } Y = 100)$$

Adams derived three Munsell value functions, V_x, V_y, and V_z. These are based on the relation between V and Y, with X_c, Y_c, and Z_c used in turn in place of Y:

$$V_x = \frac{X_c}{98.04}, \qquad V_y = \frac{Y_c}{100.0}, \qquad V_z = \frac{Z_c}{118.10}$$

where X_c, Y_c, and Z_c are the tristimulus values of the sample under illuminant C.

As the relation between V and Y is nonlinear, this gives a nonlinear transformation of the CIE 1931 diagram. Adams found that when he plotted $(V_X - V_Y)$ against $0.4(V_Z - V_Y)$, colors of the same Munsell chroma plotted reasonably close to circles. The radii of the circles changed very little at different values of Y.

More refined lightness equations have been proposed: for example (Newhall et al. 1943),

$$\frac{100R}{R_{MgO}} = 1.2219V - 0.23111V^2 + 0.23951V^3$$

$$- 0.021009V^4 + 0.0008404V^5$$

where R_{MgO} is the reflectance of the magnesium oxide standard. The Glasser cube root approach is another example (Glasser et al. 1958, quoted in Judd and Wyszecki 1975).

$$L = 25.29Y^{1/3} - 18.38$$

Adams chromatic value space is mentioned occasionally in the food color literature.

Color Difference Calculation

A difference between two colors can be calculated in any of the above coordinate systems. The equation is based on the square root of the sum of the squares of the differences in each axis. For example:

$$\Delta E_{(CIELAB)} = [(\Delta L^*)^2 + (\Delta a^*)^2 + (\Delta b^*)^2]^{1/2}$$

The near-linear nature of (*Lab*), CIELUV, and CIELAB spaces makes color difference calculation more meaningful through the whole gamut of color space. Commercial paint or textile color tolerances may be, typically, 1 to 2 ΔE units.

Any color difference can be split into three components:

lightness difference:

$$\Delta L^* = L^* \text{ sample} - L^* \text{ standard}$$

chroma difference:

$$\Delta C^* = C^* \text{ sample} - C^* \text{ standard}$$

hue difference:

$$\Delta H^* = [(\Delta E)^2 - (\Delta L^*)^2 - (\Delta C^*)^2]^{1/2}$$

or (Sève 1996)

$$\frac{a_1^* b_2^* - a_2^* b_1^*}{[0.5(C_1^* C_2^* + a_1^* a_2^* + b_1^* b_2^*)]^{1/2}}$$

The numerical value of each component is in CIELAB units (CIE 1986).

Color difference in Adams chromatic value space is (Nickerson and Stultz 1944):

$$\Delta E = [(0.23 V_Y^2 + (V_X - V_Y)^2 + 0.4 (V_Y - V_Z)^2]^{1/2}$$

More complex formulas, such as the CMC(1:c) and CIE 1994, take further account of the unevenness of measured color space. Relative weightings of the contributions of the differences in L^*, C_{ab}^*, and H_{ab}^* are varied according to the position of the color in CIELAB space. Color difference calculations are not widely relevant within the industry because of the importance of color change direction as opposed to its size and because of the large difference between color perceptibility and acceptability of most foods.

Measurements and Perceptual Correlates

Relationships between the perceptual attributes and the CIE 1976 colorimetric measures outlined above are summarized in Table 8–2. The table shows no correlate for colorfulness. It is an object of color measurement that it relates to the color as seen. However, objects in everyday life have surrounds, and the appearance of the object color is affected by other colors in the scene. Specification techniques described so far have not taken the surround into account. Color vision models have been proposed for predicting the appearance of colors under a wide range of viewing conditions. The color appearance model CIECAM97s uses the color perception model described in Chapter 4 to derive correlates for brightness and colorfulness as well as for hue, saturation, lightness, and chroma of the CIELUV and CIELAB systems. Chromatic and luminance adaptations are included within the model (Hunt 1998). Data for these models have been obtained from visual judgments of computer screens.

Table 8–2 CIE Recommended Colorimetric Measures and Their Perceptual Correlates (Hunt 1991, reproduced from *Measuring Colour*, with permission from Ellis Horwood Ltd., Chichester)

Nonuniform Measures	Approximately Uniform Measures (CIE 1976)	Approximate Perceptual Correlates
Luminance L	None	Brightness
Chromaticity x, y	Chromaticity u, v	Hue and saturation
Dominant wavelength λ_d	Hue angles h_{uv} or h_{ab}	Hue
Excitation purity p_e	Saturation s_{uv}	Saturation
Luminance factor L/L_n	Lightness L^*	Lightness
None	Chroma C^*_{uv} or C^*_{ab}	Chroma
None	None	Colorfulness

Models have not yet been applied to three-dimensional food scenes.

The Helmholtz-Kohlrausch effect is the effect whereby chromatic object colors appear lighter than achromatic object colors of the same luminance factor (L^*). A corrected luminance factor (L^{**}) that predicts perceived lightness to within interobserver variability can be calculated (Fairchild and Pirotta 1991):

$$L^{**} = L^* + f_2(L^*)f_1(h^\circ)C^*$$

where L^*, C^* and h° are CIELAB parameters and

$$f_2(L^*) = 2.5 - 0.025L^*$$

$$f_1(h^\circ) = 0.116 \sin\left(\frac{h^\circ - 90^\circ}{2}\right) + 0.085$$

Most color measurements reported in the food literature involve one of the (Lab) spaces. It is stressed that visual assessments must where possible accompany instrumental measurements. The coordinate system best related to the quality and extent of color differences observed can then be chosen for that particular food. However, X, Y, and Z, and hence (xyY) space, are the fundamental units of color measurement. All other spaces are derived from them.

Machine Vision Color Measurement

Conventional instrumental methods result in a measurement of the average color of the sample. That is, they have a low spatial resolution. This can be overcome using machine vision and image processing. The charge-coupled detector (CCD) camera can evaluate each of many thousand elements of the light reflected from an object or scene. These picture elements (pixels) are detected in terms of red, green, and blue (R, G, B). Hence, an attempt can be made to derive X, Y, Z values and detailed distributions of, for example, lightness, chroma, and hue (L, C, H) from a suitably calibrated camera. Advantages include its noncontact, nondestructive nature, disturbance to the sample being minimal. The technique can be used for product sorting and monitoring dynamic color changes taking place on processing and storage. Excellent agreement can be obtained between results obtained by remote sensing methods and standard reflectance spectrophotometer (Raggi and Barbiroli 1994). Machine color vision has huge potential for food industry applications.

Apparatus consists of a source of illumination, such as a lighting cabinet, a suitable camera and stand, a vision processor board (frame grabber) for acquisition, a computer and software for processing and storing data, and a video monitor and printer for display. Once the image is obtained, it is digitized so that features can be extracted and analyzed.

As for all color measurement, the geometry of the measurement system and the components within the system should be well defined and reproducible. For illumination and measurement of static objects, most workers use a commercial lighting cabinet or dual sources at 45° to the object. This represents a natural realistic view of the scene, specular reflection forming part of the image.

Specular reflection can be substantially reduced using the diffuse/0° geometry illustrated in Figure 8–20. The cylindrical chamber has a diameter of 59 cm and is 45 cm long. Inside, a plate, 16 cm wide and 48 cm long, is mounted between the upper wall of the chamber and the lamps so that the sample of tomatoes is exposed only to indirect lighting. Six 100-watt incandescent lamps are mounted 15 cm apart along both sides of the bottom of the chamber. The inside is coated with flat white latex paint to provide uniform and diffuse illumination. The camera views the sample through a 6-cm hole bored in the top center of the chamber (Choi et al. 1995).

Illumination geometry is similarly important for on-line measurements. Front lighting is best suited for obtaining surface characteristics of an object, while back lighting is suitable for translucent materials. Subsurface cracks in corn kernels and watercore in apples can be detected in this way (Gunasekaran 1996).

Biscuit baking provides an example of the results produced by the machine vision method. Samples of plain cracker biscuit

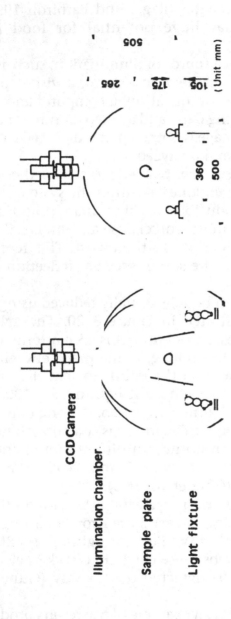

Figure 8–20 Schematic diagram of the tomato illumination chamber. *Source:* Reprinted with permission from K. Choi, G. Lee, Y.J. Han, and J.M. Bunn, Tomato Maturity Valuation Using Color Image Analysis, *Transactions of the American Society of Agricultural Engineers,* Vol. 38, pp. 171–176, © 1995, American Society of Agricultural Engineers.

taken from each zone of an industrial baking oven were imaged. Results can be displayed in terms of visual sensation, such as hue, chroma, and lightness, or in terms of *R, G, B*. The latter is in the form of a cube with three mutually perpendicular axes representing red, green, and blue. When $R = G = B = 0$, the object is black; when $R = G = B = 1$, it is white. Each pixel measurement from the biscuits is shown plotted in (*RGB*) space (Figure 8–21). The mean path clearly defines the *baking curve* of the

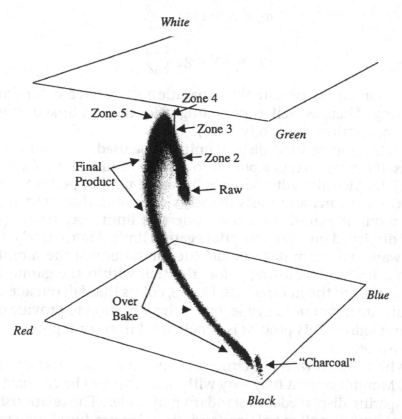

Figure 8–21 The baking curve for a plain cracker biscuit. *Source:* Reprinted with permission from L.G.C. Hamey, J. Yeh, and C. Ng, Objective Bake Assessment Using Image Analysis and Artificial Intelligence, in Cereals '97, *Proceedings of the 47th Australian Cereal Chemistry Conference*, A.W. Tarr, A.S. Ross, C.W. Wrigley, eds., p. 181, © 1997, Royal Australian Chemical Institute.

biscuit through color space. Similar characteristic color development paths are found for a wide variety of biscuit and cookie products. It is hypothesized that the chemical processes involved in color development are very similar and are independent of color development rate (Hamey et al. 1997).

Light reflected by a three-dimensional object with a uniform surface color has constant spectral balance but varies in intensity. This is caused by the variation of illumination incident on the surface of the object. To overcome this, the CIELAB cube root function is replaced by the natural logarithm. Thus:

$$a_L = 500 \log_e \frac{XY_0}{X_0 Y}$$

$$b_L = 200 \log_e \frac{YZ_0}{Y_0 Z}$$

These equations permit the separation of surface color from lighting effects as well as providing color metrics linked to human perception (Connolly 1996).

A television or VDU display unit can be used to perform two tasks. It can be used as a plotter to display results in the form of, say, CIELAB units. Alternatively, it can be used to perform critical comparisons and measurements on colors that are within the monitor gamut. Colors outside this limit may be clipped and displayed on the appropriate gamut limit. Alternatively, the software may attempt to model the limitations of the monitor gamut limit by adjusting colors that fall within the gamut inwards toward the neutral axis. Hence, colors that fall outside the gamut are reproduced inside, but with a variation to provide discriminability. This process is widely used in color reproduction industries.

When critical work is being performed, various errors can occur. Monitor output may vary with time; this can be checked by comparing displayed and standard gray scales. The relationship between the applied voltages and the color produced may vary over the area of the screen. As the eye is not sensitive to colors that are spatially gradual, this can be checked by using a comparator that enables colors of different areas to be seen or measured side by side. Colors produced may alter with screen usage.

This can be checked with a spectroradiometer (Hunt 1998). It is also necessary to calibrate the screen. This can be done using a set of eight test colors. Errors usually less than one CIELAB ΔE unit can be achieved over the gamut of displayed colors (Berns et al. 1993).

Reasonable agreement can be obtained between conventional color-measuring instruments and machine vision systems when comparing Munsell standards (Raggi and Barbiroli 1994; Ling et al. 1996). Contributing to disagreement will be the fact that the camera response characteristics are not identical to the CIE color matching functions (Figure 8–6). Also, differences in response certainty are dependent on the particular color. On the Munsell samples tested, standard deviations of replicate measurements varied from 1.1 CIE units for the 3.8BG samples and 1.7 units for 3.1YR. Color resolution is also color dependent. At a 95-percent confidence interval, 0.7 CIE unit could be resolved in the 3.8BG group of samples, but only 3.1 units could be resolved in the 9.0YR group. The good agreement obtained when using opaque, flat standards does not necessarily occur when food materials are measured. For example, L^* values of carrot purees from the machine vision system were on average 14 units greater than those from a conventional colorimeter. Values of a^* were 27 units higher, and values of b^* were 44 units lower (Ling et al. 1996). Other factors of puree appearance such as translucency, gloss, and surface texture may account for these differences.

Within the food industry, although reproducibility is demanded, accuracy may not be important. Of positive benefit, however, is the high spatial resolution of the machine vision measurement system. This includes the ability to isolate and specify appearance features such as color pattern, gloss, and surface texture, attributes not delivered by conventional instrumentation. Applications of this technique are discussed in Chapter 9.

Subtractive Colorimetry

The Tintometer is an instrument based on subtractive colorimetry. It was originally developed for the assessment of beer by

Lovibond in 1887. Colored glasses are used to subtract light from a standard white lamp until the resulting light visually matches that transmitted by the transparent sample. The combination of colored glasses needed provides a specification for the color. Opaque colors can be measured in a similar way. The same lamp is used to illuminate the sample surface and the surface of a white diffuse reflector. The glass filters are again used to produce a matching color.

The glasses are magenta, called *red*, to absorb green light; *yellow* to absorb blue; and cyan, called *blue*, to absorb red light. In the Tintometer, each color is arranged in a numerical scale from practically colorless to highly saturated. After matching, the sample color can be specified by the number of units of *red*, *yellow*, and *blue*. Using combinations of the scales, nine million colors can be produced.

The Lovibond Schofield Tintometer was developed later. With this instrument, a visual match may be made using two of the scales and a light intensity control. It is possible to convert readings from this instrument into X, Y, Z tristimulus values.

Metamerism

Metamerism occurs when two colors match under one set of viewing conditions but fail to match under a different set. There are four types: *illuminant*, *observer*, *geometric*, and *field size*.

Illuminant metamerism occurs when two samples match under one illumination but not under another. This arises from the reflectance curves of the samples not being identical. Because the visual process is an integration of three independent factors— light output, reflectance properties, and visual characteristics— it is possible for two colors to have identical X, Y, Z values under one light source but different values under another. An example of metamerism has occurred with lime juice, some types of which look green under normal white light but pink under triphosphor lamps. Metamerism normally occurs in other industries because different pigment or dye mixes have been used in the two samples. This phenomenon might have become important in foods if the industry had progressed into making food analogues, such as real beef steak look-alikes from vegetable protein. Artificial colorants would have been needed to match natural food pigments under a wide range of illumination conditions. However, the industry has developed in other directions, and illuminant metamerism is not a common problem for food itself.

Illumination can cause two types of problem in packaging and printing: adjacent packages on a shelf may not match under the particular store lighting conditions, and the picture on the pack may not match the color of the food inside. The former problem is caused by nonidentical pigments in different batches of color printing. The second problem arises because the spectral characteristics of the pack pigments are different from the natural pigments existing in the food. For example, no stable artificial pigments exist that produce the same reflectance curves as the blood pigment hemoglobin. Therefore, a nonmetameric match cannot yet be obtained for meat colors.

The method recommended for quantifying the metamerism of objects involves calculation of tristimulus values X, Y, Z for all the illuminants under which the objects are to be critically judged. The color difference between the two objects is then calculated for each of the illuminants. If this is markedly high under one of the illuminants, the closeness of the match under this illuminant may prove to be unacceptable (McLaren 1980).

Observer metamerism occurs when a pair of samples match for one person but not for another. This arises from the two observers' having different color-matching functions. This can occur even among people having normal color vision.

Geometric metamerism occurs when two samples match when viewed at one angle but do not match when the observing angle is altered. This is obvious when viewing metallic paints and is due to differences in the structure of the paint surface. It can also be obvious in foods. For example, the sheen of interference colors in bacon slices is caused by the uneven cutting of surface fibers.

Field size metamerism occurs when the size of field is changed from, say, 2° to 10°. This arises from the change in structure of the retina as the subtended angle is altered.

FLUORESCENCE

A material is fluorescent if it absorbs energy at one wavelength and re-emits it at another. This is not normally a problem with foodstuffs, although it does occur. Beef connective tissue absorbs ultraviolet radiation and re-emits between 400 and 550 nm (Swatland 1997). Monitoring of fluorescence is a rapid way of assessing lipid changes during storage of sardines (Aubourg et al. 1997) and mackerel (Maruf et al. 1990).

The most detailed color measurement of fluorescent materials is undertaken using a spectrophotometer having a dual monochromator. The sample is illuminated with monochromatic radiation, and a detector records the energy reflected plus that re-emitted at each wavelength. In this way, a total radiance (reflected plus emitted energy) curve characteristic of the material can be built up. The spectral emission distribution of a standard illuminant, such as D_{65}, can then be used and values of X, Y, and Z calculated in the normal way.

A simpler approach to the color measurement of fluorescent materials involves illumination of the sample with a white light source that includes the near-ultraviolet wavelengths. Tristimulus values of the resulting radiance can then be measured using the normal tristimulus or spectral techniques. Difficulties arise with the choice of white light illuminant, but it is possible to correct the results to daylight illumination (Hunt 1998).

TRANSLUCENCY

Translucency is a key perceptual attribute of foodstuffs. The translucency, or opacity, of diffusing films of paper and paint is indicated by its contrast ratio: that is, by the extent to which a patterned background is hidden. The Technical Associates of the Pulp and Paper Industry (TAPPI) opacity is defined as

$$C_{0.89} = R_0 / R_{0.89}$$

where R_0 and $R_{0.89}$ are the reflectances when the film is backed by a material of reflectance 0 (black) and 0.89. The normal minimum value for paint and paper is greater than 0.90. The measurement can be related to Kubelka-Munk scattering and absorption coefficients. TAPPI opacity is a practical measurement related to the visually perceived opacity of white materials (Hunter and Harold 1987). The Kubelka-Munk scatter coefficient is linearly related to panelist estimates of the opacity of model nonabsorbing emulsions (Birkett 1985).

Color measurement of translucent samples has always been a problem. Mackinney et al. (1966) were the first to realize that food translucency is a property in its own right. However, little appears to be known about the way we see it, other than that it involves perception of color contrast.

In model system studies, colored suspensions of titanium dioxide were presented either in white shallow dishes having a solid black letter stencilled inside the bottom or between glass plates that could be placed onto a contrasting background. Translucency occurs, by definition, on a scale between *transparent* and *opaque*. It was found that to some panelists, an increase in *translucency* meant that the samples became more opaque, whereas to others, it meant that the samples became more transparent. Hence, scaling was limited to *opacity* and *transparency*. As solution opacity was increased, opacity perception could be divided into three merging regions, which could be called *transparent*, *translucent*, and *opaque*. The initially sharp edge of the stencil became blurred, then totally obliterated. This was accompanied by a graying of the black stencil area until any further increase in light scatterer was judged on the increasing whiteness of the suspension. Blurring of the stencil outline, or using a gray stencil, led to higher opacity scores from panelists. It was found that care should be taken over the words used for scaling. For example, it was found that, using Stevens (1975) criteria, *opacity* and *transparency* were not equal and opposite, although on subsequent discussion most panelists had assumed that they were.

Two types of response were found among a panel of 21 observers when scaling 15 green suspensions differing in concentrations of dye and scattering material. At very low scatterer concentrations, 14 panelists scored *opacity* according to the dye concentration, and 7 gave the same score irrespective of the dye concentration. That is, some scored according to the magnitude of the contrast and others according to line sharpness. For samples containing higher levels of scatterer, an increase in dye concentration resulted in an increase in *opacity* for all panelists.

Magnitude estimations were also made of the *opacity* of red (two samples), blue (one), and green (five) soluble dyes with titanium dioxide in suspension. Transreflectance spectrophotometric measurements were made on the samples. Values of X, Y, and Z were calculated from the Kubelka-Munk–derived internal transmittance spectra. They are shown plotted against the panel *opacity* in Figure 8–22.

The three values of X, Y, and Z for each sample have been plotted. As would be expected, X is greater than Y or Z for a red sample, Y is greater for a green, and Z for a blue. It can be seen that a continuous line can be drawn through these highest tristimulus values. It seems reasonable that translucency is perceived according to the color of the sample by the receptors most sensitive to the peak transreflected wavelengths of the color. These conclusions require confirmation using other model systems and colors (Hutchings and Scott 1977; Hutchings and Gordon 1981).

Some combination of scattering and absorption in translucent fluids can be measured with a version of the Secchi disc. This instrument was developed for the measurement of daylight penetration into the sea (Nyquist 1979). The laboratory version consists of a disc that can be raised gradually through the fluid until it can just be seen. The depth at which this occurs is an indication of sample translucency.

GLOSS

Gloss is an attribute characteristic of the particular food. In the kitchen, hot vegetables are coated with butter, or a joint of meat may be glazed, but no attempt is made to control the gloss of natural foods with any subtlety. However, the degree of gloss of manufactured foods can be chosen and manipulated. Yellow spreads and cheeses, for example, range from matte to high gloss.

Faced with a number of foods covering a range of glossiness, one might assume that gloss is a continuum and that the same

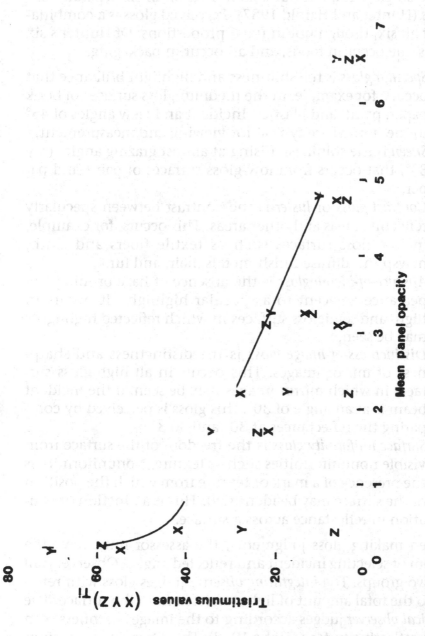

Figure 8–22 Perception of colored translucent suspensions. Relationship between the tristimulus values derived from the internal transmittance spectral curves and the mean panel opacity (Hutchings and Gordon 1981).

type of observation is being used for all samples. However, gloss is not a single sensation but a combination of six separate sensations (Hunter and Harold 1987). Perceived gloss is a combination of all six, though not in fixed proportions. Of Hunter's six types, some occur in foods, and all occur in packaging.

1. *Specular gloss* is the shininess and highlight brilliance that occurs, for example, in the medium-gloss surfaces of book paper, paint, and plastics. Incident and view angles of 45° to the normal are typical for viewing and measurement.
2. *Sheen* is the shininess arising at almost grazing angles (say 85°) that occurs from low-gloss surfaces of paint and paper.
3. *Contrast gloss* or *luster* is the contrast between specularly reflecting areas and other areas. This occurs, for example, in low-gloss surfaces such as textile fibers and cloth, newsprint, diffuse-finish metals, hair, and fur.
4. *Absence-of-bloom gloss* is the absence of haze or milky appearance adjacent to a specular highlight. It occurs in high- and semigloss surfaces in which reflected highlights may be seen.
5. *Distinctness-of-image gloss* is the distinctness and sharpness of mirror images. This occurs in all high-gloss surfaces in which mirror images may be seen. If the incident beam is at an angle of 30°, this gloss is perceived by comparing the reflectances at 30° and 30.3°.
6. *Surface uniformity gloss* is the freedom of the surface from visible nonuniformities such as texture. Nonuniformity is the presence of a mark or texture from which the position of the surface may be identified. This is a function of variation in reflectance across a surface.

When making gloss judgments, the assessor must have the freedom of selecting incident and reflected angles. Observers fall into two groups. The *integrating observer* judges gloss with reference to the total amount of light reflected from the surface. The *analytical observer* judges according to the image sharpness seen via the reflecting surface (Tighe 1978). These two sets of findings are sufficient to account for disagreement among observers when scaling or ranking gloss. The visual response to specularly reflected light is logarithmic (Kosbahn 1964, quoted in Tighe 1978).

Surface reflectance can be described in terms of goniophotometric reflectance patterns, using different angles of illumination (see Figure 3–8). The relationship between the goniophotometric curves and the perception, for samples of the same material but differing in gloss, is normally straightforward. This has led to the many American Society for Testing and Materials (ASTM) and TAPPI standards based on the above classifications related to specific materials. Goniophotometers are expensive, and simpler gloss meters designed to service these standards offer the high precision required for plastics and coated materials. None of these standards concerns food materials, but some are designed for packaging.

Unfortunately, the relationship between the goniophotometric curves of different materials exhibiting a wide gloss range and the perceptions of gloss is not simple. This is because gloss is a combination of different sensations, because observers use different criteria in judging, and because gloss is of different types.

The *gloss factor* can be derived from a goniophotometric curve for materials having moderate gloss levels. This has found many applications in a number of industries, including foods. The gloss factor (GF) is

$$GF = (I_s - I_d)/W_{1/2}$$

where I_s is the peak height at specular reflection angle, I_d the diffuse reflectance intensity, and $W_{1/2}$ the specular peak width at half height. The last enables the image sharpness as well as the extent of specular reflection itself to be included in the measurement.

The GF is useful for following changes in reflectance characteristics occurring during the weathering of polymer film. There is a quantitative relationship between GF and the average height of surface irregularity of such polymer films (Tighe 1978). For dried food polymer films used for reducing water loss in fruits, however, the situation appears more complex. Increases in concentration of agar and agarose lead to greater roughness and lower gloss. On the other hand, increases in concentration of gelatin, gellan, xanthan, and alginate also lead to increasing roughness but higher gloss. This may be caused by differing amounts of sedimentation, by the size and shape of surface particles, or by the different chemical nature of the polymers (Ward and Nussinovitch 1997).

Surface color can be compensated for. For example, for a series of papers covering a wide range of gloss and color, the gloss value (GV) related well with subjective rankings of gloss. The empirical relationship used for GV is

$$GV = S(0.2 + 10/L)$$

where S is the percent reflection at 45°, relative to the reflection from magnesium oxide (MgO) at 0°, and L is the percent reflection at 0°, relative to the MgO reflection at 0°, for an incidence angle of 45°. The L term is included to compensate for differences in color depth (Harrison and Poulter 1951).

Commercial glossmeters are designed to measure extended flat surfaces. They cannot readily be used for food materials because of their uneven, often curved, surfaces. To overcome this, a curved-surface glossmeter has been designed (Figure 8–23). The object to be measured sits on a rotatable platform on the base of this abridged goniophotometer. Two arches attached to the base are lined with photodiodes, and a half arch holds a laser that can be beamed directly at the sample. Examples of the recorded output, in terms of reflected light intensity versus distance, are shown in Chapter 9. The width of the curve at 50-percent intensity was taken as a measure of the gloss. Measurements on whole fruits and vegetables using this instrument were compared with measurements made on peeled flattened skins from the same samples using a commercial flat surface glossmeter. A linear correlation was found between them (Nussinovitch et al. 1996).

A goniospectrophotometer can be used to measure wavelength-related gloss. At 90° to the incident light and at high visible wavelengths, a much lower intensity of light is detected compared to that at 15° (Swatland 1994).

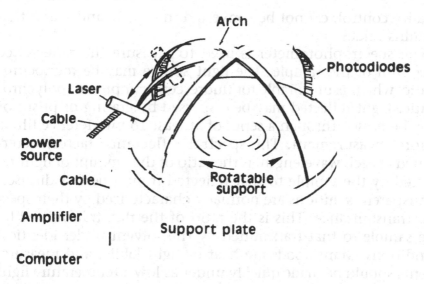

Figure 8–23 The curved-surface glossmeter. *Source:* Reprinted with permission from Nussinovitch, Ward, and Mey-Tal, Gloss of Fruits and Vegetables, Lebensmittel Wissenschaft und Technologie, Vol. 29, pp. 184–186, © 1967, Academic Press, Ltd.

COLOR-MEASURING INSTRUMENTS USED FOR FOODS

Tristimulus colorimeters and spectrophotometers designed for use in other industries can be used for color measurement of foods. In the former instruments, the sample is irradiated with polychromatic light, and that reflected is passed through three (sometimes four) filters and onto a photocell. The combination of the spectral characteristics of the light source, filters, and photocell is such that the three (or four) signals can be processed simply into X, Y, and Z and values. The tristimulus colorimeter is often less accurate but more precise than the spectrophotometer. Hence, it is particularly suitable for measuring the difference between two samples measured sequentially. One instrument (the Agtron), which measures reflectance at wavelengths determined by three spectral lines, is sometimes used for the measurement of large particulate samples, such as breakfast cereals. These measurements, although valuable for quality control, cannot be converted into X, Y, and Z and tristimulus values.

The spectrophotometer is used to measure the reflectance spectrum of the sample. The light source may be monochromatic, which is unsuitable for fluorescent samples, or polychromatic. Light reflected may be dispersed by grating or prism or may be passed through a series of at least 16 interference filters before measurement. The spectral reflectance factor, determined at each wavelength, is the ratio of the amount of light reflected by the sample to that reflected from a perfect diffuser. Transparent solutions are normally characterized by their spectral transmittance. This is the ratio of the flux transmitted by the sample to that transmitted by the solvent under identical conditions. Many foods are heat or light labile, and measurements should be made quickly under as low a temperature light source as possible.

Spectrophotometers and colorimeters can be adapted for food applications: examples are the Hunter Citrus and Hunter Tomato Colorimeters. These instruments have been developed to relate tristimulus measurements and visually defined orange juice or tomato product scales. Some instruments have been developed specifically for food materials. These include a wide-range transmission spectrophotometer for detecting internal quality, a small-area coherent fiber-optics spectrophotometer, and the white wine colorimeter. Time-dependent changes in noodles can be monitored using a computer-controlled automated sample platform attached to a reflectance spectrophotometer (Corke et al. 1997). All use the principles described in this chapter. Derivative spectral analysis might be useful to the industry. Calculations of the first and second derivatives of a spectral absorption curve are sensitive methods of examining changes in slopes of reflectance and absorption curves. The technique is applicable to clear as well as turbid samples (Lewandowska 1987). Difference spectra can be used to magnify minor changes to reflectance spectra (see Figure 9–43).

Instruments depending on subtractive and additive color matching are available. An example of the former is the Tintometer, the principles of which are described in this chapter.

Both visual matching and photoelectric forms are used for color grading a range of food materials, including edible oils and fats, sugars, syrups, beer, and caramel (Belbin 1993). The Munsell color spinning disc is an additive method that has had wide use in the industry, particularly in the specification of awkwardly shaped whole foods such as fruit and vegetables.

Color control is essential to the brewing industry. The European Brewery Commission (EBC) recommended method involves the use of EBC colored discs. Spectrophotometer measurements of clarified worts at 430 nm have become standard (White 1995), and a continuous-flow spectrophotometer can be used for monitoring (Wili 1996). As well as single wavelength monitoring, tristimulus ΔE values can be used to distinguish between beers having small color differences (Smedley 1995).

Several instruments have been developed specifically for use with food materials or have involved modification of existing commercial instruments. A tristimulus colorimeter has been adapted to measure the color of veal. This yields results in terms of the Canada color grade requirements (Canada Agricultural Product Standards Act 1984). This instrument incorporates a hinged white standard that when not used for calibration, can be secured away from the light beam and the sample (Brach et al. 1987). A scanning reflectometer for fruit assessment in the field (Reeves et al. 1997) and instruments for apple color measurement have been designed (Colosie and Breeze 1973; Girard and Lau 1995). The Kent-Jones reflectometer and the Leucometer are used to measure brightness of flour and bakery goods pastes (Zimmerman 1969).

An instrument for measuring the color of meat from freshly slaughtered animals incorporates a lance containing an optical system. This consists of a pair of probes, one carrying a microlamp situated near to a window in its side. This illuminates the meat between the window and the other probe. Light transmitted through the meat is reflected onto a photocell, also carried within the probe. Alternatively, the light can be reflected back along the probe axis for visual assessment (Slagteriernes Forskningsinstitut 1966). The fiber-optic probe is a robust, portable instrument that can be calibrated in terms of the Kubelka-Munk scattering coefficient. It was developed to investigate light-scattering properties of meat. Light transmitted by a fiber-optic bundle illuminates the meat from the side of the probe tip. The intensity of the returned light is a measure of scatter—for example, in beef and bacon (MacDougall and Jones 1980). It can also be used to study the dynamics of fat crystallization in spreads. Fiber optics in the industry has been described by Swatland (1992). Applications are numerous and include fish (Swatland et al. 1997), beef (Gariépy et al. 1994), pork (Irie and Swatland 1993), animal tissue (Tsuruga et al. 1994), and protein gels (Barbut 1996). The use of fiber-optic spectrophotometry on meat and meat products has also been described (Swatland 1985; Osawa 1995).

To overcome the problem of measurement of strongly scattering media such as orange juice, Rummens (1970) designed a hollow diffusing sphere, with the sample housed in a cylindrical cell at its center. The sphere is twice the diameter of the cell and is coated with magnesium oxide. The sample is illuminated by two tungsten lamps 90° apart, and the scattered light is sampled for measurement at a point midway between them. Good agreement can be obtained between the color coordinates of the juice measured in this way and those of a Munsell chip matched with the juice in diffuse daylight.

The color-determining properties of bulk translucent materials can be investigated by examining the behavior of a pencil beam of light incident on the surface. The spectral radiance decreases with radius. The value at the center is mainly due to scattering, while the value at the edge is determined by the scattering and absorption properties. A fiber-optic instrument has been developed to measure center and edge radiances as a function of wavelength (Borsboom and Ten Bosch 1982). Birth et al. (1978) used a laser beam and a detector mounted on a micrometer to map the geometrical distribution of light scatter in pork muscle.

Color is important to the sugar industry. The first-ever quantitative color measurement was made in 1822 to determine the capacity of finely divided bone char to remove color from cane sugar. The Stammer Colorimeter has been a widely used visual method, in which the color of a sugar solution is expressed in terms of standard filters designated in degrees Stammer. Problems with filter variability and sample color range led to its being supplanted by the Lovibond Tintometer. The Pfund Color Grader, an instrument similar in principle to the Stammer, incorporates a wedge-shaped cell used to grade honey and syrup (Francis and Clydesdale 1975).

REFERENCES

Adams EQ. (1942). X-Y planes in the 1931 ICI system of colorimetry. *J Opt Soc Am* 32:168–172.

Aubourg SP, Sotelo CG, Gallardo JM. (1997). Quality assessment of sardines during storage by measurement of fluorescent compounds. *J Food Sci* 62:295–298, 304.

Barbut S. (1996). Use of fibre optics to study the transition from clear to opaque whey protein gels. *Food Res Int* 29:465–469.

Belbin A. (1993). Color in oils. *INFORM* 4(6).

Berns RS, Motta RJ, Gorzinski ME. (1993). CRT colorimetry. *Color Res Appl* 18:299–330.

Birkett RJ. (1985). The appearance of concentrated colloidal dispersions. In *Proceedings of the Fifth Congress of the International Colour Association*, vol. 1. Monte Carlo.

Birth GS, Davis CE, Townsend WE. (1978). The scatter coefficient as a measure of pork quality. *J Animal Sci* 46:639–645.

Borsboom PCF, Ten Bosch JJ. (1982). Fibre-optic scattering monitor for use with bulk opaque material. *Appl Opt* 21:3531–3535.

Brach EJ, Fagan WE, Raymond DP. (1987). Instrument for grading veal carcasses. *J Food Qual* 19:91–99.

Canada Agricultural Product Standards Act, Veal Carcass Grading Regulations. (1984). *Can Gazette* 118:2699–2770.

Choi K, Lee G, Han YJ, Bunn JM. (1995). Tomato maturity valuation using color image analysis. *Trans Am Soc Agric Eng* 38:171–176.

Colosie SS, Breeze JE. (1973). An electronic apple redness meter. *J Food Sci* 38:965–967.

Commission Internationale de l'Eclairage. (1986). *Colorimetry*. 2nd ed. Pub. no. 15.2, Paris: CIE.

Connolly C. (1996). The relationship between colour metrics and the appearance of three-dimensional coloured objects. *Color Res Appl* 21:331–337.

Corke H, Lun AY, Xiaofang C. (1997). An automated system for the continuous measurement of time-dependent changes in noodle color. *Cereal Chem* 74:356–358.

Fairchild MD, Pirrotta E. (1991). Predicting the lightness of chromatic object colors using CIELAB. *Color Res Appl* 16:385–393.

Francis FJ, Clydesdale FM. (1975). *Food colorimetry: theory and applications.* Westport, Conn: Avi.

Gariépy C, Jones SDM, Tong AKW, Rodrigue N. (1994). Assessment of the Colormet fiber optic probe for the evaluation of dark cutting beef. *Food Res Int* 27:1–6.

Girard B, Lau OL. (1995). Effect of maturity and storage on quality and volatile production of "Jonagold" apples. *Food Res Int* 28:465–471.

Gunasekaran S. (1996). Computer vision technology for food quality assurance. *Trends Food Sci Technol* 7:245–256.

Hamey LGC, Yeh JC-H, Ng C. (1997). Objective bake assessment using image analysis and artificial intelligence. In *Cereals '97: proceedings of the 47th Australian Cereal Chemistry Conference*, ed. AW Tarr, AS Ross, CW Wrigley, 180–184. Melbourne: Royal Australian Chemical Institute, Cereal Chemistry Division.

Harrison VGW, Poulter SRC. (1951). Gloss measurement of papers: the effect of luminance factor. *Brit J Appl Phys* 2:92–97.

Hill AR. (1987). How we see colour. In *Colour physics for industry*, ed. R McDonald, 211–282. Bradford, UK: Society of Dyers and Colourists.

Hunt RWG. (1998). *Measuring colour*. 3rd ed. Chichester, UK: Ellis Horwood.

Hunter RS, Harold W. (1987). *The measurement of appearance*. New York: John Wiley.

Hutchings JB, Gordon CJ. (1981). Translucency specification and its application to a model food system. In *Proceedings of the Fourth Congress of the International Colour Association*, vol. 1, ed. M. Richter, C4. West Berlin: Deutscher Verband Farbe.

Hutchings JB, Scott JJ. (1977). Colour and translucency as food attributes. In *Proceedings of Third Congress of the International Colour Association*, ed. FW Billmeyer Jr., G Wyszecki, 467–470. Bristol, UK: Adam Hilger.

Irie M, Swatland HJ. (1993). Prediction of fluid losses from pork using subjective and objective paleness. *Meat Sci* 33:277–292.

Judd DB, Wyszecki G. (1975). *Color in business, science and industry*. 3rd ed. New York: John Wiley.

Kent M, Calvert A, MacDougall D, Malkin F, Witt K. (1993). The use of calibrants in food colour measurement: an international cooperative study. *Color Res Appl* 18:80–88.

Lewandowska C. (1987). Controlling colour quality in the drinks industry. *Lab Equip Digest* 25(10):111–113.

Ling PP, Ruzhitsky VN, Kapanidis AN, Lee T-C. (1996). Correlation between color machine vision and colorimeter for food applications. In *Chemical markers for processed and stored foods*, ed. T-C Lee, H-J Kim, 253–278. Washington, DC: American Chemical Society.

MacDougall DB, Jones SJ. (1980). Translucency and colour defects of dark-cutting meat and their detection. In *The problem of dark-cutting in beef*, ed. DE Hood, PV Tarrant. The Hague: Martinus Nijhoff.

Mackinney G, Little AC, Brinner L. (1966). Visual appearance of foods. *Food Technol* 20:1300–1306.

Maruf FW, Ledward DA, Neale RJ, Poulter RG. (1990). Chemical and nutritional quality of Indonesian dried-salted mackerel (*Rastrelliger kanagurta*). *Int J Food Sci Technol* 25:66–77.

McLaren K. (1980). Food colorimetry. In *Developments in food colours-1*, ed. J Walford, 27–45. London: Applied Science.

Melgosa M, Hita M, Poza AJ, Alman DH, Berns RS. (1997). Suprathreshold color-difference ellipsoids for surface colors. *Color Res Appl* 22:148–155.

Newhall SM, Nickerson D, Judd DB. (1943). Final report of the OSA subcommittee on the spacing of the Munsell colors. *J Opt Soc* 33:385–420.

Nickerson D, Stultz KF. (1944). The specification of color differences. *J Opt Soc Am* 34:350–354.

Nussinovitch A, Ward G, Mey-Tal E. (1996). Gloss of fruit and vegetables. *Lebensm Wiss Technol* 29:184–186.

Nyquist G. (1979). Relationships between Secchi disk transparency, irradiance attenuation and beam transmittance in a fjord system. *Marine Sci Commun* 5:333–359.

Osawa M. (1995). The measurement of meat pigments by fibre-optic reflectance spectrophotometry using the Kubelka-Munk equation. *Meat Sci* 40:63–77.

Priest IG, Gibson KS, McNicholas HJ. (1920). *An examination of the Munsell color system*. Washington, DC: National Bureau of Standards.

Raggi A, Barbiroli G. (1994). Problems in the use of non-contact colour measuring instruments. *Farbe* 40:217–226.

Reeves SG, Nock JF, Ludford PM, Hillman LL, Wickham L, Durst RA. (1997). A new inexpensive scanning reflectometer and its potential use for fruit color measurement. *Hort Technol* 7:177–182.

Rummens FHA. (1970). Color measurement of strongly scattering media, with particular reference to orange-juice beverages. *J Agric Food Chem* 18:371–376.

Sève R. (1996). Practical formula for the computation of CIE 1976 hue difference. *Color Res Appl* 21:314.

Slagteriernes Forskningsinstitut. (1966). Improvements in and relating to methods of determining the quality of meat and devices for carrying out said methods. London patent specification 1,193,844, 3 June 1970, based on Denmark patent 4301.

Smedley SM. (1995). Discrimination between beers with small colour differences using the CIELAB colour space. *J Inst Brewing* 101:195–201.

Stevens SS. (1975). *Psychophysics*. New York: John Wiley.

Swatland HJ. (1985). Color measurements of variegated meat products by spectrophotometry with coherent fiber optics. *J Food Sci* 50:30–33.

Swatland HJ. (1992). Fiber-optics in the food industry. *Food Res Int* 25:227–235.

Swatland HJ. (1994). Physical measurements of meat quality: optical measurements pros and cons. *Meat Sci* 36:251–259.

Swatland HJ. (1997). Relationship between the back-scatter of polarised light and the fibre-optic detection of connective tissue fluorescence in beef. *J Sci Food Agric* 75:45–49.

Swatland HJ, Haworth CR, Darkin F, Moccia RD. (1997). Fibre-optic spectrophotometry of raw, smoked and baked Arctic char (*Salvelinus alpinus*). *Food Res Int* 30:141–146.

Tighe BJ. (1978). Subjective and objective assessment of surfaces. In *Polymer surfaces*, ed. DT Clark, W Feast, 269–286. New York: John Wiley.

Tsuruga T, Ito T, Kanda M, Niwa S-I, Kitazaki T, Okugawa T, Hatao S. (1994). Analysis of meat pigments with tissue spectrophotometer TS-200. *Meat Sci* 36:423–434.

Ward G, Nussinovitch A. (1997). Characterising the gloss properties of hydrocolloid films. *Food Hydrocolloids* 11:357–365.

White FH. (1995). Spectrophotometric determination of malt colour. *J Inst Brewing* 101:431–433.

Wili B. (1996). Measurement of bitterness and colour of beers in the laboratory by automation. *Brewer* 82:340–342.

Wright WD. (1969). *The measurement of colour*. 4th ed. London: Hilger & Watts.

Zimmerman K. (1969). The brightness determination of flour and bakery goods with the Leucometer. *Jena Rev* 1:183–187.

CHAPTER 9

Color Specification of Foods

There are several reasons for specifying color and appearance. These include the need to develop realistic quality control methods, to help in unraveling the chemistry and physics of appearance changes taking place during processing and marketing, to quantify and back up sensory assessment, and to understand and possibly predict consumer response. The emphasis of this chapter is on the principles of color measurement as applied to different physical forms of food. Examples are given to illustrate the problems and the solutions sought by various workers.

The case for an instrumental specification of color is clear. The visual observation of samples may be inadequate because of eye fatigue, the unavailability of uniform lighting and suitable environment for the assessment, the poor color memory of observers, and the unavailability of sufficiently trained graders. Also, detailed analytical visual assessments can be time consuming and labor intensive.

Using light reflectance or transmission measurements, we may hope to make rapid physical measurements relating to visually perceived attributes and possibly chemical properties. Each food product, however, is individual in its physical properties and in the way it interacts with light to form an object for visual assessment or instrumental measurement. For materials from other industries, much understanding of material properties and behavior can be gained from instrumental reflectance or transmission measurement. This may involve the use of color-measuring instruments manufactured specifically for that industry. Color measurements can be made of constant thicknesses of paint films on standard backing materials. Textile samples can be piled up and compressed, and measurements can be made on a sample thickness equivalent to an infinite optical depth. Also, chemical determinations can be made on clear solutions that have been concentration adjusted to bring their transmissions within the optimum sensitivity range of the spectrophotometer.

Care is necessary when attempting to transfer these methodologies to the specification of food materials. Successful measurement depends greatly on sample preparation and presentation. Once this is realized and adequate techniques specific to the material under investigation have been developed and mastered, it is possible to make worthwhile measurements. Problems include the fact that foods exist in many physical forms and that foods undergoing processing often change physical form with time. During cooking, fish may change from a largely translucent form into one that is virtually opaque. Wine color changes with depth. The hue of colored solutions can significantly change with dilution because the pH may be altered or because there is a change in the relative solubility of the pigments present. Each food has its own individual material properties and should be considered individually.

Reliable instruments are available, but the sample properties are critical. Incorrect sample treatment before or during measurement or incorrect treatment of the results obtained can easily render the exercise of color measurement meaningless. Therefore, this chapter starts with the samples and their presentation to the measuring instrument. The different objectives of color measurement are then considered. These are to achieve understanding of the sensory quality, the product and its behavior during processing, and the chemistry.

SAMPLE AND INSTRUMENT

Normally, when we look at an object, interactions occur between object and perception. Perceptions alter when angles of viewing and illumination are changed, when the illumination itself is changed, and when the sample is rotated. The color changes, there is more or less gloss, or a surface texture pattern may emerge. When an instrument is designed, the quality and direction of illumination are specified, the viewing geometry is made constant, the specular component is consciously included or excluded, and the total geometry is established. At first glance, we now have a thoroughly standard instrument into which a material can be sited to yield a reliable measurement immediately relatable to a visual impression. Unfortunately, this is not so. To fix geometries, sample sizes are standardized through the introduction of apertures. If the samples are flat, perfectly opaque, and of a single uniform color, there should be little trouble. However, very many foods are translucent. That is, there is a complex continuous relationship between sample thickness, diameter, light intensity and direction, and light absorption and scattering properties (i.e., the color, reflection, and transmission properties). These interactions between sample and instrument make it essential that they be considered together.

Sample preparation is critical. Every food possesses its own individual potential handling problems. Some samples may need time for preparation. Manipulations such as slicing, grinding, pureeing, powder sizing, compressing, or extracting may be deemed necessary before measurement is commenced. Inconsistency in procedure affects reliability. Some preparation methods will not be appropriate for the many foods, such as green peas and cooked meats, that are colored differently outside and in the center. Many foods are uneven in color. The areas of each color can be estimated, and it may be possible to measure them separately. Air bubbles trapped inside samples increase light scattering, so for many materials they should be eliminated by careful manipulation or vacuum. For products such as ice cream, however, air bubbles contribute massively to appearance, and care is needed to prevent change in size. Sampling devices may have to be developed—for example, for ice cream (Dolan et al. 1985).

Measurement of difficult samples may call for ingenuity in the design of sample holders and may require even greater than normal discipline in the preparation and commission of the measurement. Examples include temperature- and time-dependent foods, such as hot brewed tea, ice cream, and beef. For quality control, and for samples that change color with time, speed and standardization of measurement are critical, and as little sample manipulation as possible is required.

There are four sources of error in colorimetry: operators uninformed of correct procedures and theoretical measurement background rules, the use and care of reference white tiles, the nonuse of chromatic standards in checking instrument performance, and measurement conditions. The last includes care and use of working standards, deterioration of integrating sphere condition, and, particularly relevant to food materials, the lack of realization of the dependence of reflectance on the ratio of illuminated area to sample port area. Reliable, reproducible results can be obtained only through scrupulous attention to the minimization of these errors (Billmeyer and Hemmendinger 1981).

Even when standard sample preparation and measurement methods have been specified, great care is required in their commission. Comparisons have been made of measurements carried out in different laboratories on different instruments and by different operators. Standard deviations of the tristimulus measurements made in a number of different laboratories were found to be 9 to 16 times greater than those obtained in a single laboratory. This may have been due to poorly informed operators, different reference standards, integrating sphere conditions, and specular inclusion/exclusion errors (Kent and Smith 1987). Measurements made on potato powder in six laboratories that were working closely to a defined measurement technique had standard deviations 6 to 8 times greater than those occurring in a single laboratory. Sample properties, such as liquid concentration as well as particle size, can also be significant contributors to differences between instruments and laboratories (Kent 1987). The relationships obtained between instruments change according to the food under examination. Large disagreements have been obtained for lower lightness jams, a notoriously difficult product to measure (Baardseth et al. 1988).

There can be many reasons for discrepancies occurring among instruments, including differences in optics (Rodgers et al.

1994). Reasons for differences among instruments of identical design include lack of attention to servicing and careless calibration or sample preparation. If good interlaboratory agreement is to be attained, it is necessary to reach agreement regarding the type of instrument used and its operation, cleaning, physical treatment, and maintenance.

Brimelow (1987) has described the work necessary before an industrywide international standard can be established for a food product. Approximately 60 percent of the world's tomato paste is produced in Europe. Its color depends on tomato variety, season and time of season, climate, storage, ripening conditions, soil condition and elements present, and processing conditions. Puree color is carried through into many products, and an instrumental color specification would be beneficial to the economics, marketing, and organization of the industry.

The following instrumental reflectance system was designed (Brimelow 1987; Hils 1987; Kent et al. 1991).

- Sample preparation procedures were based on a 12-percent tomato soluble solids (TSS) content, measured at 20°C on the sugar scale. Fewer problems associated with sample translucency were found with this concentration of puree than with the lower 8.5-percent concentration normally used in the United States.
- Aperture sizes for sample and illumination and the relationship between them were recommended. Aperture settings can be between 30 and 50 mm and illumination settings between 6 and 50 mm, provided that the illumination setting is less than or equal to the aperture setting. The instrumental geometries recommended are 45°/0° or 0°/45°, together with illuminant C, 2° observer viewing protocol.
- The use of hitching post versus white standards was compared. A hitching post reference standard, in the form of a centrally produced tomato paste red, glazed ceramic tile calibrated by the National Physical Laboratory, was recommended. This tile should be maintained at the temperature at which it was calibrated (i.e., 25°C). The calibration data are in terms of Hunter (*L*, *a*, *b*) coordinates. Instructions for care, cleaning, and use of the tile are issued with it (Kent et

al. 1991). The tiles are thermochromic and must not be allowed to warm up during the calibration procedure. No glass plate is included between the standard and the instrument during calibration, which should be repeated at least every two hours.

- The sample is contained in an optically clear glass cell with a diameter of 62.5 mm, a height of 62.5 mm, and a base thickness of 2 mm, capable of containing a test sample with a minimum depth of 20 mm. Measurements should take place within 10 minutes of sample preparation and, because of sample heating effects, within 30 seconds of placing the sample under the instrument lamp.
- A light shield should be used to cover the sample/instrument interface during measurement. This is to counteract the effects of variable ambient lighting conditions. The shield is in the form of a container with an internal diameter of 100 mm and a height of 140 mm, painted matte black internally.

Such methods can be used in conjunction with sensory standards, such as the Campden tomato paste and puree specification (Rodway 1993).

In summary, whatever experimental methods are used, instrumental selection, sample preparation and its presentation to the instrument, and the instrumental conditions must be fully under control and shown to be reproducible. That is, the total methodology must be controlled and understood. In addition, detailed procedures for sample preparation and measurement should be given when reporting color analyses.

Food materials are of three broad types: those that are effectively opaque, those that are transparent and scatter light minimally, and translucent materials, which fall between these extremes. Different instrumental geometries should be selected according to material properties. Table 9–1 lists the alternatives.

Opaque Samples

The ideal sample for measuring reflectance characteristics is a flat, homogeneously pigmented, opaque, light-diffusing material. For truly opaque materials, the sample viewing area can be the same as the beam area. Very few food materials are fully

Table 9–1 A Guide to Food Material Properties and the Choices of Instrumental Geometry for Color Measurement (adapted from Hunter and Harold 1987, with permission, copyright John Wiley and Sons, Inc.)

The Sample	Information Desired	Sphere (d/8°) Specular Included[a]	0°/45° (45°/0°)
Opaque, equal texture & gloss	Color difference	Suitable	Suitable
Opaque, texture & gloss differences	Color difference only	Preferred geometry	Readings affected by gloss/texture
Opaque, texture & gloss differences	Appearance differences, including color and gloss variances	Less effective	Preferred— better agreement with visual judgments
Transparent & translucent	Transmittance	Suitable via transmission or trans-reflectance techniques	Suitable via trans-reflectance techniques
Translucent	Color by reflection	Translucency errors possible by light trapping	Preferred, if illumination/sample/mea-surement areas suitable

[a] Spheres can exclude only the specular reflected component from mirror like surfaces; hence, only the specular included version of sphere geometry has been included in the table.

opaque, but compressed powders; powders themselves; compressed wafers of, for example, lyophilized egg whites and yolks; instant coffee; and sweet potato flakes may be approximations (Berardi et al. 1966). Nevertheless with powders, measurements can be significantly affected by particle size and distribution. For example, increasing the particle size of potato powder from less than 250 to 1,000 μm results in an increase in L value from 71 to

88.6 (Kent 1987). Hence, ground roasted coffee must be thoroughly standardized to attain uniform particle size distribution and controlled conditions of pellet formation (Little et al. 1958). Nonhomogeneous and large samples can be dealt with by using spinning techniques in which the food is rotated in front of the measuring aperture. This technique produces an average color of the surface. This is often unhelpful, since the average color of a two-color object, such as a red and green ripening tomato, is meaningless. It relates to no actual feature of the material. For vegetables like turnips (Perkins-Veazie 1991) and for fruit such as peaches, a more meaningful approach involves separate measurements on the differently colored areas. Spinning (Polesello et al. 1974) and large-area techniques are satisfactory for producing an average color for a nominally unicolor material, such as some green vegetables, or nominally flat samples, such as some breakfast cereals. Errors may occur when measurements of different-sized materials are compared. Sometimes, large particle–sized materials may be gently squashed or cut into smaller pieces to present a more uniform surface. However, differently colored material that may be inside the sample should be not revealed.

Most nominally opaque food materials are translucent. Such samples are prepared so that they have an infinite optical thickness: that is, where any further thickness increase results in no change to the reflectance measurement. Slices of raw meat, for example, can be stacked until this is achieved. Where this cannot be done, an attempt should be made to determine the effect of the sample translucency. Problems caused by changes in translucency during processing can then be minimized.

Large-area measurements of large samples such as apples and tomatoes and soft materials such as raw meat present geometrical problems through *pillowing*. When a sample projects through the colorimeter aperture, the reflected beam expands. The instrument geometry is then successfully destroyed through displacement of the beam/sample/detector spatial relationships. The instrument is no longer calibrated properly because the measurement is not being made in the plane of calibration. Significant effects of sample curvature on colorimeter reading can be avoided if the aperture size is reduced so that the ratio of the area of curved sample surface to that of the flat aperture surface is less than 1.023. If this figure cannot be achieved, lightness and chroma values can be affected, although hue angles may not (Hung 1990). Thin, optically clear glass can be placed between sample and instrument to prevent contamination of the optics and to present a flat sample to the aperture. Plastic overwrap may influence readings. If they are used, for example, to limit gas and vapor transfer to and from the surface of meat, the same wrap should be used for all samples. Again, care is needed because pillowing can occur.

Textured samples, such as meats, may present a problem when shadows are created as the light beam illuminates the sample at an angle to the fiber direction. Two measurements made with the samples at 0° and 90° to the direction of a 45° incident beam will reveal this phenomenon. Shadows may also occur during measurements of discrete samples such as berries.

Measurements of nonhomogeneous surfaces often exhibit poor reproducibility. This occurs when replicate measurements are made on areas having different light-scattering properties. During measurements on meat, this can arise from variations of the lean-to-fat ratio in the path of the incident light beam. The variability can be reduced by using a multiplicative scatter correction (MSC) on samples assumed to be identical. The method is based on the assumption that physical light scattering has wavelength dependencies different from that of absorbed light. For a set of samples considered to be members of the same population, the scatter of each sample is estimated relative to the scatter of a representative or ideal sample. The spectra are then corrected to give the same amount of scatter as the ideal sample, which may be the average of the population. The correction greatly reduces variability in the reflectance spectra and derived color coordinates (Iversen and Palm 1985). This treatment was successful in measurements on homogenized fat and meat mixtures. A correlation was indicated between the relative scatter coefficients obtained from the MSC and the fat content (Geladi et al. 1985).

Transparent Samples

Techniques for handling the transmission properties of transparent materials are well established. Francis (1972) has cited examples from the literature, including the work on nonturbid tea solutions by Staples (1972) and on model wine systems by Robinson et al. (1966). Many nominally transparent materials change in volume color with depth. This is *dichroism*, which occurs when materials change in hue with change in some element of examination, such as lighting or sample dilution or depth. This complicates the description of volume colors.

For example, color changes in cyanidin solutions are concentration dependent. Figure 9–1A shows the unequal effect of change of pigment concentration on the tristimulus values X, Y, and Z. At high concentrations, Z is almost zero, and the solution hue is red. At lower concentrations, Z increases at a faster rate than X and Y, and the solution changes to a purple hue. The reason for this can be seen in Figure 9–1B, which shows the transmission spectra for these solutions. As concentration is increased, the relative contribution of shorter wavelengths decreases (Eagerman et al. 1973a).

Presence of slight turbidity can profoundly affect transmission color measurement. Figure 9–2 shows the effect of slight turbidity on the b^* values of tea. The extent and direction of the effect is sensitive to clear tea solids' content and cell length. Turbidity has a profound effect at all three path lengths (Joubert 1995). Hence, transmission measurement problems are likely unless scattering within the product can be eliminated, as, for example, when water is filtered through a membrane filter of pore size 0.45 μm before color measurement (Hongve and Åkesson 1996). Very few food materials are truly transparent, and even high-quality wines possess subliminal scattering. This can totally negate the results obtained from conventional transmission measurements (Little 1977a). Reliable determinations of color can be made using techniques appropriate for translucent materials.

Translucent Samples

Most foods are translucent to some degree. However, except for special applications, some can be treated as opaque if the sample can be presented at an effectively infinite optical thickness. This occurs when the depth is such that any increase produces no change in the measurements observed. For example, 1.5 cm represents an infinite optical thickness of chicken breast. At a

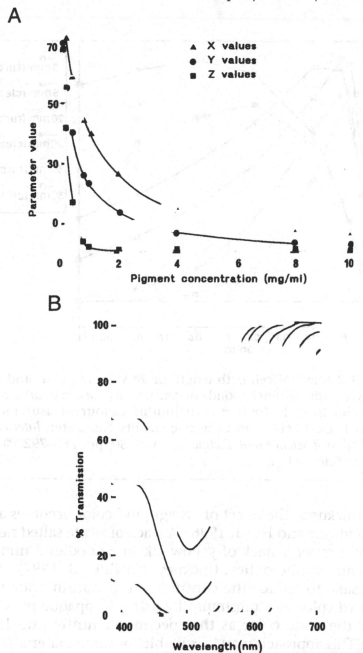

Figure 9–1 (A) The behavior of *X*, *Y*, and *Z* values with concentration of cyanidin-3-glucoside. (B) Spectral curves of cyanidin-3-glucoside solutions of concentration 0.10, 0.17, 0.69, 1.73, 3.45, 6.9, and 17.3 mg of pigment per milliliter (Eagerman et al. 1973a, with permission).

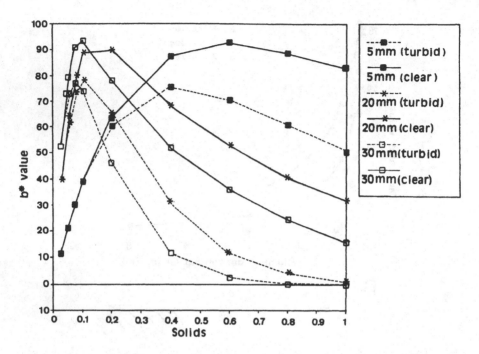

Figure 9–2 Effect of cell path length on *b** values of clear and turbid rooibos teas with different solid concentrations. *Source*: Reprinted with permission from E. Joubert, Tristimulus Colour Measurement of Rooibos Tea Extracts as an Objective Quality Parameter, *International Journal of Food Science and Technology*, Vol. 30, pp. 783–792, © 1995, Blackwell Science Ltd.

lower thickness, the effect of background color becomes apparent (Sandusky and Heath 1996). A stack of white salted noodles 10 mm deep or a stack of yellow alkaline noodles 7 mm deep represents infinite optical thickness (Solah et al. 1997). It may be possible to reduce the contribution of translucency to the measured color to a minimum by using an opaque backing of nearly the same color as the specimen (Hunter and Harold 1987). This approach may be suitable for some materials, but it should be used with caution, particularly for food materials that change in translucency during the process being studied. An infinite optical depth should be maintained for near-opaque materials. The infinite optical thickness of a material depends on light intensity, so it may change with instrument.

A common occurrence with translucent materials is that the eye and the instrument do not *see* the same thing when each views a sample. Neither do they view the same sample. The color of many foods, including meat, changes with depth. The light intensities available during subjective assessments and objective measurements are likely to be different. This leads to different proportions of light being reflected from different depths in the sample. Hence, reflected light will contain a greater or lesser component from the underlying pigment. So the stronger light of the colorimeter can be used to detect bruises beneath the skin of apples. Light penetration depth is characteristically different for different apple cultivars (Hother et al. 1995).

The relationships between illumination, sample, and measuring-area diameters are very important to the successful color measurement of translucent materials (Hunter, cited in Little 1964). The problems are shown schematically in Figure 9–3. Figure 9–3A indicates that when the cell, illumination, sample, and measuring areas are similar, a dark area can be seen around the measured area. This arises from light being reflected into the detector from the cell side. Increasing the cell diameter decreases the dark ring, but light is trapped behind the aperture and lost to the detector (Figure 9–3B). Light trapping is decreased when the aperture area is increased with respect to the illumination area (Figure 9–3C). Light trapping is reduced to a minimum as the exposed area is further increased (Figure 9–3D). The best results for color specification of translucent materials can be obtained from an instrument having a small incident beam area relative to the areas of the sample port and photoreceptor. In addition, a sphere collection system should be used, or the receptor should be close to the sample (Eagerman 1978).

When light is trapped, selective absorption of the light occurs. For example, a greater proportion of the red light is reflected with red tomato juice, and if trapping occurs, not only will a reduced intensity be recorded, but the hue measurement will not indicate the correct degree of redness (Francis and Clydesdale 1975).

Complex interactions occur within translucent samples involving internal light transmittance, absorbance, internal scat-

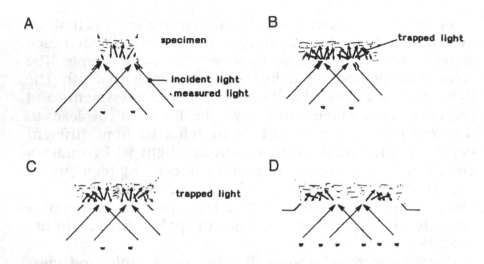

Figure 9–3 Errors possible when measuring the color of translucent samples of infinite thickness (based on Little 1964, with permission). (A) Cell area, exposed area, and illuminated area approximately equal; error through reflection from cell walls. (B) Exposed area and illuminated area approximately equal, smaller than cell area; error from light trapping. (C) Exposed area increased relative to illuminated area; error from light trapping reduced. (D) Exposed area greatly increased relative to illuminated area; error through light trapping eliminated.

tering, and light lost laterally into the sample. These lead to poor agreement among instruments. Correlation coefficients between different instruments were between 0.71 and 0.81 when chicken breasts were compared, whereas they were 0.98 to 0.99 when colored paper swatches were compared (Millar et al. 1995).

Two materials may yield similar deep-layer reflectance values but have very different pigment concentrations and colors. The sample with the lower pigment concentration will have a higher internal transmittance; this will lead to greater loss of light through trapping. The higher pigment sample will be more opaque and more absorbing. Hence, although the eye will be able to tell the difference, the instrument may not. Direct measurement of deep-layer samples of applesauce and red coloring is not as successful in ranking samples as the visual panel. The solution to these problems is transreflectance measurement. Light is transmitted through the sample, then reflected by a white tile back through the sample and into the measuring instrument. The collection of scattered light is maximized by having a relatively large aperture and placing the sample close to the photocell. This enables the Kubelka-Munk colorant-layer approach to be used. Measurements of thin layers of applesauce and red coloring backed by a standard tile lead to a reduction in internal transmittance and an increase in the separation of samples in color space. These relate more closely to visual judgments (Little 1964).

Using careful experimental procedures, Elliott (1967) defined the optical system and sample preparation technique necessary for optimal measurement of the reflectance spectrum of raw pork. He found that, of the three geometries used, the best spectral resolution was achieved using integrating spheres. In agreement with Little, Elliott found that thin layers of sample with a white rear reflector provided increased sensitivity and consequently increased resolution of muscle pigments. Thin glass cover slips free from chromatic impurities were recommended. The presence of a drop of isotonic saline placed between the glass and the sample was used to reduce refraction due to trapped air. Measurement sensitivity was further increased by decreasing the reflectance difference between the sample and the standard reference white using a neutral wedge or hitching post standard. Presence of fat in the surface affects the measurements. Elliott's results suggested that cross-section samples should be used for measurement. This minimizes the visibility of adipose tissue deposited along the fiber in a thin sheath.

UNDERSTANDING SENSORY QUALITY

Color specification of foods consists of a series of distinct clearly defined stages. The approach is relevant to the measurement of all appearance properties. The stages are preparation of the sample, the synthetic process, the analytical process, and use of the analytical process to predict the synthetic process (Little 1973).

The sample, as has been discussed, is a material having definite physical and optical properties. Each sample must be prepared and treated appropriately if the measurement is to be successful and the results are to be unambiguous. The synthetic process is the judgment of a color property arising from a visual perception of the sample. The process may involve the ranking of samples for some specific visually perceived attribute, such as browning of white wine (Little 1971) or applesauce (MacKinney et al. 1966) or lightness of canned tuna (Little 1969a). The synthetic process does not include judgments of preference, liking, or acceptability.

The analytical process involves physical measurement of the spectral reflectance and/or transmittance characteristics of the sample. Techniques include reflectance for opaque materials and thin-layer transreflectance for transparent and translucent samples. The behavior of the sample under closely controlled chemical stress can be examined at this stage. An example is the examination of the behavior of tuna under oxidizing and reducing conditions (see later in this chapter) (Little 1969a, 1969b). Use of the analytical process to predict the synthetic process is the linking of sensory evaluation and physical measurement to produce some understanding of the system under examination. Chemical information, such as pigment content, and physical information, such as particle size, can be included to produce an overall picture.

When attempting to use physical measures as predictors of appearance, it is always necessary to confirm that the one does in fact relate to the other. Measurements are not a substitute for looking at, and comparing side by side, samples representative of the experiment. The eyes should be used not just to observe that there are differences but also to assess those differences critically. Only in this way can the task to be performed by the instrument be realistically defined and an estimation be made of the efficiency of the instrumental method in fulfilling that task. This was emphasized by Little (1976) when comparing the performance of different methods for the calculation of instrumental results. For example, she found that (L, a, b) coordinates could be used to predict visual appearance. But she also found that it was necessary to use (X, Y, Z) coordinates to relate instrumental measurements to specific visual observations. The findings applied to the two experimental systems studied: (1) a pea plus carrot model system and (2) a red wine system in which browning and pigment contents were changing with storage.

Using rainbow trout, visual assessments were compared with the CIE correlates Y, λ_d, and p_e, and with the derived Hunter (*Lab*), CIELUV, and CIELAB coordinate systems. Visually perceived differences in hue and saturation were correctly indicated only by Y, λ_d, and p_e. These inconsistences were not caused by failures in the visual system or by properties of the samples but were related to the colorimetric system. An increase in pigment level causes an increase in blue absorption and an increase in the value of Z (blue reflectance), with a concomitant increase in p_e. Hence, there is a good correlation between the visual assessment of redness and the value of Z (and p_e). It was concluded that the best relationships with visual assessments are to be found with the colorimetric system most directly related to the spectral characteristics of the sample (Little et al. 1979).

Also, instrumental results contain information on structural elements within the food. Judges may ignore this factor when making visual assessments. Again, the instrument and the eye are responding differently to the same sample. For example, color measurements of white and yellow corn were found to be so affected by variation of pericarp thickness and glossiness that trained panelist assessment is used in corn breeding programs (Floyd et al. 1995). Also, structural elements may or may not remain the same—for example, when a sample is measured several times during a storage experiment. Patterning or structure changes with cut of meat and variety of apple. These factors can greatly affect the relationship between visual judgments and objective measurements. They are usually sufficient to prevent reliable relationships from being obtained across muscle type or species, and possibly even during storage of a single muscle type. However, the advantages of objective reflectance measurement are at their greatest during time trials when, through its low powers of retention, the human visual memory system is at its weakest.

The convenient axial alignments of the Hunter (*Lab*) system to red/green and blue/yellow dimensions have led to the tendency to describe an increase in *a* as an increase in *redness* or a decrease in *b* as an increase in *blueness*. This emphasizes an independence of these dimensions that does not exist. The point (*L, a, b*) is a coordinate in three-dimensional space related to perceived lightness, hue, and saturation. It is convenient sometimes to use one dimension to follow a color change, but its relevance within the three-dimensional color space must always be borne in mind (Trant et al. 1981).

The visual evaluation of the sample and the results from measurements should be critically examined. If used to their fullest extent, reflectance characteristics provide basic information about composition, the state of the product, and its visual appearance.

Visualizing Tristimulus Coordinates

The word of the commercial taster has far-reaching consequences for trade and consumer. Expert wine tasters are renowned for the delicacy of the subjective judgments that they apply to all sensory quality attributes (Bakker and Arnold 1993). Color is an important sensory property providing considerable immediate information about wine type, condition, age, and quality. Scientifically, much is known about the color of wine, and effort has been put into defining standard, sensitive procedures for its quality control (Little 1980).

The colorimetric basis of the expert wine taster's judgments has received a little attention. Using the technique pioneered for measurement of wine color (Little 1964; Joslyn and Little 1967), transreflectance measurements were made on four types of red wine. These were compared with the judgments of two expert tasters from Grants of St. James's, a large British wine shipper (Hutchings and Suter 1983). The tasters were asked to use their normal methodology and vocabulary. The written judgments included only appearance properties of wine in the glass.

The tasters' comments on a claret, a ruby port, a chianti, and a red table wine could be divided into four groups, as indicated in

Table 9–2. The first group of comments described the *depth* of color; the second described the *edge* or *rim*, the color that becomes apparent when the wine is swirled in the glass; and the third, using the adjectives *clear* and *bright*, appeared to be related to clarity and freedom from haze. The fourth group described qualities particular to the type of wine. For example, the port was described as "not ruby enough." This group of properties was not considered further.

Transreflectance spectrophotometric measurements were made on each wine at cell depths of 0.25, 1, 2, 5, and 10 mm. Tristimulus properties were calculated from the transreflectance spectra for the 2° Observer and Illuminant C. Using these measurements alone, it is difficult to visualize either the subtleties of color change with depth or the color differences among the wines. However, Figure 9–4 shows that this situation can be remedied by superimposing Munsell constant Hue lines (Judd and Wyszecki 1975) onto the chromaticity locus of the wines.

Table 9–2 Tasters' Descriptions of the Appearance of Four Red Wines (Hutchings and Suter 1983)

Appearance Attribute Group		Claret	Port	Chianti	Red Table Wine
1	Depth	Good	Medium	Medium	Fairly light
2a	Edge, rim	Slightly purple	Tawny	Slight browning	Youngish, slightly blue
2b	Browning		y	Slight	
3	Clear	y		y	y
	Bright	y	y	y	y
4			Light style because not ruby enough	Attractive	Fairly lively

Note: y denotes the presence of that attribute; all words included in the table are those used by the tasters.

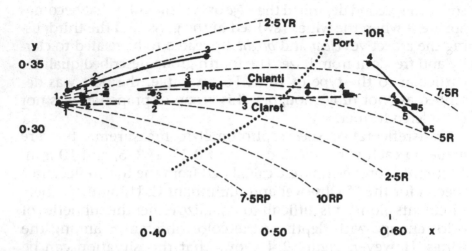

Figure 9–4 Change of chromaticity with cell depth for four wines. Munsell constant Hue and Chroma lines are included (5°/d, Illuminant C, 2° Observer). Cell 1 (depth 0.25 mm), 2 (1 mm), 3 (2 mm), 4 (5 mm), 5 (10 mm) (Hutchings and Suter 1983).

Two lines of constant Chroma are also shown. The change of color with cell depth is indicated by the number of constant Hue lines crossed as cell depth is increased. The claret and the red table wine change more than the port. It would be expected that thin layers of the claret and the red would look bluer or more purple than the chianti and port because the depth-change chromaticity loci cross into the red purple (RP) Munsell zone. It would be expected that the port would be browner because the locus enters the yellow red (YR) zone. The chianti in the thin layer is noted as having slight browning on the rim, and, in agreement, the hue of the thin layer is not as purple as that of the red or the claret. This agrees with the observed order of purpleness of the thin layer when the wines are swirled in the glass—claret (most), red, chianti, port.

It can also be seen from Figure 9–4 that at a cell depth of 10 mm, the wines were approximately the same color. When the wines were compared stationary in the glass, the port looked different from the other three, which were very similar. It is more likely that this judgment may have taken place at an equivalent transreflectance depth of 5 mm. Here the Hue of the port was different from that of the other three wines. Studies on translucent materials, such as orange juice, have concluded that the best agreement with the visual assessment of samples contained in a glass is obtained with measurements on depths of "a few mm" (Francis and Clydesdale 1975).

The last wine tasters' assessment that may possibly be interpreted using Figure 9–4 is *depth*. The rank order was red (lightest) first, then both chianti and port, and finally claret. This is the same as the rank orders of Chroma for cell depths of 2 mm and 5 mm. Again, it seems that measurements made on depths "of a few mm" are valuable. It is also possible that *depth* may be a judgment of transparency. In this case, we can consider relating it to perceived opacity. This is a combination of light scattering (perhaps related to *clear* and *bright*) and light absorption. The perceived opacity of a colored material is related to its maximum tristimulus value (X, Y, or Z), calculated from the Kubelka-Munk derived internal transmittance spectrum (see Chapter 7) (Hutchings and Gordon 1981). The values of X calculated from the internal transmittance spectra of the wines at a cell depth of 5 mm are claret, 28.3 percent; port, 36.5 percent; chianti, 37.7 percent; red, 40.0 percent. That is, the *depth* of these nominally transparent fluids may be a judgment more of visible light transmission than of chromaticity. One of these two approaches may be valid.

As is evident, it is not possible to draw categoric conclusions from this work, as only four wines were used. However, the techniques are available for an understanding to be gained of relationships between color measurement and expert taster judgments (Hutchings and Suter 1983).

Three-Dimensional Nature of Color Change

Color itself is three dimensional, and this is also true of food color change. For example, fresh veal is seen to be poorer if it is too dark (i.e., low luminance), too deep a red (high dominant wavelength) or too saturated a red (high purity). Figure 9–5 is a set of color discrimination diagrams applying to veal based on (uvW^*) space. The numbers denote the expert panel visual scores bounded by the contours indicated (Hutchings 1969). This shows the measurements set out in a logical manner that can be easily interpreted. However, a multidimensional approach can be taken, and principal components (Z_1, Z_2) were calculated to produce a "predicted" visual score (S). For example, using (xyY) space for the veal:

$$S = 24.62 + 0.405Z_1 - 0.309Z_2$$

where

$$Z_1 = -84.1x + 167.5y + 0.1483Y$$
$$Z_2 = 124.1x + 181.0y + 0.0238Y$$

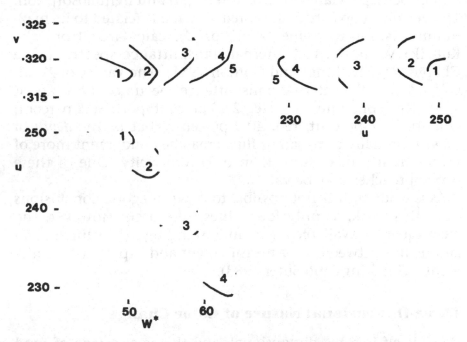

Figure 9–5 The color of fresh veal; discrimination diagrams based on (u, v, W^*) dimensions (5°/d, Illuminant C, 2° observer). Numbers indicate visual score (adapted from Hutchings 1969).

These relationships gave 73 percent of the scores to within half a visual score unit and 98 percent to within 1 unit. In any trial, the visual scores could be in error by at least 0.5 units.

Taking a multidimensional approach leads to the loss of the identity of individual color dimensions. The latter have been firmly established and are easily imagined and interpretable. Changes in the visual/measured relationships occurring, say, from season to season can more easily be realized if the relationship is in the familiar form of the color specification dimensions. Such changes in the positions of points can be more easily seen when plotted on a graph than when they are in the form of a number predicting visual impression. Later veal production trials revealed the relative nonabsolute nature of the expert visual judgments. As the work progressed, the quality of the veal improved. Consequently, samples in earlier trials had been scored high relative to samples of later trials, causing a change in the visual score/instrumental relationship (Hutchings 1978).

Chocolate color also moves in three-dimensional space. The color is affected by the fermentation, alkalization, and roasting during the processing of cocoa beans into cocoa liquor, but the color potential is determined by the origin of the beans. Final chocolate color—white, milk, or plain—is controlled by cocoa liquor, cocoa butter, sugar, and milk contents and particle size before the tempering and molding production stages. Maximum and minimum L^*, a^*, and b^* values are shown in Table 9–3. White and plain have smaller limits than milk chocolate, for which manufacturers have their own tight specification within the limits. This situation is driven by different consumer color preferences such as occur within countries of the European Economic Community (O'Carroll 1995).

Multimodal Populations

Whatever sensory work is being performed, the experimenter should always be on the lookout for the panelist behaving differently. When this occurs, the decision can be taken whether to include his or her responses. Many statistical treatments will pick out the odd individual scoring in an idiosyncratic manner.

Table 9–3 Maximum and Minimum Limits of L^*, a^*, and b^* Values of Chocolate

	L^*	a^*	b^*
White	80 to 85	−1.1 to −0.3	18 to 20
Milk	40 to 55	8 to 10	12 to 16
Plain	34 to 38	6 to 8	5 to 9

Source: Reprinted with permission from P. O'Carroll, Defining Chocolate Color, *World of Ingredients*, January/February, pp. 34–37, © 1995, C&S Publishers.

However, mathematical approaches often cannot indicate the presence of different populations within a panel. Hence, in consumer or panel work, the nonexistence of such population differences should be positively confirmed. Otherwise, totally meaningless results may be obtained.

Effects of age are significant. Sensory perceptions of the elderly differ from those of a younger population. The elderly rely more on visual cues to evaluate foods because they are less sensitive to changes in flavor (Clydesdale 1994). Panelists over 60 were more sensitive to visual cues of a cherry drink (Philipsen et al. 1995). Older panelists were more strongly influenced by color manipulation of orange juice and showed a clear preference for the flavor of the control sample as compared with an adulterated sample (Tepper 1993).

Adults, teenagers, and children were asked about the *darkness* and *liking* of orange beverages. The liking profile of children is different from that of adults, with the behavior of the teenagers somewhere between. The children appear to have preferences for two types of juice, one light and one dark. These possibly correspond to fresh orange juice and to the darker orange drink. As adulthood is reached, it seems that the liking for the darker product lessens and greater preference is expressed for the lighter colored juice (Moskowitz 1985).

Work on an ice cream product has also provided evidence for a split of population by age. The formulation was varied using three concentrations each of color, fruit, acid, flavor, and sweetener. Using the results, we can attempt answers to two questions. Do children (aged approximately between 8 and 14 years) perceive various qualities of appearance in the same way as adults, and do they have the same preferences? We can use two attributes to answer the perception question. Both groups viewed the same samples and made estimates of *darkness* and *amount of fruit*. For *darkness*, the relationship between the response of adults (A) and that of children (C) is

$$A = -0.26 + 0.91C$$

The correlation coefficient (r) is 0.94. This is a well-correlated relationship indicating a straight line of slope almost unity and going practically through the origin. Hence, adults responded to changes in *darkness* in the same way as children.

However, an equivalent plot involving *amount of fruit* yielded the relationship

$$A = 15.2 + 0.72C$$

with an r of 0.88. This is also a highly significant relationship, but the line does not go through zero and its slope is not unity, indicating that the groups did not perceive this attribute in the same way. It appears that the children gave higher *amount of fruit* scores to those samples containing the highest concentration of fruit. Adults did not appear to distinguish between the two highest concentrations. For those samples containing lowest amounts of fruit, there appeared to be some confusion among both adults and children in distinguishing fruit from other coloring material (HR Moskowitz, personal communication, 1990).

A well-designed experiment planned on demographic lines will readily yield such conclusions. However, the split response to tomato soup appearance cannot be so readily determined. A total of 20 panelists were asked to rank 12 tomato soups for the visual impression of *strength of tomato flavor*. The soups had been selected to cover the wide range commercially available in the United Kingdom but also included three formulations designed to increase the color range. The Rosenthal and Ferguson test for pairwise differences gave the mean rank order of sample numbers:

11 9 6 10 1 5 2 12 3 7 4 8

With regard to the actual samples presented to the panel, these results seemed surprisingly coarse and undiscriminating. A close examination of the responses, however, revealed two populations. One group consisted of 15 panelists, the other of 5. When they were separated, the mean rank order obtained for the group of 15 was

11 9 6 10 1 5 2 12 3 7 4 8

For the group of 5, it was

2 6 5 3 9 7 11 1 10 12 4 8

The reason for the division is immediately apparent upon examination of the tristimulus data obtained from infinite-depth measurements. A three-dimensional (xyY) plot is shown in Figure 9–6, on which the above groupings are ringed. The arrows indicate the direction of increasing *strength of tomato flavor*. The letters indicate the color appearance of each sample—orange (O), red (R), brown (B), watery orange (WO), and pink (P). Figure 9–6A contains the data from the group of 15 who judged the darkest red samples as possessing the strongest apparent tomato flavor. Figure 9–6B shows the rank-order direction for the minority group, who judged the orange samples as having strongest flavor. Both groups agreed that the pink and brown samples had the least strength of flavor.

This disagreement within a British population may have been expected. Many people are brought up to regard the orange Heinz canned cream of tomato soup as the normal color for a tomato soup. The other type of tomato soup available in the United Kingdom is the dark red packet product based on tomato powder.

The color of tomato products also differs across national boundaries. These differences, caused by tradition, legislation, and taste, face the producer with the inconvenience of controlling different formulations according to the market (Hutchings 1978). Differences are caused not only by the grade, source, and concentration of the tomato powder or puree used in the formulation but by tradition and the addition of colorants. Sensory

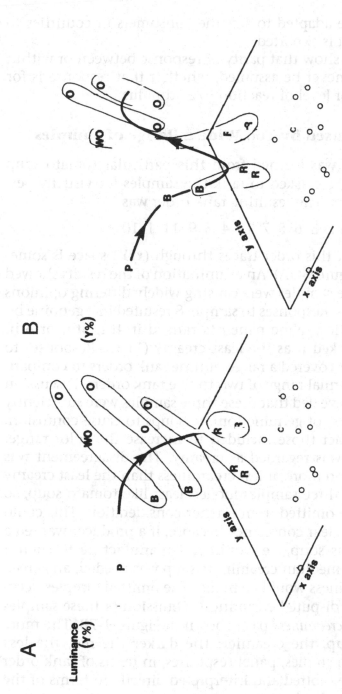

Figure 9–6 Tomato soup color, deep-layer measurements (5°/d, illuminant C, 2° observer). Soup color: O = orange, WO = watery orange, R = red, B = brown, P = pink. Linked samples show no significant differences. (A) Results of the "red equals strong" population. (B) Results of the "orange equals strong" population. Arrows denote increasing *strength of tomato flavor.*

scales should be adapted to suit the consumers in countries to which a product is exported.

These studies show that parity of response between or within groups should never be assumed, whether that response is for preference or for level of reaction to a stimulus.

Confusion Caused by Too Wide a Range of Samples

Another lesson was learned from this particular tomato soup trial. Panelists were asked to rank the samples for visually perceived *creaminess*. The resulting rank order was

$$2\ 3\ 6\ 5\ 7\ 12\ 4\ 8\ 9\ 11\ 1\ 10$$

The path that this order traces through (xyY) space is somewhat erratic (Figure 9–7A). An examination of the results showed that three of the samples were causing widely differing opinions among panelists. Responses to sample 8 resulted in a genuine bimodal distribution. Nine panelists ranked it 4th, 5th, or 6th, while seven ranked it as the least creamy (11th). Responses to samples 4 and 7 covered a range of nine rank orders in comparison with the normal range of two to five rank orders. Discussion with panelists revealed that these three samples were sufficiently outside the range of genuine tomato soups to cause confusion. They were in fact those included to increase the color range. None of them was regarded as creamy. The disagreement was whether they had more or less creaminess than the least creamy samples. These three samples looked least like tomato soup, so the results were omitted from further consideration. This could be done with a clear conscience because, if a producer wanted a high-creaminess soup, he would not manufacture to such a color. Also, if a medium-creaminess soup was needed, an agreed medium creaminess would be made. The omitted samples occupied an area of disputed creaminess. Omission of these samples resulted in the *creaminess* path shown in Figure 9–7B. The more orange the soup, the creamier; the darker the red, the less creamy. In such studies, panel responses, in terms of rank order or score, can be plotted and interpreted directly in terms of the three-dimensional color measurement properties.

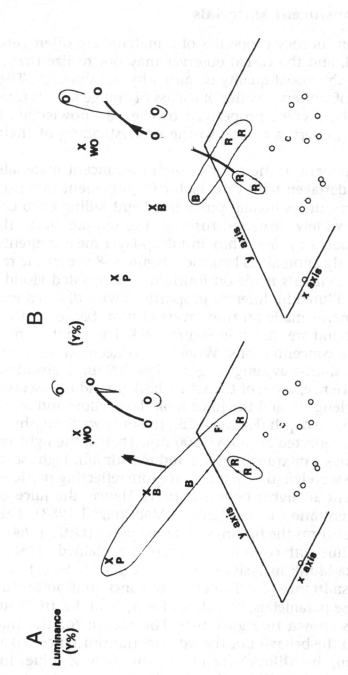

Figure 9–7 Tomato soup color (see Figure 9–6). (A) Arrow denotes increasing *creaminess*, with all samples included in the analysis. (B) Arrow denotes increasing *creaminess*, with X samples omitted.

Color of Translucent Materials

Color and translucency properties of a material are often very closely linked, and the casual observer may not realize that a major part of the visual quality is caused by translucency. The look of a cup of whitened coffee or a glass of orange juice is usually seen to be a color property. However, a knowledge of translucency properties is vital to the understanding of their quality.

The measurement of the color of such translucent materials should be undertaken with care. Color measurement of citrus fruit juices presents particular problems. Light falling onto orange juice is widely dispersed through the sample, and this causes the product to *glow*. Thin and deep-layer measurements show apparently anomalous behavior. Figure 9–8 shows the results of measurements made on fourfold concentrated Florida orange juice, diluted in different proportions with distilled water. Measurements made on thin layers (2 mm) backed with a white background are shown in Figure 9–8A. The numbers represent relative concentrations. When the concentration is increased, more short-wavelength light below 500 nm is absorbed by the juice. Hence, more of the white background is obscured at these wavelengths, and the juice looks less yellow and more red. At an infinite depth (Figure 9–8B), the only wavelengths at which light is reflected are above 500 nm. That is, the light reflected is always a mixture of green and red stimuli. Light scattered at these wavelengths increases as more reflecting particles become present at higher concentrations. Hence, the juice of greatest concentration is the lightest (MacDougall 1983). This conclusion confirms the findings on apple puree (Little 1964).

Change of hue with concentration can be explained in terms of the Kubelka-Munk analysis of absorbance (K), scatter (S), and internal transmittance (T_i). The effect of concentration on the values of these parameters, calculated for X, Y, and Z tristimulus values, is shown in Figure 9–9. The reason for the hue change lies in the behavior of the internal transmittance (T_i) at short wavelengths. Dilution results in the K for Z (blue) increasing more than the K for X and Y. The value of S increases

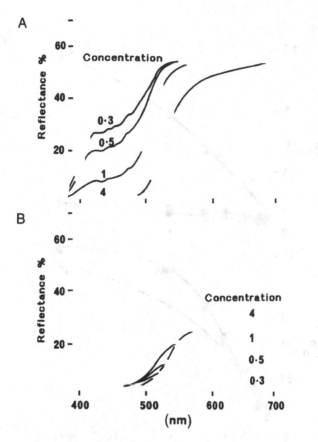

Figure 9–8 Reflectance spectra of orange juice. Numbers denote relative concentration. (**A**) 2-mm path length on white background. (**B**) 4-mm path length equal to infinite thickness (MacDougall 1983, with permission).

for X, Y, and Z. The combination of K and S behavior results in a relatively much larger increase in T_i for Z. Thus, proportionally more light from wavelengths less than 500 nm is added to the stimulus on dilution, and the color changes from orange to yellow. The visual appreciation of colored light-scattering materials in low-concentration suspensions is governed by internally scattered light emerging multidirectionally, as well as by directly reflected light (MacDougall 1983).

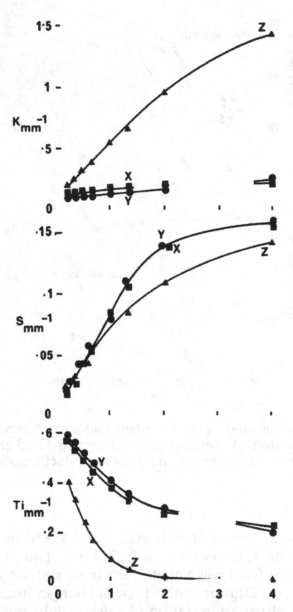

Figure 9–9 Orange juice properties. Kubelka-Munk absorbance (K), scatter (S), and internal transmittance (T_i) for the tristimulus values X, Y, and Z versus relative concentration (MacDougall 1983, with permission).

Brewed tea containing a constant volume of milk may be regarded as a system that, although translucent, can be monitored instrumentally at infinite depth, using a tristimulus color measurement. The speed of color change with time and the difficulty of preparing and handling thin layers of hot brewed tea for translucency specification call for the simplest possible measurement procedure. An appropriate approach in such cases is a deep-layer tristimulus measurement. This is relevant only in restricted cases: in this case, the level of added scatterer in the tea was large but constant. It is not applicable for products such as orange juice.

With some products, human assessors learn to ignore the presence of a certain amount of light scattering when making color judgments. This is the case with unmilked tea, although the product can be downgraded if haze is too great (Joubert 1995). When one is making instrumental measurements, haze should not be ignored. It causes light to be lost through scattering additional to light absorbed by the pigment. This is recorded as greater absorption, and hence the material is measured as having a stronger color. Gelatin is another material having turbidity as well as color. Some standards recommend filtering before color measurement. As long as the light scattered is small compared with that of the white tile, transreflectance measurements can be used for color measurement independent of turbidity (Cole and Roberts 1997).

Hedonic Assessments and Instrumental Measurements

Simple relationships claimed between hedonic responses and instrumental measures should be treated with caution. For a particular food system, there may be a continuous increasing relationship between pigment content and the visual impression of color depth. However, the link between pigment content and the liking of the color is often parabolic.

An experiment demonstrating this involved increasing the depth of color of a pale orange juice by the addition of pigment. Tristimulus color measurements were made, and panelists used their visual sense to judge *intensity* and *liking*. Examples of the results are shown in Figure 9–10. There is a clear relationship between the perceived color intensity and the measured Hunter *a* dimension ($r = 0.99$, $p < 0.001$). The figure also shows the presence of a peak in the hedonic liking response as pigment content increases. However, this relationship is such that there are still high correlation coefficients between the hedonic response and the perceived intensity ($r = 0.94$, $p < 0.05$) and between the hedonic response and the Hunter *a* dimension ($r = -0.89$, $p < 0.05$). Clearly, these last two "significances" are totally misleading. The true relationships cannot be revealed without examination of the plotted variables. This type of error occurs when no attempt is made to understand the cause-and-effect relation-

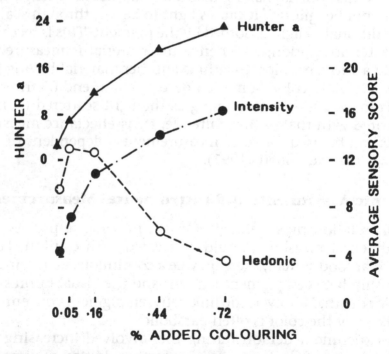

Figure 9–10 Color of orange juice: visual intensity, hedonic response, and Hunter *a* value as functions of added coloring (adapted from Trant et al. 1981, with permission).

ships within a system (Trant et al. 1981). A typical hedonic function has, not the general linear form,

$$\text{Liking} = k_0 + k_1x$$

but the quadratic form

$$\text{Liking} = k_0 + k_1x + k_2x^2$$

This represents a parabolic curve with a maximum at the point of maximum liking, the *bliss point* (Moskowitz 1977). Similarly, *fatness* (*F*) judgments of beef were related in a quadratic manner to *overall attractiveness* (OA).

$$\text{OA} = 2.32 + 0.324F - 0.716F^2$$

This equation has a maximum very close to the panelists' ideal fatness score (Rhodes 1970).

More complex pigment systems such as those occurring in raw lean meats bring other problems. When asked to judge *desirability*, panelists tended to think in terms of how the viewed sample differed from a mental impression of ideal meat color. The deviations from ideal could occur in any of the three dimensions. There were also indications that panelists were using different criteria when making their own individual judgments of quality (Eagerman et al. 1977).

Use of Color Difference Equations

Color difference (ΔE) equations, useful for defining the distance of a color from an ideal, have found some application in the food industry. The more of the fish used in the preparation of mince, the more discolored it is. It becomes progressively redder (higher *a*), less yellow (lower *b*), and darker (lower *L*). Minces were prepared using various flesh separation techniques. Values of ΔE (*Lab*) for these fish minces are listed in Table 9–4. Degree of whiteness controls product application. Values of ΔE of 1 or less were regarded as satisfactory for the best products. A value greater than 3 was unsatisfactory (Young and Whittle 1985).

Just-noticeable differences of 1 to 2.5 ΔE units, calculated using the Adams chromatic value equation, were obtained for

Table 9–4 Color Difference of Haddock Minces (Young and Whittle 1985, with permission, Crown copyright reserved)

Source of Mince	1	2	3	4
1 Fillet skin off				
2 Split without bone	7.2			
3 Cutlet without bone	7.6	0.6		
4 Headed and gutted	11.0	3.8	3.5	
5 Split bone	15.3	8.2	7.9	4.6

blends of spinach, carrot, and pear puree. The visual judgments were made on samples contained in unlabeled, 5-oz baby-food jars on a stainless steel bench (Yuson and Francis 1962).

Carrot and pea vegetable purees having large chromaticity differences were not separated using ΔE based on (Lab) space (Little 1963). However, panelists were able to rank squash puree samples differing in approximately 0.2 ΔE (Lab) units. Optimal visual rankings were found with samples 5 mm deep or more on a black background (Huang et al. 1970b). With carrot puree containing added green food coloring, panelists were able to rank samples differing by 0.10 to 0.15 ΔE (Lab) units. Optimal viewing conditions were a sample thickness of 6 mm or more on a black background (Huang et al. 1970a). Oxidation of nitric oxide myoglobin in bacon can be monitored using ΔE^* (MacDougall 1981) (see Chapter 11).

In pale beers, color differences of 1.5 ΔE CIELAB units can be perceived. The ΔE value can be used as the basis of a quality assurance test or for an in-line dosing-control system in which colorant is added to meet a predetermined target color (Figure 9–11). In a typical example, a beer may be produced slightly paler than required so that the color can be adjusted at the end of the process by adding a particular amount of any one of a number of available colorants. In the figure, the initial color difference between the beer and its specification is shown as approximately 2.2 ΔE units. Addition of caramel will make little

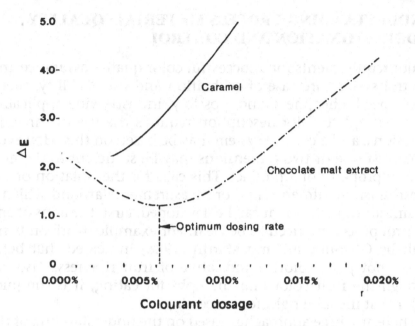

Figure 9–11 Effect on the beer color of adding chocolate and caramel extracts. ΔE is the color difference between beer and specification. The optimum dosing rate of the chocolate gives the best color match. *Source*: Reprinted with permission from S. Smedley, Towards Closed Colour Control in the Brewing Industry, Part 3, *Brewers' Guardian*, Vol. 125, No.1, pp. 15–19, © 1995, Hampton Publishing, Ltd.

improvement to the color. However, chocolate malt extract can be added at the dosing rate indicated to reduce ΔE from 2.2 to 1.00 (Chandley et al. 1995).

Values of ΔE have been used to quantify the color uniformity of processed coffee beans. Values between 3.217 and 7.25 reflect effects of variety, production method, and presence of defects (Raggi and Barbiroli 1993).

Any reduction or pooling of tristimulus data, such as calculation of ΔE, calls for caution. It ought to be justified with experimental and visual evidence and with an understanding of the theoretical basis of the manipulation. It should then be used only within the limits of the materials from which the original data were taken (Setser 1983).

UNDERSTANDING PROCESS MATERIAL: QUALITY DETERMINATION AND CONTROL

Major requirements for a successful color quality assurance and control system are ease of calibration and use, stability, precision, speed, cheapness, and, possibly, industrywide applicability. A complete color description requires the use of three dimensions, and a control system may be based on this. However, often just one or two dimensions may be sufficient to define a critical property of a product. This calls for the isolation of the most sensitive dimension(s) or measurements around which a technique or instrument can be developed. First, the aims of any control procedure need to be clear. For example, work on berry fruit by O'Donoghue and Martin (1988) indicated that berry color (not juice color) is indicative of fruit ripeness. Also, although the fruit color must be right for eating, it is the juice color that must be right for processing.

There are three approaches based on the understanding of the food system under consideration. These are discussed under the headings of tristimulus measurements, measurements made at specific wavelengths, and measurements designed to replace established visually based scales.

Pitfalls are possible with any control procedure. One is that the dimension chosen for quality control may respond to changes other than those for which the test was designed. Milks and creams may provide an example. Heat treatment and sterilization discolors milk, but there is a clear relationship between instrumental color measurements and visual judgments of quality. Processes result in browning through an increase in yellowness and redness, with consequent higher values of b and possibly a. As the value of L plays little part in the discrimination between the milks, b might be used as a basis for monitoring or control (Andrews and Morant 1987). However, changes in the value of b also correspond with changes in carotene content and to the perceived *creaminess* of milks and creams (JB Hutchings, unpublished data). Hence, an increase in the value of b might indicate an increase in browning or an increase in carotene. An injudicious extension of the use of b to monitor all milks and creams for browning might lead to incorrect conclusions. If a test is designed on the basis of changes taking place in a series of experimental samples, it will be valid only for that series of samples. It may be necessary to demonstrate, say, that browning and carotene level can be separated using one, two, or even three dimensions. A color universe for milk, however, can be established. Figure 9–12 is a plot of *L* and *b* for a number of milks. Values of *a* varied little among the samples of whole, 2-percent milk-fat, fermented, heat-treated, and nonfat milks. A plot of *L* versus *b* separated the sample whiteness into bluish, yellowish, grayish, and brownish (Rankin and Brewer 1998).

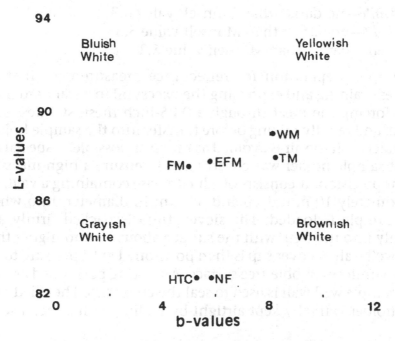

Figure 9–12 Positions of different types of milk on a lightness–yellowness diagram. Nonfat (NF), heat-treated control (HTC), fermented (FM), exopolysaccharide fermented (EFM), 2-percent milk (TM), and whole milk (WM). *Source*: Reprinted with permission from S. A. Rankin and J. A. Brewer, Color of Nonfat Fluid Milk as Affected by Fermentation, *Journal of Food Science*, Vol. 63, pp. 178–180, © 1998, Institute of Food Technologists.

Isolating the Most Sensitive Tristimulus Dimension

One, two, or three color dimensions may be needed for a control system. The number depends on the nature of the color change being monitored.

Control Based on One Dimension

Evaluation of canned tuna color on an industrywide basis provides a classic application of the use of objective tristimulus color measurements for industrial control (Little 1969a). The original scale defined tuna color in terms of the Munsell value function as measured with an optical comparator. Color quality designations corresponded to

- *white*—no darker than Munsell value 6.3
- *light*—no darker than Munsell value 5.3
- *dark*—darker than Munsell value 5.3

Sample preparation for reflectance measurement first involves draining and expressing the excess oil in a standard manner, forcing the meat through a 0.25-inch mesh stainless steel sieve, and rapidly mixing before transfer into the sample holder. Contact with the air is avoided as much as possible. A special airtight sample holder was constructed to ensure a high measurement precision. It consists of a lucite base containing a well, approximately 10 mm deep and 55 mm in diameter, into which the sample is loaded. The sieved tuna is packed firmly and evenly into the well, with the surface about 10 mm higher than the well wall. A cover slip is then positioned with pressure to obtain a uniform bubble-free surface. An O-ring positioned around the sample well wall is used to seal the cover slip. The filled sample holder is finally kept airtight by sealing with a metal screw cap.

Measurements were made using a Colormaster Differential Colorimeter on 42 coded samples of meat, obtained from a wide range of tuna species supplied by the Bureau of Commercial Fisheries. Luminous reflectance (Y) values yielded a Spearman rank correlation coefficient of 0.962 ($p < 0.001$) when compared with visual scores. Only four samples fell into different color quality groups according to whether they were ranged on the basis of Y value or visual score. The instrumental measurements provided a markedly increased sensitivity over the visual evaluations, which had been provided by three judges acting independently. The values of Y equivalent to the three grades were

- *white*—Y greater than 33.7 percent
- *light*—Y between 22.7 and 33.6 percent
- *dark*—Y lower than 22.6 percent

A strict sample preparation procedure is essential. It was found possible to use regression equations to reconcile differences among instruments. A more practical method of achieving agreement would involve the use of physical hitching post standards covering the reflectance values of commercially canned tuna (Little 1969a). Variations in presentation method have been found to result in a difference in Y of up to 4.0 percent, while delays of 20 minutes before measurement can lead to a fall of 2.6 percent (Khayat 1973). However, precanning handling practices, such as whether the animal was stressed or rested before death, temperature, time delay, or freezing before canning, did not significantly affect the postcanning color (Little 1972).

One tristimulus value may be used for control of wine (Joslyn and Little 1967). Conventional transmission spectrophotometers are designed on the assumption that the solution can be diluted if it is too deep a color and its concentration increased if it is too light a color. Such instruments are inadequate for measuring wines, for which the situation is complicated by the presence of subliminal colloidal turbidity. With thin-layer (4-mm), white background, transreflectance measurement of rosé wines, objective color measurement can be used as a sensitive indicator of anthocyanin destruction and browning development. On acidification of the wine, the chromaticity shift, as defined by blue transreflectance, is a valid measure of anthocyanin concentration. The pigment remaining unresponsive to pH change is caused by browning pigments (discussed later in this chapter). A reliable, rapid, stable, high-precision transreflectance measurement instrument, involving determination of the Commission Internationale d'Eclairage (CIE) Z function, has been designed for the specification of the color of white wine (Little and Simms 1971). The instrument and method were accepted by the Association of Official Agricultural Chemists (Horwitz 1965). Transreflectance measurements, involving CIE X, Y, and Z measurements, may also be used in stock quality control, as well as in the blending of red and rosé wines (Little and Liaw 1974).

A major quality factor of flour is *whiteness*. When measured dry, hard and soft wheat flours yield different regression lines. Soft wheats appear whiter than hard wheats at equal yellow pigment levels, probably because of ultrastructural differences. This can be overcome by measuring flour and water slurries when single regression lines are obtained between L^* and ash content and between b^* and yellow pigment content (Oliver et al. 1993). Again, consistency in sample treatment and preparation is essential. For flour, particle size has a greater influence on whiteness than extraction level or bleach treatment, which can reduce the carotene level by half (Skarsaune and Shuey 1975). Loaf crumb Z value correlates with flour Z regardless of changes in loaf volume or texture (Scanlon et al. 1993).

Yellow coloration of the papaya is used traditionally for monitoring maturity. Yellow intensity increases during ripening, with yellower fruits tending to develop their total soluble solids content more rapidly. The initial value of Hunter b is related to the time needed to ripen. Fruits with initially high values of b (>20) ripened normally in 5 to 7 days. Fruits having values between 18 and 20 ripened in 8 to 10 days, but the majority of fruits having lower values did not ripen normally (Peleg and Gomez 1974). The yellow to orange maturity of buttercup squash fruit can be monitored using the CIELAB a^* value (Harvey et al. 1997).

A yellowness index (Y_i) can be used to monitor Emmental cheese color.

$$Y_i = [100(aX - bZ)]/Y$$

where $a = 1.301$ and $b = 1.149$ for the instrumental conditions employed. Figure 9–13 shows the effect of production time on the color. Yellow cheese is produced in summer months when the cheese is produced in remote alpine pastures (see Chapter

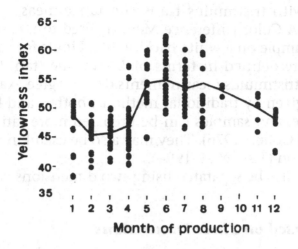

Figure 9–13 Effects of production time on the color of Vorarlberge Berkäse. *Source*: Reprinted with permission from Rohm and Jaros, Colour of Hard Cheese, *Z Lebensm Unters Forsch*, A 204:259–264, © 1997, Springer-Verlag GmbH & Co. KG.

10). During the remainder of the year, it is produced in valley plants (Rohm and Jaros 1996).

Values of Y or CIELAB L^* can also be used to monitor mushroom cap color (Lopez-Briones et al. 1992; Moquet et al. 1997), paleness in chicken breast meat (Boulianne and King 1995), and military ration packs of applesauce and cheese spread (Ross et al. 1997).

Control Based on Two Dimensions

Specifications of honey color, based on tristimulus measurements, are now taking the place of visual techniques. For the Canadian domestic market, honey is classified into the four grades **white, golden, amber,** and **dark** (Canada Agricultural Standards Act 1958). Traditional methods of specifying these grades are the Pfund Classifier using visual matching (Sechrist 1925) and a technique involving optical density measurement (Townsend 1969). The former suffers from lack of precision and agreement among observers and the latter from sample presentation problems. Many of these drawbacks have largely been overcome with tristimulus transreflectance measurement. A Hunter D25A Color Difference Meter is used to measure a 15-mm–deep sample on a white backing tile. Limits for each grade boundary are defined in terms of L and hue ($\tan^{-1}a/b$). Although the tristimulus measurements do not agree exactly with the grades given by traditional methods, both L and hue were more precise, and samples can be prepared more satisfactorily (Bowles and Gullett 1976). They may also be useful in detection of adulteration (Tisse et al. 1994).

Milks can also be separated using two dimensions (see Figure 9–12).

Control Based on Three Dimensions

The original color scoring system for tomato products was based on colored papers made by the Munsell Company. This was the sole U.S. Department of Agriculture (USDA) standard method for assessment from the time of its introduction in 1938 until 1977, when photoelectric colorimeters became reliable enough to be accepted as a basis for commercial quality–grade assignment throughout a large multilocation industry.

In terms of Hunter (*Lab*), the distribution of tomato juice colors is shown in Figure 9–14. The color gamut may be visualized as occupying the surface of the approximate triangle ABC. In terms of the third dimension, point A ($L = 21.9$) is below B (28.3) and C (27.0). Side AC shows changes in hue and lightness; BC, changes in hue and saturation; and AB, changes in saturation and lightness, with a small change in hue. In a good growing season, color might be marked by changes in hue and saturation as lycopene replaces chlorophyll. In this case, a chromaticity change paralleling CB may be expected. In a poorer season, less lycopene synthesis might occur, and the fruit might parallel CA. This model helps to reconcile conflicting data of seasonal color differences in tomatoes (MacKinney and Little 1962).

To obtain successful judgments, tomato juice graders require all three color dimensions. Similarly, all three dimensions are necessary to obtain scores from the Hunter Tomato Colorimeter, widely used in the industry to monitor tomato products (Hunter 1961). The tomato color (TC) formula consists of two

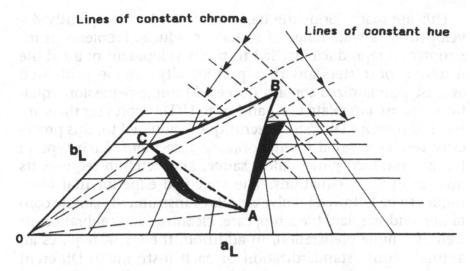

Figure 9–14 Distribution of tomato juice colors (adapted from Yeatman et al 1960, via MacKinney and Little 1962, with permission).

parts. The ratio a_L/L differentiates among samples lying along sides CA and CB but not among samples lying along side AB. This is effectively the distance from the origin—that is $(a_L^2 + b_L^2)^{1/2}$, a measure of saturation.

$$TC = \frac{a_L}{L} \frac{1}{(a_L^2 + b_L^2)^{1/2}}$$

It can be seen from Figure 8–11 that

$$\frac{a_L}{(a_L^2 + b_L^2)^{1/2}} = \cos\theta$$

where θ is the hue angle. When a scaling factor of 2,000 is used, the formula becomes

$$TC = \frac{2,000 \cos\theta}{L}$$

In CIE terms:

$$TC = \frac{21.6}{Y^{1/2}} - \frac{3.0}{Y^{1/2}} \frac{(Y - Z)}{(X - Y)}$$

where X, Y, and Z refer to source C.

This approach forms the basis of methods subsequently developed for a wide range of tomato products. Problems of instrumental reproducibility led to the development of a red tile hitching post standard that periodically can be calibrated against standardized tomato purees. Multiple-regression equations are used to relate L, a_L, and b_L to USDA scores for the sample. If a rigorous laboratory technique is adhered to, this procedure results in good interlaboratory agreement. Each type of tomato product, puree, juice, sauce, and ketchup requires its own calibration equations. The specified experimental techniques to be followed involve details of instrument, sample container, and standard tile setup, care, cleaning, and calibration, as well as sample preparation. In addition, the USDA requires an annual central standardization of each instrument. Different makes of instrument can be used, but each must be certified, and each has its own calibration equations (Zuyus, personal commication, 1991). Rigorous standardization of technique is absolutely essential for color measurement. This applies to any series of color quality measurements, whether within one laboratory, one company, or one industry.

Measurements at Specific Wavelengths

Two successful applications of color monitoring by measurement at specific wavelengths concern ripening and browning. For an example of the latter, see the section "Measurement of Browning" later in this chapter.

Spectral changes accompanying ripening of many fruits and vegetables are characterized by an increase in reflectance at the red end of the spectrum (between approximately 650 and 700 nm). Examples are shown in Figures 9–15 and 9–16. The position of the rapidly rising part of the reflectance curve in the middle of the spectrum controls whether the ripened fruit is yellow, orange, or red. As this part of the curve moves to longer wavelengths, the fruit is redder.

Apples are graded in the packing house according to their color. Visual color comparators have not proved successful because a uniform color interpretation on which several ob-

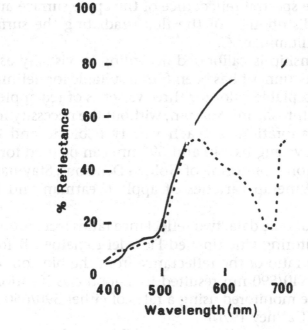

Figure 9–15 Spectral reflectance curves of the skin of extra fancy–grade Golden Delicious apples (Lott 1964, with permission).

servers can agree has not been achieved. A fully instrumental system to monitor apple redness has been constructed (Colosie and Breeze 1973). The spectral reflectance curve showing the range in skin color of extra fancy–grade Golden Delicious apples is shown in Figure 9–15. The quality control instrument has been designed to use the difference in reflectance between the total light reflected and that reflected from the red end of the spectrum. The one number grade (V) is proportional to the ratio of the radiant flux reflected between wavelengths 640 and 760 nm to that reflected throughout the visible range (380–760 nm).

$$V = \frac{\sum_{640}^{760} RE\Delta\lambda}{\sum_{380}^{760} RE\Delta\lambda}$$

where R is the spectral reflectance of the apple surface and E is the spectral distribution of the flux irradiating the surface in terms of CIE illuminant C.

The relationship is calibrated according to visually assessed grades. The instrument has been found suitable for defining the minimum acceptable color for three varieties of red apples, Red Delicious, McIntosh, and Spartan, without the necessity for separate color calibration for each variety (Colosie and Breeze 1973). The wavelengths 740 and 695 nm can be used for internal transmission monitoring of Golden Delicious, Stayman, Red Rome, and Winesap varieties of apple (Yeatman and Norris 1965).

Using similar base data, two reflectance ratios were found suitable for monitoring vine-ripened Floradel tomatoes. Before the red stage, the ratio of the reflectance from the blossom end at wavelengths 540/590 nm resulted in a good classification. Red tomatoes were monitored using a ratio of either 590/650 nm or 675/760 nm (Gaffney 1970).

Reflectance spectra of a green-to-yellow model system were made in work designed to lead to a method for monitoring banana ripeness. The system consisted of green peas and corn, blended in different proportions. The spectra are shown in Figure 9–16.

Although the overall shape of the banana curve is the same as that of the model, the 665-nm minimum is shifted to 675 nm. A correlation coefficient $r = 0.973$ was obtained when the color index values were compared with log $R675$ nm. This was higher than that obtained using other wavelengths and compared favorably with results from conventional tristimulus techniques. It is possible that this method will be useful for monitoring green color development of peas (Ramaswamy and Richards 1980).

Control Methods Reproducing Established Visual Scales

The economic value of orange juice is based on color. USDA standards for color quality are defined by the colors of six 1-

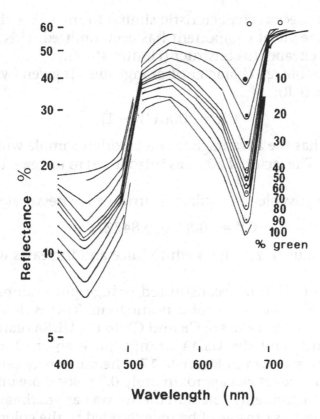

Figure 9–16 Reflectance spectra of mixtures of peas and sweet corn (Ramaswamy and Richards 1980, with permission).

inch–diameter plastic cylinders. Points are awarded on the basis of the visual color match of samples contained in tubes of the same dimensions (see Chapter 6). These colored tubes are designated OJ1 to OJ6. An orange juice matching OJ2 is given a score of 40, while OJ5 is equivalent to 36. The Hunter Citrus Colorimeter is an instrumental method capable of specifying these color grades. Light is projected onto the sample, which is contained in tubes of the type used for visual color grading. Reflected light is channeled via light pipes through three filters, A (amber), Y (green), and Z (blue), onto three photo detectors. The USDA cylinders themselves are used for standardization. The A filter has a spectral characteristic similar to the CIE X filter, except that the violet component has been omitted. This made it easier and cheaper to construct (Hunter 1967).

The desirable red character of orange juice is given by the citrus redness (CR):

$$CR = 200(A/Y - 1)$$

This scale has the advantage that a colorless sample will yield a zero score. The factor of 200 was introduced to increase the scale spacing.

It is also possible to calculate a citrus yellowness scale (CY):

$$CY = 100(1 - 0.847Z/Y)$$

CY has a value of zero for a white juice and increases with yellowness.

A number (179) of reconstituted orange juice samples were used in the development of a nomogram. This is designed to convert measured values of CR and CY to the UDSA orange juice OJ standards and the USDA orange juice color scores. The nomogram is shown in Figure 9–17. The range between means obtained by judges was approximately 0.7 color score units. The range for instrumental measurements was approximately 0.3. This approach is capable of being extended to the color specification of other citrus juices (Hunter 1967).

Comparisons have been made of results from a number of commercially available colorimeters with those from the Citrus Colorimeter (Eagerman 1978). Transmission measurements were not reproducible because of the presence of varying quantities and sizes of pulp particles. For reflectance style measurements, a sample thickness of between 10 and 20 mm was found to be optimum. Instruments yielding the highest correlation coefficients ($r > 0.98$) had a number of specific design features in common. They had a large sample port and photoreceptor area relative to that of the incident beam illuminated area, and either the receptor was close to the sample or the instrument had a sphere collector. These features minimize scattered light trapped and lost from the system.

All instruments tested, except the Citrus Colorimeter, used the transreflectance technique, with the sample contained in an

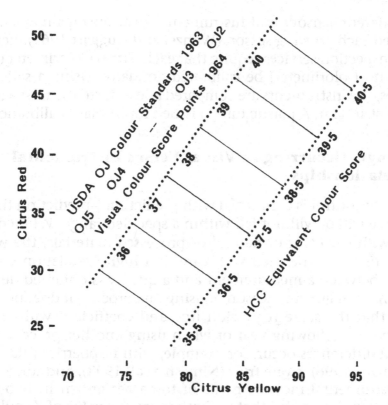

Figure 9–17 A nomogram for the conversion of Hunter Citrus Colorimeter Citrus Red and Citrus Yellow values to USDA grades (Francis and Clydesdale 1975, with permission).

optical glass cell backed with a reflecting tile. A gray reflective backing (approximately 50 percent reflection) gave higher correlations than white (90 percent) or black (5 percent) backings. This was probably because the Citrus Colorimeter has a sandblasted steel backing for the sample. This has a similar reflectance to the medium-gray painted backing used when visual judgments are made. The transreflectance method is a more expensive, awkward experimental setup to use than the tube of the Citrus Colorimeter. However, improvements, such as a special-purpose cell with a built-in reflector, are possible (Eagerman 1978).

Different sensory and instrumental relationships may be expected each growing season (Wenzel and Huggart 1969); hence, the inspection services require that each Citrus Colorimeter (and Tomato Colorimeter) be calibrated annually. Using a series of juices, the instruments are compared and adjusted against a master instrument. A plastic tube is issued for in-plant calibration.

Changes Occurring in Visual Versus Instrumental Relationships

Quality control is concerned with prediction—prediction that a sample will or will not fall within a specified range. When dealing with the color of natural or processed materials, the word *prediction* must be used with care. If a high correlation coefficient between a measurement and a quality is obtained during one natural growing season or using one process, it does not follow that the same regression line and coefficient will be obtained the following year or when using another process. Seasonal differences occur, for example, with raspberries (Riaz and Bushway 1996), stone fruit (Nimesh et al. 1993), and tomatoes.

Instrumental measurements of tomatoes grown in two successive years revealed that for both years, the value of Y could be used as an accurate marker of *maturity*. However, different relationships occurred between visual maturity score and Y. The difference was caused by the occurrence of greenback during year 2. This condition reduces the lycopene/chlorophyll ratio during ripening (Hutchings 1969). Ratios of tomato beta carotene to lycopene concentration can differ significantly from season to season. For example, in **pink** and **firm-ripe** tomatoes, the ratio was lower in 1958, by a factor of one third to one half, than in the previous year. The pHs also tended to be lower by 0.2 to 0.3 units. The differences, attributed to climatic and soil conditions (Yamaguchi et al. 1958), are derived from a change in chromaticity locus (Figure 9–14). Visual and instrumental systems respond differently to changes in chemical and structural constitution. This produces a different relationship between them and confirms the necessity for the annual calibration of colorimeters used for USDA specification.

Machine Vision Applications

The technique of machine vision is described in Chapter 8. A major advantage of color image analysis is spatial sensitivity, which allows detailed examination of differences occurring across surfaces. Application and potential application to the nondestructive examination of foods are numerous. They include laboratory study of pigment behavior and beef marbling, in-the-field studies of color development during fruit ripening, in-plant studies of heating effects, on-line detection and classification of product properties and defects, and in-home and in-store studies of consumer behavior.

Using the diffuse illumination instrument shown in Figure 8–20, colors of tomatoes of different ripeness levels were measured. Pixel hue distributions of the six USDA maturity stages are shown in Figure 9–18.

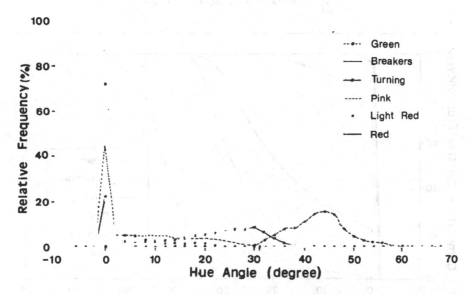

Figure 9–18 Hue distribution of six maturity stages of fresh tomatoes. *Source*: Reprinted with permission from K. Choi, G. Lee, Y. J. Han, and J. M. Bunn, Tomato Maturity Valuation Using Color Image Analysis, *Transactions of the American Society of Agricultural Engineers*, Vol. 38, pp. 171–176, © 1995, American Society of Agricultural Engineers.

Hues have different types of distribution according to the degree of ripeness. Distributions of **green** and **breaking** tomatoes are wide; all other stages have a significant proportion of pixels—that is, surface area—at a zero hue angle (red). It is difficult to discriminate maturity stages from this information because of the significant overlap of the distributions. However, all six maturity stages can be separated using the cumulative relative frequency distributions shown in Figure 9–19. Hence, a tomato classification system can be based on the aggregated percent surface area below certain hue angles. Such a continuous ripeness index agreed with manual grading in 77 percent of tested tomatoes, all samples being classified within one maturity stage difference (Choi et al. 1995).

The pixel level of detail revealed in images means that the process of ripening can be studied in terms of specific areas of tomato surface. The color change process can be analyzed in terms of the equation

$$\frac{dH}{dt} = k(H - H_{ini})(H - H_{fin})$$

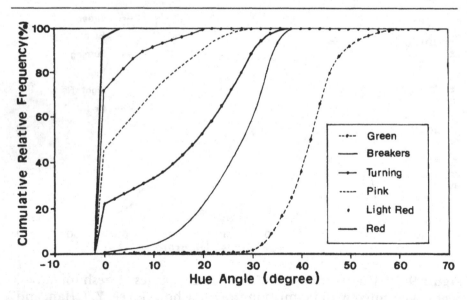

Figure 9–19 Cumulative hue distribution of six maturity stages of fresh tomatoes. *Source*: Reprinted with permission from K. Choi, G. Lee, Y. J. Han, and J. M. Bunn, Tomato Maturity Valuation Using Color Image Analysis, *Transactions of the American Society of Agricultural Engineers*, Vol. 38, pp. 171–176, © 1995, American Society of Agricultural Engineers.

where H_{ini} is an initial hue value, H_{fin} is a final hue value, and $k [h^{-1}]$ is a rate constant. Each area of the tomato possesses a particular rate constant, the highest occurring along the radial walls of the pericarp from the center of the tomato. Thus, tomato color development is related to inner structure as well as to metabolism.

Surface color change of tomatoes during ripening is caused by changes in chlorophyll, in situ carotenoids, and synthesized lycopene, which are temperature sensitive, as shown in Figure 9–20. During ripening at 25°C, the hue angle changes from 60° to 15° as chlorophyll fades and the deep red lycopene is formed (open circles). At 35°C (black squares), lycopene synthesis is inhibited, and the hue angle stabilizes at 35° until the storage temperature is reduced by 10°C, when lycope can be synthesized and the hue angle falls to 15°. This indication that hue angles had additive properties was confirmed by color measurements of chlorophyll, carotene, and lycopene mixtures (Motonaga et al. 1997).

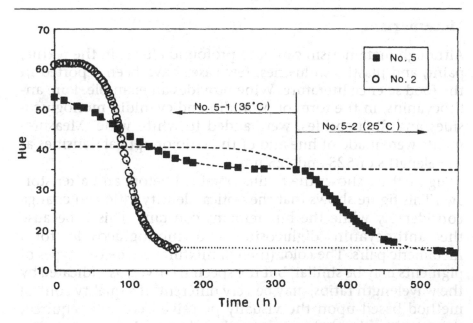

Figure 9–20 Changes of hue of ripening tomatoes: the effect of temperature. *Source*: Reprinted with permission from Y. Motonaga, T. Kameoka, and A. Hashimoto, Color Development of Tomato During Post-Ripening, in *Proceedings of Color '97, Kyoto*, p. 931, © 1997.

The image analysis technique has been used to sort bell peppers by statistical classification of hue (Shearer and Payne 1990); to extract size, shape, and color data from tomatoes (Edan et al. 1994); to assess fresh apples for color, defect, shape, and size using a hue histogram and linear discriminant analysis (Varghese et al. 1991); to measure color and marbling of beef (Gerrard et al. 1996); and to detect defects in potatoes using color and shape (Grenander and Manbeck 1993). Results from machine vision measurements have found that expert judges of dry peas respond primarily to the average luminance of the seed when ranking pea color for the market. However, intrasample variation and median seed size condition their response (Coles 1997). Extruded food products were characterized for color and surface texture, the latter using fast Fourier transforms of the hue, saturation, intensity (HSI) data (Tan et al. 1994). Other investigations involving machine vision have been made of beef and apples (Kyung et al. 1996), broccoli (Ishikawa et al. 1997), tomatoes (Choi et al. 1995; Edan et al. 1997), stone fruit (Nimesh et al. 1993), and cucumbers (Schouten et al. 1999).

Metamerism

Although metamerism can have profound effects in the textile, paint, and plastics industries, few cases have been reported in the food science literature. Wine provides an example. Pure anthocyanins, in the form of peonidin and cyanidin monoglucosides and diglucosides, were added to white wine. Measurements were made of hue and of the ratio of optical densities at wavelengths of 525 and 425 nm.

Figure 9–21 shows the results obtained before and after storage. This figure shows that the optical density ratio can change considerably while the hue remains constant. This is because the anthocyanin diglucoside and monoglucoside form metameric pairs. The color (hue) of mixtures of the two types of pigments may be similar, but the spectral curves, as indicated by the wavelength ratios, may be very different. If a quality control method based upon the visually perceived color is required, then the hue should be monitored. Only on solutions of similar

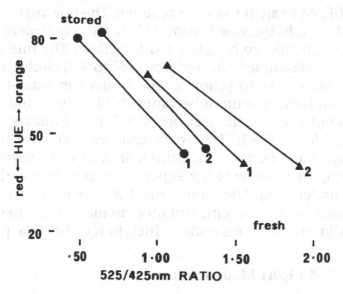

Figure 9–21 Effect of pigment color change in wine during storage at 120°F for three weeks, shown in terms of Hunter hue and the 525/425-nm optical density ratio. Pigments added to a white wine base are 1, peonidin; 2, cyanidin monoglucoside (▲) and diglucoside (●) (adapted from Robinson et al. 1966, permission received from *Am J Enol Vitic* 17:178).

chemical composition should ratio measurements be used for this purpose (Robinson et al. 1966).

Another example of metamerism is the greenness that can occur in uncooked beef. There are two causes of green color. One is due to the formation of sulfmyoglobin in putrid meat; the other occurs in beef fat. The latter is a transient phenomenon that depends on the illumination. Greenness becomes visible in thin layers of fat under intense, blue, sunless, in-shadow illumination. The color of the reduced myoglobin becomes just perceptible through the translucent collagen where the fat is less than 2 mm thick. In an investigation of this, color measurements were made between a correlated color temperature of 5,000K, approximately equivalent to average noon sunlight, and 20,000K, equivalent to a clear blue sky. The hue angle of the normal fat sample increased from 34° to 68°, equivalent to a Munsell hue change from yellow red to yellow. The hue angle of the abnormal sample increased from 75° to 90°, equivalent to a change from yellow to yellow green. Under some day-to-day lighting conditions, greenness becomes visible. This could result in an apparently serious spoilage problem for the industry (Mac-Dougall and Jones 1983). This work again emphasizes the fact that the eye must be the final arbiter in any work involving color. Again, the purpose of the experiment must be clearly defined and understood. The sample must receive great attention in its preparation, assessment, and measurement, and attention must be paid to conditions under which judgments take place.

Transmitted Light Measurements

Light transmission measurements are useful for monitoring the internal quality of many whole fruits and vegetables. Such measurements reveal information about the state of internal pigment development and maturity, as well as indicating the presence of internal defects caused by mechanical damage and disease. Such attributes are not revealed by reflectance measurements. Kramer and Smith (1947) made transmission spectrophotometry measurements on extracts from fruit macerates, but others have since applied the technique to many whole fruits and vegetables (Birth and Norris 1965). The internal-quality literature has been reviewed by Deshpande et al. (1984).

The food to be measured is clamped between a light source and photodetector. Seals prevent loss of light on entering and leaving the sample. Measurement wavelengths are carefully chosen to indicate, for example, the decay of chlorophyll during ripening and to minimize sample size effects. Measurements are made in terms of the ratio of the intensities of transmitted to incident light at single or multiple wavelengths. An alternative is to use optical density measurements. The emphasis of transmission measurements is on pigment analysis; hence, the wavelengths employed are not restricted to the visible range.

An example of spectral transmittance curves showing progressive stages in maturation of peaches in terms of optical density is given in Figure 9–22. Curve 4 represents the tree-ripe stage. The decrease in absorbance at 675 nm indicates a decrease in chlorophyll content as maturation and ripening progress. The maturity of the fruit in terms of eating quality is independent of chlorophyll content and fruit size. Maturity was related ($r = 0.957$) to the difference between optical densities at 700 and

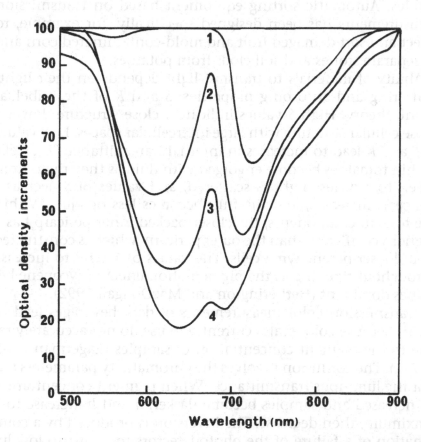

Figure 9–22 Optical density curves of four intact Elberta peaches at progressive stages of maturation. 1, typical of fruit harvested August 3; 2, August 17; 3, August 21; 4, August 26, tree-ripe stage. (Sidwell et al. 1961, with permission).

740 nm (Sidwell et al. 1961). In similar studies, the value of the transmittance peak ratio at wavelengths of 730 and 850 nm has been used to separate JH Hales peaches into predominantly mature and immature groups (Romani et al. 1962).

Such studies have been carried out on a number of foods. Examples include apples (e.g., Yeatman and Norris 1965), plums (Romani et al. 1962), papaya (Birth et al. 1978), pears, peaches, and apples (Bittner and Norris 1968), and tomatoes (e.g., Birth et al. 1957). Also, this technique has been applied to the detection of bruising and disease in a wide range of fruits and vegetables. Automatic sorting equipment based on transmission measurements has been designed specifically, for example, to select and sort damaged fruit and mold-contaminated corn and to separate stones and soil clods from potatoes.

Ability of materials to transmit light depends on their light-scattering and absorbing properties, S and K of the Kubelka-Munk theory. High S values indicate a close structure, low S a loose cellular structure with large intercellular spaces. Low values of S and K lead to increases in internal transmittance (T_i). Cells within tomatoes become engorged with fluid as they ripen from green to red, less light is scattered, and values of S decrease; hence T_i increases, and the fruit becomes less opaque. Within the tomato components, the closer packed inner pericarp has a higher S coefficient than the outer pericarp, which is constructed from looser parenchyma cells. The values of K tend to increase throughout ripening as the pigmentation effect of lycopene becomes dominant (Hetherington and MacDougall 1992).

Transmission color measurements of dark beverages lead to errors because color scales currently in use do not accurately relate to the pigment concentration of samples (Eagerman et al. 1973a). The confusion involves the chromaticity parameters but not the luminous transmittance. When pigment concentration is increased and samples become darker, a and b increase to a maximum, then decrease. The inversion is produced by a combination of a failure of the photodetectors to adjust to low luminosities and the decreased color gamut in the dark part of color space. An attempt can be made to use a color scale expansion of (Lab) space to compensate for low light transmission properties of dark liquids. This was done so that a color scale could be developed suitable for on-line measurements of liquids containing a number of soluble colorants. The type of scale expansion used is shown in Figure 9–23.

The figure is a representation of an *L, a', b'* and the *L, a₂, b₂* scale expansion. The chromaticity parameters at lower *L* values

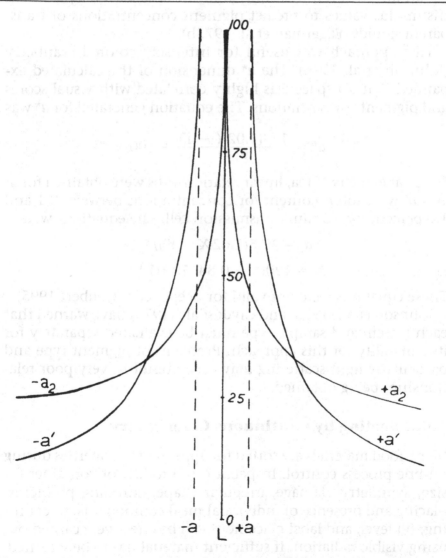

Figure 9–23 A representation of L, a', b' and L, a_2, b_2 color scale expansion (Eagerman et al. 1973a, with permission).

have been expanded relative to their values at higher L values. Unfortunately, it was found that different formulas relating the two scales were necessary for each color. It is not feasible to develop a colorimeter yielding accurate readings related to pigment concentration for all systems. If a single product is being worked with, it may be possible to develop a perfectly linear equation. However, this study indicates the feasibility of using tristimulus values to predict pigment concentrations of transparent liquids (Eagerman et al. 1973b).

This approach was useful for intensely colored cranberry (Johnson et al. 1976). The a^* dimension of the calculated expanded ($L^*a^*b^*$) space was highly correlated with visual scores and pigment concentrations. The equation generated for a^* was

$$a^* = \frac{170(1.02X - Y)}{Y^{1.60}} + 200$$

For clear extracts of tea, linear relationships were obtained for a' and b' with solids content for concentrations between 0.1 and 1.0 percent for a 5-mm transmission cell. The equations were:

$$a' = 17.5(1.02X - Y)/Y$$

$$b' = 120(Y - 0.8467Z)/Y^{1.5}$$

These equations were not valid for deeper cells (Joubert 1995).

Johnson et al. (1976) and Clydesdale (1976) have warned that each system and sample type must be evaluated separately for its suitability for this approach. Problems of pigment type and particularly light scattering may easily result in very poor relationships being obtained.

Color Sorting by Continuous Colorimetry

Many food materials are sorted for appearance properties during on-line process control. Inspection of product or container for size, symmetry, damage, irregular shape, enrobing problems, placing and presence of individual meal components, or counting, fill level, and label placement can be effectively carried out using visible radiation. If sufficient material has to be screened, the advantages of machine sorting are many, including lower labor costs; less wastage and variation in sorted product quality through human error; more efficient and reliable sorting into possibly four or five groups of product category; stable standards that do not drift through, for example, operator fatigue; and the ability to sort in inhospitable environments.

The operational sequence of sorting food materials such as grains, fruit, and vegetables by color involves a number of steps:

1. *Feed and spread*: As the sorter acts on single discrete units, the material must be flowing at a consistent rate and spread across the width of the unit.
2. *Accelerate and singulate*: Acceleration of the product (grains probably under free fall) contributes to singulation, the object being to bring the product to constant velocity as it enters the viewing section.
3. *Illuminate and view*: Each unit has to be maneuvered in front of the sensors, where it is illuminated and evaluated by photodetector or line scan against the response from a standard.
4. *Eject and separate*: A targeted particle is removed by the ejector, which works on the assumption that all the product is moving with the same velocity. Ejected material is physically separated from the bulk.

Food process environments can be hostile to optical and electronic equipment, and long-term repeatability cannot be assumed. Spiked electrical supplies, heat, and dirt from moisture, oil, vapor, and dust reduce reliability and accuracy of the instrumentation. The head must be protected from the environment by double enclosure and temperature control, and adequate checks for cleanness and calibration are required. Where products vary in thickness or size, it may be necessary to correct reflectance data using an on-line sonic bed depth measurement (Coatney 1986). Throughput is limited by mechanical rather than optical factors. Rate depends somewhat on product size, but speeds of between 200,000 pieces of fruit per hour to 200,000,000 grains of rice per hour can be achieved (Philpotts 1983).

Monochrome vision systems can be used to separate materials by size and shape but not by color (McClure 1988, cited in

Tao et al. 1995). Color is needed for sorting according to ripeness. Combined size, shape, and color monitoring is needed for sorting vegetables from foreign objects of the same color—for example, separating green beans from stem pieces ("Bean Sorter" 1995). On-line measurement can be used to provide feedback for process control of foods such as bakery products (Scott 1995). For example, reflectance measurements on the end product can be used, via a closed-loop control system, to alter the temperature profile within a baking oven (McFarlane 1985). However, frying of fried potato chips (crisps) is a more complex problem. Color depends on the potato sugar/starch ratio, and drying properties depend on the moisture content. Both color and moisture require control to obtain an optimum product. High-moisture chips have poor texture and oxidize faster, while low-moisture chips contain more oil and are more expensive to produce. The lighter the chip color, the more desirable. Dark chips are seen to be burned, although the darkness may be caused by sugar content. A dark, low-moisture chip indicates overcooking relative to chip thickness, and a lower temperature or faster fryer belt speed is required. A high-moisture, light color indicates the opposite. A high-moisture, dark chip needs longer cooking at a lower temperature. A low-moisture, light-colored product needs shorter cooking at a higher temperature. Reflectance and moisture content monitoring may be used to alter the process variables of slice thickness, feed rate to the fryer, fryer temperature, and fryer speed and/or agitation accordingly (Coatney 1986).

Four fundamental procedures are necessary in any color specification technique, including those applicable to continuous colorimetry. These are consideration of the object itself, the viewing technique, the analytical procedures, and the relationship with subjective response. Cranberry cocktail is a dark, deeply colored, non–light-scattering juice. Tristimulus colorimetry was found to be unsuitable, as insufficient light intensity is available for transmission through 4 mm, the minimum sample depth for practical on-stream installation. The desired control criterion (the visually perceived color of cranberry juice, as determined by visual matching using a Lovibond Schofield

Tintometer) is related to its pigment content. This in turn can be monitored by an optical density difference (515–420 nm) measurement, using high-intensity illumination. Hence, the final color can be manipulated by adding a quantity of lighter or darker juice, controlled by the on-stream optical density value. On-line refractometry measurements could also be installed in such a system to control sugar content (Clydesdale 1978).

Older methods of mechanical and electromechanical machine sorting are being replaced by machine vision inspection. This easily interfaces with the operator via a keyboard, it can be more easily calibrated, it is faster, and it is "intelligent." Such systems are capable of screening very many product units at the same time against many criteria. Their portability means that they can be moved around the factory according to the product in season. Discrimination can be on a monochromatic (gray scale), bichromatic (using two filters), or full-color (three filter) basis. More than one camera may be needed to detect defects on the sides or bottom of the product, but cleaning may become a problem in this case. Alternatively, the product can be turned. Whatever sorting system is used, adequate maintenance of the system and training of the operators are essential if maximum usefulness is to be achieved and retained (Heffington and Nilsson 1990).

For such sorting, a CCD color camera (see Chapter 7) can be used to generate *RGB* pixel values for conversion into hue, saturation, and intensity (HSI) coordinates. Good Russet potatoes possess hue values ranging from 16 to 32, while those greened through exposure to light possess hues between 16 and 45. A pixel summation graph of hues 27 to 40 plotted against hues 23 to 26 (Figure 9–24) almost separates green from good potatoes. These clusters can be fully separated in multidimensional feature space. Excellent agreement can be obtained using machine color vision systems with conventional expert visual inspection techniques for potatoes and apples (Tao et al. 1995). Golden Delicious apples can be sorted for size and shape using Fourier descriptors, as well as for color and ripeness using machine color vision (Heinemann et al. 1995). A prototype automated inspection station is almost adequate for sample sorting, but move-

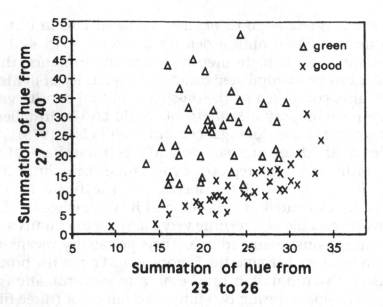

Figure 9–24 A plot of hue clusters of green and good potatoes in two-dimensional feature space. The clusters are separable in multidimensional feature space. *Source*: Reprinted with permission from Y. Tao, P. H. Heinemann, Z. Varghese, C. T. Morrow, and H. J. Sommer III, Machine Vision for Color Inspection of Potatoes and Apples, *Transactions of the American Society of Agricultural Engineers*, Vol. 38, pp. 1555–1561, © 1995, American Society of Agricultural Engineers.

ment of the potatoes during inspection at present reduces classification accuracy (Heinemann et al. 1996).

Determination of the color class of wheat is necessary, as functionality traits of genetically red and white wheats cause them to mill and bake differently. This can be achieved using automated color classification (Dowell 1998), as can recovery of broken grains during screening (Haas 1995). Algorithms have been developed for real-time maturity grading of fresh market peaches (Nimesh et al. 1993).

Machine vision is used for robotic harvesting of fruit from the tree. A gray-level thresholding method of distinguishing between object and background cannot be used to detect oranges against a background of leaves. This is because oranges in the sun are brighter than leaves in the sun, but oranges in the shade are darker than leaves in the sun. Hence, color information is needed to discriminate fruit from other likely background objects. With *RGB* information, a multivariate statistical pattern classification technique is used to determine the probability of a pixel's belonging to the orange or the background. If sufficient pixels indicate that an orange has been detected, the centroid of the fruit is located, and the robotics controlling picking can be suitably directed (Slaughter and Harrell 1989).

Light rays from points on an object surface form a planar cluster in color space. The directional vectors within the space can be used to perform image segmentation. By means of a training algorithm, parameters used in color segmentation can be calculated. Fruit location is taken from the color-segmented image. Citrus robotic harvesting can locate 95 percent of the visible fruits, failures being caused by the presence of soil of similar color on the oranges. The system can be used during the day under natural lighting or at night using artificial illumination (Pla et al. 1993).

UNDERSTANDING PROCESSING

Applications of color measurement to the understanding of the product during processing are many and varied. Examples are grouped under three headings: effect of processing on color, browning kinetics, and mathematical modeling.

Effect of Processing on Color

Processing affects product color through changes in pigment nature or content and/or the physical state. Changes to pigments on baking can be seen in Figure 8–21. Physical state changes occur when salmon is processed. Canned salmon with higher oil contents tends to have a more desirable color (Schmidt and Idler 1958). This is because a high dietary lipid content improves the utilization of carotenoids by the fish (Foss et al. 1984). Also, there is a greater refractive index difference between air and muscle than between oil and muscle. Thus, a lower oil content will lead to increased light scattering and lower redness. Salmon intake of oil and pigment varies with season and environment. This leads to different relationships occurring between carotenoid determinations and visual redness.

Light scattering also affects salmon color during cooking and canning. On heating, protein precipitation occurs and opacity increases, with a consequent rise in the value of lightness. The opacity increase reduces the optical path length in the flesh, and light entering the surface has a lower chance of becoming selectively absorbed. Hence, the perceived redness falls, and consequently so does the value of a^*. A similar process takes place when salmon is subjected to fluctuating temperatures while in frozen storage. The color difference on baking has been found to be rather greater for farmed fish (containing canthaxanthin) than for wild (containing astaxanthin) (Skrede and Storebakken 1986). It can be appreciated that the decolorizing effect of cooking on badly stored farmed fish, containing perhaps lower concentrations of the less intensely red canthaxanthin, can lead to a product more resembling white fish than salmon.

The color of vegetables, such as green beans, is also affected by both physical structure and pigment content. Color and pigment measurements were made on raw and blanched samples. Blanching methods were microwave, convection oven, steam, and water (Muftugil 1986). A continuous relationship was produced involving chlorophyll and Hunter $-a$ values for the water-blanched and steam-blanched (greenest) samples and the microwaved and raw (least green) samples. Samples with the lowest chlorophyll were the most green. The convection oven sample, although retaining a reasonable value of $-a$, had the least chlorophyll because of the high blanching time necessary to inactivate the peroxidase. Hence, a knowledge of the pigment content would not be a predictor of the perceived color. Steam- and water-blanched samples were a more intense green color because water had replaced air inside the beans. The change of relative refractive index reduced light scattering and the vegetables looked greener. There is no free water available during convection oven heating. Similarly, changes taking place on blanching and frozen storage of asparagus (Begum and Brewer 1997) are probably caused by both pigment changes and structural damage. Cells are ruptured with consequent release of cell contents, and water is substituted for air in the structure. Both sensory and properly carried out instrumental color measurements detect color changes occurring in green vegetables. However, the sensitivity of detection should be determined before such changes are measured during storage (Gnanasekharan et al. 1992).

A more complex example of color change on processing concerns tuna. For the quality control of tuna, the luminous reflectance Y can be used as an indicator of visually perceived quality (see earlier in this chapter). Further insight may be gained into the quality of the processed product by quantifying the chromaticity change on chemical reduction. In the homogenized sample prepared for reflectance measurement, it can be assumed that the hemepigment is essentially in the ferri-hemochrome state. This pigment can be converted to the ferro-hemochrome state by treatment with reducing agents such as sodium dithionite. To do this, the sample is prepared and measured in the airtight cell as already described in this chapter. The cell is opened, and 0.5 ml of 5 percent freshly prepared sodium dithionite is applied to the sample surface. The cell is then recapped, and a second measurement is made after 30 minutes.

The chromaticity shift, x and y, is plotted as shown in Figure 9–25 (Little 1969b). Measurements were made on 67 samples of canned tuna, including albacore, yellowfin, skipjack, and bluefin. The measurements were divided into three groups according to federal standards of identity criteria. Group I (open circles) contained all the white albacore samples and all samples graded I. The 37 samples in grades II and the 10 in grade III (closed circles) cannot be grouped unequivocally on the basis of chromaticity shift. However, there is a marked tendency for the magnitude of the shift to increase with decreasing Y value in the direction of $+x$, $-y$. High values of $+x$ are characteristic of group III samples. The converging lines indicate the direction of chromaticity change with decrease in Y.

The magnitude of the chromaticity shift can be used as a measure of relative concentration of reducible pigment. The advantage to the understanding of processing lies in the consistency of the chromaticity shift–versus–Y relationship. A sample having a low Y value with a small chromaticity shift is evidence that the dark color is the result of something other than the normal

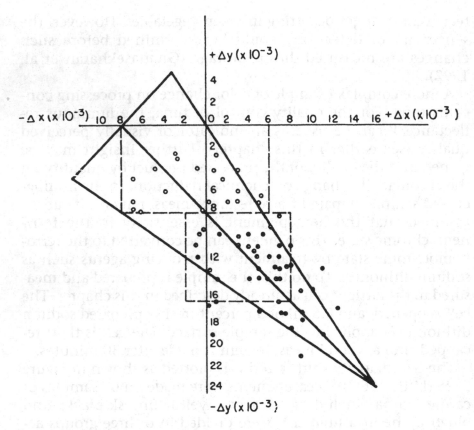

Figure 9–25 Color of tuna. Relationship of chromaticity shift to Y value. Group 1 (○), $Y \geq 33.7$; group 2 (◎), Y = 33.6–23.7; group 3 (●), $Y \leq 23.6$ (Little 1969b, with permission).

pigment. This might be browning, scorching, or oxidative degradation (Little 1969b).

Prediction of the Effect of Processing on Product Color

Much pilot plant work might be avoided if the color of a processed food could be predicted from the color of its ingredients. An attempt was made to predict strawberry jam color from the color of the strawberries using six varieties (Skrede 1982). The attempt failed because changes produced by the processing are large in terms of raw material properties and not constant. Also, in the process, pH, acid, sugar content, and soluble solids levels are adjusted far beyond the levels found in fresh fruit. Good correlations produced one year were not repeatable in subsequent seasons. Hence, where foods undergo significant processing, it is likely that there are few cause-and-effect relationships involving appearance. Skrede concluded that the main emphasis of investigations on such foods should be given to the final products rather than to the characteristics of the fresh ingredients. That is, from an industrial viewpoint, the most important factors are the color, consistency, and taste of the final product at the time of production and the changes occurring during handling and storage.

For other foods, however, it is possible to establish consistent relationships between raw material and finished product. For example, for canned salmon, there is a good relationship across species between carotenoid content, visual redness, and $+a$ value (Saito 1969). The a value of cooked sockeye salmon can be predicted from the value of the a/b ratio obtained on the raw salmon. Figure 9–26A shows how this relationship has been used to define the three grades of canned sockeye (Schmidt and Idler 1958). Automatic sorting of raw fish using reflectance properties is therefore possible. However, it can be seen from Figure 9–26B that the relationship between the a value and the visual FIL color rating depends on species (Schmidt and Cuthbert 1969). The color of cooked salmon is generally related to that of the raw muscle (Swatland et al. 1997). Using consistent processing, it has been found that redder raw fish leads to redder cooked or baked fish (Skrede and Storebakken 1986), to redder canned fish (Saito 1969), and to redder smoked slices. For Arctic char (*Salvelinus alpinus*), there is a negative relationship between external skin color and internal musculature ($r = -0.65, p < .001$). Muscle scores can be predicted from fiber-optic probe measurements using reflectance data at 520, 580, and 530 nm ($r = 0.905, p < 0.01$) (Swatland et al. 1997).

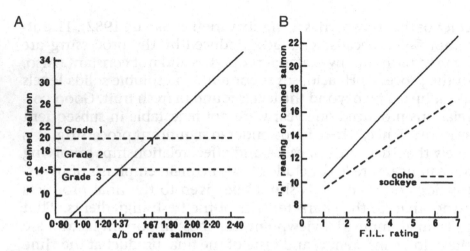

Figure 9–26 Canned salmon color. (A) Prediction of canned sockeye color from its raw color (Schmidt and Idler 1958, with permission). (B) Regression lines for *a* values versus FIL color ratings for canned salmon (Schmidt and Cuthbert 1969, with permission).

Measurement of Browning

Browning development caused by active processing or storage is of vital interest to the industry. Colors generated in model studies include blood red and green as well as the yellows, oranges, and browns. These can be quantified using tristimulus measurements (MacDougall and Granov 1997). However, the colored products of Maillard reactions are often measured by selected wavelength absorption.

Conventional extraction methods for the measurement of browning in citrus products consist of three main stages: extraction of soluble coloring materials (if the food is in a solid form), clarification, and colorimetric measurement. An extra stage is necessary to eliminate carotenoid interference. This involves centrifugation, followed by an ethyl alcohol dilution to flocculate the remaining cloud, which is filtered off. The browning index is defined, using a conventional spectrophotometer, as the absorbance measured at 420 nm using a 10-mm–path length cell (Meydev et al. 1977). For storage-abused whole orange juice, a good correlation has been found between this browning index and the tristimulus measurements Y ($r = -0.956$), and X ($r = -0.940$) ($p < .001$) (Robertson and Reeves 1981). Browning in blackberry juice and wine can be measured as the absorbance at 420 nm after the sample has been bisulfite bleached (Rommel et al. 1990).

Absorption measurements at 420 nm can be used to determine browning pigments extracted from stored hygroscopic whey powders. Samples stored in sealed pouches released water when crystallization of the amorphous lactose occurred. Storage at 35°C resulted in an increase in local moisture content and a consequent increase in the extent of browning compared with that of samples stored in an open system. Further, the kinetics were not the same for both conditions, browning being greater for the closed system. Caution is therefore required when using open-system reactions to characterize and predict reaction kinetics that will occur in closed systems (Kim et al. 1981).

The color of processed black currant juices after storage influences sensory acceptance. Cyanidin-rich juices brown in a few weeks; hence, for a longer shelf life, black currants of high pigment content, containing particularly delphinidins and acylated anthocyanins, are required (Brennan et al. 1997). Measurements at specific wavelengths can also be used as indicators of black currant syrup quality. Three samples having different degrees of visually judged browning were diluted 1:4 with water. A conventional spectrophotometer was used to measure absorbance spectra. Examples are shown in Figure 9–27A.

Two groups of wavelengths appear sensitive to changes in browning. Initial browning results in a fall in absorbance at approximately 520 nm, while later stages produce changes around 400 nm. The authors chose to use the absorbance ratio of A_{520}/A_{420}, and this is shown plotted against visual judgments of *browning* in Figure 9–27B. This ratio has the advantage of being independent of initial pigment content. The standard error of prediction (SEP) was approximately 0.8 sensory units. Hunter a values also produced a consistent relationship with visual scores, the SEP being approximately 0.7. It was found possible to set acceptability limits in terms of absorbance; this may lead to

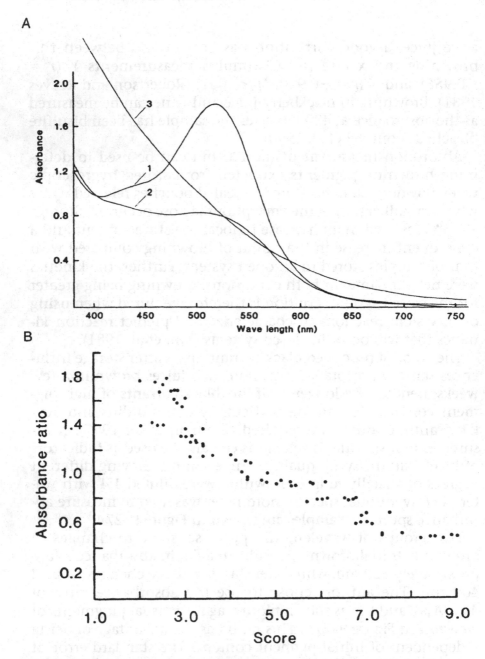

Figure 9–27 Color deterioration of black currant syrup. (A) absorbance spectra of syrup diluted 1:4 with water. The visual scores for *brown color* are 1, 2.0; 2, 4.8; and 3, 9.9. (B) absorbance ratios A_{520}/A_{420} versus visual estimates of *browning* (Skrede et al. 1983, with permission).

a method for predicting storage life for juices (Skrede et al. 1983). The absorbance ratio A_{max}/A_{420} has been used to measure color quality in a variety of anthocyanin-containing samples, including strawberry jam (Spayd and Morris 1981).

The susceptibility of apples to browning depends on phenolic compounds and polyphenol oxidase (PPO) activity. Apple fruits contain different classes of phenol with differing susceptibilities to enzymic oxidation. The degree of browning can be determined by simultaneous measurements of soluble (absorbance at 400 nm) and insoluble (lightness L^*) brown pigments. The sum of the two expresses the degree of browning, closely related with the amount of phenolics degraded. Maturity did not appear to influence the development of browning (Amiot et al. 1992). Enzymic browning in Golden Delicious apple pulp can be monitored using L^* values. Pulp made with unripe apples browned at the highest rate. This was attributable to ascorbic acid content and PPO activity. Browning rate increased with temperature but was strongly nonlinear (Lozano et al. 1994). Because of fruit-to-fruit variability in the extent of enzymic browning, multilevel treatments with browning inhibitors should be compared using several plugs from the same apple. Tristimulus L and a values can be used to monitor inhibitor effectiveness on apple and pear plugs (Sapers and Douglas 1987).

Reflectance ratios have been used to quantify the extent of browning in green vegetables. Spinach browning causes changes in reflectance between 500 and 650 nm. As browning progresses, reflectances at 590 and 636 nm increase. The *browning number* (BN) is calculated from

$$BN = \frac{R_{590} - R_{400}}{R_{530} - R_{400}} \times 100$$

Inclusion of the reflectance at 400 nm compensates for differences in chlorophyll content. A regression coefficient of $r = 0.962$ was obtained for 11 samples (Loeff 1974). A brown index (BI), defined as the brown color purity, was developed as an indicator of browning in sugar-containing products:

$$BI = 100(x - 0.31)/0.172$$

where x is the chromaticity coordinate (Buera et al. 1986). The index can be used, for example, to determine processing conditions leading to minimum browning development in banana puree (Guerrero et al. 1996).

In red wine, the percentage brown component and the relative loss of anthocyanin can be followed during storage (Little 1977a) (see below). Presence of browning or other degradation in processed tuna can be deduced from measurements of the relative concentration of reducible pigment (Little 1969b) (see above). The work of Loeff (1974) and Skrede et al. (1983) at specific wavelengths (above) suggests methods for the study of the browning in green vegetables and fruit products.

Mathematical Modeling

The effects of processing on the color of many products are complex. However, empirical mathematical modeling techniques available for linking cause and effect include multidimensional analysis (see Chapter 5).

Neural Networks

The formation of an appearance image can be understood either in terms of a knowledge of the cause and effect of the system or by using mathematical and statistical analysis of the suspected causes. For the latter, multivariate methods such as principal component regression analysis and canonical analysis have gained wide use in the food industry (e.g., Piggott 1988). The success of such systems depends on the presence of linear relationships among the data (see Chapter 5). Multilayer neural networks, on the other hand, are designed to deal with nonlinear data. A neural network is an information-processing technique that attempts to simulate the functions of living nervous systems (Beale and Jackson 1990; Finlay and Dix 1996).

The neural network is built of neurons in three layers: the input, hidden, and output layers. Each input-layer neuron takes a piece of input data. Each neuron in the output layer stores or generates an output number, but the layers between are hidden (Figure 9–28).

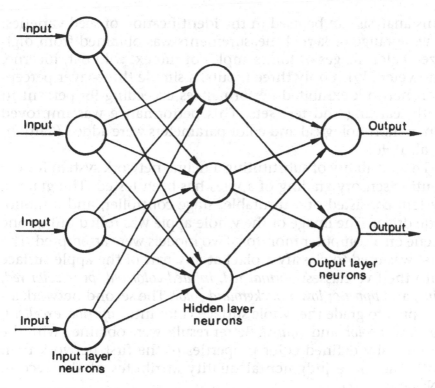

Figure 9–28 A typical neural network.

The strength of connection between two neurons is the weight, and the input is multiplied by its weight to give a signal between the two neurons. A transfer function in the hidden layer determines whether inputs are sufficient to produce an output: the neuron output is either on or off. The output depends solely on the inputs: a certain number must be on at any one time to make the neuron fire. Errors are used to adjust the weights until errors between predicted and experimental results are minimized. Thus, the network can be "trained." There are several types of transfer function (e.g., Luedeking 1991; Wythoff 1993).

Neural networks have been used to manipulate a number of chemical, physical, or social parameters into yes/no decisions.

This analysis can be used in the identification of seed varieties. A wide range of kernel measurements was obtained from digitized color images of four samples of rumex, wild oat, lucerne, and vetch. With only three inputs, a simple three-layer perception network exhibited performances exceeding 99 percent in both learning and test sets. This performance was improved when morphological and color parameters were added (Chtioui et al. 1996).

The possibility of substituting a neural network system for the routine sensory grading of apples has been tested. The grading system consisted of a turntable, stage controller, and a mono-axle driver. The image of the whole apple was rolled out as one scene on a computer monitor. Two models were attempted. The first was used to classify a pixel at any part of the apple surface into the five classes: *normal red*, *injured color red*, *poor color red*, *vine*, and *upper or lower background color*. The second network attempted to grade the whole surface color into *superior*, *excellent*, *good*, *poor color*, and *injured*. Better results were obtained with the more easily defined color properties of the first network than with the more judgmental quality attributes of the second (Nakano 1997).

Color development of biscuits during baking yields characteristic color dimension curves (see Figure 8–21). Neural networks can be used to represent and classify products according to these curves and a 7-point expert judgment of bake level. From day to day, the trained networks performed more consistently than human judges (Yeh et al. 1995).

Most young Rioja red wines that fail to qualify for an Origin Denomination label do so because of their color. A trial involving 126 young Rioja wines showed that neural network analysis involving both color coordinates and routine chemical analyses gave a correct pass/fail classification above 95 percent, sufficient to justify confirmatory studies involving a greater number of samples (Ortiz-Fernandez et al. 1995). Color coordinates can be used in a similar way to model the aging of port wine (Ortiz-Fernandez et al. 1996).

Other neural network application examples include the prediction of foam characteristics from physicochemical properties

(Arteaga and Nakai 1993), loaf volume from flour physical and chemical properties (Horimoto et al. 1995), sensory quality of olive oil (Angerosa et al. 1996), and milk shelf life from gas chromatograph data (Vallejo-Cordoba et al. 1995).

A number of studies involved comparison of results using neural nets and discriminant analysis. The neural network approach outperformed multidimensional analysis in work on seed identification (Chtioui et al. 1996), prediction of flour loaf volume (Horimoto et al. 1995), prediction of turkey meat color from pH and color coordinates (Sante et al. 1996), and work on sensory-instrumental comparisons (Bardot et al. 1994).

The Scheffé Model

Polynomials can be used to relate mixture composition directly to tristimulus color properties (Scheffé 1963). For example, a mixture having two colorant components will describe a line in three-dimensional space. Three components may result in a two-dimensional surface, while four components results in a three-dimensional volume. Empirical models based on three-dimensional mixture space may be solved, theoretically uniquely, for component values as a function of color dimension. The sum of the components of the mixture must be constrained at 1. That is, in terms of (L, a, b) coordinates:

$$L = f_L(C_1, C_2, C_3, C_4)$$

$$a = f_a(C_1, C_2, C_3, C_4)$$

$$b = f_b(C_1, C_2, C_3, C_4)$$

$$1 = C_1 + C_2 + C_3 + C_4$$

where f_i is a general (continuous differentiable) functional relationship for each color response and each C is a color-active ingredient within the mix. Polynomial equations can be used to describe these relationships. The precision obtained is related to the number of samples tested within the mixture space. Color response surfaces can be visualized using contour plots that depict color variation for a series of compositions (Alman and Pfeifer 1987). These polynomials have been applied to food formulation in general (Hare 1974), to colorant mixture models

(Alman and Pfeifer 1987), to wine blending (Ayala et al. 1996; Negueruela et al. 1988), and to meat processing (Ellekjaer et al. 1996; MacDougall et al. 1988; MacDougall and Allen 1984).

Scheffé models and triangles were constructed for the lightness, hue angle, and chroma of a raw and cooked homogenized meat product. The color-active ingredients in the product were ground chicken (M), ground pork fat (F), water (W), and blood solution. The same concentrations of salt, tripolyphosphate, and sodium nitrite were included in each sample. Seven basic formulations were used to determine the polynomials leading to the construction of each triangle. These formulations were located at the vertices, midsides, and centroid of the triangle. The samples located at the vertices of the equilateral triangles corresponded to

$$Z_1 (0.9M, 0.0F, 0.1W);$$

$$Z_2(0.6M, 0.3F, 0.1W); Z_3 (0.6M, 0.0F, 0.4W)$$

The Scheffé diagrams constructed provided a systematic and sensitive method of examining the effects of formulation on product color. Figure 9–29, for example, shows the effects of composition on the hue h^* of the raw and cooked product. Samples plotted in row A triangles contained no blood; those in row B contained 2 percent commercial blood solution substituted for meat (MacDougall et al. 1988).

A similar approach with sensory data was used when optimizing the color of commercial port blends. Blends of tawny, ruby, and white ports were ratio scaled for their *color intensity*, *redness*, *brownness*, and *color* relationships to an ideal port. The results were plotted in a triangle, the vertices of which corresponded to 100 percent tawny, ruby, and white respectively. For example, Figure 9–30 shows the results of the *brownness* scaling. The size of each rectangle represents the deviation of the sample from an ideal tawny port. Hollow rectangles indicate less brown than ideal; the filled-in rectangles, browner than ideal. The shaded area corresponds to the blend composition closest to the ideal value of *brownness* (Williams et al. 1986).

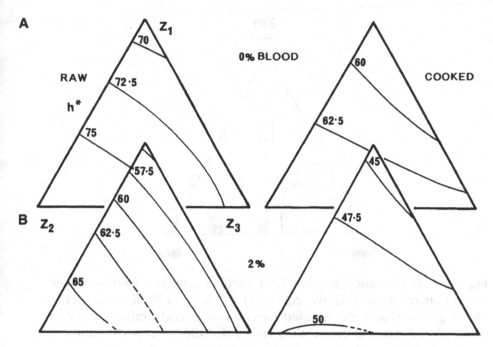

Figure 9–29 Models of the interaction of meat (M), fat (F), and water (W) on hue angle (h^*) in a raw and cooked meat product, with 0% blood substituted for meat (row A) or 2 percent blood substituted (row B). Z_1 (0.9M, 0.0F, 0.1W); Z_2 (0.6M, 0.3F, 0.1W); Z_3 (0.6M, 0.0F, 0.4W) (from MacDougall et al. 1988, with permission).

Reaction Kinetics

Minimization of quality loss and maintenance and improvement of food processes depend upon knowledge of the relationships between attribute or attribute change and treatment severity. This applies to quality loss, perhaps through heating or photodegradation, as well as to quality improvement, perhaps through the development of optimum browning. Available techniques predict the time taken for a product to change to a given acceptability or attribute level.

Equations have been developed to link surface color and firmness of Nemato tomatoes with time and fluctuating storage temperature. The ratio Hunter a_L/b_L was used to monitor color, and

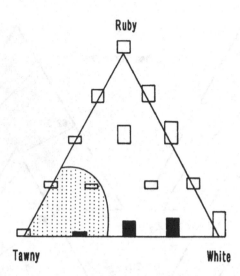

Figure 9–30 Deviation of ports from ideal *brownness* for assessors preferring tawny ports. Hollow rectangles = less than ideal; solid rectangles = greater than ideal. Shaded area denotes blend composition closest to ideal *brownness* (from Williams et al. 1986, with permission).

a deformation measurement was used for firmness. The method is based on the time-temperature-tolerance hypothesis developed for the storage of frozen foods (van Arsdel 1957). The product quality is first monitored at a series of different constant temperatures. For any one temperature, the deterioration rate can be expressed as the reciprocal of the storage life. A deterioration curve corresponding to the time-temperature storage history can be constructed. The area under the curve is a measure of the total deterioration occurring. For this hypothesis to be valid, the assumptions made are that the changes taking place must be additive and commutative and that there are no effects caused by the temperature change itself. These assumptions were found to be valid between 12 and 27°C for tomatoes. Below 12°C, chilling injury occurs, and above 27°C, the ripening process becomes disrupted (Thorne and Alvarez 1982). Dynamically changing storage temperatures can be applied to such models (Tijskens and Evelo 1994).

Hothouse tomatoes may take days to reach market because of the distances involved. A method of selecting the appropriate market for a particular batch of fruit would therefore be commercially advantageous. Light transmission measurements at the wavelengths 675 and 575 nm have been used to produce an index of ripening. Using the index, tomatoes can be sorted, the storage life under various conditions can be calculated, and the preferred market can be identified (Tchalukova and Krivoshiev 1978). The tomato handling and distribution management system uses computer models to convert (*L*, *a*, *b*) measurements to sensory equivalent scores. Models developed for Sunny tomatoes were used the following season to predict behavior in the Flora-Dade variety. Changes in vine-ripened Flora-Dade tomatoes ripened at 15, 21, and 25°C were reasonably well estimated, but less success was achieved for ethylene-treated fruit (Shewfelt et al. 1988). When tomatoes are wrapped in polymeric film, respiration and color development are affected. Computer modeling has been developed to predict the ripening of the fruit under constant temperature conditions from knowledge of the oxygen, carbon dioxide, and water vapor permeabilities of the film (Yang and Chinnan 1988).

The reaction kinetics used to determine heat treatments for product sterilization can also be used to predict appearance shelf life of, for example, sterilized fish pudding. Samples were prepared in thermal death time cans of diameter 73 mm, outer height 1 mm, and inner height 4 mm. Figure 9–31 shows the relationships obtained. As cooking time is increased, the product darkens and becomes less acceptable. Panel scores corresponding to **excellent, good,** and **acceptable** are also shown. Calculations of *z* values, equivalent to thermal death times, can be made by plotting the common logarithm of the rate of *appearance* deterioration against temperature. Figure 9–32 shows the logarithm of the heating time required for the panel score to go from **good** to **acceptable**, plotted against temperature (Ohlsson 1980).

Examples of *z* values and the slope of the line, with regression coefficients, determined from panel scores and instrumental *L* or *a* measurements, for a range of products are given in Table

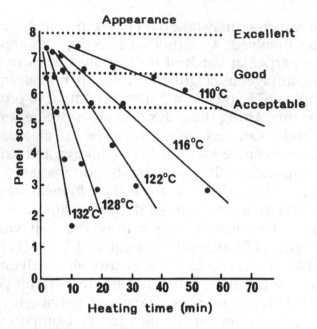

Figure 9–31 Appearance of sterilized fish pudding. Average values and linear regression line of the panel scores are plotted against the corrected heating time. The panel score levels were determined to correspond to **excellent, good,** and **acceptable** appearance (Ohlsson 1980, with permission).

9–5. The reason for the variation between products may be that the chemical and physical changes of major importance to the appearance change are different. The differences between the results of the sensory and instrumentally based determinations may be due to sample presentation problems of translucent samples.

It may be possible to develop Ohlsson's work to produce a classification of products according to the appearance behavior on storage. Exponential curves have been fitted to published temperature/time/quality data for fresh and frozen fruits and vegetables (Thorne and Meffert 1978). The two constants p and q of the exponential equation

$$t = pe^{qT}$$

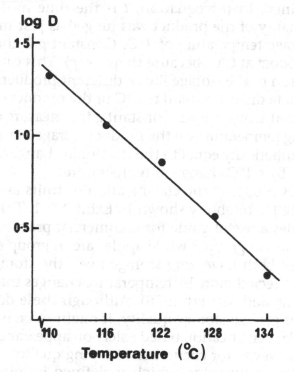

Figure 9–32 Appearance of sterilized fish pudding: the common logarithm of the heating time required for the panel score to go from **good** to **acceptable**, as a function of the heating temperature. The z value = 23°C (Ohlsson 1980).

Table 9–5 Appearance z Values (°C) and Linear Regression Coefficients for a Range of Products (from Ohlsson 1980, with permission)

	From Appearance Panel Scores	From Color Measurement Parameters
Fish pudding	23 (0.99)	25 (0.99) *L*
Liver paste	33 (0.86)	21 (0.96) *L*
Strained beef	21 (0.96)	22 (0.97) *L*
Strained vegetables	21 (0.72)	21 (0.97) *L*
Tomato sauce	16 (0.79)	28 (0.90) *a*
Vanilla sauce	21 (0.99)	20 (0.99) *L*

were determined. In the equation, *t* is the time in days after which the quality of the product was judged as **just unacceptable** at a storage temperature of *T*°C. Constant *p* is the storage life of the product at 0°C (because then $t = p$). This can be used as a comparison of the storage life of different products, even if the original data did not extend to 0°C or the product could not be stored at that temperature. Constant *q* is a measure of the effect of varying temperature on the product storage life. For small values, *q* is numerically equal to the fractional change in storage life produced by a 1°C change in temperature.

The authors proposed the classification of fruits and vegetables according to storability shown in Exhibit 9–1. This classification provides a useful guide for commercial practice. For example, pears are in group 3 while apples are in group 4: that is, although they both have long storage lives, the storage life of pears will be affected more by temperature changes than that of apples (Thorne and Meffert 1978). Although these determinations were for general storage quality, it ought to be possible to derive a similar classification based solely on appearance change properties. However, for cucumbers, keeping quality cannot be predicted using only color, which is defined by machine vision–derived average red (*R*) intensity divided by the average blue (*B*) intensity. Photosynthetic measurement data are also required (Schouten et al. 1999). This model has been validated under dynamic storage conditions for minimally processed endive (Vankerschaver et al. 1996).

Reaction kinetics can be studied in terms of individual color attributes. For example, the kinetics of color coordinates (*L**,*H**,*C**) have been investigated for heat-treated comminuted meat (Palombo and Wijngaards 1990a). Empirical models were fitted to plots of each color attribute versus time. During comminution, both *L** and *H** show a rapid increase and then, on storage, a gradual decline. The following form of nonlinear model was used for these two attributes:

$$H^* = m - n(1 - e^{-ct})$$

where *m* is the initial value, *n* the extent of the decrease, $m - n$ the final value, *c* the rate constant (min^{-1}), and *t* the time (min).

Exhibit 9–1 Proposed Classification of Fruits and Vegetables According to Storability (Thorne and Meffert 1978, with permission)

- Group 1, Delicate, $p < 10$; $q < 0.1$
  - raspberries
  - strawberries
  - mushrooms
  - plums
  - purslane
  - soft fruits

- Group 2, Half Hardy, $10 < p < 40$; $0.1 < q < 0.2$
  - lettuce
  - chicory
  - asparagus
  - endive
  - Victoria plums
  - cauliflower

- Group 3, Hardy, $p > 40$; $q > 0.1$
  - cucumber
  - tomatoes
  - carrots
  - pears

- Group 4, Long Storage, $p > 40$; $q < 0.1$
  - iceberg lettuce
  - oranges
  - potatoes
  - apples
  - cabbage
  - onions

The value of C^* is almost a mirror image of H^*, so the model used was

$$C^* = m - n(e^{-ct})$$

where m is the final value, n the extent of the increase, and $m - n$ the initial value.

Although these models were adequate for a temperature of 15°C, only H^* could be modeled at the higher temperatures of 40 to 60°C. The Arrhenius model was applied to H^*:

$$c = c_0\, e^{E/RT}$$

where c is the rate constant (min^{-1}), c_0 a constant (min^{-1}), E the activation energy (cal/mol), R the gas constant (1.986 cal/mol °K), and T the temperature (°K).

A mathematical model for the prediction of L^* values of a comminuted porcine lean meat was constructed using kinetic data collected at temperatures of 50 to 100°C and heating times of up to three hours (Palombo and Wijngaards 1990b). Model constants were obtained from regression analysis of L^* versus time. The nonlinear equation produced a reasonable fit to the experimental data:

$$L^* = g - je^{-ct}$$

where g is the final value (L^* units), j the extent of increase (L^* units), $g - j = L^*$ at zero heating time, c the rate constant (min^{-1}), and t the time (min). j was found to depend on temperature:

$$j = d + uT$$

where d is a constant (L^* units) and u a rate constant (L^* units/K). These derived constants were subjected to an Arrhenius equation treatment to produce a predictive equation for L^*. The latter equation was substantially verified using a replicate experiment.

A statistical evaluation of the Arrhenius model and its applicability to prediction of quality loss has been made (Cohen and Saguy 1985). Zero-order and first-order reactions were considered. The classic method of estimating Arrhenius parameters has two stages, similar to the above treatment by Ohlsson (1980). The first involves regression of the color parameter versus time at a series of temperatures to determine the rate constant. The second involves regression of the logarithm of the rate constant versus the reciprocal of the absolute temperature. From this, the activation energy and the pre-exponential factor can be deduced. A disadvantage of this method is the need to estimate many intermediate values. The Cohen and Saguy (1985) preferred method was a single nonlinear regression on all data points.

Temperature has a major effect on browning rate. At temperatures up to 60°C, browning is normally a zero-order reaction (Labuza and Saltmarch 1981). That is, it obeys the relationship

$$B = B_0 + kt$$

where B and B_0 are measures of the browning at time t and time zero and k is the rate or reaction constant. At higher temperatures, a plot of brown pigment versus time curves upwards in a first-order reaction: that is, when log B is proportional to time. Then there will be a maximum as the substrate becomes exhausted (Warmbier et al. 1976; Labuza and Schmidl 1986).

This was confirmed by Petriella et al. (1985), who studied the kinetics of nonenzymic color reactions in high–water activity model systems. The model systems consisted of the reactants lysine and glucose. Sodium chloride was added to control the water activity between 0.90 and 0.95. Added phosphates were sufficient to buffer the system during heating at 35, 45, and 55°C. Experiments without the lysine were performed to confirm that any contribution to the browning from caramelization could be neglected and that only Maillard reactions were the cause of brown pigment formation.

Values of X, Y, and Z were calculated, for illuminant C and the 2° standard observer, from spectral transmittance measurements made in 10-mm glass cells. The metric saturation (S_{uv}) has been used to illustrate the results. Figure 9–33A shows the effect of heating time, at 45°C, $a_w = 0.95$ and pH = 6, on the difference in metric saturation $(S - S_0)_{uv}$. S and S_0 are the values obtained from the heated and nonheated samples respectively. The upper line was obtained for the glucose-lysine system, the lower line for a similar system without lysine. The glucose-lysine model system consists of an initial induction phase, followed by one of rapid linear increase, then by a slowing of the color development. The linear phase rate of browning is a zero-order reaction. There were indications that a more complex model may need to be applied at the highest temperature.

The Arrhenius plot showing the effect of temperature on the rate of color development is shown in Figure 9–33B. Activation

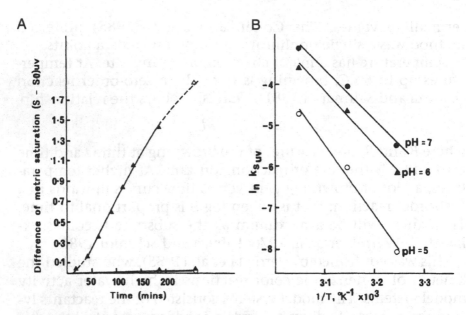

Figure 9–33 (A) Effect of heating time at 45°C on the increase in the color function $(S - S_0)_{uv}$ for a glucose-lysine model system at $a_w = 0.95$ and for a similar system without lysine. (B) Arrhenius plot for the rate of color development in a glucose-lysine model system at $a_w = 0.95$ (from Petriella et al. 1985, with permission).

energies (E_a) obtained from the slopes can be seen to be independent of pH at higher pHs. The increase at the lowest pH probably indicates a change of browning mechanism. The E_a values obtained of 20 to 30kcal/mol were in agreement with those found by others working with a wide range of model and real food systems. From a survey of the values of E_a found in the literature, the authors concluded that the possibly expected dependence of the reactants on the chemical identity did not occur. The Arrhenius plot showing the effect of pH can be seen in Figure 9–34. This plot illustrates the adverse effects of high temperature and high pH on the color deterioration. The effect of a_w was small (Petriella et al. 1985).

The extent of browning occurring on accelerated storage of glucose/aspartame and glucose/glycine model systems has been determined (Stamp and Labuza 1983). Temperatures from 70 to 100°C and an a_w of 0.80 were used. Extrapolation using zero-order kinetics was successful in predicting shelf life at lower temperatures and longer times. The authors found that the relationships obtained were applicable to reduced-calorie syrups and soft drinks.

A novel approach to accelerated shelf life testing has been suggested through a comprehensive kinetic equation for the browning of a pure glucose-glycine model system. A high acceleration ratio may be obtained by the combined effect of a relatively large number of factors in the equation, yet with only a

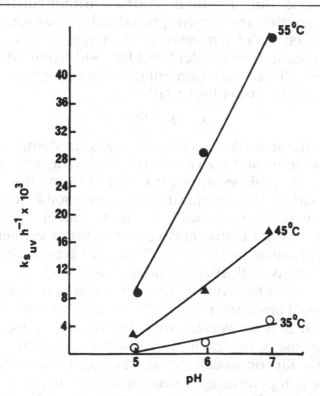

Figure 9–34 Effect of pH value on the rate of color development in a glucose-lysine model system at $a_w = 0.95$ (Petriella et al. 1985, with permission).

small departure from normal conditions (Weissman et al. 1993).

The mobility of small compounds in glassy and rubbery states is different, and the glass transition theory helps to explain kinetics of reactions requiring some mobility. Browning formation kinetics has been investigated for polyvinylpyrrolidone model systems of different molecular weights. Pigment formation rate was influenced significantly more by glass transition temperature than by water activity (Bell 1996). Crystallization of the matrix and structure collapse also affect mobility (Buera and Karel 1995).

Nonenzymic browning of grapefruit juice has been modeled as a material exhibiting two rate processes. A slower, exponential lag phase, during which colorless intermediates of the browning reaction are formed, precedes the zero-order reaction phase. The effects of temperature (between 60 and 96°C) on both rate constants were calculated for solids contents between 11.2 and 62.5°Brix. Activation energies for each phase were calculated from the Arrhenius equation

$$k = Ke^{-E/RT}$$

where E is the activation energy, R the gas constant, T the absolute temperature, and K a constant found to depend on E. The length of the lag phase ranged from 10 to 60 minutes. This behaved linearly with temperature at constant solids content. Using these parameters, it was possible to model the extent of browning during thermal and concentration treatments. Successful application has been made to two typical commercial processes (Saguy et al. 1978b; Cohen et al. 1994).

Similar studies have linked kinetics of pea puree with quality of the canned processed product (Shin and Bhowmik 1995) and have found that conversion of chlorophyll to pheophytin is solely responsible for the loss of visual green color (Steet and Tong 1996). Kinetic models can also be used to simulate and optimize thermal processing of canned white tuna, leading to optimal retention of surface lightness (Banga et al. 1993). Also developed are conditions for maximum storage stability of cheddar cheese powder (Kilic et al. 1997) and a generalized kinetic model of onion browning as a function of time, water activity, and temperature (Rapusas and Driscoll 1995). Photodegradation of beta carotene in model dispersions fits a first-order kinetic model, the relative amounts of the *cis* isomer increasing with time of exposure (Pesek and Warthesen 1988). Fluorescent light produces photocatalyzed autoxidation of carotene, but the loss from a fat solution is less as fat saturation is increased. The loss can be further retarded by the presence of antioxidants (Carnevale et al. 1979). Color of meatball crusts changes during frying, with the lightness, redness, and yellowness decaying exponentially with time (Ateba and Mittal 1994).

Attempts have been made to extend mathematical modeling to the correlation of storage conditions and shelf life of dehydrated vegetable products, defined by off flavors or color changes (Villota et al. 1980). An empirical equation was produced:

$$\ln t_f = a_0 + E/R(1/T) + a_2(m - \text{BET})$$

where t_f is the time of failure (days), m the moisture content, BET the monolayer moisture content (grams of water per grams sample), and a_0 and a_2 constants. The temperature function is based on the Arrhenius equation. Using data taken from the literature, the authors found that the correlation was lower when color rather than flavor was the cause of failure. It is possible that authors of the different studies from which the results were taken were more agreed on the criterion for flavor failure and more tolerant of the color accept/reject boundary.

There are two possible uses for determining reaction kinetics of a process. They are to deduce information about mechanisms, chemical or other, involved in the reactions, and to derive information that can be used to link process optimization with final product quality. Statistical and inferential implications of reaction kinetics are critically considered by van Boekel (1996).

UNDERSTANDING CHEMISTRY

Pigment Content Prediction

Many workers have attempted to use reflectance or light absorption measurements to predict pigment content. Such measurements can be quicker, easier, and more consistent than chemical analysis. However, attempts often fail because sample structure plays a significant part in how a product looks but does not affect its pigment concentration. Also, seasonal, climatic, and geographical conditions affect both structure and balance of pigments. Failure to demonstrate significant relationships between a reflected property and pigment content have been reported for meat products, due to structure differences (Trout et al. 1990); green beans, due to structure changes (Muftugil 1986); fruit, due to complex pigment mixtures (Lancaster et al. 1997); wine, due to metameric effects of anthocyanin pigments (Robinson et al. 1966); and salmon, due to oil content and structure (Foss et al. 1984).

Another reason for such failure may be the uneven distribution of pigment within a batch of food. Fresh cranberry samples, consisting of pale and deeply colored berries from two varieties of cranberry, were picked, and the reflectance was measured. The poor relationship produced between Agtron G reading and pigment content is shown in Figure 9–35. When the berries were hand sorted into uniformly colored samples, a much bet-

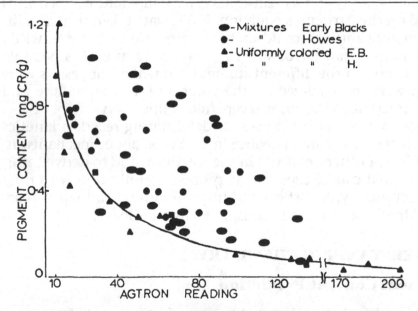

Figure 9–35 Plot of Agtron reading against pigment content for fresh cranberries. The triangles and squares refer to uniformly colored berries. The circles and ellipses refer to mixtures of well-colored and pale berries (Francis 1969, with permission).

ter relationship ($r = 0.961$) was obtained. This is shown by the continuous line. There are two effects that need comment. The relationship for the uniformly colored berries is log-log. This arises because a small amount of pigment added to a very dark berry will have no effect on the surface color, whereas the same amount of pigment in a very pale berry will have a large influence. The large scatter for the mixed-color berries arises because it is possible for a mixture to have the same overall surface color, but very different pigment contents. For example, the same overall color results from a mixture of 70 percent dark red with 30 percent white berries and a mixture of 80 percent medium red with 20 percent white berries. However, the pigment contents were 0.81 and 0.23 mg/g respectively (Francis 1969).

Attempts to use reflectance/pigment relationships may be divided according to the approach used, the tristimulus or the spectral.

The Tristimulus Approach

Thin-layer transreflectance measurements of wine can provide analytical data for both anthocyanin degradation and browning pigment development (Joslyn and Little 1967). This involves making thin-layer (4-mm) R, G, B transreflectance measurements of wine at two pH levels. Anthocyanins exist as equilibrium mixtures of the colored oxonium ion (the F+ form), and the colorless pseudobase. The proportion of the F+ form, present in the equilibrium mixture at pH 3.9, is approximately 15 percent. At pH < 1.0, the proportion increases to 100 percent. As pH is decreased, the visual color intensifies because the molar extinction coefficient markedly increases. When pure cyanidin chloride solutions and blends of rosé and white wines were used, it was found that there was a consistent relationship between anthocyanin content and both the chromaticity shift on acidification $(\Delta r^2 + \Delta g^2)^{1/2}$ and the difference in the blue absorbance, $\Delta - \log B$ (Little 1977b). Further work included real red wines. It was found possible to follow the brown (pH nonresponsive) component present, as well as the anthocyanin (pH responsive) degradation during storage.

Values of $-\log B$ are obtained for the wine adjusted to pH values of 3.4 (A) and 1.0 (B). $\Delta \log B(C)$ is calculated by subtraction:

$C = B - A$. Using the assumption that 100 percent of the pigment is anthocyanin, the value of C is used to calculate the corrected value of B (B_{cor}) at pH < 1.0. This is done using Figure 9–36, which was obtained using cyanidin solutions. The brown component (D) is determined from

$$D = B - B_{cor}$$

The percent brown component in the wine is:

$$(D/A) \times 100$$

The fate of the anthocyanins during storage can be followed by using the experimentally determined value of C. The relative concentration at day 1 is 100 percent. Subsequent determinations of C (C_n) are divided by C and multiplied by 100 to convert them to percentages. For a dark burgundy red wine stored

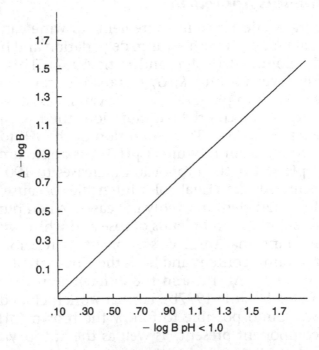

Figure 9–36 Cyanidin Cl, relationship of $\Delta - \log B$ to $-\log B$ at pH1 with increase in pigment concentration (Little 1977a, with permission).

at 37.7°C, this value fell to zero after 30 days (Little 1977b). The percentage polymeric color in blackberry wine has been measured as the pigment's resistance to bleaching (Rommel et al. 1992).

The transreflectance technique should be used for detailed studies of wine browning development because their high pigmentation and slight light scattering render the normal analytical absorption spectrophotometric methods inoperative. This general method can also be extended to other anthocyanin-containing foods (Little 1977a).

A relationship between tristimulus values and pigment content has been found for parsley. Freshly dried parsley leaves, as well as samples dried at two temperatures and stored for two years, were used in the investigation. The hue function $\tan^{-1}(b^*/a^*)$ was highly correlated with chlorophyll content ($r = 0.973$) and visual ranking ($r = 0.979$). It is possible that the simple linear regression obtained could be used to measure the degree of chlorophyll degradation and color quality during drying and storage (Berset and Caniaux 1983).

The Spectral Approach

The diffuse reflectance physics of a color-containing material may be summarized in terms of a reflectance-versus-wavelength spectrum. Because of nonlinear relationships, this cannot be used directly to indicate pigment concentration. However, the optical signature, a plot of log K/S against wavelength, has wide application to colorant formulation (Judd and Wyszecki 1975).

$$K/S = \frac{(1 - R_\infty)^2}{2R_\infty}$$

where K and S are the Kubelka-Munk absorption and scattering coefficients and R_∞ is the reflectance of an optically infinitely thick layer of sample (see Chapter 3).

The technique is used in various color-measuring industries to identify pigments and determine their concentrations. The optical signatures of samples of minced white cod to which suspensions of canthaxanthin have been added are shown in Figure 9–37. The proportions of pigment added were 1:2:4. The differ-

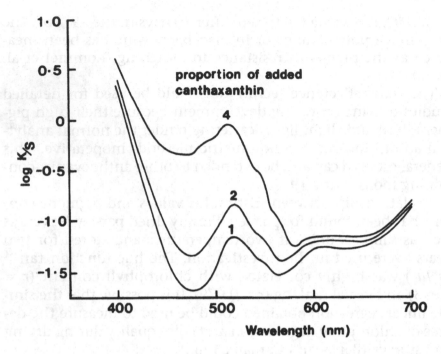

Figure 9–37 Optical signatures of cod with added canthaxanthin.

ences in spacing of the curves are in proportion to the differences in pigment concentration. That is, the log K/S peak values, at approximately 510 to 520 nm, have a linear relationship with the proportions of pigment added. The cod/canthaxanthin formulations were used as a model system for salmon (JB Hutchings, unpublished data). However, pigments do not usually appear singly in natural systems. For example, pigments in wild sea salmon have been found to contain approximately 93 percent astaxanthin, and wild river salmon approximately 85 percent astaxanthin (Skrede and Storebakken 1986).

Methods have been reported for determining the concentrations of dyes present in mixtures. Separation and isolation of components from a dye mixture are expensive and time consuming. The procedure is based on a nonlinear curve fitting of the visible spectrum of the pigments with a predicted function of the individual dyes. This has been used successfully with mixtures of food colorants (Saguy et al. 1978b) and in the continuous monitoring of the major beetroot pigments and browning during drying and storage (Saguy et al. 1978a).

The most common pigments present in fresh meat are oxymyoglobin (MbO_2), myoglobin (Mb), and metmyoglobin (MMb). Their relative proportions, together with scattering from the structure, govern the perceived color. Care is needed with sample presentation, as, for example, the presence of fat on the surface will negate the results. Figure 9–38 shows the reflectance spectra of the three forms of the pigment.

The K/S function from spectrophotometric reflectance measurement can be used to determine the percentages of each fresh

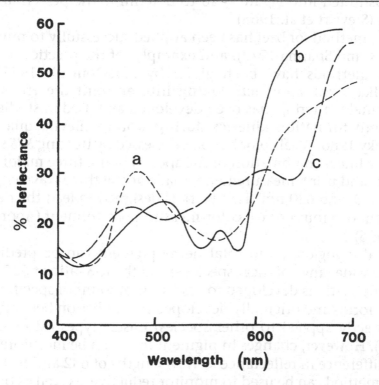

Figure 9–38 Reflectance spectra of the meat pigments: a, myoglobin; b, oxymyoglobin; c, metmyoglobin.

meat pigment present in a sample (Dean and Ball 1960). Percentage reflectance measurements are made at four wavelengths, and the values of two ratios are calculated:

$$\frac{K/S \text{ at } 507 \text{ nm}}{K/S \text{ at } 573 \text{ nm}} \quad \text{and} \quad \frac{K/S \text{ at } 473 \text{ nm}}{K/S \text{ at } 597 \text{ nm}}$$

The wavelengths were chosen on the basis of the extraction technique developed by Broumand et al. (1958). The K/S ratios are used, via Figure 9–39, to determine the percentages of MbO_2, Mb, and MMb present. Calculation of total pigment content can be based on the reflectance at the isobestic point of all three pigments. This occurs at 525 nm. The ratio determined

$$\frac{K/S \text{ at } 572 \text{ nm}}{K/S \text{ at } 525 \text{ nm}}$$

is substituted into Figure 9–40 to determine the percentage of MMb (Stewart et al. 1965).

This method for beef has been applied successfully to mutton (Attrey and Sharma 1979), and examples of the practical use of these methods have been given by MacDougall (1977). A Kubelka-Munk approach, taking into account the effects of structural scattering, has been developed and used in studies of pigment formation kinetics during storage (Bevilacqua and Zaritzky 1986). Veal color has been assessed by treating K/S spectra as a linear combination of the spectra of the three major pigments and using measurements made at the three wavelengths 480, 570, and 620 nm. This work is designed to lead the development of a routine veal color-measuring instrument (Aporta et al. 1996).

Total myoglobin and total heme pigment can be predicted over a wide range of meat species using the K/S ratio at 525 nm. Other equations developed to assess the amount of specific pigment forms and originally developed for studies on beef cannot be directly applied to other species/muscle types (Moss et al. 1994). However, changes in pigment form can be monitored by the difference in reflectance at wavelengths of 632 and 614 nm. This method can be used to monitor reductive, as well as oxidative, changes between Mb and MMb occurring in a single piece of meat (Eagerman et al. 1978).

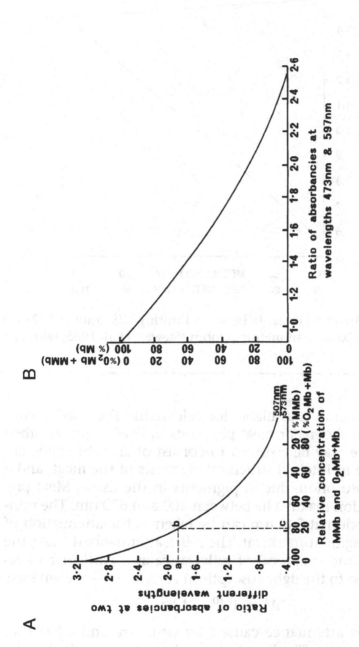

Figure 9–39 (A) Curve for determination of the relative concentration of MMb and (O_2Mb + Mb) after measuring the absorbancies at 507 and 573 nm. (B) Curve for determination of the relative concentration of Mb and (O_2Mb + MMb) after measuring the absorbancies at 473 and 597 nm (Broumand et al. 1958, with permission).

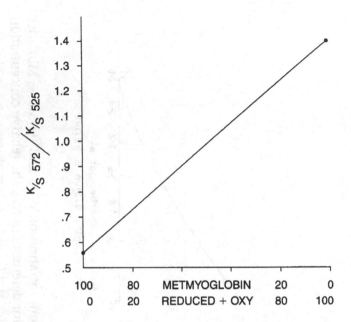

Figure 9–40 Assumed linearity between limiting *K/S* ratios at 572/525 nm for 0 and 100 percent metmyoglobin (Stewart et al. 1965, with permission).

Another method is available for calculating the relative concentrations of the major meat pigments in beef. Light absorbed by the surface may be assumed to consist of an achromatic absorption, due to internal structural elements of the meat, and a chromatic absorption, due to pigments in the tissue. Meat pigment absorption maxima lie between 400 and 630 nm. The minimum absorption at 730 nm can be taken as the attenuation of the light by pigment-free meat. The reflex attenuance (i.e., A^λ, the logarithm of the reciprocal of reflectance at a particular wavelength, related to the light absorption) consists of two elements:

$$A^\lambda = A^{730} + A_p^\lambda$$

where A^{730} is attenuance caused by structure and A_p^λ is that caused by pigment. The Beer-Lambert law can be applied to the pigment absorption element,

$$A_p^\lambda = e^\lambda c d$$

where e^λ is the molar absorption coefficient (MAC) at λ wave-length, c is molar concentration of pigments, and d is the path length of light in the surface layer. For a mixture of three pigments, the MAC comprises the fractional contribution, x, y, z from each, where $x + y + z = 1$. Using attenuation values at the isobestic points, the contribution of the three pigments can be calculated. All three MAC values at 525 nm are equal, and

$$A_p^{525} = A^{525} - A^{730}$$

depends only on the total concentration of all myoglobin derivatives and independent of their proportions. For equal path lengths, the difference in their A_p^{525} absorbances is directly proportional to the difference in pigment concentration in the sample. At the 572-nm isobestic point, the MACs of myoglobin and oxymyoglobin are equal. The MAC of a mixture of the three derivatives at 572 nm is

$$e^{572} = e_{myo\,=\,ox}^{572} (1 - y) + e_{met}^{572} y$$

To calculate y, the metmyoglobin fraction, consider the ratio:

$$a_1 = \frac{A_p^{572}}{A_p^{525}}$$

The ratio of two absorbencies of the same meat sample is equal to the ratio of the MAC values:

$$a_1 = \frac{e_{myo}^{572} (1 - y) + e_{met}^{572} y}{e^{525}}$$

After substituting the MAC values from Table 9–6, the value of y can be calculated:

$$y = 1.395 - a_1$$

where

$$a_1 = \frac{A^{572} - A^{730}}{A^{525} - A^{730}}$$

Table 9–6 Millimolar Absorbance Coefficients of Myoglobin, Oxymyoglobin, and Metmyoglobin at Some Isobestic Points

	Wavelength (nm)		
Pigment Form	473	525	572
Myoglobin	4.4	7.6	10.6
Oxymyoglobin	7.6	7.6	10.6
Metmyoglobin	7.6	7.6	3.0

Source: Reprinted with permission from H. Broumand, C.O. Ball, and E.F. Stier, Factors Affecting the Quality of Prepackaged Meat, II: Determining the Proportions of Heme Derivatives in Fresh Meat, *Food Technology*, Vol. 12, pp. 65–77, © 1958, Institute of Food Technologists.

By analogy, and at 473 nm, where the MACs of oxymyoglobin and metmyoglobin are equal, the fraction x of myoglobin can be calculated as

$$a_2 = \frac{A^{473} - A^{730}}{A^{525} - A^{730}}$$

$$a_2 = \frac{A_p^{473}}{A_p^{525}} = \frac{e_{met=ox}^{473}(1 - x) + e_{myo}^{473}x}{e^{525}}$$

$$x = 2.375 \, (1 - a_2)$$

Finally, the oxymyoglobin fraction may be obtained from (Krzywicki 1979):

$$z = 1 - (x + y)$$

Reflectance measurements are also used to follow fading in cooked cured meats. Subjective judgments of fading are complicated because a darker sample will appear less faded than a pale sample when exposed to the same illumination. Hence, an objective measurement must take into account the limits of fading as well as the fading itself. The difference between these parameters represents the pigment change from nitrosomyochrome to metmyochrome. This change has been monitored successfully using the ratios R_{570}/R_{650}, R_{540}/R_{560} (Barton 1967), and R_{650}/R_{570} (American Meat Science Association [AMSA] 1991). Reference values for the R_{650}/R_{570} nm ratio and cured color intensity are no cured color, 1.1; moderate fade, 1.6; less intense by noticeable cured color, 1.7 to 2.0; and excellent cured color, 2.2 to 2.6. A tan or brownish color may develop during cured-meat illumination. This discoloration and fading decreases *a* and increases *b* values. The *a/b* ratio is sensitive to shifts from pink to tan, or red to maroon, or red to brown (AMSA 1991). For cooked meat, there are significant changes at the wavelengths 545, 580, and 630 nm. Hue angle measurements are useful for measurement of cooked pork (AMSA 1991).

It is possible to increase the sensitivity of reflectance spectra by eliminating the regularly reflected component of the reflected light or by using a multiplicative scatter correction (see Chapter 3).

Pigment Behavior

Reflectance spectrophotometry and tristimulus colorimetry may be used to study pigment behavior in food materials. Figure 9–41 shows two reflectance spectra of homogenized tuna.

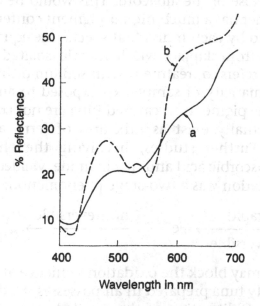

Figure 9–41 Reflectance spectra of homogenized canned tuna. (a) Untreated; (b) after treatment with Na dithionite (Little 1969c, with permission).

Tuna prepared in air has the spectrum shown in curve a. It was originally assumed that this was caused by pigment in the oxidized ferrihemochrome state (Brown et al. 1958). All attempts to protect the sample from oxidization by preparation under nitrogen failed to yield curve b, the highly differentiated spectrum characteristic of ferrohemochromes. In fact, a spectrum very similar to curve a was obtained. Curve b, exhibiting the strong absorption maxima at 560, 530, and 425 nm, was obtained only after application of sodium dithionite to the sample surface. It was observed that upon exposure to air, curve a was once again obtained. This was studied further using tristimulus methods (Figure 9–42).

Plots 1 and 2 were obtained with canned albacore and skipjack tuna respectively. Points A refer to the chromaticities of samples homogenized in air, and points N to those of samples homogenized in nitrogen. R_A and R_N refer respectively to samples A and N treated with sodium dithionite. The lengths of vectors A to R_A and N to R_N indicate the relative magnitudes of the chromaticity shifts. The shifts produced by the skipjack are longer than those of the albacore. This would be expected because the former has a much higher pigment content. However, shifts produced by each individual species are equal. Points 3N were obtained from skipjack whole muscle isolated under nitrogen. R_N again refers to treatment with sodium dithionite, while A is the chromaticity of sample R_N exposed to air. Little concluded that the pigments in canned tuna are not completely reduced but actually exist as mixtures of ferri- and ferrohemochromes. Further studies, involving the effects of the antioxidants ascorbic acid and niacinamide, yielded the conclusion that oxidation was a two-stage phenomenon:

$$Fe^{++} \xrightarrow[\text{reversible}]{\text{rapid}} Fe^{+++} \xrightarrow{\text{nonreversible}} \text{oxidative degradation}$$

Antioxidants may block the oxidation sequence at Fe^{+++}.

Good-quality tuna prepared in air possesses the differentiated spectral form. Obliteration of the detailed spectral characteristics of the homogenized sample is caused by the altered physi-

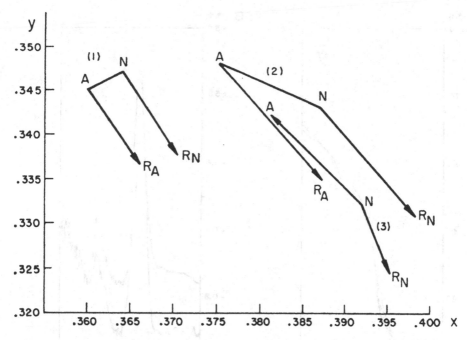

Figure 9–42 Comparison of chromaticity coordinates and chromaticity shifts for samples of canned tuna prepared under N_2 and in air (Little 1969c, with permission).

cal state of the sample. Pigments exhibit an absorption coefficient K, and the structural components of the matrix contribute a light-scattering coefficient S. If the value of the ratio S/K is large, then the effect of the pigment is masked. The effect of the sodium dithionite is to increase K, thus reducing the ratio and the masking effect of the background (Little 1969c).

Pigment changes often result in minor changes to reflectance spectra. If the spectra are reproducible, they can be examined as difference spectra. For example, Figure 9–43 shows reflectance and difference spectra of ground-carrot samples dried at temperatures of 40, 70, and 100°C. High-temperature drying causes a *trans/cis* isomerization of the alpha and beta carotenes. This is accompanied by changes particularly well shown by reflectance spectra difference plots (Marty et al. 1988).

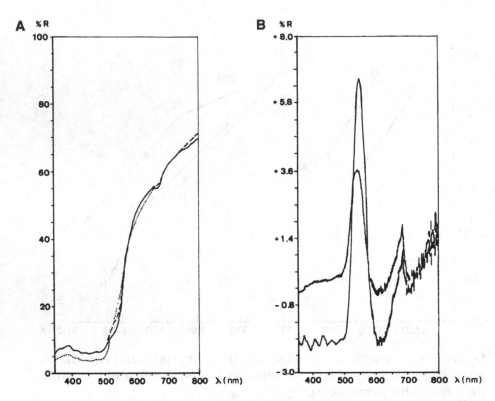

Figure 9–43 (A) Reflection spectra of ground dried carrots. (B) Difference spectra between ground samples dried at 70°C and 100°C and the sample dried at 40°C. *Source*: Reprinted with permission from C. Marty, A. Lebert, E. Lesellier, C. Berset, and J. Bimbenet, Color Control of Carrots Subject to Different Drying Experimental Conditions, in *Proceedings of the Congress on Progress in Food Preservation Processes*, pp. 81–88, © 1989, ENSIA.

REFERENCES

Alman DH, Pfeifer CG. (1987). Empirical colorant mixture models. *Color Res Appl* 12:210–222.

American Meat Science Association Committee. (1991). *Guidelines for meat color evaluation*. Chicago: National Live Stock and Meat Board.

Amiot MJ, Tacchini M, Aubert S, Nicolas J. (1992). Phenolic composition and browning susceptibility of various apple cultivars at maturity. *J Food Sci* 57:958–962.

Andrews GR, Morant SV. (1987). Lactulose content, colour and the organoleptic assessment of ultra heat treated and sterilized milks. *J Dairy Res* 54:493–507.

Añgerosa F, Giancinto LD, Vito R, Cumitini S. (1996). Sensory evaluation of virgin olive oils by artificial neural network processing of dynamic headspace chromatographic data. *J Sci Food Agric* 72:323–328.

Aporta J, Hernández B, Sañudo C. (1996). Veal colour assessment with three wavelengths. *Meat Sci* 44:113–123.

Arteaga GE, Nakai S. (1993). Predicting protein functionality with artificial neural networks: foaming and emulsifying properties. *J Food Sci* 58: 1152–1156.

Ateba P, Mittal G. (1994). Dynamics of crust formation and kinetics of quality changes during frying of meatballs. *J Food Sci* 59:1275–1278, 1290.

Attrey DP, Sharma TR. (1979). Estimation of fresh mutton color by reflectance spectrophotometry. *J Food Sci* 44:916–917.

Ayala ZF, Echavarri GJF, Negueruela SAI. (1996). Comparison of colour calculation methods of three rose wine mixtures. *Bull OIV* 69:973–986.

Baardseth P, Skrede G, Naes T, Thomassen MS, Iversen A, Kaaber L. (1988). A comparison of CIE (1976) L*a*b* values obtained from two different instruments on several commodities. *J Food Sci* 53:1737–1742.

Bakker J, Arnold GM. (1993). Analysis of sensory and chemical data for color evaluation of a range of port red wines. *Am J Enol Viticulture* 44: 27–34.

Banga JR, Alonso AA, Gallardo JM, Perez Martin RI. (1993). Kinetics of thermal degradation of thiamine and surface colour in canned tuna. *Z Lebens Unter Forsch* 197: 127–131.

Bardot I, Bochereau L, Martin N, Palagos B. (1994). Sensory-instrumental correlations by combining data analysis and neural network techniques. *Food Qual Pref* 5:159–166.

Barton PA. (1967). Measurement of colour stability of cooked cured meats to light and air. *J Sci Food Agric* 18:298–307.

Beale R, Jackson T. (1990). *Neural computing: an introduction.* Bristol, UK: Adam Hilger.

Bean sorter shapes up. (1995). *Food Processing* 56(11):105–106.

Begum S, Brewer MS. (1997). Microwave blanching effects on color, chemical and sensory characteristics of frozen asparagus. *J Food Qual* 20: 471–481.

Bell LN. (1996). Kinetics of non-enzymatic browning in amorphous solid systems: distinguishing the effects of water activity and the glass transition. *Food Res Int* 28:591–597.

Berardi LC, Martinez WH, Boudreaux GJ, Frampton VL. (1966). Rapid reproducible procedure for wafers of dried foods. *Food Technol* 20:124–127.

Berset C, Caniaux P. (1983). Relationship between color evaluation and chlorophyllin pigment content in dried parsley leaves. *J Food Sci* 48: 1854–1957, 1877.

Bevilacqua AE, Zaritzky NE. (1986). Rate of pigment modifications in packaged refrigerated beef using reflectance spectrophotometry. *J Food Processing Preservation* 10:1–18.

Billmeyer FW, Hemmendinger H. (1981). Instrumentation for colour measurement and its performance. In *Golden Jubilee of Colour in the CIE: proceeding of a Colour Group (GB) symposium*. Bradford, UK: Society of Dyers and Colourists.

Birth GS, McGee JB, Cavalette C, Chan H. (1978). *An optical technique for measuring papaya maturity*. Paper no. 78-3071. St. Joseph, Mich: American Society of Agricultural Engineering.

Birth GS, Norris KH. (1965). *Difference meter for measuring interior quality of foods and pigments in biological tissues*. Tech. Bull. 1341. Washington, DC: U.S. Department of Agriculture.

Birth GS, Norris KH, Yeatman JN. (1957). Nondestructive measurement of internal colour of tomatoes by spectral transmission. *Food Technol* 11:552–555.

Bittner DR, Norris KH. (1968). Optical properties of selected fruits vs maturity. *Trans Am Soc Agric Eng* 11:534–536.

Boulianne M, King AJ. (1995). Biochemical and color characteristics of skinless boneless pale chicken breast. *Poultry Sci* 74:1693–1698.

Bowles LH, Gullett EA. (1976). Color classification of honey using reflectance measurements. *J Inst Can Sci Technol Aliment* 9:125–129.

Brennan RM, Hunter EA, Muir DD. (1997). Genotypic effects on sensory quality of blackcurrant juice using descriptive sensory profiling. *Food Res Int* 30:381–390.

Brimelow CJB. (1987). Measurement of tomato paste colour: investigation of some method variables. In *Physical properties of foods 2*, ed. R Jowitt, F Escher, M Kent, B McKenna, M Roques. London: Elsevier Applied Science.

Broumand H, Ball CO, Stier EF. (1958). Factors affecting the quality of prepackaged meat, II: determining the proportions of heme derivatives in fresh meat. *Food Technol* 12:65–77.

Brown WD, Tappel AL, Olcott HS. (1958). The pigments of off-color cooked tuna meat. *Food Res* 23:262–268.

Buera MP, Karel M. (1995). Effect of physical changes on the rates of nonenzymatic browning and related reactions. *Food Chem* 52:167–173.

Buera MP, Lozano RD, Petriella C. (1986). Definition of colour in the nonenzymatic browning process. *Farbe* 32/33:318–322.

Canada Agricultural Standards Act and the honey regulations. (1958). Report 1967. 1779.

Carnevale J, Cole ER, Crank G. (1979). Fluorescent light catalyzed autoxidation of beta-carotene. *J Agric Food Chem* 27:462–463.

Chandley P, Smedley SM, Hammond RV. (1995). The application of new technology in pursuit of quality. In *Proceedings of the European Brewing Convention, 25th Congress*, Brussels, 684–694.

Choi K, Lee G, Han YJ, Bunn JM. (1995). Tomato maturity evaluation using color image analysis. *Trans Am Soc Agric Eng* 38:171–176.

Chtioui Y, Bertrand D, Dattée Y, Devaux M-F. (1996). Identification of seeds by colour imaging: comparison of discriminant analysis and artificial neural network. *J Sci Food Agric* 71:433–441.

Clydesdale FM. (1976). Instrumental techniques for color measurement of foods. *Food Technol* 30 (Oct):52–59.

Clydesdale FM. (1978). Colorimetry: methodology and applications. *CRC Crit Rev Food Sci Nutr* 10:243–301.

Clydesdale FM. (1994). Changes in color and flavor and their effect on sensory perception in the elderly. *Nutr Rev* 52(8):S19–S20.

Coatney CW. (1986). Color sensing for automatic process control. *Cereal Foods World* 31:404–405.

Cohen E, Birk Y, Mannheim CH, Saguy IS. (1994). Kinetic parameter estimation for quality change during continuous thermal processing of grapefruit juice. *J Food Sci* 59:155–158.

Cohen E, Saguy I. (1985). Statistical evaluation of Arrhenius model and its applicability in prediction of food quality losses. *J Food Processing Preservation* 9:273–290.

Cole CGB, Roberts JJ. (1997). Gelatine colour measurement. *Meat Sci* 45:23–31.

Coles GD. (1997). An objective dry pea "colour" scoring system for commercial and plant breeding applications. *J Sci Food Agric* 74:435–440.

Colosie SS, Breeze JE. (1973). An electronic apple redness meter. *J Food Sci* 38:965–967.

Dean RW, Ball CO. (1960). Analysis of the myoglobin fractions on the surfaces of beef cuts. *Food Technol* 14:271–286.

Deshpande SS, Gunasekaran S, Paulson MR. (1984). Nondestructive optical methods of food quality evaluation. *CRC Crit Rev Food Sci Nut* 21:232–379.

Dolan KD, Singh RP, Wells JH. (1985). Evaluation of time-temperature related quality changes in ice cream during storage. *J Food Processing Preservation* 9:253–271.

Dowell FE. (1998). Automated color classification of single wheat kernels using visible and near-infrared reflectance. *Cereal Chem* 75:142–144.

Eagerman BA. (1978). Orange juice color measurement using general purpose tristimulus colorimeters. *J Food Sci* 43:428–430.

Eagerman BA, Clydesdale FM, Francis FJ. (1973a). Comparison of color scales for dark colored beverages. *J Food Sci* 38:1051–1055.

Eagerman BA, Clydesdale FM, Francis FJ. (1973b). Development of new transmission color scales for dark colored beverages. *J Food Sci* 38:1056–1059.

Eagerman BA, Clydesdale FM, Francis FJ. (1977). Determination of fresh meat color by objective methods. *J Food Sci* 42:707–710, 724.

Eagerman BA, Clydesdale FM, Francis FJ. (1978). A rapid reflectance procedure for following myoglobin oxidative or reductive changes in intact beef muscle. *J Food Sci* 43:468–469.

Edan YH, Pasternak DG, Ozer N, Shmulevich I, Rachmani D, Fallik E, Grinberg S. (1994). *Multisensor quality classification of tomatoes*. Paper no. 946032. St. Joseph, Mich: American Society of Agricultural Engineering.

Edan Y, Pasternak H, Shmulevich I, Rachmani D, Guedalia D, Grinberg S, Fallik E. (1997). Color and firmness classification of fresh market tomatoes. *J Food Sci* 62:793–796.

Ellekjaer MR, Naes T, Baardseth P. (1996). Milk proteins affect yield and sensory quality of cooked sausages. *J Food Sci* 61:660–666.

Elliott RJ, (1967). Effect of optical systems and sample preparation on the visible reflection spectra of pork muscle. *J Sci Food Agric* 18:332–338.

Finlay J, Dix A. (1996). *An introduction to artificial intelligence*. London: UCL Press.

Floyd CD, Rooney LW, Bockholt, AJ. (1995). Measuring desirable and undesirable color in white and yellow food corn. *Cereal Chem* 72:488–490.

Foss P, Storebakken T, Schiedt K, Liaaen-Jensen S, Austreng E, Strieff K. (1984). Carotenoids in diets for salmonids. *Aquaculture* 41:213–226.

Francis FJ. (1969). Pigment content and color in fruits and vegetables. *Food Technol* 23:32–36.

Francis FJ. (1972). Colorimetry of liquids. *Food Technol* 26(11):39–42, 46, 48.

Francis FJ, Clydesdale FM. (1975). *Food colorimetry: theory and applications*. Westport, Conn: Avi.

Gaffney JJ. (1970). *Photoelectric color sorting of vine-ripened tomatoes*. Marketing Research Rep. no. 868. Washington, DC: U.S. Department of Agriculture.

Geladi P, MacDougall DB, Martens H. (1985). Linearisation and scatter-correction for near-infrared and reflectance spectra of meat. *Appl Spectrosc* 39:491–500.

Gerrard DE, Gao X, Tan J. (1996). Beef marbling and color score determination by image processing. *J Food Sci* 61:145–148.

Gnanasekharan V, Shewfelt RL, Chinan MS. (1992). Detection of color changes in green vegetables. *J Food Sci* 57:149–154.

Grenander U, Manbeck KM. (1993). A stochastic shape and colour model for defect detection in potatoes. *J Comput Graph Stat* 2:131–151.

Guerrero S, Alzamora SM, Gerschenson LN. (1996). Optimisation of a combined factors technology for preserving banana puree to minimise colour changes using the response surface methodology. *J Food Eng* 28:307–322.

Haas R de. (1995). Colour sorting as a wheat screening stage. *Getreide, Mehl Brot* 49:200–201.

Hare LB. (1974). Mixture designs applied to food formulation. *Food Technol* 28(3):50–56, 62.

Harvey WJ, Grant DG, Lammerink JP. (1997). Physical and sensory changes during the development and storage of buttercup squash. *NZJ Crop Hort Sci* 25:341–351.

Heffington JC, Nilsson RH. (1990). Sorting technology for food processing plants. *Food Technol Int Eur*, 117–120.

Heinemann PH, Pathare NP, Morrow CT. (1996). An automated inspection station for machine-vision grading of potatoes. *Machine Vision Appl* 9:14–19.

Heinemann PH, Varghese ZA, Morrow CT, Sommer HJ. III, Crassweller RM. (1995). Machine vision inspection of "Golden Delicious" apples. *Appl Eng Agric* 11: 901–906.

Hetherington MJ, MacDougall DB. (1992). Optical properties and appearance characteristics of tomato fruit (*Lycopersocon esculentum*). *J Sci Food Agric* 59:537–543.

Hils AKA. (1987). Farbmessung von Tomatenmark: eine Massnahme zur Qualitatssicherung. *Ind Obst-Genuseverwert* 72:267–276.

Hongve D, Åkesson G. (1996). Spectrophotometric determination of water colour in Hazen units. *Water Res* 30:2771–2775.

Horimoto Y, Durance T, Nakai S, Lukow OM. (1995). Neural networks vs principal component regression for prediction of flour loaf volume in baking tests. *J Food Sci* 60:429–433.

Horwitz W. (1965). *Official methods of analysis*. 10th ed. Washington, DC: Association of Official Agricultural Chemists.

Hother K, Herold B, Geyer M. (1995). Detecting quality defects in apple tissue using spectral reflection measurement. *Gartenbauwiss* 60:162–166.

Huang I-L, Francis FJ, Clydesdale FC. (1970a). Colorimetry of foods: carrot puree. *J Food Sci* 35:771–773.

Huang I-L, Francis FJ, Clydesdale FC. (1970b). Colorimetry of foods: color measurement of squash using the Kubelka-Munk concept. *J Food Sci* 35:315–317.

Hung Y-C. (1990). Effect of curvature and surface area on colorimeter readings: a model study. *J Food Qual* 13:259–269.

Hunter RS. (1961). Direct reading colorimeter. *J Opt Soc Am* 51:552–554.

Hunter RS. (1967). Development of a citrus colorimeter. *Food Technol* 21:906–911.

Hunter RS, Harold RW. (1987). *The measurement of appearance*. New York: John Wiley.

Hutchings JB. (1969). Tristimulus colour measurement in the food industry. In *Proceedings of the First International Colour Association Congress*, 581–589. Stockholm: Swedish Colour Centre Foundation.

Hutchings JB. (1978). Psychophysics of colour and appearance in product development. In *Proceedings of a symposium on food colour and appearance*, 46–55. London: University of Surrey, the Colour Group (GB).

Hutchings JB, Gordon CJ. (1981). Translucency specification and its application to a model food system. In *Proceedings of the Fourth Congress of the International Colour Association*, vol. 2, ed. M Richter. West Berlin: Deutscher Verband Farbe.

Hutchings J, Suter H. (1983). The perception of wine colour. Paper presented to the Colour Group (GB). April. London.

Ishikawa Y, Makino Y, Satoh H, Hirata T. (1997). Evaluation of color changes in broccoli packaged with plastic films by computerized image analysis. *J Jpn Soc Food Sci Technol* 43:1170–1175.

Iversen AJ, Palm T. (1985). Multiplicative scatter correction of visible reflectance spectra in color determination of meat surfaces. *Appl Spectrosc* 39:641–646.

Johnson LE, Clydesdale FM, Francis FJ. (1976). Use of expanded color scales to predict chemical and visual changes in solutions. *J Food Sci* 41:74–77.

Joslyn MA, Little A. (1967). Relation of type and concentration of phenolics to the color and stability of rose wines. *Am J Enol Viticulture* 18:138–148.

Joubert E. (1995). Tristimulus colour measurement of rooibos tea extracts as an objective quality parameter. *Int J Food Sci Technol* 30:783–792.

Judd DB, Wyszecki G. (1975). *Color in business, science and industry*. 3rd ed. New York: John Wiley.

Kent M. (1987). Collaborative measurements on the colour of light-scattering foodstuffs. In *Physical properties of foods 2*, ed. R Jowitt, F Escher, M Kent, B McKenna, M Roques, 277–294. London: Elsevier Applied Science.

Kent M, Malkin F, Verrill JF, Henshall DJ, (1991). *The certification of a tomato paste colour reference tile*. Rep. EUR 13392. Luxembourg: Commission of the European Communities.

Kent M, Smith GL. (1987). Collaborative experiments in colour measurement. In *Physical properties of foods 2*, ed. R Jowitt, F Escher, M Kent, B McKenna, M Roques, 251–276. London: Elsevier Applied Science.

Khayat A. (1973). Some observations on the colour measurement of canned tuna. *J Food Sci* 38:716–717.

Kilic M, Muthukumarappan K, Gunasekaran S. (1997). Kinetics on nonenzymatic browning in cheddar cheese powder during storage. *J Food Processing Preservation* 21:379–393.

Kim MN, Saltmarch M, Labuza TP. (1981). Non-enzymatic browning of hygroscopic whey powders in open versus sealed pouches. *J Food Processing Preservation* 5:49–57.

Kramer A, Smith HR. (1947). Electrophotometric methods for measuring the ripeness and color of canned peaches and apricots. *Food Technol* 1:527–539.

Krzywicki K. (1979). Assessment of relative content of myoglobin, oxymyoglobin and metmyoglobin at the surface of beef. *Meat Sci* 3:1–10.

Kyung MK, Dong WS, Jae KC. (1996). Image processing system for color analysis of food. *Korean J Food Sci Technol* 28:786–789.

Labuza TP, Saltmarch M. (1981). The nonenzymatic browning reactions as affected by water in foods. In *Water activity: influences on food quality*, ed. LB Rockland, GF Stewart, 605–650. New York: Academic Press.

Labuza TP, Schmidl MK. (1986). Advances in the control of browning reactions in foods. In *Role of chemistry in the quality of processed food*, ed. OR Fennema, WH Chang, C-Y Lii, 65–95. Westport, Conn: Food and Nutrition Press.

Lancaster JE, Lister CE, Reay PF, Triggs CM. (1997). Influence of pigment composition of skin colour in a wide range of fruits and vegetables. *J Am Soc Hort Sci* 122:594–598.

Little AC. (1963). Evaluation of single-number expressions of color difference. *J Opt Soc Am* 53:293–296.

Little AC. (1964). Color measurement of translucent food samples. *J Food Sci* 29:782–789.

Little AC. (1969a). Reflectance characteristics of canned tuna, 1: Development of an objective method for evaluating color on an industry-wide basis. *Food Technol* 23:1301–1318.

Little AC. (1969b). Reflectance characteristics of canned tuna, 2: the relationship of pigment concentration to luminous reflectance and color evaluation. *Food Technol* 23:1466–1468.

Little AC. (1969c). Reflectance characteristics of canned tuna, 3: observations on physical and chemical properties of the pigment system. *Food Technol* 23:1468–1472.

Little AC. (1971). The color of white wine. *Am J Enol Viticulture* 22:144–149.

Little AC. (1972). Effect of pre- and post-mortem handling on reflectance characteristics of canned skipjack tuna. *J Food Sci* 37:502.

Little AC. (1973). Color evaluation of foods: correlation of objective facts with subjective impressions. In *Sensory evaluation of appearance of materials*, ASTM STP545, 109–127. Philadelphia: Am Society for Testing and Materials.

Little AC. (1976). Physical measurements as predictors of visual appearance. *Food Technol* 30(10):74–82.

Little AC. (1977a). Colorimetry of anthocyanin pigmented products: changes in pigment composition with time. *J Food Sci* 42:1570–1574.

Little AC. (1977b). Pigment analysis in red and rose wines by tristimulus colorimetry. *Am J Enol Viticulture* 28:166–170.

Little AC. (1980). Colorimetry of wines. *Color Res Appl* 5:51–56.

Little AC, Chichester CO, MacKinney G. (1958). On the color of coffee II. *Food Technol* 12:505–508.

Little AC, Liaw MW-Y. (1974). Blending wines to color. *Am J Enol Viticulture* 25:79–83.

Little AC, Martinsen C, Sceurman L. (1979). Color assessment of experimentally pigmented rainbow trout. *Color Res Appl* 4:92–95.

Little AC, Simms RJ. (1971). The color of white wine, III: the design, fabrication and testing of a new instrument for evaluating white wine color. *Am J Enol Viticulture* 22:203–209.

Loeff HW. (1974). Instrumental colour measurement of processed spinach and other leaf vegetables. *Confructa* 19:120–130.

Lopez-Briones G, Varoquaux P, Chambroy Y, Bouquant J, Bureau G, Pascat B. (1992). Storage of common mushroom under controlled atmospheres. *J Food Sci Technol* 27:493–505.

Lott RV. (1964). Variability in color and associated quality constituents in Golden Delicious apple packs. *Proc Am Soc Hort Sci* 83:139–148.

Lozano JE, Drudis-Biscarri R, Ibarz-Ribas A. (1994). Enzymatic browning in apple pulps. *J Food Sci* 59:564–567.

Luedeking S. (1991). *Introduction to neural networks*. Nevada City, Calif: California Scientific Software.

MacDougall DB. (1977). Colour in meat. In *Sensory properties of foods*, ed. GG Birch, JG Brennan, KJ Parker, 59–69. London: Applied Science Publishers.

MacDougall DB. (1981). Visual estimate of colour changes in meat under different illuminants. In *Proceedings of the Fourth Congress of the International Colour Association*, ed. M Richter. West Berlin: Deutsche Verband Farbe.

MacDougall DB. (1983). Instrumental assessment of the appearance of foods. In *Sensory quality in foods and beverages*, ed. AA Williams, RK Atkin, 121–139. London: Society of Chemical Industry.

MacDougall DB, Allen RA. (1984). Mathematical modelling in meat processing. In *Procedures of the 30th European Meeting of Meat Research Workers*, 298–299. Bristol: Meat Research Institute.

MacDougall DB, Brace J, Allen RA, Robinson JM. (1988). Model of the effect of the four major components of a homogenised meat product on its colour. In *Trends in modern meat technology 2*, ed. B Krol, PS van Roon, JH Houben, 15–21. Wageningen: Pudoc.

MacDougall DB, Granov M. (1997). Relationship between ultraviolet and visible spectra in Maillard reactions and CIELAB colour space and visual appearance. In *The Maillard reaction in foods and medicine*, ed. J O'Brien, HE Nursten, MJC Crabbe, JM Ames. Cambridge, UK: Royal Society of Chemistry.

MacDougall DB, Jones SJ. (1983). Elucidation of a transient green colour in beef fat. In *Uncommon applications of colour measurement: conference proceedings*. 2–4. London: Colour Group (GB).

MacKinney G, Little AC. (1962). *Color of foods*. Westport, Conn: Avi.

MacKinney G, Little AC, Brinner L. (1966). Visual appearance of foods. *Food Technol* 20:1300–1306.

Marty C, Lebert A, Lesellier E, Berset C, Bimbenet J. (1988). Color control of carrots subject to different drying experimental conditions. *Procedings of the International Symposium: Progress in food preservation processes*, vol. 2, 81–88. Brussels: CERIA.

McFarlane I. (1985). On-line color measurement and baking oven control. *Cereal Foods World* 30:386–388.

Meydev S, Saguy I, Kopelman IJ. (1977). Browning determination in citrus products. *J Agric Food Chem* 25:602– 604.

Millar SJ, Moss W, MacDougall DB, Stevenson MH. (1995). The effect of ionising radiation on the CIELAB colour co-ordinates of chicken breast meat as measured by different instruments. *Int J Food Sci Technol* 30:663– 674.

Moquet F, Guedes-Lafargue MR, Vedie R, Mamoun M, Olivier JM. (1997). Optimum measure of cap color in *Agricus bisporus* wild and cultivated strains. *J Food Sci* 62:1054–1056, 1079.

Moskowitz HR. (1977). Magnitude estimation: notes on what, how, when and why to use it. *J Food Qual* 3:195–227.

Moskowitz HR. (1985). *New directions for product testing and sensory analysis of foods*. Westport, Conn: Food and Nutrition Press.

Moss BW, Millar SJ, Stevenson MH. (1994). The use of reflectance measurements at selected wavelengths to predict the amount and proportion of myoglobin in a range of meat species. Paper presented at the 40th International Congress of Meat Science and Technology, The Hague, August/ September.

Motonaga Y, Kameoka T, Hashimoto A. (1997). Color development of tomato during post-ripening. In *Proceedings of the 8th Congress of the International Colour Association*, vol. 2, 929–932. Tokyo: Color Science Association of Japan.

Muftugil N. (1986). Effect of different types of blanching on the color and the ascorbic acid and chlorophyll contents of green beans. *J Food Processing Preservation* 10:69–76.

Nakano K. (1997). Application of neural networks to the color grading of apples. *Comput Electronics Agric* 18:105–116.

Negueruela AI, Echavarri JF, Los Arcos ML, Lopez de Castro MP. (1988). Contribution to the study of wines' color: application of Scheffe's design to calculus of color of three red wines mixtures. *Opt Pura Apl* 21:45–51.

Nimesh S, Delwiche MJ, Johnson RS. (1993). Image analysis methods for real-time color grading of stonefruit. *Comput Electronics Agric* 9:71–84.

O'Carroll P. (1995). Defining chocolate color. *World Ingredients* (Jan/Feb): 34–37.

O'Donoghue EM, Martin W. (1988). Juice loss as a simple method for measuring integrity changes in berry fruit. *J Hort Soc* 63:217–220.

Ohlsson T. (1980). Temperature dependence of sensory quality changes during thermal processing. *J Food Sci* 45:836–839, 847.

Oliver JR, Blakeney AB, Allen HM. (1993). The colour of flour streams as related to ash and pigment contents. *J Cereal Sci* 17:169–182.

Ortiz-Fernandez MC, Herrero-Gutierrez A, Sanchez-Pastor MS, Sarabia LA, Iniguez-Crespo M. (1995). The UNEQ, PLS, and MLF neural network methods in the modelling and prediction of colour of young red wines from the Denomination of Origin "Rioja." *Chemometrics Intell Lab Syst* 28: 273–285.

Ortiz-Fernandez MC, Sarabia LA, Symington C, Santamaria F, Íñiguez M. (1996). Analysis of aging and typification of vintage ports by partial least squares and soft independent modelling class analogy. *Analyst* 121: 1009–1013.

Palombo R, Wijngaards G. (1990a). Kinetic analysis of the effect of some processing factors on changes in color of comminuted meats during processing. *J Food Sci* 55:604–612.

Palombo R, Wijngaards G. (1990b). Predictive model for the lightness of comminuted porcine lean meat during heating. *J Food Sci* 55:601–603, 612.

Peleg M, Gomez B. (1974). External color as a maturity index of papaya fruits. *J Food Sci* 39:701–703.

Perkins-Veazie PM. (1991). Color changes in waxed turnips during storage. *J Food Qual* 14:313–319.

Pesek CA, Warthesen JJ. (1988). Characterisation of the photodegradation of beta-carotene in aqueous model systems. *J Food Sci* 53:1517–1520.

Petriella C, Resnik SL, Lozano RD, Chirife J. (1985). Kinetics of deteriorative reactions in model food systems of high water activity: color changes due to nonenzymatic browning. *J Food Sci* 50:622–626.

Philipsen DH, Clydesdale FM, Griffin RW, Stern P. (1995). Consumer age affects response to sensory characteristics of a cherry flavored beverage. *J Food Sci* 60:364–368.

Philpotts P. (1983). Eyes on the production line. *Food Processing* 52 (Jun): 27–29.

Piggott J. (1988). *Statistical procedures in foods research*. London: Elsevier.

Pla F, Juste F, Vicens M. (1993). Colour segmentation based on a light reflection model to locate citrus fruits for robotic harvesting. *Comput Electronics Agric* 9:53–70.

Polesello A, Gorini FL, Bertolo G. (1974). Valutazione del colore mediante riflettanza dei frutti sferici. *Agricoltura* 11:173–177.

Raggi A, Barbiroli G. (1993). Colour uniformity in discontinuous foodstuffs to define tolerances and acceptability. In *Proceedings of the Seventh Congress of the International Colour Association*, 199. Budapest: Technical University of Budapest.

Ramaswamy HS, Richards JF. (1980). A reflectance method to study the green-yellow changes in fruits and vegetables. *Can Inst Food Sci Technol J* 13:107–111.

Rankin SA, Brewer JA. (1998). Color of nonfat fluid milk as affected by fermentation. *J Food Sci* 63:178–180.

Rapusas RS, Driscoll RH. (1995). Kinetics of non-enzymatic browning in onion slices during isothermal heating. *J Food Eng* 24:417–429.

Rhodes DN. (1970). Meat quality: the influence of fatness of pigs on the eating quality of pork. *J Sci Food Agric* 21:572–575.

Riaz MN, Bushway AA. (1996). Effect of cultivars and weather change on Hunter 'L', hue angle and chroma values of red raspberry grown in Maine. *Fruit Varieties J* 50:131–135.

Robertson GL, Reeves MJ. (1981). Relationship between colour and brown pigment concentration in orange juices subjected to storage temperature abuse. *J Food Technol* 18:535–541.

Robinson WB, Weirs LD, Bartino JJ, Nattick LR. (1966). The relationship of anthocyanin composition to color stability of New York State wines. *Am J Enol Viticulture* 17:178–182.

Rodgers J, Wolf K, Willis N, Hamilton D, Ledbetter R, Stewart C. (1994). A comparative study of color measurement instrumentation. *Color Res Appl* 19:322–331.

Rodway E. (1993). *Specification for tomato paste and puree*, L80/1. Leatherhead, UK: Campden Food and Drink Research Association.

Rohm H, Jaros D. (1996). Colour of hard cheese, part 1. *Z Lebens Unter Forsch A* 203:241–244.

Romani RJ, Jacob FC, Sprock CM. (1962). Studies on the use of light transmission to assess the maturity of peaches, nectarines and plums. *Proc Am Soc Hort Sci* 80:220–225.

Rommel A, Heatherbell DA, Wrolstad RE. (1990). Red raspberry juice and wine: effect of processing and storage on anthocyanin pigment composition, color and appearance. *J Food Sci* 55:1011–1017.

Rommel A, Wrolstad RE, Heatherbell DA. (1992). Blackberry juice and wine: processing and storage effects on anthocyanin composition, color and appearance. *J Food Sci.* 57:385–391,410.

Ross EW, Shaw CP, Friel M. (1997). Color measurement as predictor of consumer ratings of military ration items. *J Food Qual* 20:427–439.

Saguy I, Kopelman IJ, Mizrahi S. (1978a). Computer-aided determination of beet pigments. *J Food Sci* 43:124–127.

Saguy I, Mizrahi S, Kopelman IJ. (1978b). Mathematical approach for the determination of dyes concentration in mixtures. *J Food Sci* 43:121–123, 134.

Saito A. (1969). Color in raw and cooked Atlantic salmon. *J Fisheries Board Can* 26:2234–2236.

Sandusky CL, Heath JL. (1996). Effect of background color, sample thickness and illuminant on the measurement of broiler meat color. *Poultry Sci* 75:1437–1442.

Sante VS, Lebert A, Pottier G, Ouali A. (1996). Comparison between two statistical models for prediction of turkey breast meat colour. *Meat Sci* 43:283–290.

Sapers GM, Douglas FW Jr. (1987). Measurement of enzymatic browning of cut surfaces in juice of raw apple and pear fruits. *J Food Sci* 52:1258–1262, 1285.

Scanlon MG, Zhou H-M, Curtis PS. (1993). Tristimulus assessment of flour and bread crumb colour with flours of increasing extraction rate. *J Cereal Sci* 17:33–45.

Scheffé H. (1963). Experiments with mixtures. *J Roy Stat Soc B* 20:344–360.

Schmidt PJ, Cuthbert RM. (1969). Colour sorting of Pacific salmon. In *Technical Conference on Fish Inspection and Quality Control*. Halifax, Canada.

Schmidt PJ, Idler DR. (1958). Predicting the color of canned sockeye salmon from the color of the raw flesh. *Food Technol* 12:44–48.

Schouten RE, Otma EC, van Kooten O, Tijskens LMM. (1999). Keeping quality of cucumber fruits predicted by the biological age. *Postharvest Biol Technol* (in press).

Scott AJ. (1995). On-line inspection of bakery and snack products. *Cereal Foods World* 40(1):15–18.

Sechrist EL. (1925). *The color grading of honey*. USDA circ. no. 364. Washington, DC: U.S. Department of Agriculture.

Setser CS. (1983). Color: reflections and transmissions. *J Food Qual* 6:183–197.

Shearer SA, Payne FA. (1990). Color and defect sorting of bell peppers using machine vision. *Trans Am Soc Agric Eng* 33:2045–2050.

Shewfelt RL, Thai CN, Davis JW. (1988). Prediction of changes in color of tomatoes during ripening at different constant temperatures. *J Food Sci* 1433–1437.

Shin S, Bhowmik SR. (1995). Thermal kinetic of colour changes in pea puree. *J Food Eng* 24:77–86.

Sidwell AP, Birth GS, Ernest JV, Golumbic C. (1961). The use of light transmittance techniques to estimate the chlorophyll content and stage of maturation of Elberta peaches. *Food Technol* 15(2):75–78.

Skarsaune SK, Shuey WC. (1975). The effect of several variables on instrument flour colour. *Cereal Foods World* 20:286–288, 292.

Skrede G. (1982). Quality characterisation of strawberries for industrial jam production. *J Sci Food Agric* 33:48–54.

Skrede G, Storebakken T. (1986). Characteristics of color in raw, baked and smoked wild and pen-reared Atlantic salmon. *J Food Sci* 51:804–808.

Skrede G, Naes T, Martens M. (1983). Visual color deterioration in blackcurrant syrup predicted by different instrumental variables. *J Food Sci* 48:1745–1749.

Slaughter DC, Harrell RC. (1989). Discriminating fruit for robotic harvest using color in natural outdoor scenes. *Trans Am Soc Agric Eng* 32:757–763.

Solah VA, Chiu PC, Crosbie GB. (1997). The measurement of colour in raw noodle sheets. In *Cereals '97: Proceedings of the 47th Australian Cereal Chemistry Conference, Perth*, ed. AW Tarr, AS Ross, CW Wrigley, 289–293. Melbourne: Royal Australian Chemical Institute, Cereal Chemistry Division.

Spayd SE, Morris JR. (1981). Influence of immature fruits on strawberry jam quality and storage stability. *J Food Sci* 46:414–418.

Stamp JA, Labuza TP. (1983). Kinetics of the Maillard reaction between aspartame and glucose in solution at high temperatures. *J Food Sci* 48:543–544, 547.

Staples LC. (1972). *Colorimetry of tea*. PhD diss., Rutgers University, New Brunswick, NJ.

Steet JA, Tong CH. (1996). Degradation kinetics of green color and chlorophylls in peas by colorimetry and HPLC. *J Food Sci* 61:924–927, 931.

Stewart MR, Zipser MW, Watts BM. (1965). The use of reflectance spectrophotometry for the assay of raw meat pigments. *J Food Sci* 30:464–469.

Swatland HJ, Haworth CR, Darkin F, Moccia RD. (1997). Fibre-optic spectrophotometry of raw, smoked and baked Arctic char (*Salvelinus alpinus*). *Food Res Int* 30:141–146.

Tan J, Gao X, Hsieh F. (1994). Extrudate characterisation by image processing. *J Food Sci* 59:1247–1250.

Tao Y, Heinemann PH, Varghese Z, Morrow CT, Sommer HJ. III. (1995). Machine vision for color inspection of potatoes and apples. *Trans Am Soc Agric Eng* 38:1555–1561.

Tchalukova R, Krivoshiev G. (1978). Color changes of hot-house grown tomatoes during storage. *J Food Sci* 43:218–221.

Tepper BJ. (1993). Effects of slight color variation on consumer acceptance of orange juice. *J Sensory Stud* 8:145–154.

Thorne S, Alvarez JSS. (1982). The effect of irregular storage temperatures on firmness and surface colour in tomatoes. *J Sci Food Agric* 33:671–676.

Thorne S, Meffert HFT. (1978). The storage life of fruits and vegetables. *J Food Qual* 2:102–112.

Tijskens LMM, Evelo RG. (1994). Modelling of tomatoes during postharvest storage. *Postharvest Biol Technol* 4:85–98.

Tisse C, Dordonnat M, Tisse C, Guerere M. (1994). Characterisation of honeys by colour analysis. *Ann Falsifications Expertise Chem Toxicol* 87(928):163–172.

Townsend GF. (1969). Optical density as a means of color classification of honey. *J Agric Res* 8:29–31.

Trant AS, Pangborn RM, Little AC. (1981). Potential fallacy of correlating hedonic responses with physical and chemical measurements. *J Food Sci* 46:583–588.

Trout GR, Chen CM, Dale S. (1990). Effect of calcium carbonate and sodium alginate on the textural characteristics, color and color stability of restructured pork chops. *J Food Sci* 55:38–42.

Vallejo-Cordoba B, Arteaga GE, Nahai S. (1995). Predicting milk shelf life based on artificial neural networks and headspace gas chromatograph data. *J Food Sci* 60:885–888.

van Arsdel WB. (1957). The time-temperature-tolerance of frozen foods: introduction. *Food Technol* 11:28–33.

van Boekel MAJS. (1996). Statistical aspects of kinetic modeling for food science problems. *J Food Sci* 61:477– 485, 489.

Vankerschaver K, Willcox F, Smout C, Hendricks M, Tobback P. (1996). Modeling and prediction of visual shelf life of minimally processed endive. *J Food Sci* 61:1094–1098.

Varghese Z, Morrow CT, Heinemann PH, Sommer HJ. III, Tao Y, Crassweller RM. (1991). *Automated inspection of Golden Delicious apples using color computer vision.* Paper no. 91-7002. St Joseph, Mich: American Society of Agriculture Engineering.

Villota R, Saguy I, Karel M. (1980). An equation correlating shelf life of dehydrated vegetable products with storage conditions. *J Food Sci* 45:398–399, 401.

Warmbier HC, Schnickles RA, Labuza TP. (1976). Effect of glycerol on nonenzymatic browning in solid intermediate moisture model food system. *J Food Sci* 41:528–531.

Weissman I, Ramon O, Kopelman IJ, Mizrahi S. (1993). A kinetic model for accelerated tests of Maillard browning in a liquid model system. *J Food Processing Preservation* 17:455–470.

Wenzel FW, Huggart RL. (1969). Instruments to solve problems with citrus products. *Food Technol* 23:147–150.

Williams AA, Baines CR, Finnie MS. (1986). Optimisation of colour in commercial port blends. *J Food Technol* 21:451–461.

Wythoff BJ. (1993). Back propagation neural networks: a tutorial. *Chemometrics Intell Lab Syst* 18:115–155.

Yamaguchi M, Howard FD, Luh BS, Leonard SJ. (1958). Effect of ripeness and harvest dates on the quality and composition of fresh canning tomatoes. *Proc Am Soc Hort Sci* 76:560–567.

Yang C, Chinnan MS. (1988). Computer modeling of gas composition and color development of tomatoes stored in polymeric film. *J Food Sci* 53: 869–872.

Yeatman JN, Norris KH. (1965). Evaluating internal quality of apples with new automatic fruit sorter. *Food Technol* 19:123.

Yeh JCH, Hamey LGC, Westcott T, Sung SKY. (1995). Colour bake inspection system using hybrid artificial neural networks. In *Proceedings of the 1995 IEEE International Conference on Neural Networks*, 37–42. Piscataway, NJ: Institute of Electrical and Electronic Engineers.

Young KW, Whittle KJ. (1985). Colour measurement of fish minces using Hunter L,a,b values. *J Sci Food Agric* 36:383–392.

Yuson SM, Francis FJ. (1962). Relation between visual color differences and tristimulus color readings for pureed carrots, spinach and pears. *J Food Sci* 27:295–302.

CHAPTER 10

Measurement of Other Appearance Properties

This chapter is devoted to the specification and measurement of appearance properties other than color. They are grouped under the headings of physical form, translucency, gloss, uniformity and pattern, temporal properties, and analysis of complex scenes. The principles of established techniques, together with relevant examples and applications, are included. The examples quoted inevitably apply to specific foods or systems, but the methodology is widely applicable.

PHYSICAL FORM

Structure is controlled by the organization of a number of elements, their binding together, and the relationships between the elements or their groupings. For example, the structural hierarchy of an animal body consists of atoms and their components, molecules, cell parts (organelles), cells, muscle parts (cell aggregates), muscles, organs, fatty tissues, bones, and animal body. With the naked eye, we perceive the grossest elements, the macrostructure. Nevertheless, the ultra- and microstructure control what we see. Cell parts consist of elementary structures built from polyhedrons, which are

- two-dimensional, flat ordered, linear, randomly branched networks
- three-dimensional, spatially ordered, randomly branched networks

405

- crystalline, symmetrical, spatially ordered, regularly branched structures

Such elementary structures can form randomly ordered aggregates, linear ordered unbranched fibers, and linear fibers with random branching. These three elements in turn form the basis of all homogeneous food structures, as shown in Figure 10–1.

Homogeneous materials consist of combinations of similar elements. Most foods, however, are embedded structures formed from heterogeneous combinations of dissimilar elements: that is, materials formed by a homogeneous structure surrounded by, or embedded with, another structure (Raeuber and Nikolaus 1980). These structures form the *visual structure* properties, including size, shape and *surface texture*, which comprises factors such as roughness, the presence of fibers, and granularity. All these attributes occur naturally in foods with which we are familiar. They concern the grower, sorter, packer, retailer, purchaser, and consumer. Product response to stress governs harvesting and packing machinery design. Surface convolution controls how a food is processed; physical complexities influence the technologist's attempts to fabricate natural-looking foods from more basic materials. Concerns of the consumer include considerations such as portion size and number of portions per unit weight.

Simple methods are often adequate for size specification. These include measurements of length, diameter, and the count per weight or volume. Measurements traditionally carried out on large volumes, such as bread loaves, are height, width (or average diameter), specific volume, and form ratio (width/height). Volumes of discrete objects, such as fried gluten balls (Chen et al. 1998) and loaves (Kühn and Grosch 1988), can be measured by rape seed displacement or laser profile (Scott 1995). Sieve analysis is appropriate for near-spherical objects, such as peas and potatoes, for which specification of the narrowest dimension may be sufficient (Szczesniak 1983). Sizes of smaller food particles such as flour can be determined using sieve analysis (Obuchowski and Bushok 1980), sedimentation rate, or laser beam diffractometry (McDonald 1994).

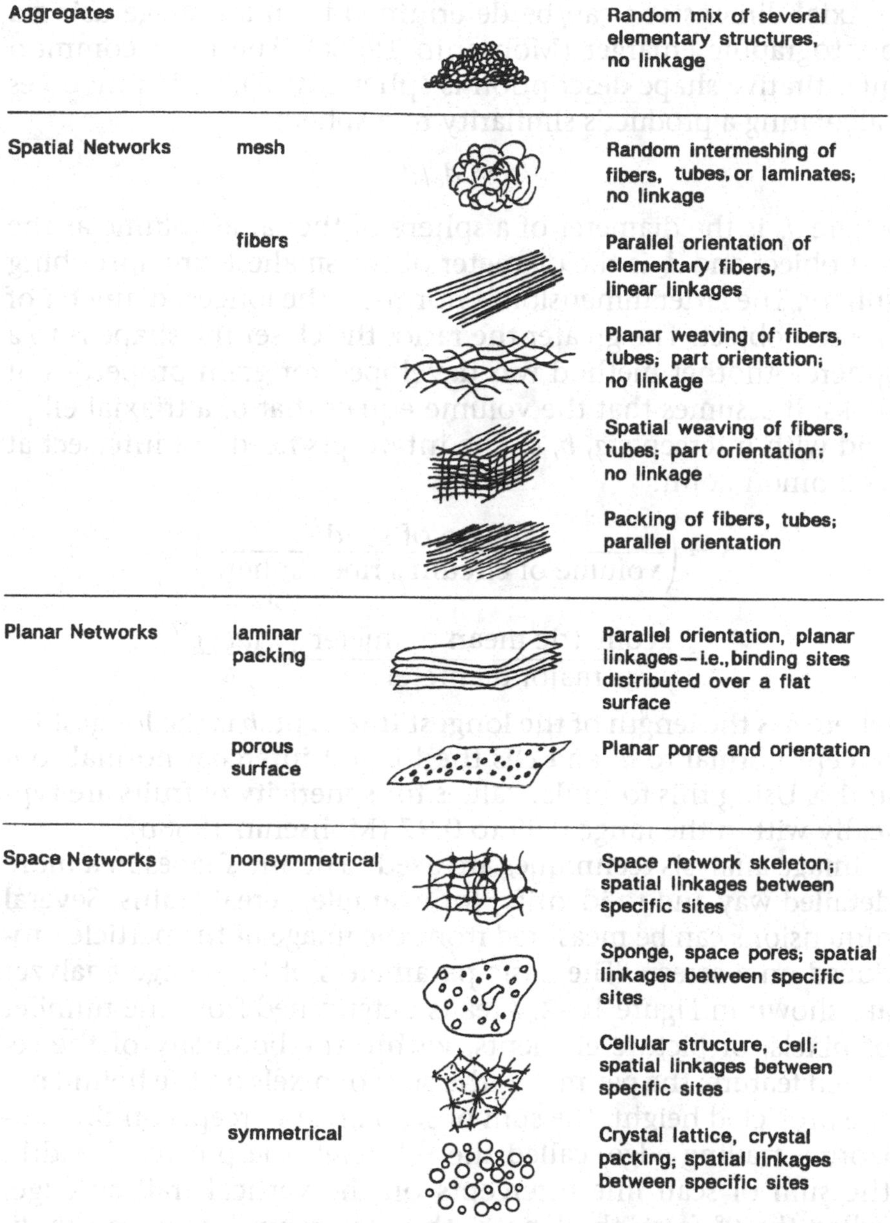

Aggregates			Random mix of several elementary structures, no linkage
Spatial Networks	mesh		Random intermeshing of fibers, tubes, or laminates; no linkage
	fibers		Parallel orientation of elementary fibers, linear linkages
			Planar weaving of fibers, tubes; part orientation; no linkage
			Spatial weaving of fibers, tubes; part orientation; no linkage
			Packing of fibers, tubes; parallel orientation
Planar Networks	laminar packing		Parallel orientation, planar linkages—i.e., binding sites distributed over a flat surface
	porous surface		Planar pores and orientation
Space Networks	nonsymmetrical		Space network skeleton; spatial linkages between specific sites
			Sponge, space pores; spatial linkages between specific sites
			Cellular structure, cell; spatial linkages between specific sites
	symmetrical		Crystal lattice, crystal packing; spatial linkages between specific sites

Figure 10–1 Homogeneous compounded structures (after Raeuber and Nikolaus 1980, with permission).

Axial dimensions can be determined from an image using a photographic enlarger (Mohsenin 1968a). The most common quantitative shape description is sphericity (S), which involves calculating a product's similarity to a sphere.

$$S = d_e/d_c$$

where d_e is the diameter of a sphere of the same volume as the test object and d_c is the diameter of the smallest circumscribing sphere. The latter dimension is normally the longest diameter of the test object. The greater the ratio, the closer the shape is to a sphere. Another method was developed for grain properties of rocks. It assumes that the volume equals that of a triaxial ellipsoid with intercepts a, b, c. The intercepts need not intersect at a common point.

$$S = \left(\frac{\text{volume of solid}}{\text{volume of circumscribed sphere}} \right)^{1/3}$$

$$= \frac{\text{geometric mean diameter}}{\text{major diameter}} = \frac{(abc)^{1/3}}{a}$$

where a is the length of the longest intercept, b is the longest intercept normal to a, and c is the longest intercept normal to a and b. Using this formula, values for sphericity of fruits are typically within the range 0.89 to 0.97 (Mohsenin 1968b).

Image analysis techniques are used to define shapes in a more detailed way and to identify, for example, cereal grains. Several dimensions can be measured from the image of the particle produced on a screen. The basic parameters of the image analyzer are shown in Figure 10–2. Area is determined from the number of pixels, or picture elements, within the boundary of the selected feature; the perimeter, the sum of pixels on the boundary; the projected height, the sum of scan line intercepts on the horizontal trailing edge, called the 90° feret; the projected width, the sum of scan line intercepts on the vertical trailing edge, called the 0° feret; the length, the maximum linear feature dimension independent of orientation; the width, the dimension at 90° to the longest dimension; the volume, the integrated optical brightness or density within the boundary; and the convex perimeter, the length around the outside of the feature. These

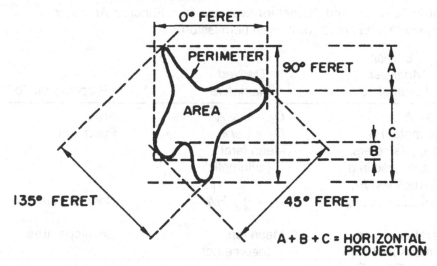

Figure 10–2 Basic parameters of the image analyzer (Lai et al. 1986, with permission).

parameters are used to produce a set of derived diameters according to Table 10–1. Examples of shape functions based on these determinations are shown in Figure 10–3. Examples of pattern parameters for various grains are given in Table 10–2 (Lai et al. 1986). Specific cultivars of wheat can be identified (Keefe and Draper 1986). Grain purity (Jakobsson 1997) and bran particle content of white flour (Evers 1993) can be quickly assessed.

Color image analysis can be used to identify seed variety using size, shape, and texture parameters. Rumex, wild oat, lucerne, and vetch seeds were identified with 99-percent success, the artificial neural network outperforming stepwise discriminant analysis (Chtioui et al. 1996). Shapes of larger specimens, such as carrot roots and chrysanthemum plant leaves, can also be specified by image analysis (Keefe and Draper 1988). Biscuit and shelled almond shape (Gunasekaran and Ding 1994), cooked french fry color and shape (Ritchie 1994), bakery and snack products (Scott 1995), and baked products such as biscuits and pizza (Perry 1992) can be monitored on line. High-speed grading and sorting are possible by size, color, shape, or defect.

Table 10–1 Derived Diameters Available from Function Analyzer Parameters (Lai et al. 1986, with permission)

Function Analyzer Parameters	Derived Diameters	Proportional To
Area A	$D_1 = 2 \sqrt{A/\pi}$	Area
Perimeter Pe	$D_2 = Pe/\pi$	Perimeter
Feret diameters, measured in n directions, F_1, F_2 .... F_n	Mean feret (arithmetic) $D_3 = \sum_{i=1}^{n} F_i/n$	Convex perimeter
Feret diameters, measured in n directions, F_1, F_2 .... F_n	Mean feret (geometric) $D_4 = n \sqrt{F_1 F_2 ...... F_n}$	Envelope area
Feret diameters, measured in n directions, F_1, F_2 .... F_n	Maximum feret $D_5 = \max(F_i)$	Length
Feret diameters, measured in n directions, F_1,	Minimum feret $D_6 = \min(F_i)$	Envelope width

Cake batter is an aerated fat in water emulsion. During mixing, large air bubbles are introduced into the mix through the lower hydrostatic pressure behind the stirring blade. Shear stress within the moving batter elongates the bubbles, which break up into smaller bubbles. The success of the cake depends on bubble size. If they are too large, buoyancy during baking carries them to the surface, from which they escape. Too many large bubbles lead to the cake's rising before egg protein coagulates and starches gelatinize. Both effects lead to a smaller cake. A narrower and smaller bubble size distribution leads to a finer texture product, as bubbles expand the structure in a uniform man-

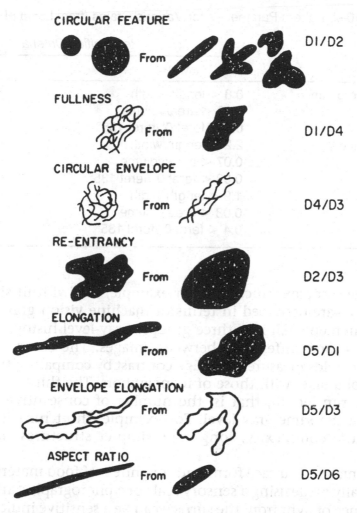

CIRCULAR FEATURE — From — D1/D2

FULLNESS — From — D1/D4

CIRCULAR ENVELOPE — From — D4/D3

RE-ENTRANCY — From — D2/D3

ELONGATION — From — D5/D1

ENVELOPE ELONGATION — From — D5/D3

ASPECT RATIO — From — D5/D6

Figure 10–3 Examples of shape functions based on diameter definitions listed in Table 10–2 (Lai et al. 1986, with permission).

---

ner. Bubble formation can be studied dynamically using hot-stage video microscopy (Pateras et al. 1994). Final bubble sizes can be measured by image analysis of xerox copies of cake or loaf cross section. Cell sizes are determined by image analysis applied to cross sections, distributions being described mathematically (Barrett and Peleg 1992).

Table 10–2 Pattern Parameters for Various Grains (from Lai et al. 1986)

Grain	Recognition Criteria	
	Min	Max
Grain sorghum	0.8 < length/width	<1.25
	0.12 < area/volume	<0.20
	0.8 < feret 0°/feret 135°	<1.2
White rice	2.2 < length/width	<3.7
	0.07 < area/volume	<0.12
	0.03 < feret 0°/feret 135°	<0.68
Barley	1.9 < length/width	<2.9
	0.08 < area/volume	<0.18
	0.4 < feret 0°/feret 135°	<0.68

More complex structures—for example, of kiwi fruit slice geometry—are described in terms of machine vision gray levels. Measurements fall into three groups. Gray-level histograms indicate coarse differences between images. The co-occurrence matrix yields measures of image contrast by comparing the gray level of a pixel with those of its neighbors. The third group involves run length; that is, the number of consecutive points having the same gray level. For example, short runs indicate structure complexity, long runs simpler structures (Roudot 1989).

Estimates of surface form and roughness of food materials are normally made using a sensory scale or a photographic atlas. Reflectance of light from the surface can be a sensitive indicator of roughness (Bennett and Porteus 1961; Graham and Choong 1981). Surface roughness testers can be used to give values of roughness (μm) in terms of the arithmetic mean deviation of the profile determined per unit length at constant speed. Measurements can be made on strips of the surface in an atmosphere maintained at 100-percent relative humidity to prevent drying. Eggplants and apples have greater surface roughness after harvesting than tomato, which does not change during the ripening process. As may be expected, stripping the wax from the surface results in an increase in roughness (Ward and Nussinovitch 1996a). Surface contact methods have been used to determine the roughness of human skin (Marks and Pearce 1975). Surface topography—for example, of onion and garlic skins—can by visualized using atomic force microscopy (Hershko et al. 1998).

Overall particle shape as well as surface morphology can be expressed in terms of fractal dimensions. This technique is used to describe the geometry of fragmented shapes of, for example, shorelines, clouds, and plants (Mandelbrot 1977). Potential applications to the specification of food shape have been reviewed (Peleg 1993). Fractal dimensions can be used to monitor fine-particle properties (Rahman 1997) and to determine effects of particle breakage and erosion—for example, in instant food powders where attrition is important (Barletta and Barbosa-Cánovas 1993).

Three-dimensional shape properties of seeds have been determined for soybean seed. Surface area and volume have been measured by an air permeability method (Jindal et al. 1974), possibly coupled with a shadowgraph and the assumption that the shape is an ellipsoid (Chuma et al. 1982). Three-dimensional shape can be measured more precisely using an extension of the binocular stereo method, incorporating laser illumination (Sakai and Yonekawa 1991).

TRANSLUCENCY

Translucency is an essential appearance element of many foods, including meat (e.g., MacDougall 1983; Little 1964), hot drinks (e.g., Hutchings and Gordon 1981), cold drinks (e.g., Mac-Dougall 1987; Francis and Clydesdale 1975; Little 1971; Staples 1969, cited in Francis and Clydesdale 1975), and rice (Te-Chen and Song 1984). Both color and translucency arise as phenomena through the same basic elements of light absorption and scattering. These occur on the surface and within the body of a material in a diffuse, nondirectional manner (see Chapter 3). This leads to their close interaction as visual properties, and both are often unfortunately classified under the single term *color*. This often leads to difficulties. The reliance on the same basic elements means that color and translucency are, in visual terms, not totally independent; this leads to problems in the color specification of translucent foods (see Chapter 8). If a translucent food is to be understood as a material, or in relationship to its structure and ingredients, or in giving rise to consumer attributes, a clear distinction must be made between color and translucency properties. The following examples illustrate interactions between them and the importance of translucency measurements to the understanding of consumer behavior toward foods.

Interaction Between Color and Translucency

The complexity of the color and translucency interaction can be illustrated with meat. The appearance of fresh meat is caused by the concentration of heme pigment, the degree of desiccation, the state of the pigment both on the surface and several millimeters below, and the light-scattering properties of the muscle proteins (MacDougall 1983). This interaction between pigmentation and scattering is shown in the color discrimination diagram in Figure 10–4.

The myoglobin content of muscle increases in the order pork, veal, beef. The color coordinates (L^*, h^*, C^*), however, depend on the scattering coefficient S, as well as the pigment content. At slaughter, muscle is dark and translucent and has a pH of 7.0. This falls to 5.5 during rigor mortis as the glycogen converts to lactic acid. The muscle protein then reaches its isoelectric point, and the light scattering increases. As handling and aging increase, so does the opacity through an increase in S. Pale, soft, exudative (PSE) muscle arises from too fast a pH fall at too high a temperature after slaughter. This rapid postmortem glycolysis causes structural changes in the muscle. The myofilaments have a distinct granular appearance, which is possibly due to precipitated sarcoplasmic proteins or to conformational changes in myofibrillar proteins. These changes result in abnormally high values of scattering, the subsequent paleness being manifest in higher values of L^*. The pH of muscle from animals suffering from glycogen depletion, through preslaughter stress or exhaus-

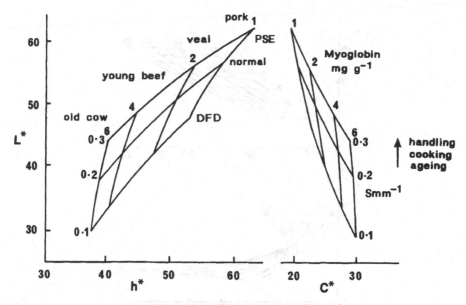

Figure 10–4 Relationship of CIELAB lightness (L^*), hue (h^*), and chroma (C^*) to light scatter (S) and oxymyoglobin in fresh meat (adapted from MacDougall 1987, with permission).

tion, does not fall below 6.0. Light scattering does not increase to its normal value, and this dark, firm and dry (DFD) muscle has a low value of L^* (MacDougall 1987). As an indicator of water-holding capacity, light scattering (Irie and Swatland 1993) is more reliable than L^* (van Laack et al. 1994).

Similarly, the color of frozen meat is governed by the freezing rate, which affects light-scattering properties. Fast freezing results in small crystal sizes. These scatter more light than the large, slow-growing crystals produced on slow freezing. Fast-frozen meat is opaque and pale, and slow frozen is translucent and dark (MacDougall 1970a, 1970b, 1982).

Color and translucency interactions change with sample depth or thickness. The interaction between values of K and S and chromaticness of bacon as a function of slice thickness is shown in Figure 10–5 in terms of the 1960 Commission Internationale d'Eclairage (CIE) Uniform Colour Spacing coordinates

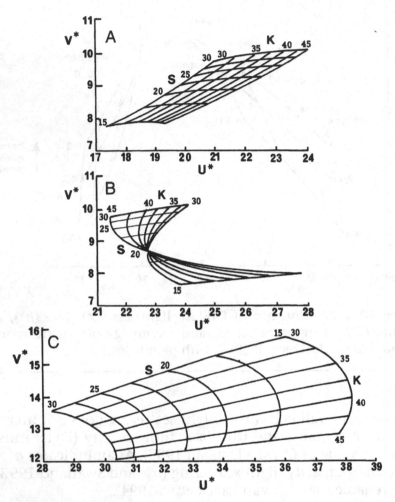

Figure 10–5 Effect of bacon slice thickness on 1964 CIE uniform chromaticness spacing (u^*, v^*). Slice thicknesses: **A**, infinite (155mm); **B**, 8mm; **C**, 3mm, on a white background (MacDougall 1971, with permission).

u^* and v^*. As slice thickness is decreased to the normal commercial thickness of 3 mm (bottom), the chromaticity space occupied is greater. The effect of increase in K is to make the hue more red. In visual terms, the redness produced is twice that seen in an infinitely thick sample. The reversal in the saturation effect, as the sample thickness is reduced, occurs when the thin-layer background starts contributing to the reflectance. At an 8-mm slice thickness, the color space occupied twists. The translucent bacon with low S values increases faster in saturation (increase in u^* and v^*) than the opaque bacon with high S values (MacDougall 1971).

Translucency of Foods

Visual perception of the color of some foods is better related to color measurements made on thin layers of the sample than to measurements made on thicker layers. This discovery led to the realization that the Kubelka-Munk theoretical approach (see Chapter 3) might be as valuable to food technology as it had been for some years to paints, plastics, and textiles (Little 1964). Internal transmittance (T_i) measurements can be used to monitor sensitively addition of brown sauce to applesauce and the mixing of skimmed and whole milks (MacKinney et al. 1966). Internal transmittance was also found to control the sugar-likeness of a range of white granular experimental products. In that work, members of a consumer group were asked to judge the appearance of a series of achromatic sweeteners in terms of similarity to white granulated sugar. Samples judged to look like sugar possessed a greater value of T_i than the more opaque non-sugarlike products (Hutchings 1978).

Coffee drink is a convenient model system for demonstrating the effect of translucency properties on sensory evaluation. Nine samples, differing in coffee solids and whitener concentration, were presented in vertical-sided, glazed white cups and were assessed for derived perceptions of visually apparent *creaminess* and *strength*. The same samples were also presented in thin layers for an assessment of the basic perception of *opacity*. It was found that these three attributes can be defined in terms of Kubelka-Munk ratio of absorption and scattering coefficients (K/S) and T_i. These are determined from transreflectance spectrophotometry measurements made on sample layers 2 mm thick. The spectral response of coffee is not highly differentiated; therefore, a wavelength in the middle of the visually sensitive spectral range was chosen. Figure 10–6 is a plot of K/S ver-

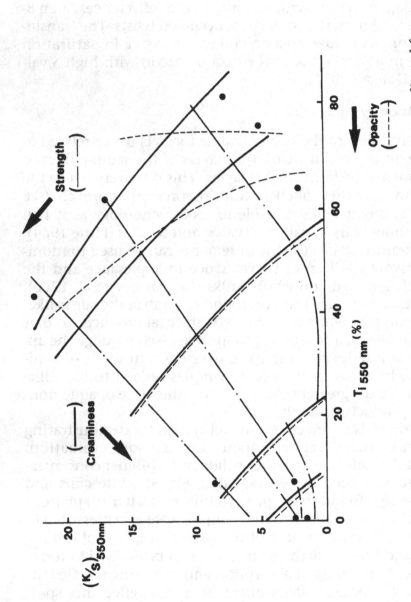

Figure 10–6 Kubelka-Munk–derived properties of nine coffee samples. Contours are indicated for *creaminess* and *strength* viewed in the cup, and for *opacity* viewed in a thin layer (from data used in Hutchings and Scott 1977, and Hutchings and Gordon 1981).

sus T_i at 550 nm, showing contours that can account for the in-the-cup judgments of *creaminess* and *strength*, as well as judgments of *opacity* as viewed in a thin layer.

The most creamy samples have high scattering and consequently low values of T_i and K/S. The impression of low creaminess can arise from high values of T_i (i.e., low S) either with or without relatively high values of K/S. If two samples have equal values of T_i, the one that is more pigmented—that is, having the higher K/S—will have a lower perceived creaminess. Perceived strength increases with increase in pigmentation (increase in K/S). For samples having equal K/S, the one with lower T_i will be seen to be the stronger. *Opacity* and *creaminess* contours follow similar patterns. The main difference between these attributes occurs with high K/S (more highly pigmented) samples, which are seen as being more opaque (Hutchings and Scott 1977; Hutchings and Gordon 1981). Coffees having the same coffee-to-milk ratio have the same values of measured K/S irrespective of actual coffee and milk concentrations. They also look alike and possess the same visually assessed *darkness*, *creaminess*, and *strength*. However, subjects were able to detect in-mouth differences of *creaminess* and *strength* (MacDougall and Lima 1999). Hence, it is instructive to regard such a product, varying widely in opacity, as both an absorbing and a scattering material.

Orange juice is another example of a translucent material for which thin-layer transreflectance measurements are useful in understanding and defining consumer attributes. Measurements on a series of 11 commercial and experimental products resulted in a conclusion that 3.7 mm was an optimal depth for orange juice. These products were visually assessed by 22 panelists for rank order of the derived perceptions of *strength of flavor*, *preference for a breakfast drink*, and *preference for a refreshing drink*. It was found that all rank orders could be explained using a plot of one translucency property (K/S) and one color property (dominant wavelength).

Figure 10–7 shows the two preference assessments as contours in terms of these variables. A preferred breakfast drink (Figure 10–7A) should have a low (yellower) dominant wavelength (λd),

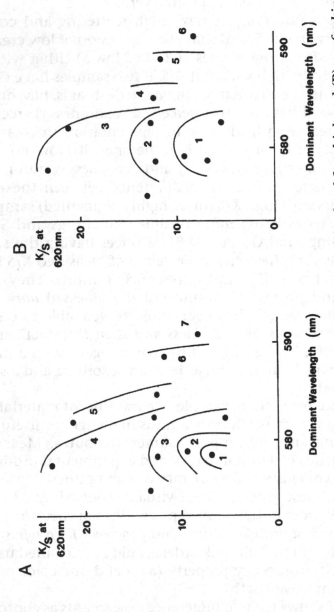

Figure 10–7 The physical basis of orange juice preference as (A) a *breakfast drink* and (B) a *refreshing drink*. Means of responses from 22 panel members. Cell depth = 3.7 mm.

coupled with a low K/S (high opacity). As K/S increases, the low λ_d is still preferred. The lowest preference is reserved for high λ_d (redder) samples. Figure 10–7B indicates a similar picture for a refreshing drink, except that low- and mid-λ_d samples of low K/S are equally preferred.

Two types of response were found in the rank orders for strength of flavor. One was given by 14 panelists, the other by 8. The contours of the responses are shown in Figure 10–8. Both groups in this bimodal response indicated that a low K/S indicated strength. However, the majority (Figure 10–8A) saw the reddest as strongest; the minority (Figure 10–8B) saw the reddest as weakest. The majority ranked strength in order of reducing K/S, but they were also influenced in favor of higher λ_d. The minority population ranked strength almost in the same way as the whole panel ranked the same samples as a refreshing drink (Figure 10–7B). The minority group perhaps considered a lower strength drink to be less preferable as a refreshing drink. Hence, orange juice is a material in which the interaction between color and translucency is important to the consumer.

An emulsion is an inherently unstable dispersion of one liquid in another, immiscible liquid that can be stabilized by the inclusion of an interfacial emulsifying agent (Parker 1987). Different mechanisms, such as creaming, flocculation, and coalescence, are involved in emulsion breakdown. These normally cause a degradation in visual quality of the food system involved (Klemaszewski et al. 1992). Visual properties, such as transparency, gloss, and whiteness, of fresh and condensed (Patel et al. 1996) milk, and hence nondairy milks, are important to the dairy industry because they relate directly to product quality and consumer acceptance. Peanut milk can be used in nondairy milk whiteners (Malundo et al. 1992; Abdullah et al. 1993). Effective fat substitutes for fat-free milk should make the color whiter, less green, and less blue. Such a substitute needs to change the appearance attributes of the milk more than its tactile attributes (Phillips, McGiff, et al. 1995a). This has been done, for example, by adding nonfat dry milk (Phillips, McGiff, et al. 1995b) or casein/titanium dioxide blends (Phillips, Barbano, and Lawless 1995) or by enzyme treatment of the milk,

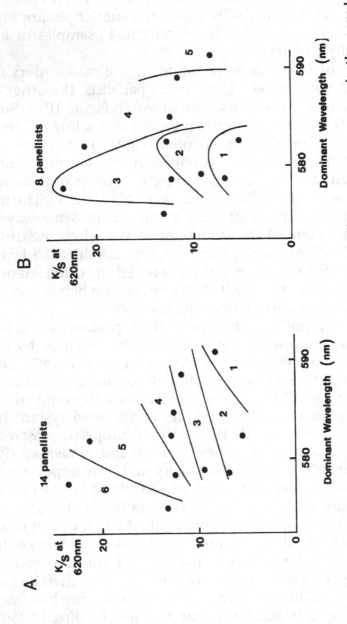

Figure 10-8 The physical basis of orange juice *strength of flavor*. There were two groups in the population tested. (A) Mean responses from 14 panel members. (B) Mean responses from 8 panel members. Numbers indicate order of *strength of flavor*. Cell depth 3.7 mm.

causing aggregation of the casein micelles (Savello and Solorio 1995).

Fruit beverages are stabilized using clouding agents. Such agents must contribute to the opacity of the drink without affecting cloud stability—that is, by creaming, wringing, or separation. Clouding agents include proteins, water-soluble gums used as emulsifiers, and oil-soluble gums. Naturally based soy proteins can also be used (Klavons et al. 1992).

Haze

Haze is an important visually perceived property of drinks. For beer, two types exist; a reversible chill haze, which forms on cooling but disappears on warming, and an irreversible haze, which usually forms slowly at normal storage temperatures. It is vital that no haze form even when the pH is suddenly lowered on the addition of lime juice. Haze is quantified visually or by nephelometric measurement of the sample under defined temperature conditions. The nephelometer is an instrument in which a tube containing the sample is illuminated from a set angle. Light scattered by any suspension present is measured at another set angle, commonly 90° to the angle of illumination. Both visual and instrumental methods involve standard suspensions against which the sample is compared. Standard suspensions include formazin, or a mixture of gum arabic and gum mastic (Francis and Clydesdale 1975). As color affects nephelometer readings (for beer, Hudson 1969), a standard the same color as the sample should be used.

Turbidity meters can be used to monitor light scattering in solids and fluids. Gelatin turbidity is caused by the molecular size properties and imperfect filtration. These effects can be reduced by enzyme treatment and submicron filtration (Cole and Roberts 1997). Apple juices contain haze-active proteins and polyphenols. Turbidimetry can be used to assess treatments aimed at haze stabilization in such products in which there is little color change. Temperature control during the haze period is critical for reproducibility (Siebert and Lynn 1997).

The accuracy of these instruments in quantifying haze or turbidity should not be overestimated. The turbidity will be quantified only if the light scattered at a set angle is a fixed proportion of the total light scattered from the path of the beam. It is evident from Figure 3–12 that measurements made at angles of 45° and 90° will place the two particle-sized suspensions in reverse order. The usefulness of the instrument is probably based on its illumination and collection angles' being wide enough to average out the scattering caused by the particle size range normally encountered in the practical production situation. It appears not ideal either for detailed use in the research and development laboratory or for strict comparison with consumer appreciation of haze (Ehmann 1972; Francis and Clydesdale 1975).

Haze is a consumer-important property of brewed tea. Staples (1969, cited in Francis and Clydesdale 1975) found that the problems of nephelometer measurement may be overcome using an integrating sphere for the collection of all forward-scattered light. Little and Brinner (1981) used the technique developed for tea by Little (1971) for the study of wines possessing subliminal scattering (see Chapter 8). They found that, with judicious application, the Kubelka-Munk theory can be used to provide physical explanations for the visually perceived attributes of this product. Kubelka-Munk absorption and scattering coefficients were determined for the tristimulus values X, Y, and Z. Darkening of color could be separated from the turbidity increase, which occurs with increasing mineral content. Turbidity, if not excessive, can normally be ignored during subjective evaluation, but color and clarity should be evaluated on separate scales to indicate their individual contributions to the overall visual appearance of tea. Dilution decreases turbidity effects but causes a significant change in hue (Joubert 1995).

Of possible application to the food industry are the findings of Billmeyer and Chen (1985), who have investigated the haze and optical clarity of plastic films and sheets. They used an integrating-sphere spectrophotometer for the accurate determination of haze and showed that valuable information regarding the origin of the haze can be gained by comparing total and diffuse transmittance spectra.

GLOSS

Goniophotometric properties—that is, change of reflected intensity with angle of illumination and/or viewing—give rise to perceptions of gloss. Food qualities relying heavily on gloss include such visually derived perceptions as *apparent softness*, *stickiness*, and *oiliness* of spreads. However, although the visually perceived basic perception might be labeled simply *gloss*, different viewing conditions are necessary for its perception, and different instrumental methods may be needed for its quantification (see Chapter 3). The many yellow spreads on the market range from high gloss to almost matte. Instrumentally, gloss may be discussed with reference to the product's goniophotometric reflectance envelope (see, e.g., Figure 10–9).

It is unlikely that one single attribute of the goniophotometric curve will relate to the visual rank order of gloss of such a range of samples (Hunter and Harold 1987). Indeed, observation indicates that the gloss of less glossy spreads may relate to the intensity of diffuse reflectance, while the gloss of highly glossy spreads may be judged by the sharpness of the reflected image. Both specular gloss and distinctness-of-image gloss measurements are required to quantify such attributes.

Sample preparation for measurements using conventional gloss meters is difficult because, for most foods, the need for a flat sample necessitates destroying the surface for which the specification is required. Measurements made on prepared surfaces of lard and dripping illustrate some of the difficulties. Goniophotometric curves for samples cooled to 10°C or heated to 25°C are shown in Figure 10–9. Lard is softer than dripping, and this markedly affects development of the specular reflectance curves with temperature (MacDougall 1970b). Machine vision techniques are potentially valuable for gloss determination.

Chocolate consists of a mixture of small solid particles of principally sugar, cocoa, and milk, suspended in a continuous fat phase. Cocoa butter is a highly polymorphic fat that can exist in six forms, I to VI, increasing in order of melting point and thermodynamic stability. When chocolate is properly tempered, it will crystallize into form V, slowly changing at low temperatures

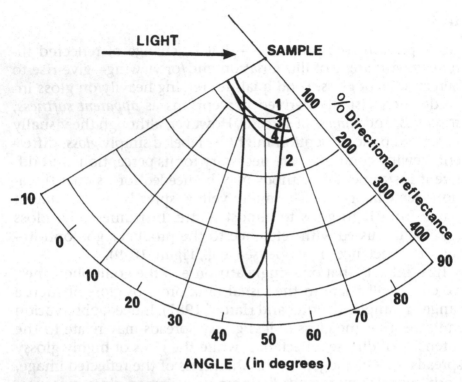

Figure 10–9 Goniophotometric reflection curves of lard and dripping at two temperatures. Lard at 1, 10°C and at 2, 25°C. Dripping at 3, 10°C and at 4, 25°C (MacDougall 1970b, with permission).

into form VI. Melting followed by cooling results in recrystallization into forms IV and V, and bloom results. Blooming also occurs slowly on low-temperature storage caused by changes from form V to VI (Talbot 1995).

Shininess is caused by the presence of a layer of glossy fat on the surface, and loss of gloss can also be migration induced. The surface of fat may withdraw into the body of the chocolate as a result of suction, or the glossy fat may be disrupted by extrusion of coarse crystalline fatty material from within the chocolate. Settling of the solids can occur at high temperatures when the fat softens. Then the intergranular space is reduced and surplus fat extruded. The nonformation of gloss on manufacture is a special case of this condition in that the surface film of fat is disrupted from the start (Koch 1978). Fat migration also occurs from migration of non–cocoa butter fats from the center of a coated product into the chocolate coating (Talbot 1995). These movements lead to an increased porosity of bloomed chocolate (Loisel et al. 1997).

Fat that has been solidified when firmly in contact with a polished metal surface, however, results in a smooth surface, as indicated in Figure 10–10A. Much of the incident light is reflected specularly. Light penetrating the smooth surface is scattered by randomly orientated particulate solids beneath. As indicated in Figure 3–13, part of the scattered light emerges from the surface and contributes to the diffuse reflectance from the sample. When chocolate is allowed to solidify freely, the surface is not flat. In this case, light reflected both from the surface and beneath is diffuse, as shown in Figure 10–10B (after Musser 1973).

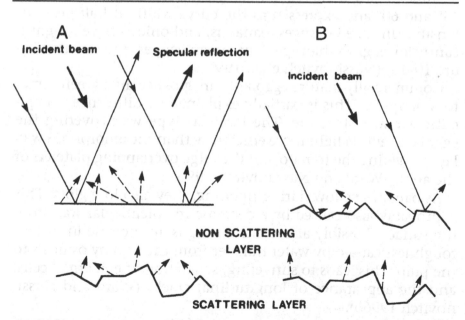

Figure 10–10 Different patterns of light reflection from chocolate with a molded surface (top) and a freely solidified surface.

These differences in structure give rise to the type of change in goniophotometric curve shown in Figure 10–11. The curves are for chocolate samples having high and medium gloss. The peak height of the high-gloss sample has been reduced so that the curves have the same peak height. The more diffuse nature of the sample having lower gloss is clearly indicated by the peak breadth.

Samples of chocolate containing different particle sizes of solids were assessed by a panel using a 0-to-10 *gloss* scale. The anchor points were matte brown paper (0) and a highly polished colored tile (10). The results are shown in Figure 10–12. Gloss depends on the way the chocolate is molded. If there is sufficient fat to give a continuous surface (the molded samples), particle size has less effect because most of the light is specularly reflected from the surface. The effect of size is greater for freely solidified samples because more light is reflected from the particles. As particle size is reduced, the gloss intensity increases as shown (Musser 1973).

Gloss characteristics of a number of intact fruits have been determined using a curved surface glossmeter of light incidence 45° and 60° and expressed as the curve width at half peak intensity. Unwaxed oranges, bananas, and onions have a significantly lower gloss than eggplant, green pepper, and tomato (Figure 10–13) (Nussinovitch et al. 1996).

Commercially mature eggplants are glossier than green tomatoes or apples. This is partially explained by differences in epicuticular wax structure. The lamellae-type wax covering the eggplant reflects light more efficiently than the amorphous wax layer covering the tomato and the large overlapping platelets of the apple (Ward and Nussinovitch 1996a).

As bananas yellow during ripening, they also lose gloss. This is probably also caused by a decrease in epicuticular wax from the surface. Possibly also contributing is an increase in surface roughness caused by water transfer from the skin by osmosis to the pulp. This leads to shriveling, separation of epidermal cells, and the appearance of longitudinal cracks (Ward and Nussinovitch 1996b).

Failure to relate panel assessments of gloss to goniophotometric measurements of processed purees and sauces has been

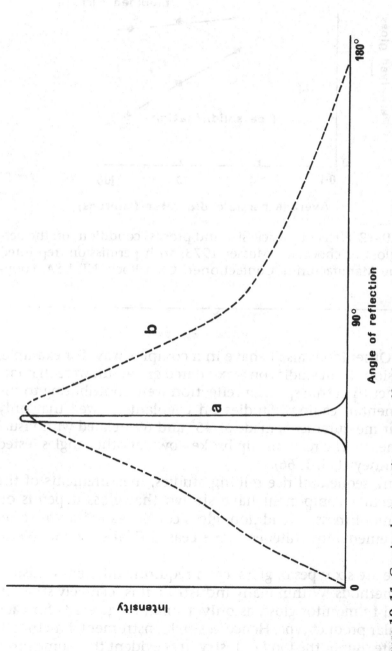

Figure 10–11 Goniophotometric curves of high-gloss (a) and medium-gloss (b) chocolate samples adjusted to equal peak heights. Angle of incidence = 60°.

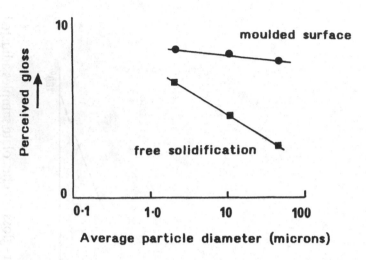

Figure 10–12 Effect of particle size and process conditions on the perceived gloss of chocolate (Musser 1973, with permission, reprinted from The Manufacturing Confectioner. Glen Rock, NJ, USA, copyright).

noted. Other foods also behave in a complex way. For example, conclusions from studies on wax-coated grapefruit were that factors other than true specular reflection were contributing to the instrumental readings. Studies on eggplant showed that only specular measurements made at 45° and 60° related with visual assessments. The relationship broke down at other angles tested (MacKinney et al. 1966).

During sequential rice milling studies, measurements of the 45° specular component have shown that gloss depends on grain smoothness. In addition, gloss continues to increase after improvements in whiteness have ceased (Blakeney and Welsh 1988).

There are six types of gloss, each requiring different measurement methods. Within many industries, it is relatively straightforward to monitor gloss, as only a small range exists for each particular product type. Hence, a single instrumental method is adequate. But in the food industry, it is evident that some products undergo too large a gloss change to be adequately specified

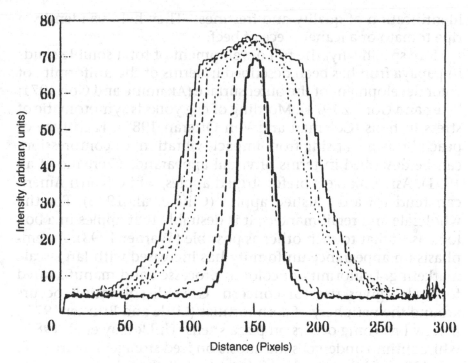

Figure 10–13 Typical goniophotometer curves obtained from a curved-surface glossmeter. The curves of the outermost to innermost represent fruits and vegetables in the order: orange, banana, onion, tomato, pepper, and eggplant. *Source*: Reprinted with permission from Nussinovitch, Ward, and Mey-Tal, Gloss of Fruits and Vegetables, *Lebensmittel Wissenschaft und Technologie*, Vol. 29, pp. 184–186, © 1967, Academic Press, Ltd.

using only one method. The relationships between the different types of gloss are not sufficiently understood to produce a continuous measurement scale relating to the apparent perceptual continuum.

UNIFORMITY AND PATTERN

Uniformity and variation in material properties, and hence the basic perceptions of visual structure, surface texture, color, translucency, and gloss, specifically characterize species and variety of natural foods. They form the basis of recognition and identification of quality and freshness. They help us identify a ripe tomato or a leaner piece of beef.

More specifically, the full development of total soluble solids in papaya fruit has been described in terms of the uniformity of color development of the outer surface (Akamine and Goo 1971; Peleg and Gomez 1974). Mottling of egg yolks is symptomatic of stress in hens (Coleman and Van Otteran 1982). Each type of peach bruising, arising from impact, vibration, or compression, can be described in terms of visual appearance (Vergano et al. 1991). Asian markets prefer striped apples, while North American tend toward blushed apples (Cliff et al. 1996). For the wholesale and retail markets, it is desirable that apples in a box look as similar to each other as possible (Warner 1993). An emphasis on appearance uniformity has increased with large-scale marketing. Uniformity of color in processed and manufactured food also gives rise for concern. Gray discoloration occurs around the periphery of corn tortillas (Little and Brinner 1977), patchy browning occurs on pizza shells (Unklesbay et al. 1983), white shrimp undergoes melanosis on iced storage (Luzuriaga et al. 1997), and the visual surface texture of chocolate is sensitive to the state of precrystallization (Tscheuschner and Markov 1986). The quality of a cup of coffee improves with more homogeneous batches of beans (Richard 1977, cited in Raggi and Barbiroli 1993). Methods by which uniformity may be specified within and among products are therefore an essential part of the food industry. Many specifications of quality involve sensory judgments. This approach is required for the setting up of raw material and product standards, such as those proposed by authorities like the U.S. Department of Agriculture (USDA). Examples are given in Chapter 6.

Throughout Europe, beef and sheep carcasses are classified by visual inspection according to the EUROP scheme regulated by the European Union. *Fatness* and *conformation* are defined on scales from 1 to 5, each of which may contain three subclasses. Since the early 1970s, the Meat and Livestock Commission has operated a carcass authentication service in Britain. *Fat color* is an important quality within the Danish industry, and this can be scored on a range from 1 (**very white**) to 10 (**very yellow**).

These fatness and conformation qualities can also be successfully measured using image analysis. A frame is used to position the half-carcass in the slaughter line, and a video camera is used to provide three-dimensional shape information. Fatness and fat color can be determined from the color image data (Borggaard et al. 1996).

Measurement of Discontinuities

In many foods, discontinuity of color is almost as important as the color itself. Patterns occur in fruit ripeness, multicomponent foods, and the brownness of baked products. Patterns of discontinuities can be linked to structure in beef marbling. Such patterns can be described using machine vision analysis and small-aperture fiber-optic colorimetry. These techniques allow pattern to be specified and provide a basis for examination of consumer response to, for example, product component distributions.

Image analysis has been used to investigate patterns on products produced by cooking. Computer images of browned pizza bases were scanned for intensity and spatial location. Each brown intensity level was multiplied by its corresponding percentage area to produce a plot of browning severity as a percentage of the total pizza area against the brown intensity. An example is shown in Figure 10–14. The presence of brown in the crust resulted in lighter areas in the image. Hence, the lower the brown intensity number, the greater the browning. The traces illustrate the effects of ingredient and of including a black coating on the pan. The effect of the latter is to produce a higher degree of browning in the crust. In a bare pan, wheat flour leads to a less brown pizza than soy flour (Unkelsbay et al. 1983). Tomato ripening patterns are discussed in Chapter 8.

Machine vision can be used to determine surface areas, for example, of histological sections of meat products. The sections were stained to identify denatured, raw, and native collagens, elastin, and bone. The results compared well with standard histology point counting and chemical reference measurements (Hildebrandt and Hirst 1985). Image analysis can also be used

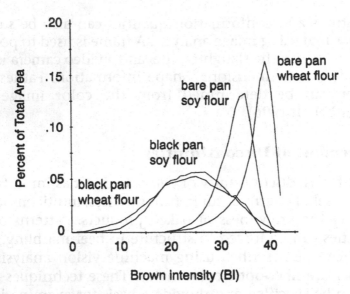

Figure 10–14 Distribution of brown intensity of pizza crusts. Note that as BI increases, the degree of browning decreases (Unklesbay et al. 1983, with permission).

for on-line quantification of fat and lean components of mince meat (Newman 1987).

Multicomponent foods may also be examined using a fiber-optic spectrophotometer. For example, visually apparent characteristics of fat and lean tend to be related to the eating quality of fresh meat. Marbling properties—that is, amount and spatial distribution of individual fat deposits—form part of quality grading for beefsteaks. USDA marbling score cards (USDA 1989, 7 CFR §§ 54.102–54.107) are traditionally used for rating, but image processing can also be used. Marbling density—that is, the size and number of marbling flecks per unit area of steak— and spatial distribution can be determined from R, G, B images. Even though the subjective evaluation of marbling appears complex, image processing can effectively predict it ($R^2 = 0.84$) (Gerrard et al. 1996). Black specks in semolina can be counted by digital imaging. Alternatively, they can be defined in terms of gray level (Harrigan and Bussmann 1998).

The coherent fiber-optic spectrophotometer can be used to measure the color of areas less than 1 mm in diameter. Results obtained from the three components of a sample of bologna are shown in Figure 10–15.

The absorbance spectra are plotted in terms of histogram bars, standard deviations being shown by notches in the top of each bar. Pork heart muscle in the bologna exhibited the characteristic 550-nm peak of nitrosylhemochrome against the less specific absorption of the surrounding matrix. The green pepper has the maximum absorption bands toward each end of the visible spectrum, typical of chlorophyll (Swatland 1985).

A microdensitometer may be used to measure variations in reflectance as the measuring head scans across the surface or across the negative of the photographic image of the surface. For example, a numerical quantification of *mottling* of pharmaceutical tablets can be specified using the standard deviation about a mean of the reflectance (Armstrong and Marsh 1974). It is possible to automate such measurements and calculations (e.g., Tobias et al. 1989). Another method involves measurement of transmitted scattered distribution resulting from laser (Birth et al. 1978; Swatland 1991) or xenon arc (Swatland 1991) illumination.

Air foams play an important role in bread, cake, whipped cream, ice cream, and toppings. Each product has its own characteristic ranges of sizes, shapes, and air cell phase volumes, and these are largely responsible for the product's particular range of appearances and textures. The basis of gas stabilization of a foam may be either or both the water-soluble protein, such as egg protein, and the fat phase (Sahi 1996). Foaming can be assessed by measurement of expansion, determined by the increase, then decrease of volume of a sample held at constant temperature, and flow properties of time-dependent viscosity (Bombara et al. 1994). Alternatively, foam can be specified by overrun, stability, and viscosity (Yankov and Panchev 1996) or foam capacity and foam stability (Arteaga and Nakai 1993).

The brewing industry has its own special consumer-important appearance attributes. For example, lager has four apparently independent groups of properties. These are *clarity, amber/color in-*

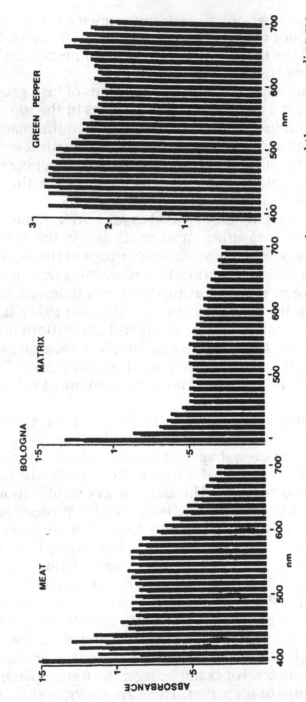

Figure 10–15 Focal absorbance spectra of lean meat and green pepper fragments and the surrounding matrix of bologna (Swatland 1985, with permission).

tensity, gray froth, and *head persistence/head quantity/rising bubbles/bubble size/head density* (Spooner 1996). Apart from *color* and *haze,* properties important to beer include *foam quality, lacing,* and *nucleation.* There are regional differences in preference. For example, in the United Kingdom, flat beer tends to be preferred in the south and foamier beer in the north. Lace is generally not important in the Midlands but becomes more liked among customers to the east and west. Beers have characteristic foam properties. Pilsner has large and transparent bubbles, Guinness smaller and more opaque. Foam quality depends to a great extent on proteins and polypeptides present, and many factors affect the ability of proteins to stabilize the interface (Bamforth 1985). At low alcohol concentration, lipids are major foam-destabilizing components in beer and wine. At high alcohol concentration, foam behavior is governed mainly by ethanol content (Dussaud et al. 1994; Brierley et al. 1996). In the industry, foam quality is measured by foaming the beer in a tube and measuring the rate of drainage of a set volume of beer from the foam (Rudin 1957; American Society of Brewing Chemists 1957, 1958, 1959, 1960, 1961). Foam collapse can be measured directly (Weyh 1988). Lacing is the adherence of the foam to the sides of the glass after the beer has been drunk. The extent of nucleation in a glass of beer is largely controlled by the carbon dioxide concentration in solution, the temperature, and the dispense pressure. Carbon dioxide concentration can be measured manometrically (Martin 1970), but gas analyzers are more convenient.

Carbonated soft drinks contain dissolved carbon dioxide gas under pressure. When pressure is released or temperature increased, the mixture becomes metastable, and the gas is spontaneously released. Each flavor has an optimum carbonation level promoting the flavor. The product is controlled by the relationship between product pressure, temperature, and dissolved gas volume (Murphy 1997).

Machine vision within the food industry is used for gauging for dimensions, verifying presence or absence of features, flaw detecting, identifying an object, recognizing the identity of object features, and locating (Gunasekaran 1996). In coordination with robot packing, individual chocolates can be checked for correct area, perimeter, centroid coordinates (for the picking-up point), and calculation of the angle of orientation. Robots equipped with vision systems are available for building calculated nonuniformity into complex meals, such as pizza, ready meals, and desserts. They are used in conjunction with cutting and portioning machines to produce attractive fish or meat cuts and to achieve effective pack weight control (Lightbody 1990). Vision robotic pick-and-place technology is in wide use in the biscuit and baked products industry (e.g. "Robots" 1995). On-line inspection of bakery and snack products can be used to monitor and control loaf profile, cookie and muffin dimensions, and color, as well as to detect pieces of burned debris (Scott 1995). For example, broken shaped cookies and shelled nuts can be identified using dimensional measurements for *wholeness* specific to the product. Using definitions obtained as a result of artificial neural net trials, damaged products can be rejected (Gunasekaran and Ding 1994). Presence of different pizza components, such as green peppers and herbs, cheese, red peppers and tomatoes, and yellow sweet corn, can be detected using computer color filters (Perry 1992). Machine vision line speeds greater than 200 units per minute can be achieved using pick-and-place action on payloads of a few grams up to 1 kg to an accuracy of less than or equal to 3 to 5 mm (Wallin 1997).

Where shapes and sizes can be characterized, as in ice crystals (Hobbs 1974), shark skinning in extruded meats (Hawkins 1971), faults in fabrics (e.g., British Standards Institute 1983), structural variations in leaves (Araus et al. 1986), epithelial cells (Engvall et al. 1981), surface irregularity (e.g., American Society for Testing and Materials 1977), chatter marks on machined surfaces (Rau and Huebner 1986), container cleanness and filling, and closure security, there is a potential for automatic analysis using machine vision analysis. Cosmetic features detectable are the presence of labels and their position, quality, and alignment (Grikitis 1987).

Distribution of Discontinuities

The means whereby the distribution of a quality can be specified has been discussed with reference to the color uniformity of canned green beans (Ross et al. 1959). A measure of disuniformity (D) is the square of the standard deviation of the distribution about the mean of measurements made on a number of randomly selected beans. The greater the value of D, the more nonuniform the material. To reduce error, a minimum number of 50 beans is necessary. A Hunter Color and Color Difference meter was used with a 12-mm aperture, small enough to make measurements on single beans. Color variation among beans was large compared with variations within a single sample. Instrumental readings of (R_d, a, b) were made on three strains harvested on three dates. Significant differences in D (using R_d, a function of L, measurements) were found between strains ($p <$ 0.01). Also, as harvesting progressed, significant increases of D (using b) with picking time were found, indicating changes in the yellow component of the color. Mean values of D (using b) across strains for each harvesting date of canned beans were

day 1, $D = 1.26$; day 5, $D = 1.60$;

day 10, $D = 1.95$; day 17, $D = 2.02$

In terms of incorporation into the U.S. Department of Agriculture (1953) specification for three grades, the grade limits for *color uniformity* (Ross et al. 1959) were calculated to be

Grade A, $D \leq 3.4$;

Grade B, $3.4 < D \leq 4.9$;

Grade C, $4.9 < D \leq 6.4$

Uniformity ratio is a useful measure of, for example, variation in the size of individuals captured in a fish catch. This is the weight ratio of the largest 10 percent of the catch of, say, shrimp, to the smallest 10 percent (Luzuriaga et al. 1997).

TEMPORAL ASPECTS

Temporal aspects of food appearance are few, but they do exist. For example, one of the properties that can reveal the identity of a reformed or synthetic product is its behavior with time during cooking. Other attributes in this class fall within the overlap region of appearance and kinesthetic properties (see Figure 2–1); that is, those structural properties estimated by the visual sense. Properties such as the wobbliness of a jelly and the viscosity of a syrup are revealed by how the product behaves with time in response to deformation. The interaction between texture perception, for example, of gel strength depends upon the sense used. When one is placing different types of gel in a rank order of strength, the estimate made by the visual sense when judging the gel's response to shaking is the same as that when estimated by depressing the surface with the finger. However, this is very different from the rank order produced when the gel is eaten. The former is governed mainly by the reciprocal of the yield strain; that is, the movement of the gel in response to a force. The eating strength is governed by the yield stress, the force required to deform the gel (Hutchings and Wood 1974, cited in Wood 1974).

ANALYSIS OF COMPLEX SCENES

Daily judgments are made of food scenes around us, and images arising from inspection of the scene form the basis of subsequent actions. Appearance profile analysis provides a method of describing food materials in terms of *basic perceptions* of structural elements, their color, translucency, and gloss properties, and *derived perceptions* of visual expectation (see Chapter 6). This is an analytical tool. However, the situation of the diner looking at the plateful in front of him or her on the table requires further consideration. Such studies are normally comparative and take the form of intensive debate using articulate panelists. A scientific approach, capable of analyzing, measuring, and comparing complex appearance scenes would help to understand mechanisms of design and choice in complex situations. This is not an established area of study, but leads exist whereby a suitable methodology might be developed. Such a methodology would apply to consideration of a simple product and its place as part of a lifestyle or in a meal and would be applicable to selection from the shelf and consumption within an eating environment.

In the real world, the eater, the food, and the environment form a continuum. The food itself can be analyzed using appearance profile analysis (see Chapter 7). Layout of the food on the plate can be discussed in terms taken from the design industry. Any structure can be considered in terms of alternation, dominance, contrast, harmony, and balance of elements (Radhakrishnan 1980). *Alternation* is a basic shape as it undergoes changes in organization, size, and color. *Harmony* occurs when one or more qualities of, say, shape, size, or color are alike. Extremes are complete repetition and discord. *Contrast* is the bringing together of two opposites of shape, line, or color. *Dominance* is the proportion by which some parts of the design are given more importance. *Balance* is the equilibrium of opposing forces; *formal balance* is symmetry about an axis, such as the human form or a leaf, whereas *informal balance* is the balancing of one or more dissimilar elements on opposite sides of an axis. Application of these design elements to, for example, pizza can be readily appreciated.

The interaction of the food with the environment consists of the decor and other aspects of food presentation. These can add or subtract from the food. Responses are dictated by, for example,

- the container—its size, color, pattern, cleanness, and temperature
- the degree of fill—the container/portion size ratio
- the tabletop elements such as cutlery—appropriateness and effectiveness
- the environment—temperature, comfort, appropriateness
- the illuminated decor and its interaction with the meal and the occasion

An analysis of the food itself includes

- whether it can be recognized and identified
- whether it is appropriate with relation to the eater's immediate environment and inherited and learned responses to specific situations (see Chapter 2)
- whether it accords with what the eater knows and is used to and with how the eater feels it ought to be cooked

- whether it contains sufficient vitamins or calories that the eater feels such a dish ought to provide
- whether the portion size is appropriate for the eater's needs
- whether the portion content and its aesthetic properties of total appearance are appropriate
- whether the eater thinks he or she ought to be seen to be associated with the product
- whether the value judgment of quality relative to cost is appropriate

Methods of assessment and measurement of scene elements and whole scenes are available, as is methodology for their application and analysis.

CONNOTATIVE MEANINGS OF SCENES OR SCENE ELEMENTS

Any product or scene can be quantified by connotative analysis. Psychophysical relationships between objective properties of surfaces and their psychological meanings have been applied to architecture (Burnham and Grimm 1973). The connotative meanings of an object or concept can be measured using the semantic differential technique (Osgood et al. 1957). Respondents are asked to indicate, on a 7-point scale between a series of polar adjective pairs, the point that best describes the relationship or association between the concept and the adjective. Factor analysis is then used to derive ratings for three independent dimensions. The dimensions, with their relevant polar adjective pairs, are

- *Evaluation:* **beautiful/ugly, pleasant/unpleasant, harmonious/dissonant, meaningful/meaningless, cheerful/sad**, and **refined/vulgar**
- *Activity*: **energetic/inert, tense/relaxed, dynamic/static, interesting/boring, warm/cold**, and **fast/slow**
- *Potency*: **rugged/delicate, hard/soft, tenacious/yielding, strong/weak, tough/tender**, and **masculine/feminine**

The dimensions can be plotted orthogonally in Cartesian coordinates, thus developing a semantic space. The technique can be used to compare the meanings of two concepts—say, a design concept and a visual surface property. The formula used for this is

$$D_{ds} = [(E_d - E_s)^2 + (P_d - P_s)^2 + (A_d - A_s)^2]^{0.5}$$

where D_{ds} is the distance between the meanings in semantic spatial units, E the evaluative coordinate, P the potency coordinate, A the activity coordinate, d the subscript indicative of the design concept, and s the subscript indicative of the visual property of the surface. This technique can be a rational basis for the selection of a particular surface from a number of alternatives. The surface selected is the one that matches the design concept in connotative meaning—that is, one for which D_{ds} is minimized.

In the architectural example used by Burnham and Grimm (1973), the connotative meanings for unglazed brick textures, colors, and sizes were determined. Judgments of texture showed that, in general, brick textures are neither **good** nor **bad** but may be **strong** or **weak**, or **active** or **passive**. Few brick colors were either **good** or **bad**. Moderately orange bricks were determined to be the most **active**, brownish pink the most **passive**. Grayish reddish brown bricks were the **strongest**, pale yellowing pink the **weakest**. The potency ratings given to size varied considerably more than the other two dimensions. The intraobserver reliability and the agreement among two groups of cooperative and well-motivated observers were found to be high.

Methods have been suggested for pretesting such market stimuli as product benefit combinations, package designs, brand-price combinations, advertisements, and special offers. Marketing problems can be broken down into specific areas: for example, measuring the relative importance of a group of product benefits, or selecting a package design that best relates to the psychological imagery of selected benefits, or determining what point-of-purchase display materials, brand name, and pricing strategy to employ in the market introduction. All these problems involve non-numerical, judgmental responses to multiattribute marketing stimuli.

This type of problem may be approached using conjoint measurement. This involves using conventional market research techniques to obtain a list of product benefits thought to be the most important to the market. Conjoint measurement algorithms require respondents to rank-order fractional factorial designs that incorporate the product benefits. Benefit interactions may also be included in the analysis (Green 1973).

Relationships between the color of a space and its psychological effects have been investigated. Judgments of the *dynamism factor* (e.g., **active/passive, exciting/calming**), the *spatial quality factor* (**cramped/spacious**), and the *emotional tone factor* (**hot/cold, lush/austere**) were found to be relatable to Munsell Chroma, Value, and Hue. The similarity of findings with those of other authors led to the conclusion that there may be some inherent, possibly biological, relationship between color and the above quality judgments. The failure to find consistent relationships between color quality and the *complexity factor* (**complex/simple, usual/unusual**) and the *evaluation factor* (**repellent/receptive, unpleasant/pleasant**) suggested that these may be more influenced by learning (Hogg et al. 1979). Similar studies of two-color combinations resulted in the isolation of the five factors *evaluation, articulation, light/heavy, warm/cold,* and *originality* (Sivic and Toft 1989, 1992).

That so many adjectives can be clustered into so few (four or five) groups may seem surprising. However, Osgood and Akuto (cited in Kobayashi 1981) have found that adjectives in the Japanese language can be classified into just four fundamental groups. These are *moral correctness, sensory pleasure, dynamism,* and *magnitude*.

Scales specific to the product can also be selected. These include, for example, scales describing the derived perceptions of visually assessed flavor and texture (see Chapter 6), portion size, ingredient variety, and whether the product is appetizing.

The approach to the analysis of complex scenes will be enhanced by knowledge of how the customer looks at the scene. The use of eye movement monitoring and virtual reality techniques has application both in the small scale of pack design and in the larger world of allowing the architect's customer to walk through rooms before the building is constructed. Computer design techniques allow images complete with logos to be created on the screen. Also, packs can be massed on the screen and set up with others for an initial assessment of design or logo effectiveness to be made against those of competitors.

These examples are sufficient to promise the successful application to foods and lead to an understanding of the appearance of a product within the framework of the image transfer process and total appearance.

REFERENCES

Abdullah A, Malundo TMM, Resurreccion AVA, Beuchat LR. (1993). Descriptive sensory profiling for optimizing the formula of a peanut milk-based liquid coffee whitener. *J Food Sci* 58:120–123.

Akamine EK, Goo T. (1971). Relationship between surface color development and total soluble solids in papaya. *Hort Sci* 6:567–568.

American Society for Testing and Materials. (1977). *Standard test method for surface irregularities of flat transparent plastic sheets*. ANSI/ASTM D637-50. Philadelphia: ASTM.

American Society of Brewing Chemists. (1957). Reports of the Subcommittee on Evaluation of Beer Foam. *Proc Am Soc Brewing Chem* 135.

American Society of Brewing Chemists. (1958). Report of the Subcommittee on Evaluation of Beer Foam. *Proc Am Soc Brewing Chem* 137.

American Society of Brewing Chemists. (1959). Report of the Subcommittee on Evaluation of Beer Foam. *Proc Am Soc Brewing Chem* 175.

American Society of Brewing Chemists. (1960). Report of the Subcommittee on Evaluation of Beer Foam. *Proc Am Soc Brewing Chem* 214.

American Society of Brewing Chemists. (1961). Report of the Subcommittee on Evaluation of Beer Foam. *Proc Am Soc Brewing Chem* 139.

Araus JL, Sabido J, Aguila FJ. (1986). Structural differences between green and white sectors of variegated *Scindapsus aureus* leaves. *J Am Soc Hort Sci* 111:98–102.

Armstrong NA, Marsh GA. (1974). Quantitative assessment of surface mottling of colored tablets. *J Pharmacol Sci* 63:126–129.

Arteaga GE, Nakai S. (1993). Predicting protein functionality with artificial neural networks: foaming and emulsifying properties. *J Food Sci* 58:1152–1156.

Bamforth CW. (1985). The foaming properties of beer. *J Inst Brewing* 91:370–383.

Barletta BB, Barbosa-Cánovas GV. (1993). Fractal analysis to characterize ruggedness changes in tapped agglomerated food powders. *J Food Sci* 58:1030–1035, 1046.

Barrett AM, Peleg M. (1992). Cell size distributions of puffed corn extrudates. *J Food Sci* 57:146–148, 154.

Bennett HE, Porteus JO. (1961). Relation between surface roughness and specular reflectance at normal incidence. *J Opt Soc Am* 51:123–129.

Billmeyer FW Jr, Chen Y. (1985). On the measurement of haze. *Color Res Appl* 10:219–224.

Birth GS, McGee JB, Cavalette C, Chan H. (1978). An optical technique for measuring papaya maturity. Paper no. 78-3071. St. Joseph, Mich: American Society of Agricultural Engineering.

Blakeney AB, Welsh LA. (1988). Objective measurement of rice grain color and gloss. *Cereal Foods World* 33:679.

Bombara N, Pilosof PMR, Anon MC. (1994). Mathematical model for formation rate and collapse of foams from enzyme modified wheat flours. *J Food Sci* 59:626–628, 681.

Borggaard C, Madsen NT, Thodberg HH. (1996). In-line image analysis in the slaughter industry, illustrated by beef carcass classification. *Meat Sci* 43:S151–S163.

Brierley ER, Wilde PJ, Onishi A, Hughes PS, Simpson WJ, Clark DC. (1996). The influence of ethanol on the foaming properties of beer protein fractions: a comparison of Rudin and microconductivity methods of foam assessment. *J Sci Food Agric* 70:531–537.

British Standards Institute. (1983). *British standard method for numerical designation of fabric faults by visual inspection*. BS 6395. London: British Standards Institute.

Burnham CA, Grimm CT. (1973). Connotative meaning of visual properties of surfaces. In *Sensory evaluation of materials*, ed. RS Hunter, PN Martin. Philadelphia: American Society for Testing and Materials.

Chen C-S, Chen J-J, Wu T-P, Chang C-Y. (1998). Optimising frying temperature of gluten balls using response surface methodology. *J Sci Food Agric* 77:64–70.

Chtioui Y, Bertrand D, Dattée Y, Devaux ME. (1996). Identification of seeds by colour imaging: comparison of discriminant analysis and artificial neural network. *J Sci Food Agric* 71:433–441.

Chuma Y, Uchida S, Shemsanga KHH. (1982). Simultaneous measurement of size, surface area and volume of grains and soybeans. *Trans Am Soc Agric Eng* 25:1752–1756.

Cliff MA, Dever MC, MacDonald RA, Flemming WW. (1996). Development of a photographic scale for the visual evaluation of apple stripe density. *J Food Qual* 19:31–40.

Cole CGB, Roberts JJ. (1997). Gelatine colour measurement. *Meat Sci* 45:23–31.

Coleman MA, Van Otteran E. (1982). The effect of crowding on the incidence of yolk mottling in WLH hens. *Poultry Sci* 61:1374.

Dussaud A, Robillard B, Carles B, Duteurtre B, Vignes-Alder M. (1994). Exogenous lipids and ethanol influences on the foam behavior of sparkling base wines. *J Food Sci* 59:148–151, 167.

Ehmann EP. (1972). The measurement of beer color. PhD diss, University of Massachusetts.

Engvall J, Greenberg SD, Spjut HJ, Estrada R, Subach J, Kimzey SL, King JF, DiTripani PM. (1981). Development of a mathematical model to analyse color and density as discriminant features for pulmonary squamous epithelial cells. *Pattern Recognition* 13:37–47.

Evers T. (1993). On line quantification of bran particles in white flour. *Food Sci Technol Today* 7(1):23–27.

Francis FJ, Clydesdale FM. (1975). *Food colorimetry: theory and applications.* Westport, Conn: Avi.

Gerrard DE, Gao X, Tan J. (1996). Beef marbling and color score determination by image processing. *J Food Sci* 61:145–148.

Graham D, Choong YC. (1981). Potential use of image analysis techniques in surface texture studies. *Precision Eng* 3:209–213.

Green PE. (1973). Measurement of judgmental responses to multi-attribute marketing stimuli. In *Sensory evaluation of appearance of materials*, ed. RS Hunter, PN Martin, 139–153. Philadelphia: American Society for Testing and Materials.

Grikitis K. (1987). Focus on quality. *Food Processing* 56(9): 25–29.

Gunasekaran S. (1996). Computer vision technology for food quality assurance. *Trends Food Sci Technol* 7:245–256.

Gunasekaran S, Ding K. (1994). Using computer vision for food quality evaluation. *Food Technol* 48(6):151–154.

Harrigan KA, Bussmann S. (1998). Digital speck counting of semolina using automated image analysis. *Cereals Foods World* 43:11–16.

Hawkins AE. (1971). Non-Newtonian technology associated with some food products. *Chem Eng* 245:19–23.

Hershko V, Weisman D, Nussinovitch A. (1998). Method for studying surface topography and roughness of onion and garlic skins for coating purposes. *J Food Sci* 63:317–320.

Hildebrandt G, Hirst L. (1985). Determination of the collagen, elastin and bone content in meat products using television image analysis. *J Food Sci* 50:568–572, 576.

Hobbs PV. (1974). *Ice physics.* Oxford, UK: Clarendon Press.

Hogg J, Goodman S, Porter T, Mikellides B, Preddy DE. (1979). Dimensions and determinants of judgements of colour samples and a simulated interior space by architects and non-architects. *Br J Psychol* 70:231–242.

Hudson JR. (1969). Institute of Brewing: Analysis Committee measurement of colour in wort and beer. *J Inst Brewing* 75:164–168.

Hunter RS, Harold RW. (1987). *The measurement of appearance*. New York: John Wiley.

Hutchings JB. (1978). Psychophysics of colour and appearance in product development. In *Proceedings of a symposium on food colour and appearance*, 46–55. London: University of Surrey, Colour Group (GB).

Hutchings JB, Gordon CJ. (1981). Translucency specification and its application to a model food system. In *Proceedings of the Fourth Congress of the International Colour Association*, ed. M Richter, Vol. 1. West Berlin: Deutscher Verband Farbe.

Hutchings JB, Scott JJ. (1977). Colour and translucency as food attributes. In *Proceedings of the Third Congress of the International Colour Association*, Troy, New York, ed. FW Billmeyer Jr, G Wyszecki, 467–470. Bristol, UK: Adam Hilger.

Irie M, Swatland HJ. (1993). Prediction of fluid losses from pork using subjective and objective paleness. *Meat Sci* 33:277–292.

Jakobsson K. (1997). A digital analysis instrument for consistent grain assessment. In *Cereals '97: proceedings of the 47th Australian Cereal Chemistry Conference, Perth*, ed. AW Tarr, AS Ross, CW Wrigley, 176–179. Melbourne: Royal Australian Chemical Institute, Cereal Chemistry Division.

Jindal VK, Mohsenin N, Husted JV. (1974). Surface area of selected agricultural seeds and grains. *Trans Am Soc Agric Eng* 17:720–728.

Joubert E. (1995). Tristimulus colour measurement of rooibos tea extracts as an objective quality parameter. *Int J Food Sci Technol* 30:783–792.

Keefe PD, Draper SR. (1986). The measurement of new characters for cultivar identification in wheat using machine vision. *Seed Sci Technol* 14:715–724.

Keefe PD, Draper SR. (1988). An automated machine vision system for the morphometry of new cultivars and plant genebank accessions. *Plant Varieties, Seeds* 1:1–11.

Klavons JA, Bennett RD, Vannier SH. (1992). Stable clouding agent from isolated soy protein. *J Food Sci* 57:945–947.

Klemaszewski JL, Das KP, Kinsella JE. (1992). Formation and coalescence stability of emulsions stabilized by different milk proteins. *J Food Sci* 57:366–371, 379.

Kobayashi S. (1981). The aim and method of the color image scale. *Color Res Appl* 6:93–107.

Koch J. (1978). Some thoughts on the gloss of chocolate. *Confectionery Prod* 44(May):182, 184, 254.

Kühn MC, Grosch W. (1988). Influence of enzymatic modification of the non-starchy polysaccharide fractions on the baking properties of reconstituted rye flour. *J Food Sci* 53:889–895.

Lai FS, Zayas I, Pomeranz Y. (1986). Application of pattern recognition techniques in the analysis of cereal grains. *Cereal Chem* 63:168–172.

Lightbody MS. (1990). New technology approaches to reducing uniformity in processed foods. *Food Sci Technol Today* 4(1):37–40.

Little AC. (1964). Color measurement of translucent food samples. *J Food Sci* 29:782–789.

Little AC. (1971). The color of white wine: evaluation by transreflectometry. *Am J Enol Viticulture* 22:144–149.

Little AC, Brinner L. (1977). Factors affecting color and appearance of corn tortillas. *J Food Qual* 1:141–146.

Little AC, Brinner L. (1981). Optical properties of instant tea and coffee solutions. *J Food Sci* 46:519–522, 525.

Loisel C, Lecq G, Ponchel G, Keller G, Ollivon L. (1997). Fat bloom and chocolate structure studied by mercury porosimetry. *J Food Sci* 62:781–788.

Luzuriaga DA, Balaban MO, Yeralan S. (1997). Analysis of visual quality attributes of white shrimp by machine vision. *J Food Sci* 62:113–118, 130.

MacDougall DB. (1970a). Characteristics of the appearance of meat—I: the luminous absorption, scatter and internal transmittance of the lean of bacon manufactured from normal and pale pork. *J Sci Food Agric* 21:568–571.

MacDougall DB. (1970b). Instrumental assessment of food appearance. *Proc Nutr Soc* 29:292–297.

MacDougall DB. (1971). Characteristics of the appearance of meat—II: uniform lightness and chromaticness spacing of the lean of sliced fresh bacon. *J Sci Food Agric* 22:427–430.

MacDougall DB. (1982). Changes in the colour and opacity of meat. *Food Chem* 9:75–88.

MacDougall DB. (1983). Instrumental assessment of the appearance of foods. In *Sensory quality in foods and beverages*, ed. AA Williams, RK Atkin, 121–139. London: Society of Chemical Industry.

MacDougall DB. (1987). Effects of pigmentation, light scatter and illumination on food appearance and acceptance. In *Food acceptance and nutrition*, ed. J Solms, DA Booth, RM Pangborn, O Raunhardt, 29–46. London: Academic Press.

MacDougall DB, Lima RC. (1999). Coffee and milk with Kubelka and Munk. *Proc Colour '98 Harrogate*. In press.

MacKinney G, Little AC, Brinner L. (1966). Visual appearance of foods. *Food Technol* 20:1300–1306.

Malundo TMM, Resurreccion AVA, Koehler PE. (1992). Sensory quality and performance of spray-dried coffee whitener from peanuts. *J Food Sci* 57:222–226, 251.

Mandelbrot BN. (1977). *The fractal geometry of nature*. New York: WH Freeman.

Marks R, Pearce AD. (1975). Surfometry: a method of evaluating the internal structure of the stratum corneum. *Br J Dermatol* 92:651.

Martin PA. (1970). The Institute of Brewing Analysis Committee: determination of carbon dioxide in beer. *J Inst Brewing* 76:344–347.

McDonald CE. (1994). Collaborative study on particle size in wheat flour by laser instrument (AACC method 50–11). *Cereal Foods World* 39:29–33.

Mohsenin NN. (1968a). *Physical properties of plant and animal materials*, vol. 1, part 1. Pittsburgh, Pa.: Pennsylvania State University.

Mohsenin NN. (1968b). *Physical properties of plant and animal materials*, vol. 1, part 2. Pittsburgh, Pa.: Department of Agricultural Engineering, Pennsylvania State University.

Murphy C. (1997). Carbonation: the science of bubbles. *Food Processing* 66(4):16–17.

Musser JC. (1973). Gloss on chocolate and confectionery coatings. In *Proceedings of the 27th Pennsylvania Manufacturing Confectioners' Association. Production Conference* 46–50.

Newman PB. (1987). The use of video image analysis for quantitative measurement of visible fat and lean in meat. *Meat Sci* 19:129–150.

Nussinovitch A, Ward G, Mey-Tal E. (1996). Gloss of fruit and vegetables. *Lebensm Wiss Technol* 29:184–186.

Obuchowski W, Bushok W. (1980). Wheat hardness: comparison of methods and its evaluation. *Cereal Chem* 57:421– 426.

Osgood CE, Succi GJ, Tannenbau PH. (1957). *The measurement of meaning.* Urbana: University of Illinois Press.

Parker NS. (1987). Properties and functions of stabilising agents in food emulsions. *CRC Crit Rev Food Sci Nutr* 25:285–316.

Patel AA, Gandhi H, Singh S, Paril GR. (1996). Shelf-life modelling of sweetened condensed milk based on kinetics of Maillard browning. *J Food Processing Preservation* 20:431–451.

Pateras IMC, Howells KF, Rosenthal AJ. (1994). Hot-stage microscopy of cake batter bubbles during simulated baking: sucrose replacement by polydextrose. *J Food Sci* 59:168–170, 178.

Peleg M. (1993). Fractals and foods. *Crit Rev Food Sci Nutr* 33(2):149–165.

Peleg M, Gomez B. (1974). External color as a maturity index of papaya fruits. *J Food Sci* 39:701–703.

Perry S. (1992). Cracking vision. *Food Processing (GB)* 61(11):27–30.

Phillips LG, Barbano DM, Lawless HT. (1995). The sensory attributes of skim milk containing a casein and titanium dioxide blend to improve whiteness. *J Dairy Sci* 78 (suppl 1): 135.

Phillips LG, McGiff ML, Barbano DM, Lawless HT. (1995a). The influence of fat on the sensory properties, viscosity and color of lowfat milk. *J Dairy Sci* 78:1258–1266.

Phillips LG, McGiff ML, Barbano DM, Lawless HT. (1995b). The influence of nonfat dairy milk on the sensory properties, viscosity and color of lowfat milks. *J Dairy Sci* 78:2113–2118.

Radhakrishnan KK. (1980). Colour and textile design: Colour 80 seminar proceedings. *Colourage* (suppl.) 43–48.

Raeuber HJ, Nikolaus H. (1980). Structure of foods. *J Texture Stud* 11:187–198.

Raggi A, Barbiroli G. (1993). Colour uniformity in discontinuous foodstuffs to define tolerances and acceptability. In *Proceedings of the Seventh Congress of the International Colour Association*, 265–346. Budapest: Techical University of Budapest.

Rahman MS. (1997). Physical meaning and interpretation of fractal dimensions of fine particles measured by different methods. *J Food Eng* 32:447–456.

Rau N, Huebner G. (1986). Optical measurement of chatter marks. *Wear* 109:225–239.

Ritchie RH. (1994). Color scanner, advanced imaging software improve product quality. *Food Processing (USA)* 55(4):15–20.

Robots hit the spot for biscuit packaging. (1995). *Confection* 2(11):11–13.

Ross E, Pauls RH, Hard MM. (1959). Uniformity of color measure in green beans. *Food Technol* 13:711–715.

Roudot A-C. (1989). Image analysis of kiwi fruit slices. *J Food Eng* 9:97–118.

Rudin AD. (1957). Measurement of the foam stability of beers. *J Inst Brewing* 63:506–509.

Sahi S. (1996). Forever blowing bubbles. *Food Manuf* 71(6):32–34.

Sakai N, Yonekawa S. (1991). Three-dimensional image analysis of the shape of soybean seed. *J Food Eng* 15:221–234.

Savello PA, Solorio HA. (1995). Increased whiteness and decreased blueness in enzyme-treated pasteurised skim milk. In *Proceedings of the IFT Annual Meeting*, 221.

Scott AJ. (1995). On-line inspection of bakery and snack products. *Cereal Foods World* 40(1):15–18.

Siebert KA, Lynn PY. (1997). Haze-active protein and polyphenols in apple juice assessed by turbidimetry. *J Food Sci* 62:79–84.

Sivic L, Toft C. (1989). Semantic variables for judging color combinations. *Göteborg Psychol Rep*, 19(5). Göteborg: University of Göteborg.

Sivic L, Toft C. (1992). Color combinations and associated meanings. *Göteborg Psychol Rep* 19(2). Göteborg: University of Göteborg.

Spooner M. (1996). Making sense of sensory analysis. *Food Manuf* 71(12): 32–33.

Swatland HJ. (1985). Color measurements of variegated meat products by spectrophotometry with coherent fiber optics. *J Food Sci* 50:30–33.

Swatland HJ. (1991). Spatial and spectrophotometric measurements of light scattering in turkey breast meat using lasers and a xenon arc. *Can Inst Sci Technol* 24:27–31.

Szczesniak AS. (1983). Physical properties of foods. In *Physical properties of foods*, ed. M Peleg, EB Bagley, 1–41. Westport, Conn: Avi.

Talbot G. (1995). Chocolate fat: the cause and the cure. *Int Food Ingredients* (1):40–45.

Te-Chen K, Song S. (1984). A spectrological method for measuring rice chalkiness and translucency. *J Agric Assoc China* 128:7–16.

Tobias PE, Ricks J, Chadwick M. (1989). Objective, reproducible measurement of printing mottle with a mottle tester. *Tappi J* 72 (May):109–112.

Tscheuschner H-D, Markov E. (1986). Instrumental texture studies on chocolate—III: processing conditioned factors influencing the texture. *J Texture Stud* 17:377–399.

Unklesbay K, Unklesbay N, Keller J, Grandcolas J. (1983). Computerized image analysis of surface browning of pizza shells. *J Food Sci* 48:1119–1123.

U.S. Department of Agriculture, (1953). *U.S. standard for grades of canned green beans and canned wax beans.* [since superseded by 37FR87]. Washington, DC: USDA.

U.S. Department of Agriculture. (1989). *U.S. standards for grades of carcass meat.* Washington, DC: USDA.

van Laack LJM, Kauffman RG, Sybesma W, Smulders FMJ, Eikelenboom G, Pinheiro JC. (1994). Is colour brightness (L-value) a reliable indicator of water-holding capacity in porcine muscle? *Meat Sci* 38:193–201.

Vergano PJ, Testin RF, Newall WC Jr. (1991). Distinguishing among bruises in peaches caused by impact, vibration and compression. *J Food Qual* 14:285–298.

Wallin PJ. (1997). Robotics in the food industry: an update. *Trends Food Sci Technol* 8:193–198.

Ward G, Nussinovitch A. (1996a). Gloss properties and surface morphology relationships of fruits. *J Food Sci* 61:973–977.

Ward G, Nussinovitch A. (1996b). Peel gloss as a potential indicator of banana ripeness. *Lebensm Wiss Technol* 29:289–294.

Warner G. (1993). Down to the blush; new cameras sort by hue. *Good Fruit Grower* 44:15.

Weyh H. (1988). Zur Reproduzierbarkeit von Schaummessungen Teil 2: NIBEM Methode. *Monatsschr Brauwiss* 11:441–445.

Wood FW. (1974). The application of psychophysics to the food industry with special reference to the textures of liquid foods and gels. *Proceedings of the Fourth International Congress of Food Science and Technology*, vol. 2, 273–276. Valencia: Instituto de Agroquímica y Technología de Alimentos.

Yankov S, Panchev I. (1996). Foaming properties in sugar-egg mixtures with milk protein concentrates. *Food Res Int* 29:521–525.

CHAPTER 11

Food Color Mechanisms

The ever-increasing volumes and varieties of fresh and pro-cessed produce being brought onto the shelves are responsible for much research into the mechanisms responsible for food ap-pearance. Driving forces arise from the vital part played by fruit and vegetables in our well-being, attempts to increase the time between food production and sell by date, and, as palettes be-come less conservative, an unceasing quest for new products and ways to upgrade raw materials as they become available.

The volume of material in the literature dealing with pigment chemistry is too large for an exhaustive treatment to be included in a volume taking a broad look at food appearance. The aim of this chapter is to outline coloring systems in food materials and changes that occur during processing. First, there is a discussion of fresh, minimally processed, and processed fruit and vegetable systems. Accounts of meat, marine systems, and browning fol-low. There is then a discussion of the contribution made to food appearance by light scattering, and the chapter ends with a sec-tion on added color.

FRUIT AND VEGETABLE COLOR

Chlorophyll molecules formed the basis of the first self-sustain-ing life on earth. They are responsible in plant leaves for chan-neling and converting radiant energy from the sun into chemi-cal energy through the process of photosynthesis. Physically, chlorophylls are held between layers of lipid and protein in the thylakoid membrane within the leaf chloroplast, along with light-absorbing carotenoid, carotene, and xanthophyll molecules. These layers are disc-shaped particles of approximate diameter 0.1 nm. The chlorophyll molecule consists of two parts: the magnesium-chelated tetrapyrrole associated with the protein and the fat-soluble phytol "tail" associated with the lipid layer. Chlorophylls a and b are the chief forms of the pigment existing in higher plants and algae. Their structures can be depicted as in Figure 11–1.

The bright green colors of fresh vegetables act as a visual cue to quality, as their appearance is affected by aging, pH, heat, metal complexes, oxidation, enzymes, and fermentation. Chlorophylls in foods have been discussed by Francis (1999), Heaton and Marangoni (1996), Schwartz and Lorenzo (1990), and von Elbe (1986a).

Carotenoids occur widely in nature, where they contribute to the color of many yellow, red, and orange organisms. They are synthesized by bacteria, fungi, and higher plants and are conspicuous in foods such as carrots, citrus fruits, and tomatoes.

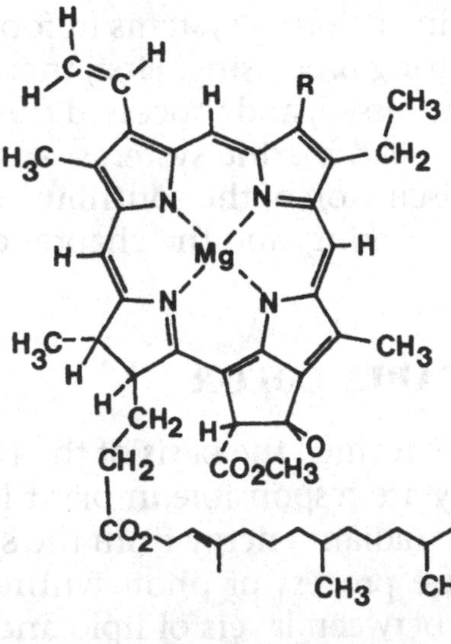

Figure 11–1 Structures of chlorophyll a (R=CH₃), and b (R=CHO).

Presence of carotenoids in many plants is initially masked by chlorophylls and only becomes apparent in autumn when greens fade and fruit ripens. Animals do not have the ability to synthesize carotenoids, but all depend upon a supply of vitamin A, of which carotene is a precursor, for maintenance of their metabolism and growth. Animal products dependent on carotenoids for their color include egg yolks, liver, body fat deposits, lobsters, salmon, and the milk fat of cattle. The few hundred naturally occurring carotenoids are generally insoluble in water but dissolve in fat solvents.

There are two classes of carotenoids, hydrocarbon carotenes and their oxygenated derivatives xanthophylls. The chromophore consists of a chain of conjugated carbon-carbon double bonds, joining, in the case of beta carotene, two beta-ionone rings. Increasing the number of these bonds progressively causes a displacement of light absorption toward longer wavelengths into the blue region of the spectrum, thus increasing apparent redness. Opening the ring, as in lycopene, further increases redness (Karrer 1962). The structures of some of the carotenoids important to food are shown in Figure 11–2.

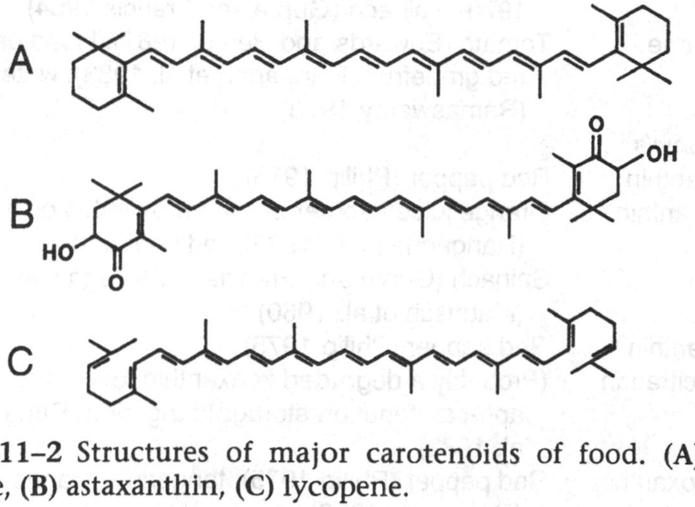

Figure 11–2 Structures of major carotenoids of food. (A) beta-carotene, (B) astaxanthin, (C) lycopene.

Examples of carotenoid pigments in plant products are listed in Table 11–1. Many carotenoids can occur in a single fruit. For example, there are more than 50 in the Israeli main crop Shamouti orange, and 16 in Marsh seedless grapefruit (Ranganna et al. 1983). Carotenoids in plant tissue are susceptible to light-induced oxidation, warm temperatures, oxygen, enzymes, and storage. Lipoxygenases are the major enzymes involved in carotenoid degradation. Carotenoids in foods are discussed by Francis (1999) and Britton (1996).

There are two types of flavonoid pigment, the widely occurring anthocyanins and the less apparent anthoxanthins. The

Table 11–1 Examples of Carotenoid Occurrence in Plant Products

Carotenoids	Occur In
Carotenes	
Alpha carotene	Carrot (Francis and Clydesdale 1970), lemon (Ranganna et al. 1983), watermelon (Ramaswamy 1973)
Beta carotene	Banana, jack fruit, maize, mango, papaya, pumpkin, watermelon (Ramaswamy 1973), red pepper (Philip 1975), carrot (Francis and Clydesdale 1970), spinach (Gupte and Francis 1964)
Lycopene	Tomato (Edwards and Reuter 1967), blood oranges, red grapefruit (Ranganna et al. 1983), watermelon (Ramaswamy 1973)
Xanthophylls	
Capsanthin	Red pepper (Philip 1975)
Violaxanthin	Orange juice (isomerizes to auroxanthin on storage) (Ranganna et al. 1983), red pepper (Philip 1975)
Lutein	Spinach (Gupte and Francis 1964), egg yolk (Maurisch et al. 1960)
Zeaxanthin	Red pepper (Philip 1975)
Beta-citraurin	(Probably a degraded zeaxanthin; decomposes to apo-carotenal on storage) tangerine (Ranganna et al. 1983)
Cryptoxanthin	Red pepper (Philip 1975), tangerine, orange (Ranganna 1983)

former are responsible for many of the wide range of red, blue, and purple hues of fruits, vegetables, and flowers. To the plant, they function as attractants in fruit dispersal and may protect cells by absorbing ultraviolet radiation. To human beings, they supply aesthetic and economic value. Flavonoids occur in the vacuolar sap of plant cells, but the amount of anthocyanin present varies with season and growing conditions. The most common forms of anthocyanidin are pelargonidin, cyanidin, delphinidin, peonidin, malvidin, and petunidin. As glycosides of anthocyanidins, they have a flavillium structure. The particular color depends on the number and orientation of hydroxyl and methyl groups. An increase in hydroxylation leads to an increase in blueness, while an increase in methoxylation leads to an increase in redness. Flavonoids are water soluble and, being highly reactive, are readily oxidized or reduced, the glycoside linkages undergoing hydrolysis. They may also form salts with acids or bases. The structures, together with that of the colorless leucocyanidin, which can take part in color-forming reactions, are sketched in Figure 11–3. These forms occur singly or in combination in foods, and examples are listed in Table 11–2. Many anthocyanidins may occur in a single fruit: for example, there are 14 in Concord grapes (Skalski and Sistrunk 1973).

Anthoxanthins (flavones) are colorless in acid but pale yellow in an alkaline medium. The most common of these easily masked

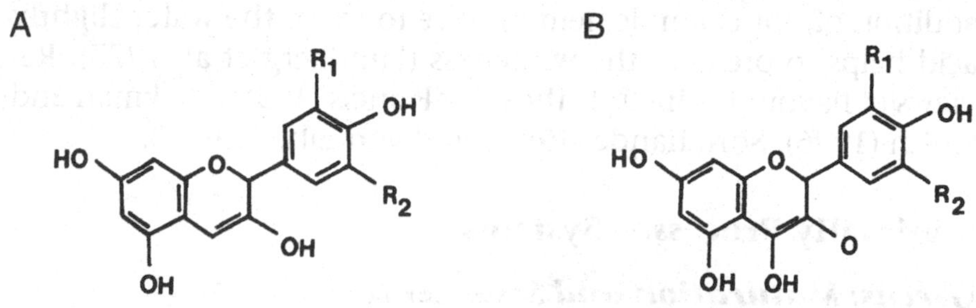

Figure 11–3 (A) Anthocyanidins. Pelargonidin (R_1=R_2=H), Delphinidin (R_1=R_2=OH), Cyanidin (R_1=OH, R_2=H), Petunidin (R_1=OCH$_3$, R_2=OH), Peonidin (R_1=OCH$_3$, R_2=H), Malvidin (R_1=R_2=OCH$_3$). (B) Leuco Anthocyanidin Leucocyanidin (R_1=OH, R=H).

Table 11–2 Anthocyanidins Occuring in Edible Fruit (compiled from Shrikhande 1986)

Anthocyanidins	Occur In
Pelargonidin	Banana, radish, potato
Cyanidin	Apple, blackberry, sweet cherry, elderberry, fig, gooseberry, mulberry, onion, peach, pear, raspberry, red cabbage, rhubarb
Cyanidin plus delphinidin	Black currant, "blood" orange juice, red currant
Delphinidin	Passion fruit, pomegranate, aubergine, french beans
Cyanidin plus peonidin	Cranberry, plum, sloe, sour cherry
Petunidin plus malvidin	Huckleberry

pigments is quercitin. First isolated from the oak, quercetin is also found in black currants, prunes, apricots, olives, corn, apples, honey, and tea leaves. The paleness of anthoxanthins restricts their contribution to overall color to one of subtle shading; hence, they have not been widely studied as constituents of fruits and vegetables. They complex with aluminium to form a yellow color and with iron to form brown. Anthoxanthins are responsible for the whiteness of vegetables such as cauliflower, onion, and potato. Boiling in alkaline water results in yellowing. However, addition of, for example, lemon juice to make the water slightly acid helps to preserve the whiteness (Lundberg et al. 1973). Reviews of flavonoids include those by Francis (1999), Jackman and Smith (1996), Shrikhande (1986), and von Elbe (1986a).

Minimally Processed Systems

Greens: Maturation and Senescence

Chlorophyll degradation is believed to result in the formation of four intermediate groups of compounds plus one end-product group. Degradation routes of chlorophyll a are summarized in Figure 11–4. There are two routes to the conversion of the blue-green chlorophyll a to the olive brown pheophorbide. Blue-green chlorophyllide is formed when the enzyme chlorophyllase cleaves the phytol chain from the chlorophyll. This is converted to pheophorbide when the magnesium ion is lost. The second route involves initial loss of the magnesium ion

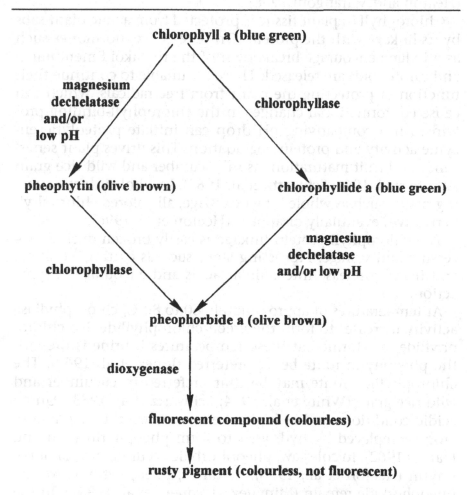

Figure 11–4 Chlorophyll degradation pathway. *Source*: Reprinted from *Trends in Food Science and Technology*, Vol. 7, Heaton and Marangoni, Chlorophyll Degeneration in Processed Foods and Senescent Plant Tissues, pp. 8–15, Copyright 1996, with permission from Elsevier Science.

when the olive brown pheophytin is formed. This is converted into pheophorbide on cleavage of the phytol chain. On cleavage of the porphyrin ring, the pheophorbide is converted into a colorless fluorescent compound, possibly through the action of dioxygenase. As what is believed to be the final step, this is converted into a nonfluorescent compound called rusty pigment (Heaton and Marangoni 1996).

Chlorophyll in plant tissue is protected from acidic plant saps by its linkage with the protein. On senescence, hormones such as ethylene encourage breakdown of the thylakoid membrane, and carotenoids are released. These are unable to continue their function of protecting the plant from free radicals, which can cause conformational changes in the chlorophyll-binding proteins. An accompanying pH drop can initiate proteolytic enzyme activity and protein degradation. This drives plant senescence and fruit maturation, as in cucumber and wild rice grain (White et al. 1964; Schwartz et al. 1983). In whole intact senescing tissue, such as whole intact cabbage, all colored chlorophyll derivatives eventually disappear (Heaton et al. 1996).

The chlorophyll-protein linkage is easily broken during processing and storage. Processing steps such as cutting, cooking, and freezing release intercellular acids and encourage enzyme action.

At temperatures of approximately 60 to 80°C, chlorophyllase activity increases to form the green chlorophyllide. No chlorophyllides are formed at these temperatures if brine is present, the pheophytin route being preferred (Jones et al. 1964). The chlorophyllide route may be that preferred in cucumber and wild rice grain (White et al. 1964; Schwartz et al. 1983). Under acidic conditions and at higher temperatures, the magnesium atom is replaced by hydrogen to form pheophytin (Tan and Francis 1962). In coleslaw, pheophorbide accumulates via pheophytin (Heaton et al., 1996); in olives, pheophytin as well as pheophorbide remain (Minguez-Mosquera et al. 1989). On the other hand, chlorophyll degradation in cabbage does not stop at pheophorbide but possibly continues to the colorless rusty pigment (Heaton et al. 1996). Black tea manufacture involving the intensified withering produced by the crush-tear-curl process also facilitates this removal of magnesium ions before the chlorophyllase becomes activated (Mahanta and Hazarika 1985). This may arise because the death of the tissue inhibits further degradation. Formation of chlorophyll intermediates is product specific. Commodities with low pH follow the pheophytin route; in other cases, both pathways seem to play a part (Heaton et al. 1996). Chlorophyll b tends to be more resistant to heat treatment than the a form (Schwartz and von Elbe 1983).

A number of possible strategies for controlling chlorophyll degradation have been suggested. These include the use of "stay-green" mutants and the selection of cultivars with modified chlorophyllase activity or chlorophyll concentration or with an increased ability to metabolize chlorophyll degradation products into colorless compounds (Heaton and Marangoni 1996).

The Ripening Process

Fruit skin color results from a dynamic process involving continual changes in pigment composition taking place through growth, maturity, and storage. Chlorophyll is lost as chloroplasts are converted to chromoplasts during ripening, when yellow and red pigments are revealed or synthesized. Violet or purple colors characterizing maturity in fruits such as olives are caused by anthocyanins, which appear toward the end of the maturation process (Minguez-Mosquera and Garrido-Fernandez 1989; Vlahov 1992).

During the ripening of San Marzano tomatoes from **full green** to **uniform red** grades, total chlorophyll content falls from 14 μg/g to 0, while lycopene concentration increases from 0 to 181 μg/g (Edwards and Reuter 1967). Normal ripening patterns occur only between 12 and 30°C. Chlorophyll degradation and lycopene formation are both reduced at temperatures below 12°C (Koskitalo and Omrod 1972). The ability to ripen at normal temperatures is lost after prolonged chilling. At temperatures higher than 30°C, chlorophyll is degraded, beta carotene is accumulated, and lycopene synthesis is inhibited. This results in yellow fruit (Sayre et al. 1953) (see Chapter 8). The spectral quality of the light falling on mature green fruit has a marked effect on color development. Red light is most effective in accelerating chlorophyll biodegradation, and blue light is more effective in enhancing the biosynthesis of carotenoids. Absorption maxima of chlorophyll and carotenoids occur in these two regions of the visible spectrum. Normal tomato ripening color is characterized by the increasing domination of lycopene. Conversely, lycopene declines in the Ruby Red grapefruit, and the pink-brown or off-yellow color of the ripe fruit becomes dominated by beta carotene (Lee 1997).

Changes in skin carotenoids of Golden Delicious apples occur during development. Those associated with photosynthesis (beta carotene, lutein, violaxanthin, neoxanthin) are replaced by those contributing to the postmaturity color (mono- and di-hydroxy xanthophylls). During ripening of these apples from **green** to **extra fancy** grades, carotenoid flavonol and flavan contents remain constant while chlorophyll concentration falls, as is shown in Figure 11–5 (Gorski and Creasy 1977). Hence, for this fruit, the ripe color is revealed rather than synthesized during ripening.

On the other hand, anthocyanin and carotenoid pigments of Cox's Orange Pippin apples increase during ripening both on and off the tree, when the xanthophyll fraction is predominant (Knee 1972). In McIntosh apples, however, the carotene fraction is the most important yellow pigment (Francis et al. 1955). Cyanidin glycosides stabilized or copigmented by quercetin glycosides are responsible for the red coloring in apples (Lancaster 1992).

Soil and cultivation conditions influence fruit skin color. For example, premature development of yellow ground color and development of red overcolor in Granny Smith apples can be reduced by application of nitrogen to the soil surface during growth (Ruiz et al. 1986). Nitrogen applied to leaves of Gala apple trees has significant effects on final fruit appearance. Application of urea to the tree during the previous season results in greener fruit of higher chlorophyll. These apples reach an acceptable color for market two weeks later than fruit from untreated trees (Reay et al. 1998).

Low temperatures during the growing season promote anthocyanin synthesis, and high temperatures inhibit it (Mazza and Miniati 1993). Hence, over-tree sprinkler irrigation can result in

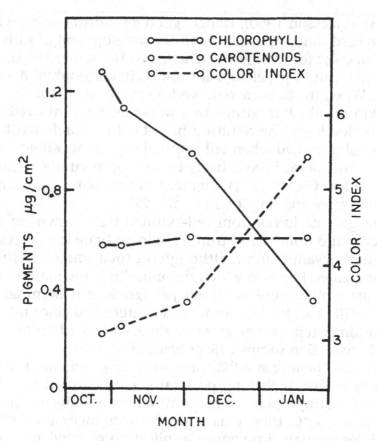

Figure 11–5 Chlorophyll and carotenoid contents and color index of Golden Delicious apple skin on storage at 4°C. An increase in color index number indicates an increase in yellowing (Gorski and Creasy 1977, with permission).

a greater proportion of pear fruit surface with red blush (Dussi et al. 1997). Upper canopy fruit is redder than fruit from the shade of the lower canopy. This occurs in peaches (Bible and Suman 1993) and apples (Warrington et al. 1996). Color development is also sensitive to latitude, tree height, and openness (Wagenmakers and Callesen 1995). Light encourages formation of anthocyanins in apples and areas exposed to the sun are redder (Siegelman and Hendricks 1958). Redness increases with light intensity (Jackson 1980). Hence, good light distribution within the orchard canopy is essential in the development of sufficient red skin color for the market (Dussi and Huysamer 1995). Covering the fruit in paper bags inhibits skin chlorophyll development. When the bags are removed several weeks before harvest, the skin is pale, but within two weeks anthocyanin red color rapidly develops. The resulting bright color is preferred to the duller color formed when chlorophyll is present (Kikuchi et al. 1997). Anthocyanin synthesis is developmentally regulated (Proctor and Creasy 1971). Pigment in ripe pears, for example, degrades after ripening (Dussi et al. 1995).

Pomegranate juices from red-skinned fruit grown on outer branches and from yellow fruit from inner branches possess the same anthocyanin profile, although the total amount in the former is greater. Juices in which delphinidin is the main component are violet; those in which pelargonidin is dominant are scarlet (Gil et al. 1995). Carotenoid content and color intensity of passion fruit juice is greater for winter-harvested fruit than for fruit harvested in summer (Sepulveda et al. 1996).

Bananas ripening at 20°C completely de-green, and the fruits become yellow, with total peel carotenoid content remaining constant during ripening. In locations with ambient temperatures above 30°C, they remain green due to incomplete chlorophyll degradation. Exogenous application of ethylene accelerates ripening but has no effect on chlorophyllase levels, chlorophyll degradation or carotenoid content at either temperature (Thomas and Janave 1992). Plantains lose greenness more quickly than bananas at high as well as low temperatures (Kajuna et al. 1998).

Dates undergo a specific ripening process. At the *kimri* stage, the fruit is green and hard and is used in pickles and chutney. At *khlal*, the fruit is firm and has developed a typical yellow, red, or yellow-scarlet color; it is used in jam, butter, and dates in syrup and for eating as fresh dates. At the *rutab* stage, the fruit is darker and softer and is used for sweeter products as well as being eaten fresh. At the *tamar* stage, the fruit is wrinkled, softer, and drier, is of maximum sweetness, and is dark gray black. It is further dried for storage and a prolonged shelf life. These stages can be clearly separated trichromatically (Al-Hooti and Sidhu 1997).

Maturation Control

The most important factors affecting the maintenance of fruit and vegetable quality are harvesting at optimum maturity, minimization of mechanical damage, use of good sanitation practice, and provision of optimum temperature and humidity after harvest. Success of the present marketing strategy to widen produce choice and extend season and convenience at minimum cost to the consumer depends largely on the extent to which fruit and vegetable maturity can be controlled. Possible technologies available are low–thermal input processes, refrigeration, additives, controlled or modified atmosphere (MA) packaging, and gas absorbers or emitters (Smith et al. 1990).

Maturity can be monitored using visual appearance scales (see Chapter 6) or instrumental techniques (see Chapter 8). Models can be used to predict ripening patterns during marketing and distribution (see Chapter 8). The relationship between tomato color and firmness can be modeled using the time-temperature-tolerance hypothesis developed for frozen foods (van Arsdel 1957). This is valid between temperatures of 12 and 27°C (see Chapter 8) (Thorne and Alvarez 1982). An on-line evaluation of tomato ripeness involving both color and mechanical properties may be possible (Hetherington et al. 1990).

Color quality and intensity are determined by genetic and environmental factors, and in the industry, color controls end use. In the United Kingdom, for example, pale peas are suitable only for canning; only dark green peas are used for quick freezing and dehydration. Yellowing in broccoli, a natural consequence of senescence, is a symptom of chlorophyll loss and a sensitive marker of perceived freshness. Hence, the use of yellowing-resistant varieties is an important factor for the wholesale market. Bioengineering of tomato plants has yielded redder fruit with increased lycopene and beta carotene (Schuch 1994; Coghlan 1995). Tomatoes that have been given a firmer texture can be allowed to ripen on the vine, thus improving flavor (Roller 1996). White strawberries and red gooseberries have been produced at the Horticultural Research International at East Malling, where work has continued on controlling apple ripening (Watson-Smyth 1996).

Pre- and postharvest treatments can extend the period of fresh vegetable acceptability. Products with long shelf lives usually have low respiration rates. Broccoli respiration rate can be reduced and senescence delayed by preharvest spraying or postharvest dipping with cytokinins (Shewfelt et al. 1983). Respiration and metabolic rates in broccoli florets are high; consequently, carbohydrate reserves are depleted. Sucrose applied immediately after harvest increases these reserves, retards chlorophyll loss, and delays yellowing. Floret tissues appear to sense postharvest decline in sucrose, and application of cytokinin may block the sensing mechanism (Irving and Joyce 1995). Water dipping of broccoli at 47°C reduces yellowing (Tian et al. 1996). Gibberellic acid can be used to retard senescence in shado benni plants before bagging (Mohammed and Wickham 1995). The greenness of field-harvested peas is affected by the length of time plants spend in windrow (piled up in a row) and by the holding time after shelling. Peas were greener after a windrow holding of six hours than after a holding of one or three hours (Freeland 1970).

Potatoes exposed to light in the field or on the shelf are subject to greening through the formation of chlorophyll. This can become a problem after three days in a small pack on the shelf. Although there is no metabolic link, chlorophyll formation is often accompanied by photosynthetically produced, bitter-tasting, poisonous glycoalkaloids (Edwards et al. 1998). Their formation is also encouraged by postharvest mechanical damage. Glycoalkaloids are not water soluble and not destroyed during cooking, so affected parts should be removed before processing (Jones et al. 1996). Preharvest regreening occurs in lemons and oranges. Postharvest regreening of pummelo (*Citrus grandis*) occurs on storage in natural or fluorescent light. This is caused by ultrastructural changes to the fruit chloroplasts (Saks et al. 1988). On exposure to light, green peas lose their greenness, but yellow peas are more stable (Heiss and Radtke 1968). Sultana grapes dried in the light lose their greenness through loss of the chlorophyll pigments, but those that are dark dried remain green (Bottrill and Hawker 1970). Light promotes carotenoid pigment losses in spinach, but none of the major carotenoids in carrots are affected (Kopas-Lane and Warthesen 1995).

During preparation for sale, fundamental changes can occur to the physiology of produce that in turn affect color and appearance. For example, fresh whole lettuce is a low–respiration rate vegetable, but after it is cut, the rate increases, markedly affecting ripening dynamics. Skins of packed fresh peaches can develop "black streak," a discoloration of the red-pigmented area. Mechanical damage resulting from defuzzing, washing, and drying procedures may alter the cyanidin pigment environment, allowing ion complexing to occur in low-pH conditions and alkaline hydrolysis at high pH (Denny et al. 1986). Shelf life of minimally processed carrots is sometimes limited by a white discoloration developing during storage. This may be caused by the formation of lignin as a wound barrier (Bolin and Huxsoll 1991) or by dehydration of the abraded surface (Tatsumi et al. 1991). Treatments include dipping in heated acid or alkali (Bolin 1992), water dipping (Cisneros-Zevallos et al. 1995), and application of steam (Howard et al. 1994) or edible coating (Howard and Dewi 1995).

Minimally processed vegetables under refrigeration have a shelf life of five to seven days (Carlin et al. 1990, cited in Paradis et al. 1996). In the store, mechanical refrigeration helps maintain chlorophyll level of broccoli heads on beds of cubed ice; cubed ice on top damages appearance (Perrin and Gaye 1986). However, refrigerated storage cannot be used for all produce. Among chill-sensitive fruits are tomatoes, melons, and bananas. Low-temperature storage (below 7°C) of potatoes converts starch to sugars. High levels of sugar are undesirable in potatoes that are to be fried because, through browning reactions, they lead to a product that is too dark a color (Thomas 1981).

In the tropics, lycopene formation in tomatoes is suppressed. Lycopene content resulting from bulk storage at 20 to 25°C and 92- to 95-percent relative humidity is doubled when the room is conditioned at 29 to 33°C and 45- to 65-percent relative humidity (Thiagu et al. 1991). Tomato ripening can be retarded by low temperature (Salunkhe and Desai 1984), but storage under refrigeration can cause chilling injury (Ryall and Lipton 1983). Storage life of tomatoes (Bertola et al. 1990) and strawberries (Gil et al. 1997) can be extended by carbon dioxide diffusion during cold storage. Ripening of mangoes can be delayed by storage in rooms at 26 to 29°C. During storage at 12 to 15°C, chilling injury in the form of pitting and black spots occurs in mature green fruit, and brown discoloration, pitting, scald, and decay occur in ripe fruit (Joseph and Aworh 1991). Peppers are normally bulk stored at 7 to 8°C. Packing in perforated polythene bags improves color development and preserves moisture levels without increasing rot (Meir et al. 1995).

Anthocyanins are sensitive to temperature and pH (Markakis 1982). They are also subject to enzyme degradation, with the damage being caused by glycosidases (Huang 1956), peroxidases (Grommeck and Markakis 1964), and phenolases (Peng and Markakis 1963). Browning of cut surfaces of fruits, such as apples and peaches, can be prevented by application of acid juice or sugar syrup. Preservation of the color and flavor of fresh fruit is traditionally carried out by adding sugar to the raw fruit before freezing. This treatment has a protective effect on the anthocyanin pigment of strawberries while retarding browning and polymeric pigment formation. This effect may be due to inhibition of the beta-glycosidase and polyphenol oxidase, or anthocyanin-ascorbate reaction, or by provision of a partial oxygen barrier (Wrolstad et al. 1990). Fructose, arabinose, lactose, and sorbose are more destructive to strawberry anthocyanin pigment than sucrose, glucose, and maltose (Markakis 1982). Addition of hexose, disaccharides, or pentose helps preserve the 520-nm absorbance of raspberry juice on storage at 5 and -20°C (Joo 1982). Pigment degradation in black currants is nonenzymic, depending on temperature and oxygen availability (Skrede et al. 1983). The anthocyanin content and hence red color intensity of the Royal Gala apple can be increased after harvest by irradiation with ultraviolet and white light (Yi et al. 1995).

Secondary factors affecting fruit and vegetable quality include maintenance or modification of suitable oxygen, carbon dioxide, ethylene, and/or water vapor concentrations in the product environment. Shelf life of many fruits and vegetables can be significantly increased by storage under refrigeration using MA conditions of high carbon dioxide, low oxygen, and high relative humidity. These conditions lower respiration rates and delay maturation, ripening, and senescence. Packaging also reduces formation of soluble protein, lipoxygenase, and hence polyunsaturated fatty acids that may contribute to postharvest deterioration (Zhuang et al. 1994). Packaging effectiveness depends on product tolerance to low oxygen and high carbon dioxide levels and on temperature, respiration rate at different temperatures, and ethylene production rate. It is also dependant on wrap permeability properties. The term *controlled atmosphere (CA) storage* implies a greater degree of control over gas levels than *MA storage*. Typical atmospheres used for fruit and vegetable produce contain 2 to 5 percent oxygen, 3 to 8 percent carbon dioxide, and 87 to 95 percent nitrogen (Phillips 1996). Synthesis of anthocyanins during ripening can be controlled by MA packaging (Goodburn and Halligan 1988; Zagory and Kader 1988). MA packaging has been reviewed by Day (1997), Phillips (1996), Cameron and Talisila (1995), Church and Parsons (1995), and Farber and Dodds (1995).

MA packaging alone can protect market quality of broccoli spears. However, appropriate temperature and relative humidity conditions are recommended for optimum shelf life (Barth et al. 1993). For example, a comparison of control, iced, and packaged fresh broccoli showed that, after air withdrawal, nitrogen flushing, and heat sealing in multilayer plastic bags, spears were still acceptable at 10 weeks compared with the 7-week life of the control (Shewfelt et al. 1983). Although MA packaging can lead to improvement in greenness retention, there can be a reduction in stem turgor (Paradis et al. 1996). Visually perceived color loss from green beans, bell peppers, and spinach is greater at 20 than at 10°C, and unpackaged vegetables lose more than packaged (Watada et al. 1987). Increase in the respiration rate of different types of lettuce after cutting can be alleviated to some degree by MA storage (Hamza et al. 1996; Morales-Castro et al. 1994a). MA storage can also be used successfully, for example, for mature tomatoes (Yang and Chinnan 1988), green tomatoes

(Bhowmik and Pan 1992), fresh green chile peppers (Wall and Berghage 1996), cut endive (Vankerschaver et al. 1996), sweet corn on the cob (Morales-Castro et al. 1994b), sweet cherries (Meheriuk et al. 1995), and strawberries (Gil et al. 1997).

Britain's major cooking apple, Bramley's Seedling, and its major eating apple, Cox's Orange Pippin, are harvested in early autumn. Most of the fruit is stored in refrigerated CA stores and is marketed up to May (Cox's) and July (Bramley's) of the following year. Ripening and color change continue after removal from the store, and poor-quality fruit can still reach the customer if MA packaging is not used. Chlorophyll loss is retarded when the oxygen concentration is reduced below 10 percent, and it is possible to slow yellowing over the entire storage season. Concentrations below 2 percent are necessary to retard softening (Knee 1980). The type of packaging film and the atmosphere necessary for optimum ripening retardation are different for each variety and are dependent on the CA storage history (Geeson et al. 1987). MA retail packaging of apples has been discussed by Smith et al. (1987).

Conference pears harvested in early autumn are stored in air at −1°C and marketed until the following May. The fruit remains hard and green in store, but when it is removed into ambient temperatures or is placed in ripening rooms at 17–20°C, it commences softening and chlorophyll loss within one day. Pears reach eating ripeness in five to seven days and become overripe in a further three to four days. MA packing effectively inhibits chlorophyll degradation and yellowing, but the fruit ripens unevenly in the pack and fails to develop normal eating characteristics. Hence, it is a technique apparently unsuitable for commercial purposes (Geeson et al. 1991). In contrast, behavior of MA-packed apples is consistent and yields more predictable ripening patterns (Geeson and Smith 1989).

Ethylene in the storage atmosphere hastens chlorophyll breakdown and carotenoid biosynthesis in orange rind. The color of the flesh and juice, however, is not improved (Houck et al. 1978). High levels of oxygen fade rind color but result in deeper flesh and juice color than is produced by storage in air (Aharoni and Houck 1980). The response of blood oranges, which contain anthocyanins as well as carotenoids, to storage atmospheres depends on the cultivar. In experiments involving three varieties, ethylene may deepen rind color, and 80 percent oxygen may result in a paler orange rind. The use of ethylene may result in deeper flesh and juice colors, but a high oxygen concentration deepens the red color of the flesh and juice of all three cultivars (Aharoni and Houck 1982).

Peaches are normally maintained at 5°C in commercial operations. Color and flavor development are arrested, but softening, although retarded, continues. Softening rate increases with increased maturity at harvesting (Shewfelt et al. 1987). Peaches and nectarines stored at low temperatures for three to four weeks fail to ripen satisfactorily when removed from storage (Lill et al. 1989). Symptoms of chilling injury are internal breakdown and internal reddening of the flesh. CA storage can prevent appearance of these symptoms, but the full development of juiciness is also prevented (Lurie 1992). The extent of quality improvement of nectarines and peaches in MA storage depends on the number of fruits per pack, with six generating a higher carbon dioxide level than two or four fruits (Lurie 1993). Intermittent warming of packs of peaches during cold storage can result in improved eating quality (Fernández-Trujillo and Artés 1998).

Use of MA packaging in the form of edible film coatings, such as sucrose polyesters and cornzein, an alcohol-soluble protein, is an alternative method for extension of postharvest life. For example, such films can be used to delay ripening in bananas (Banks 1984), pineapples (Nisperos-Carriedo et al. 1990), apples (Chai et al. 1991; Park, Chinnan, and Shewfelt 1994), pears (Sümnü and Bayindirli 1994), and tomatoes (Park, Bunn, Vergano, and Testin 1994). Peeled carrots coated with edible cellulose are more orange and have reduced white surface discoloration during storage at 2°C (Howard and Dewi 1995). Edible waxes such as carnuba wax, polyethylene wax, beeswax, rice bran wax, and shellac have also been used for mangoes (Castrillo and Bermudez 1992) and citrus fruits (Kaplan 1986), such as mandarins (Bayindirli et al. 1995) and limes (Sierra et al. 1993). The subject of edible film coatings has been reviewed by Debeaufort et al. (1998) and Baldwin et al. (1997), and minimal processing has been reviewed by Wiley (1994).

Ionizing radiation can be used to delay ripening and senescence of climacteric fruit—that is, fruit harvested before ripening, which may include bananas, mangoes, papayas (Thomas 1985), avocados (Thomas 1986a), apples, pears, peaches, nectarines, and plums (Thomas 1986b). Visible radiation damage of citrus fruit can be affected by such factors as time of season and stage of ripeness (Thomas 1986a). Minimally processed refrigerated fruits and vegetables have their shelf life limited by spoilage microorganisms. The colors of unpasteurized chopped mixed salad (pico de gallo) are not affected by gamma radiation at levels that substantially reduce populations of microflora (Howard et al. 1995). Irradiation of shredded carrot in microporous plastic bags also results in reduced microflora and prevents losses of carotene orange color (Chervin and Boisseau 1994). Radiation used to eliminate fungal pathogens and insects can affect pigment formation and development. For example, pink anthocyanins can be formed in peaches in areas normally yellow, and the radiation dose necessary to destroy most of the microorganisms on bilberries destroys 50 percent of the anthocyanins (Thomas 1986b). Mangoes and capsicums can be irradiated at doses necessary for disinfestation without significant loss of carotene (Mitchell et al. 1990). Irradiation of frozen and refrigerated sweet corn results in a loss of beta carotene, cryptoxanthin, and other carotenoids, but packaging under nitrogen tends to improve their retention (Tichenor et al. 1965). Irradiation is an alternative to fumigation with methyl bromide to combat cherry insect pests. The fruit can be irradiated as soon as the quality has reached commercial harvest levels (Drake et al. 1994).

Greening of potato tubers can be delayed for nine days by gamma irradiation. Three types of discoloration can, however, take place after irradiation. Black spot, attributed to oxidation of phenolics, tissue and vascular browning, and after-cooking darkening, can be induced or enhanced. Darkening, which occurs soon after cooking, is caused by ferrous-phenolic complexes formed on heating. These convert to bluish, dark, ferric-phenolic complexes on exposure to air. Various factors such as iron, orthodiphenols, organic acids contents, and pH contribute to these reactions (Thomas 1981).

Processed Systems

Chlorophyll-Dominated Systems

There are several processing strategies for controlling chlorophyll degradation. They include heat inactivation of chlorophyllase with minimal conversion of chlorophyll to pheophytin, prevention of tissue pH fall by the addition of magnesium and zinc hydroxides, addition of antioxidants, and controlling of pH, temperature, and ionic strength to modify chlorophyllase and dioxygenase activities or to decrease magnesium ion losses (Heaton and Marangoni 1996).

Blanching and Heat Treatment. Fresh greens deteriorate on frozen storage, with lower storage temperatures leading to more rapid loss of color (Klimczak et al. 1993). Such deterioration can be greatly reduced by blanching the vegetables prior to freezing and selecting the conditions such that relevant enzymes are inactivated without undue heat initiation of other systems. The degree of blanch treatment influences chlorophyll deterioration rate during processing.

Loss during blanching is related to blanch severity. Chlorophyll b is lost more slowly than a initially, but this relative rate appears to increase after four minutes' blanch. Blanch time also affects chlorophyll degradation on subsequent frozen storage. Losses continue through residual enzyme activity. Thermal treatment greater than that required for enzyme inactivation results in increased chlorophyll loss rates (Katsaboxakis and Papanicolaou 1984). This is probably due to oxidation and related to fat peroxidation (Buckle and Edwards 1970b). It is customary prior to freezing to blanch vegetables until there is zero peroxidase activity. This is a very heat-resistant enzyme not shown to be directly responsible for quality deterioration. Halpin and Lee (1987) suggested using lipoxygenase as a marker. Lipoxygenase rather than chlorophyllase is involved in the decoloration of chlorophyllic pigments in olive oil (Gandul-Rojas and Minguez-Mosquera 1996).

Blanching in buffered water at pH 7.0 gives improved chlorophyll retention. Spinach, green peas, broccoli, lima

beans, brussels sprouts, and green beans can retain chlorophyll if they are soaked before cooking in a citrate-phosphate
buffer at pH 6.8 to 7.0 for 10 minutes (Sweeney 1961). Sodium
sulfite dipping before freezing retards chlorophyll and
carotenoid degradation in unblanched capsicums (Rahman
and Buckle 1981).

Heat shock treatment involving exposure to boiling water or
steam for 5 to 15 seconds is a substitute for normal blanching of
leguminous vegetables, notwithstanding residual peroxidase,
catalase, and lipoxygenase activities. Color as well as flavor and
texture properties are better preserved (Steinbuch 1984). Steam
and water blanching produce a more intense green color in
beans than microwave and convection oven treatments. Microwave blanching preserves the color of frozen green beans as
well as conventional boiling water treatment (Brewer et al.
1994). Microwave treatment results in beans of the highest
chlorophyll content and convection oven the lowest (Muftugil
1986). The reason for the intense green of the water- and steam-
blanched samples is physical. Air is removed from the vegetable
surface and intercellular spaces (Adams 1978); also, chloroplasts
swell and tend to burst, producing a more uniform green
throughout the cell (Lee 1958).

Blanching and heating processes providing faster heat transfer lead to improved chlorophyll retention. Oil blanching of
vegetables involving temperatures above 175°C for less than
one minute results in a brighter, greener, crisper product than
steam or water blanching (Wu 1986). Chinese stir-fry methods
of cooking broccoli lead to high (69 percent) chlorophyll retention. Comparative figures resulting from other cooking methods are microwave (38 percent), and conventional (30 percent)
(Eheart and Gott 1965). Aseptic processing and packaging similarly lead to greater chlorophyll retention; 68-percent retention
has been reported for spinach puree (Schwartz and Lorenzo
1989).

The green chlorophyllide molecule is formed through the naturally occurring enzyme chlorophyllase, and considerable
quantities can be formed (Buckle and Edwards 1970a; Schwartz
and Lorenzo 1990). During heat processing, chlorophyllides

may be more susceptible to magnesium ion loss than the parent chlorophyll molecule. Some vegetables, such as snap beans, are devoid of chlorophyllase (Jones et al. 1964).

Heating causes denaturation of the chlorophyll-binding protein, and the chlorophyll dissolves in the lipids. In this form, it is easily degraded at high temperatures, and conventional high-temperature, short-time (HTST) blanching carried out close to the boiling point of water causes undesirable tissue damage. HTST processing without pH control (Schwartz and Lorenzo 1990) or with alkali treatment (Gupte and Francis 1964) reduces the extent of this undesirable conversion. Magnesium carbonate was found to be the most effective compound in retaining chlorophyll during HTST processing. Color improvement of spinach by this process was due primarily to decreased degradation of chlorophyll a. Although short-term protection is given to the greenness, degradation normally occurs rapidly on storage (Gupte and Francis 1964). Two-stage blanching involving low-temperature, long-time processing followed by HTST processing with vacuum infiltration of calcium can lead to greener, firmer green beans and peppers (Seow and Lee 1997). Green vegetables to be stored at chill temperatures retain their greenness when water or steam blanched, subjected to a vacuum, contacted by deaerated water, and immersed in a pH 8.7 alkaline solution. The vacuum removes the surface microlayer of air, which causes chlorophyll oxidation at refrigerator temperatures (Rogers et al. 1987).

The green color of canned vegetables can be retained by using either (1) alkaline hydroxides or carbonates or (2) metallic salts such as the chlorides and acetates of zinc and copper. In the former instance, the green color intensity decreases on storage at room temperature. In the latter, the zinc ion replacing the magnesium in the chlorophyll molecule may result in a stable green compound (von Elbe 1989). Chlorophyll loss is reduced by shorter processing times resulting from can rotation during retorting. However, degradation continues in the can on storage at room temperature (Steele 1987), as it does in bottled vegetables exposed to the light (Segal 1971). The green of canned brussels sprouts and vegetable purees is caused by bright green, stable, water-soluble pigments formed from pheophytins a and b and by pheophorbide complexing with copper (Swirski et al. 1969). The "Veri-Green" patent describes a process that combines a metal ion blanching treatment with a can that incorporates metal ions such as copper or zinc into the lining (Segner et al. 1984). The green color improvement is apparently due to the formation of the more stable zinc complexes, mainly of the a derivatives (von Elbe et al. 1986; von Elbe 1989). Color of endive and leeks is improved after blanching at 50 to 80°C in water containing zinc and calcium ions (Lin and Schyvens 1994). Effects are variable. Zinc cation concentration greatly affects the color of sterilized green beans and endive, has less effect on the color of frozen green beans after cooking, and has no effect on the color of endive after heating (Lin and Schyvens 1995).

Acidified bulk storage of blanched green beans was successful when hydrochloric acid was used to achieve a pH less than or equal to 1.5 (Basel 1982). This method allows the beans to be stored for processing after harvest has finished. After canning, the color is better than that achieved with the conventionally canned product (Basel 1983). Containerized vegetables retain their greenness after processes involving blanching, storage in an aqueous solution of zinc and copper ions, and sterilization (LaBorde and von Elbe 1996).

Green chlorophyll pigments are slightly water soluble and are converted to the undesirable olive green pheophytin in an acid medium. In an alkaline solution, chlorophyll is a bright green. However, alkali addition during cooking results in a vegetable of low vitamin content and limp appearance. Different schools of thought exist in the kitchen on how to keep the greenness of vegetables during cooking. Certainly, to retain maximum greenness, green vegetables should be plunged into a large volume of already boiling salted water. If they are not to be eaten immediately, quick cooking should be followed by draining and rinsing in cold water. This arrests color degradation. There are those who advocate keeping the lid on the pan during cooking to reduce cooking time and retain vitamins. Unfortunately, this encourages dissolved acids to remain in solution and hastens color degradation. Conversely, encouraging the volatile organic acids to evaporate by leaving the lid off will increase the time needed to return the water to boiling point and hence result in more color degradation. The lid-off method tends to lead to a more desirable color. Copper vessels might be used to retain the color of greens, but this could lead to an undesirable level of contamination (Pike 1974). Whatever the cooking route chosen, we may bear in mind advice given in the *Finchley Manual for the Training of Servants*, published in 1800 (quoted in Hartley 1954):

> Q. How do you boil peas?
> A. Briskly, ma'am.

Other Aspects. Olives are picked at different stages of ripeness. Green olives are picked and processed with sodium hydroxide when they are unripe. Violet anthocyanin-colored olives are picked when partially ripe. Black shiny olives are picked fully ripe before fermentation with polyphenols and are oxidized. This achieves a jet black finish that is fixed with an iron salt (Romero et al. 1995). Moroccan and French black olives are "dry-cured"—that is, packed with salt until the skin shrivels and has a chewy texture. Fermentation and brine storage of olives result in the destruction of chlorophylls a and b to form pheophytins a and b and pheophorbides a and b. Chlorophyllides a and b produced in the earlier stages of the process are also converted. However, the total pigment content does not change during the process, demonstrating the absence of other types of oxidative reaction, such as contact with oxidized lipids or lipoxidase which might form uncolored products (Minguez-Mosquera et al. 1989). The process of fat peroxidation was relatable to chlorophyll conversion to pheophytin during the anaerobic initiation stage of fat peroxidation, as well as to the destruction of both pigments during the aerobic propagation stage (Walker 1964).

Chlorophyll degradation can be retarded in environments of low water activity (a_w). At an a_w greater than 0.32 and a temperature of 37°C, chlorophyll is transformed to pheophytin in spinach puree and other food systems. At lower values of a_w, the degradation rate is lower, and other compounds in addition to pheophytin are formed (Lajolo et al. 1971). The reaction follows first-order kinetics even in regions of "firmly bound water." First-order reactions also occur with pH and water content. Addition of glycerol enhances degradation rate, and the critical value of minimum a_w for stability is shifted to lower values. The glycerol may act as an aqueous medium dissolving or aiding diffusion of the reaction species (Lajolo and Marquez 1982).

Dehydrated vegetables lose color when exposed to air. On dehydration, air temperature is the most important variable involved in the color degradation of green herbs. Color is better conserved when the greens are blanched prior to drying though a decrease in drying time is required (Rocha et al. 1992). High-pressure treatment has been tested for its effectiveness as an alternative to heat in blanching. High pressures combined with freezing and dehydration resulted in no marked deterioration in the color of green beans (Eshtiaghi et al. 1994). Chlorophyll stability of dried systems during storage is related to product glass transition temperature. Incorporation of different sugars into kiwi fruit by osmodehydration modifies low-temperature phase transitions. Kiwi fruit samples osmodehydrated in maltose had a lower glass transition temperature and a lower chlorophyll degradation rate during storage at $-10°C$ than those containing other sugars (Torreggiani 1995).

Carotenoid-Dominated Systems

Blanching and Heat Treatment. Carotenoids are oil soluble, and cooking has little effect on the carotene content of fresh carrots. Blanching by hot water, steam, or microwaves results in yellowing, but this has been attributed to structural changes occurring during processing (Mirza and Morton 1977). Loss of carotenoids from peas has been linked with lipoxygenase, which occurs widely in vegetables containing chlorophyll (Rhee and Watts 1966). However, carrots possess little lipoxygenase, and color changes due to blanching are usually small (Edwards and Lee 1986).

Some high-carotenoid vegetables such as squash show a distinct color change when heated in water. This has been partly attributed to isomerization of the *trans*-carotenoids to the less highly colored *cis* forms (DellaMonica and McDowell 1965), to degradation of chromoplasts, and to solution of carotenes in other cellular lipids (Purcell et al. 1969). Carotenoid droplets from chromoplast degradation also occur during the lye peeling of sweet potatoes (Walter and Giesbrecht 1982). The color shift of tomato is not as great as with other high-carotenoid material. This is due to the formation of lycopene crystals after heating, the natural lipid content being insufficient for solution (Purcell et al. 1969).

Freezing followed by thawing and immediate cooking results in no loss of carotene in carrots, broccoli, or spinach. However, after thawing for six hours before cooking, the carotene content is reduced to 84 percent for broccoli and 79 percent for spinach, there being no significant differences between microwave and conventional cooking. After the same thaw time, carrot carotene contents are 87 percent with conventional cooking and 72 percent with microwave (Park 1987).

Effects of Water and Oxygen. In model cellulose and starch food systems, water exerts a protective influence on the carotenoids beta carotene, apo-8'-carotenal, and canthaxanthin. The effect is evident both below and above the monolayer coverage for both systems, and the extent of the protection increases with moisture content (Ramakrishnan and Francis 1979). Higher water activities also have a protective role in freeze-dried salmon (Martinez and Labuza 1968). The free radicals involved in pigment loss may be produced during the oxidation process, and their activity may be reduced by contact with the increasing amounts of water available at higher water activities (Labuza et al. 1970). Carotenoids in freeze-dried papaya are most stable in the region of $0.33a_w$. Below and above this, the destruction rate is higher (Figure 11–6).

If the availability of oxygen is nonlimiting, pigment loss follows first-order reaction kinetics (Ramakrishnan and Francis 1979). Two mechanisms may contribute to the protective effect of water. The first is the reduction of free-radical activity by hydrogen bonding with the water on the surface of the food. The second involves the reduction of metal catalyst activity through hydration or through the formation of insoluble metal hydroxides (Labuza et al. 1970). Mathematical models have been de-

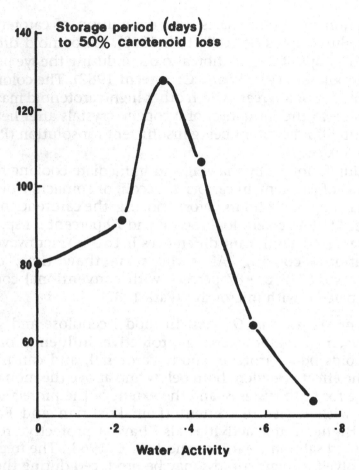

Figure 11–6 Effect of water activity on the stability of carotenoids in freeze-dried papaya (plotted from data of Arya et al. 1983).

veloped to describe the effect of water activity on the rate of loss of beta carotene and ascorbic acid in dehydrated real and model systems (e.g., Saguy et al. 1985). Carotene loss is sensitive to drying method. Carotene contents of fresh, vacuum-dried, and microwave-dried carrots were found to be 989, 459, and 368 μg/g respectively (Park 1987). The poor stability of carotenoids in stored, dehydrated pepper products constitutes a serious economic problem. Carotene destruction in powdered paprika is also sensitive to a_w, with bleaching consisting of an induction period followed by a stable period, then a period of revived oxidation. The length of the stable period can be extended by the inclusion of an antioxidant system of ascorbic acid and copper. This overcomes the pro-oxidant effect of the peroxidase. However, ascorbic acid itself may act as a pro-oxidant or antioxidant, depending on concentration and water activity (Kanner et al. 1978). Figure 11–7 illustrates these effects (Kanner and Budowski 1978).

The mean carotene contents of carrot, broccoli, and spinach decline significantly ($p < 0.01$) on drying after a freeze-thaw cycle. This indicates that drying or a combination of freezing and drying can reduce the content of chemically unstable provita-

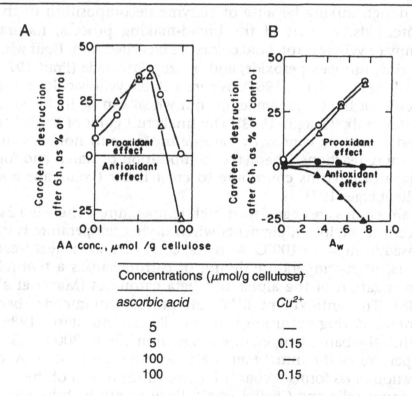

Concentrations (μmol/g cellulose)	
ascorbic acid	Cu^{2+}
5	—
5	0.15
100	—
100	0.15

Figure 11–7 (A) Effect of ascorbic acid concentration with and without $CuSO_4$ on carotene oxidation in a cellulose model at $a_w = 0.75$: (O) $CuSO_4$; ($\triangle$) 0.15 μmol of $CuSO_4$ per gram of cellulose. (B) Effect of water activity on carotene bleaching in the cellulose model in the presence of ascorbic acid and copper ions (Table from Kanner and Budowski 1978, with permission).

min A in vegetables (Park 1987). Prior blanching with (Eshtiaghi et al. 1994) or without (Lesellier et al. 1988) high-pressure and freezing pretreatment has a protective effect on the color of dehydrated carrots. Sulfite addition can increase the protective effect, possibly by reducing the formation of browning compounds during drying and subsequent storage (Arya et al. 1982). Sulfite and starch treatments prior to drying lead to an increase in the redness of the dry product (Zhao and Chang 1995).

The degree of yellowness in bread mainly depends on wheat carotenoid content but is also influenced by the milling regime, extraction rate, and possibly pigments from wheat kernel impurities. The natural yellow color of wheat flour is partly lost during dough mixing because of enzyme decomposition of the carotenoids. As part of the bread-making process, natural creamy or yellow carotenoid colors are bleached from flour with chlorine, nitrogen peroxide, and benzoyl peroxide (Pratt 1971; Kent-Jones and Amos 1967). Where a stable yellow color is important for the bread, amber durum wheat can be included in the flour mix (Quaglia 1988). The final crust color of the baked bread is mainly caused by caramelization and nonenzymic browning reactions. Thermal decomposition of starch and formation of dextrins contribute to crust luster (Pomeranz and Shellenberger 1971).

Carotenoids are degraded at high temperatures. There is a 24-percent reduction in pigments when drying temperature is increased from 40 to 100°C. As is the case with other heat treatments, high-temperature drying of carrots causes a *trans/cis* isomerization of the alpha and beta carotenes (Marty et al. 1988). The antioxidant BHT can protect commercial beta carotene during extrusion cooking (Berset and Marty 1986). Within the barrel temperature range from 125 to 200°C, 38 to 73 percent of the initial *trans*–beta carotene can be destroyed. Creation of *cis* forms accounts for two thirds to half of this loss (Guzman-Tello and Cheftel 1990). Drying with high-humidity air gives some protection (Lesellier et al. 1988). During dehydration of shelled corn, grain color degradation depends more on temperature than on moisture content. It is necessary to dry at a low temperature to prevent discoloration (Aguerre and

Suarez 1987). In dried carrots, carotenoids can be relatively stable at an a_w between 0.32 and 0.57 (Arya et al. 1982). Carotenoid loss in dried red pepper products is affected by a_w, package atmosphere, storage temperature, and treatment of the pepper. High carotenoid retention and low browning can be achieved by storing the coarse powder, including seeds, at an a_w less than 0.3 in a nitrogen atmosphere (Lee et al. 1992).

Carotenoids are stable in the absence of oxygen, but the high degree of unsaturation in their structure makes them susceptible to oxidative changes. Loss of color in dehydrated carrot powder, caused by nonenzymic oxidation, is greatly reduced by storage under nitrogen (Mackinney et al. 1958). In dehydrated systems, oxygen can be a crucial factor in beta carotene degradation. Even a concentration of 1 percent oxygen can lead to a significant loss of pigment. A free-radical–initiated degradation mechanism was indicated. Presence of antioxidants or lower a_w slows pigment loss (Goldman et al. 1983). Beta carotene content of dried carrots appears to be chemically stabilized by the presence of high levels of tocopherol, although tocopherol itself can react with an oxidant in the presence of iron to produce a red-brown color (DellaMonica and McDowell 1965).

Dairy product color is dominated by carotenoids obtained during grazing. Animals fed on the alpine pastures of Europe during the summer months of May to September produce milk higher in carotenoids, resulting in cheese of significantly greater yellowness (see Chapter 8) (Rohm and Jaros 1997). Surface discoloration of cheeses is mainly caused by specific microorganisms. For example, in Italian Taleggio cheeses, a plain white surface indicates the presence of *Geotrichum candidum*; a diffuse yellow, *Corynebacterium flavescens*; yellow-green patches, the *Arthrobacter globiformis/citreus* group; and a uniform pink-red surface, *Brevibacterium linens* (Piantanida et al. 1996).

Flavonoid-Dominated Systems

Blanching and Heat Treatment. Mechanisms contributing to loss of anthocyanin color include enzymic and nonenzymic browning, ascorbic acid degradation, and polymerization with other phenolics. Enzymes used to assist cranberry processing can decompose anthocyanins (Wightman and Wrolstad 1995), but enzymes that destroy anthocyanins may be inactivated by heat. For example, red tart cherries blanched before freezing show minimal anthocyanin loss (Siegel et al. 1971). Tannic acid has been used to stabilize anthocyanins in pasteurized orange juice. This was attributed to the formation of molecular complexes more stable to photochemical and enzymic degradation (Maccarone et al. 1987). In general, HTST processes are recommended for pigment retention (Markakis et al. 1957).

Anthocyanins degrade on heating, and up to 90 percent of the red pigment in strawberries is lost in jam making. Pigment concentrations in strawberry jam can be reduced to less than 50 percent after three months' storage, but five to seven months are required for similar losses to occur in black currant syrups (Skrede et al. 1992). The deep color of these products indicates the comparatively large amount of pigment present in the fruit. In this respect, an acceptable frozen strawberry product can contain up to 40 percent of green (unripe) strawberry puree (Sistrunk et al. 1983). Loss of the pigment in jam is reasonably rapid: the half-life in strawberry preserves is approximately eight weeks at 20°C. Leucoanthocyanins and flavanols, which are reactive phenolics, may play a major role in the deterioration (Abers and Wrolstad 1979). Fortunately, even zero–anthocyanin content strawberry jam is reasonably attractive because the degradation products, as for red wine, are reddish brown. Similar red-brown pigments contribute to the pink skin of sun-dried *mei* (Chinese plum) (Huang 1986). Thermal degradation of the pigments follows first-order kinetics in strawberry preserves and juice (Markakis 1982; Speers et al. 1987). Processing involving high pressures up to 1,000 MPa can be used as an alternative to heat treatment for fruit juice and jams. Colors are brighter and flavors more like fresh fruit than in conventionally processed products. Brown color development is significantly inhibited (Ames 1995).

Maturation of black-eyed peas is accompanied by chlorophyll degradation and the formation of a black-red anthocyanin in the eye. Discoloration of the canned peas is due to the absorption of anthocyanin precipitate by other can constituents. Controlled processing can prevent this discoloration (Sistrunk and Bailey 1965). The major pigment in discolored canned pears is a purple-pink insoluble tin-anthocyanin complex. Some pear varieties are more susceptible to discoloration than others, but the basis of this is leucoanthocyanidin content (Chandler and Clegg 1970a, 1970b, 1970c). The pathway of pink discoloration of lychee is suggested to involve enzymic conversion of phenolic compounds to leucoanthocyanidin and the formation of a colorless intermediate. During canning and subsequent storage, the intermediate combines nonenzymically with protein, leucoanthocyanidin, pectin, tin, and iron to form a colored cyanidin complex. Methods of preventing the formation of the pink discoloration include providing proper sterilization conditions, inhibiting enzyme reactions by reducing production delays, or immersing the flesh in sodium bisulfite solution (Hwang and Cheng 1986).

Effect of pH. Anthocyanins are more stable in an acid environment than in neutral or alkaline environment. When raw red cabbage is cooked, its acids tend to dissipate, and the color changes to bluish purple. This change can be prevented, for example, in dishes that combine the cabbage with acid apples (Hillman 1983). Molecular structural transformations of pelargonidin chloride occurring with pH are shown in Figure 11–8 (adapted from Harper 1968). At pH 1 to 3, the pigment exists as the red oxonium ion; the hydrated form between pH 3 and 7 is a colorless pseudobase. At higher pHs, the purple anhydro base is formed, but this ionizes above pH 10 and becomes blue (Harper 1968). Similar structural transformations have been proposed for the change of red cyanidin chloride at acid pHs, through the neutral colorless pseudobase to the violet anhydro base and the high pH blue anhydro base salt (Bentley 1960). The hyperchromic effects of organic acids in raspberry juice increase (Park and Joo 1982) in the order

formic acid > acetic acid > n-butyric acid > propionic acid

The mechanism is poorly understood, but one involving the formation of coumarin from the anhydro base has been proposed (Markakis 1982). Equilibrium reactions among anthocyanin

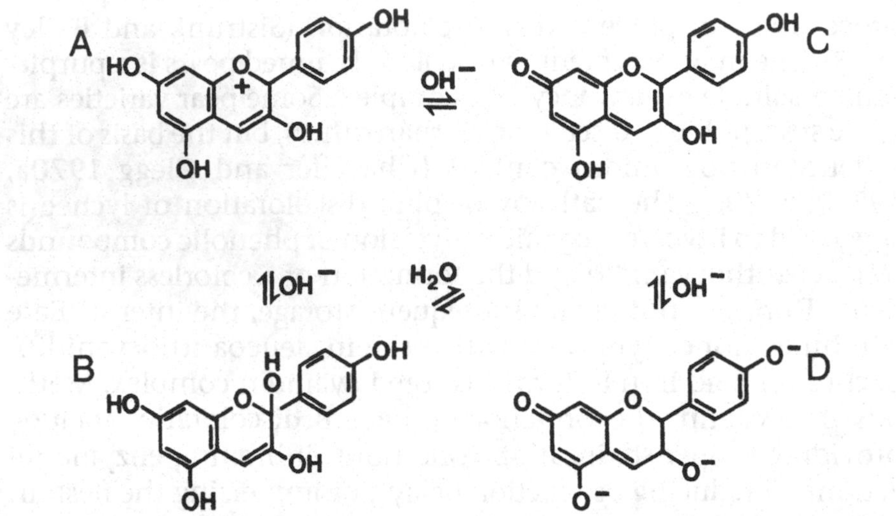

Figure 11–8 Behavior of pelargonidin chloride with pH. **(A)** pH 1 to 3, red oxonium ion. **(B)** pH 3 to 7, colorless pseudobase. **(C)** pH 7 to 10, purple anhydro base. **(D)** pH 10, blue anhydro base, ionized (from Harper 1968, with permission).

structures in solution at 20°C are believed to be in the form of (Brouillard and Delaporte 1977):

high pH	low pH		
quinoid	flavylium	carbinol	chalcone
base ↔	cation ↔	pseudo base ↔	
blue	red	colorless	colorless

Heating moves the equilibrium to the chalcone, with the reversion taking longer than the decoloration. The distribution of these four pH-dependent forms can be seen in Figure 11–9. This was derived from studies on a weak solution of malvidin-3-glucoside (Timberlake and Bridle 1983). The distribution may not apply to natural beverages containing much higher concentrations of anthocyanin.

Sulfur Dioxide. Enzymes that destroy anthocyanins may be inactivated by sulfur dioxide. The resulting bleaching may be reversible or irreversible. Very low concentrations (30 ppm) can inhibit enzyme degradation of anthocyanins in red tart cherries without bleaching (Goodman and Markakis 1965). Low concentrations of sulfur dioxide have also been used to stabilize anthocyanins in canned cherries, strawberries, and plums (Adams 1973). Addition of sulfur dioxide slows the loss of anthocyanins and formation of polymeric pigment in strawberry puree and juice (Bakker and Bridle 1992) and wine (Allen 1983). Moderate concentrations (500–2,000 ppm) of sulfur dioxide assist in extraction of anthocyanins from plant tissues and can be used to preserve fruit and nuts. However, to restore the anthocyanin coloration, the preservative must be removed before further processing. High concentrations of sulfur dioxide in the region of 10,000 ppm are used to bleach irreversibly the red cherries used for the production of maraschino, candied, or glacé cherries (Markakis 1982). Sulfur dioxide is also used during the production of red, green, or yellow glacé cherries from the white

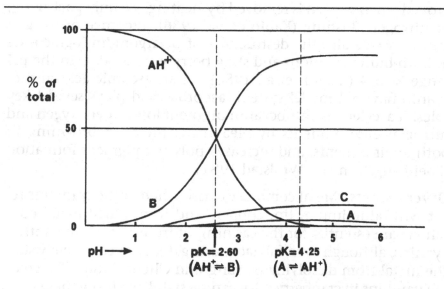

Figure 11–9 Distribution of anthocyanin structures with pH: malvidin-3-glucoside, 25°C. A, blue quinidal base; AH^+, red cation; B, colorless carbinol base; C, colorless chalcone. (Timberlake and Bridle 1983, reproduced with permission from Ellis Horwood Ltd, Chichester).

cherries of Provence (Hewitt 1995). Hydrogen peroxide, used to suppress bacterial growth, causes bleach damage to prunes, raspberries, and strawberries (Sapers and Simmons 1998). Also, hydrogen peroxide produced by ascorbic acid oxidation in the presence of copper decolorizes anthocyanins (Sondheimer and Kertesz 1953).

Oxygen and Ascorbic Acid. Strawberries are lower in pigment and ascorbic acid (vitamin C) than black currants. Anthocyanins in strawberry syrup are less stable than those of black currant syrup, and color stability depends upon total anthocyanin content rather than qualitative pigment composition. Addition of ascorbic acid further decreases pigment stability in strawberry syrup, and strawberry anthocyanin pigment fortification has a slight protective effect on the ascorbic acid (Skrede et al. 1992). Presence of oxygen or ascorbic acid contributes to increased rates of anthocyanin degradation. Loss of anthocyanins in the presence of oxygen is pH dependent and related to the amount of pseudobase present. Retention of color is improved when oxygen is removed by heating, vacuum packaging, or nitrogen flushing (Kallio et al. 1986). Presence of oxygen greatly accelerates the destruction of pelargonidin 3-glucoside in both buffer solutions and strawberry juice at 45°C in the pH range 2 to 4 (Lukton et al. 1956). In an ascorbic acid–anthocyanin-flavonol model system, ascorbic acid plays several key roles. It accelerates anthocyanin pigment loss under oxygen and nitrogen environments by bleaching, increases browning in both environments, and increases polymer pigment formation (Poei-Langstron and Wrolstad 1981).

Other Aspects. Metal complexes arise when anthocyanidins react with aluminium, tin, copper, and iron. Lined metal containers are essential for the canning of fruits containing anthocyanins, although even full lacquering does not completely stop the metal from dissolving (Hwang and Cheng 1986). The color of cyanidins in cranberry juice can be stabilized using these metals, but unfortunately other metal and tannin complexes are produced. These yield undesirable blue and brown colors (Starr and Francis 1973). Addition of Fe^{3+} and Al^{3+} improves the stability of anthocyanins in the crowberry (Kallio et al. 1986). Aluminium and cupric ions accelerate anthocyanin degradation in raspberry juice more than other metal ions (Park and Joo 1982). Discoloration of the red pigmented areas of fresh peach skin may be caused by ion complexing with the cyanidin pigment. Model reaction studies involving various metals and chlorine with cyanidin-3-glucoside result in a range of colors through yellow, orange, and pink (Denny et al. 1986).

Figure 11–10 shows the effects of pH on the visible spectrum of cyanidin-3-glucoside when complexed with aluminium chlo-

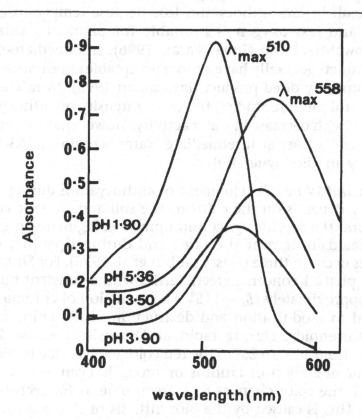

Figure 11–10 Effect of pH on visible spectra of cyanidin-3-glucoside $(3.5 \times 10^{-8}$ mol/l) + $AlCl_3$ $(8.3 \times 10^{-4}$ mol/l). Spectra measured in aqueous buffer solutions after one minute (Jurd and Asen 1966, reprinted with permission from Phytochemistry, Pergamon Press).

ride. The wavelength of peak absorbance increases as pH is increased. The maximum complex formation occurs at approximately pH 5.5, when it is reasonably stable in diffuse light (Jurd and Asen 1966).

Drying process and storage time affect color quality of dried flavonoid-containing material. Microwave-vacuum drying ensures that the product spends less time at high temperatures. Hence, it results in a redder, softer product than that produced by conventional oven drying. Similarly, continuous-mode microwave drying, as opposed to pulsed-mode drying, results in a redder product because drying time is shorter. Lower air pressures result in less anthocyanin loss because temperatures are higher and less oxygen is available for pigment oxidation (Yongsawatdigul and Gunasekaran 1996). Intermediate-moisture products generally have a more acceptable appearance than the completely dried product (Jayaraman 1995). In model water/glycerol mixture studies, the color intensity of anthocyanin increases with decrease in water activity. However, the color loss on storage is least at intermediate water activities (0.63–0.74) (Kearsley and Rodriguez 1981).

Grapes and Wine. Development of anthocyanins during grape ripening depends on the cultivar, the soil and climatic conditions, and the specific agricultural practice. Pigments in grapes are located in or near the skin, and during ripening, color changes occur in three phases (Robin et al. 1996). For Shiraz red grapes, phase 1 concerns green berries with a constant hue angle of approximately $H_0 = 115°$. The reduction of chroma is attributed to modification and destruction of the chlorophylls and carotenoids. During rapid and transitory phase 2, H_0 changes from 115° to 25° as the red color evolves due to biosynthesis or migration of tannins or proanthocyanidins. During phase 3, the color changes to a green blue as H_0 decreases to $-100°$. This is caused by the biosynthesis of skin anthocyanidins or by glycosylation of tannins and proanthocyanidins. The value of L^* decreases throughout these phases (Robin et al. 1996). Throughout the ripening of red grapes, malvidin-3-glucoside is the most abundant anthocyanin. A Colour Index for

Red Grapes (CIRG) has been proposed for mature red table grapes (Carreño, Martinez, et al. 1995),

$$CIRG = \frac{(180 - H_0)}{(L^* + C^*)}$$

The index is highly related ($r = 0.935$, $p < 0.01$) with maturity index (°Brix/titratable acid) (Carreño, Almela, et al. 1995).

Anthocyanin content of grapes depends on local conditions. High-quality grape juice is associated with a typical purple red, poor quality with brown. Anthocyanin concentration and composition responsible for red wine color are governed by fermentation conditions. Individual wineries producing within the homogeneous standards of a particular Appellation of Origin nevertheless yield wines of their own particular attributes (Gómez Cordovés et al. 1995). The complex pigment system accounts for the wide range of red wine colors from dark rose to deep purple. Although white wine is produced from white grapes, normal champagne is made from black grapes, the skins being excluded from the fermentation process. For red wine, fermentation of the skins is permitted. Rosé wine is made by blending red and white wines, by not allowing the full extraction of color from red grapes, or by macerating the grapes and allowing the pink-tinged juice to be fermented. The manufacture of rosé from blends of red and white wines is forbidden in Europe, except for rosé champagne. Differences can be detected, but not by color measurement (García-Jares 1993).

Red wine color is also affected by maturation and aging, during which grape anthocyanins are gradually replaced by more stable polymeric pigments. These may account for 50 percent of the color density of one-year-old wine. The reduction in pH sensitivity that follows the decrease in anthocyanins forms the basis of a sensitive method of tristimulus assessment of browning extent (see Chapter 9). The two groups of compounds, anthocyanin and polymeric, undergo different oxidation, condensation, and polymerization reactions, resulting in changes from bluish to orange tones. These alterations are more intense during oxidative aging—that is, in wood—than during aging in the bottle, a reducing condition. Browning reactions, oxidation products, and pigments extracted from wooden containers also contribute to the gradual change on storage of the rim color of red wine from purple red to brown red. Stabilization of oxidation products can be achieved with added sulfur dioxide. This acts as a hydrogen peroxide scavenger and inhibits oxidative enzymes such as polyphenol oxidase (Allen 1983). Presence of sulfur dioxide in wine reduces anthocyanin loss, but not in the presence of acetaldehyde (Pinicelli et al. 1994).

In some areas—for example, the Ribera del Duero region of Spain—wines retain the purple tones of young wine, either through a maintained equilibrium between the flavium (red) and quinone (violet) forms or because the compounds formed by condensation with the tannins are only slightly oxidized, thus preventing formation of orange pigments (Gómez Cordoves et al. 1995). The process of color change in wine is sensitive to such factors as oxygen, sulfur dioxide, acetaldehyde, temperature, pH, and the concentration of molecules with the ability to act as copigments. Anthocyanins in grapes and grape products have been reviewed by Mazza (1995).

MEAT COLOR

The color of meat is influenced by a number of factors, including genetic inheritance, diet, animal stress susceptibility, myoglobin concentration, and meat processing and packaging environment. An example of genetic inheritance is poultry green muscle disease (hereditary myopathy). This is possibly caused by interference to the blood supply to deep pectoral muscle, causing muscle death and greening. Capillary fragility, which leads pork fat to turn brown/green during storage and curing, is caused by an excessive number of weak capillaries rupturing on slaughter, thus increasing the fat blood content. In muscle steatosis, fat replaces lean, causing large areas of meat to appear pale (Church 1998b).

At the point of purchase, a bright cherry red color typical of oxymyoglobin is the paramount indication of fresh beef quality. An increasing proportion of the brown metmyoglobin form of pigment is associated with deterioration, and there is a consequent reduction in sales (Hood and Riordan 1973).

In fresh meat, myoglobin occurs with hemoglobin. Hemoglobin, present in erythrocytes, is bled out on slaughter to a concentration between 12 and 30 percent in the final meat (Rickansrud and Hendrickson 1967). In general, both pigments undergo the same color change reactions, although rate differences can occur (Elliott 1968). Hemoglobin carries oxygen in the blood for diffusion into the muscle fibers, where it is bound by the myoglobin molecule. Myoglobin is responsible for storing and carrying oxygen needed for aerobic metabolism. All colors characteristic of raw, cooked, and cured meats are caused by changes to this molecule. There are two parts, the protein (globin) and the nonprotein (heme). Figure 11–11 is a schematic representation.

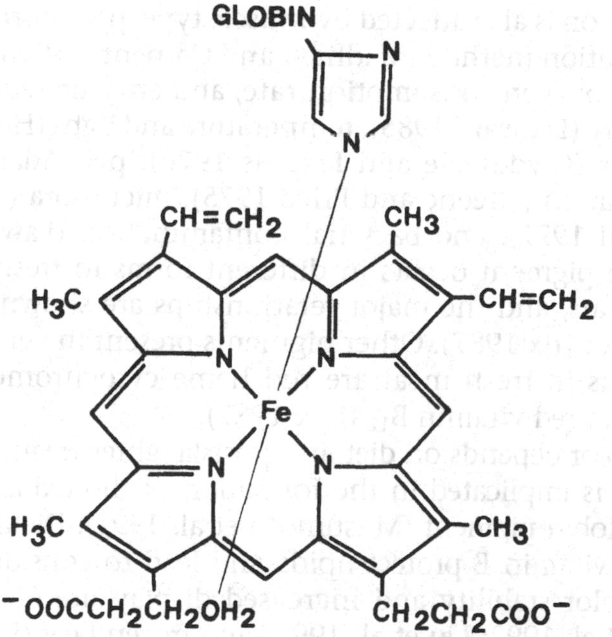

Figure 11–11 Structural representation of myoglobin.

Major changes can involve the iron atom in the heme, which can be in the oxidized or reduced form, and the protein globin, which can be in the native state or denatured. In raw meat, the complex occurs in three forms—myoglobin, the oxygenated oxymyoglobin, and the oxidized metmyoglobin. In hemoglobin and oxymyoglobin, the iron is in the ferrous (Fe^{2+}) state, and the oxygen occupies the free bonding site of the iron atom. In met-myoglobin, the iron atom is in the ferric (Fe^{3+}) state and is un-able to bind the oxygen. Red meat is red because muscle fibers contain high myoglobin and mitochondria contents. Both are red. Mitochondria organelles use oxygen to supply energy for muscle contraction. Hence, active, energy-using, and fatigue-re-sistant muscles have higher oxidative capacities and are redder. Myoglobin concentration depends on a number of biological factors, such as type and function of muscle, age, sex, species, and breed. The characteristic ascending order of redness of poul-try, pork, veal, and young beef and old beef muscles is caused by increasing concentrations of myoglobin. Surface metmyoglobin accumulation is also affected by muscle type, postmortem aging, and fabrication method (Madhavi and Carpenter 1993), temper-ature, pH, oxygen consumption rate, and enzyme-reducing sys-tem activity (Ledward 1985), temperature and light (Hood 1980), metal ions (Clydesdale and Francis 1976), pH (Adams 1976), lipid oxidation (Greene and Price 1975), microflora (Ockerman and Cahill 1977), and bacterial contamination (Lawrie 1985). The heme pigment occurs in different forms in fresh and pre-served meats, and the major relationships are shown in Figure 11–12 (after Fox 1987). Other pigments present in very low con-centrations in fresh meat are red heme cytochromes, yellow flavins, and red vitamin B_{12} (Fox 1987).

Flesh color depends on diet and postslaughter treatment. Lipid oxidation is implicated in the formation of the oxidized brown metmyoglobin pigment (Mitsumoto et al. 1991). Dietary supple-ments of vitamin E protect lipids and lead to considerably im-proved color stability and increased display life of fresh beef (Arnold et al. 1992; Liu et al. 1996) and frozen beef (Lanari et al. 1995). Evidence for improvement in pork and poultry is less con-clusive. However, prolonged storage and processing may accelerate

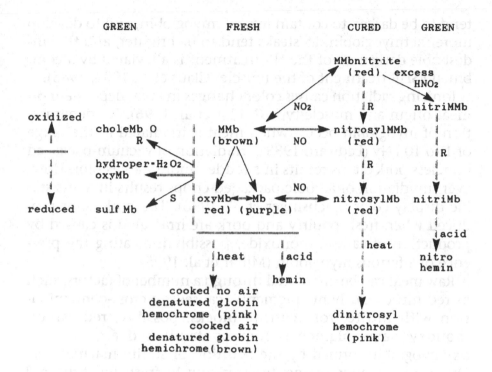

Mb = myoglobin
MMb = metmyoglobin

Figure 11–12 The heme pigments of meat (after Fox 1987, with permission).

lipid oxidation, and end-product quality seems related to dietary vitamin levels (Jensen et al. 1998). Improvements do not necessarily occur when tocopherol is added after slaughter (Sprouls and Brewer 1997), but combination of α-tocopherol and beta-carotene may increase delay in oxidation (Yin and Cheng 1997).

Electrical stimulation (ES), hot boning (HB), and chilling rate of carcasses affect meat color. High temperatures encourage pigment instability; low temperatures promote oxygen penetration and solubility in tissue. Rapidly chilled meat tends to be darker in color, perhaps due to higher oxygen consumption rate and an inhibition of oxymyoglobin formation (Renerre 1990). HB steaks tend to be darker, to contain less oxymyoglobin, and to develop more metmyoglobin. ES steaks tend to be brighter, and the undesirable darkening of the HB treatment is alleviated by a combined ESHB treatment of the muscle (Claus et al. 1984, 1985).

Ionizing radiation causes color changes in meat dependent on meat origin and muscle type (Millar et al. 1996). No denaturation of myoglobin occurs when meat is irradiated in the range of 1 to 10 kGy (Ledward 1983). Irradiation of vacuum-packaged boneless pork chops results in a redder, more stable color. However, irradiation of aerobic packaged chops results in a less stable display color (Luchsinger et al. 1996). The pink color produced when fresh poultry and pork are irradiated is caused by production of carbon monoxide, possibly indicating the presence of a ferrous myoglobin (Millar et al. 1995).

Raw meat can be discolored through a number of factors, such as red nitric acid heme pigments, caused by cross-contamination with nitrite from cutting blocks; possibly, red carbon monoxy heme pigments from contaminated air; and red oxymyoglobin formed by the addition of sulfite niacinamide. Other endogenous pigments occurring in fresh meat are red hemoglobin blood splashes; greenish choleglobin, caused by oxidation under mild reducing conditions; red peroxymetmyoglobin, from peroxides from catalase negative bacteria and radiation; green hydroperoxymetmyoglobin, formed by peroxides in acid; and green sulfmetmyoglobin, from sulfides produced by bacteria and radiation and by oxidation to the ultimate degradation product brown bile pigments (Fox 1987).

Endogenous pigments occurring in normal fat are yellow carotenes in grass-fed animals; the brown "bronze" shoulder fat catecholamines; the brown fat/protein pigments in aged animals; and the yellow fluorescent fat/protein lipofuscin pigment. Fat can be discolored by the endogenous red hemoglobin pigment resulting from capillary rupture and by brown Maillard reaction carbonyl-amino pigments (Fox 1987). The opalescent or iridescent sheen sometimes occurring in meat is discussed later in this chapter in the section "Light Scattering."

Reviews of different aspects of fresh meat color have been written by Gray et al. (1996), Froning (1995), Church and Parsons

(1995), Renerre (1990), Fox (1987), Lawrie (1985), and Ledward (1983).

Raw Meat

The freshly cut surface of a piece of beef is the purple color of the uncomplexed myoglobin molecule. Oxygen in the air oxygenates the myoglobin, and beef blooms to the bright red oxymyoglobin (Lanari and Zaritski 1991). Oxygen diffuses into the beef but is consumed by enzymes present for a considerable time postmortem. Hence, a point of balance is set up where the oxygen diffusion rate is equal to the rate at which it is consumed by the enzymes. The depth of redness is governed by the oxygen diffusion rate, which varies with muscle type. Since the oxygen diffusion coefficient decreases less than the oxygen consumption coefficient for a given fall in temperature, the layer of oxymyoglobin will be deeper at 0 than at 20°C (Brooks 1929). Oxymyoglobin oxidizes most rapidly at low oxygen tensions, and a brown layer of metmyoglobin is able to develop beneath the oxymyoglobin layer (George and Stratman 1952). On storage, the oxidation proceeds nearer to the surface, and a red-to-brown transition results. These deeper layers influence the appearance of the red oxymyoglobin surface. However, the red coloration at the surface may be maintained for several days by exposure to high oxygen tensions (MacDougall and Taylor 1975). There is continual interconversion between these three forms of the molecule in fresh meat, but in all cases the heme nucleus remains intact and the globin remains native. Beef blooms rapidly, but although oxymyoglobin is slowly formed in pork, the formation in chicken is limited (Millar et al. 1994).

Light scattering also plays a large part in the appearance of meats through differences in the biochemistry and fiber structure. Short-term stress in pigs at slaughter can lead to pale, soft, and exudative muscle through rapid glycolysis and a rapid drop in pH. Muscles reaching a pH of 6 while above 30°C are affected. The increase in light scattering is caused when sarcoplasmic proteins denature and come into solution between muscle fibers. As only certain muscles may be affected, this can result in the production of two-tone cuts. On the other hand, young cattle and pigs that have undergone long-term stress have low glycogen levels at slaughter. The resulting muscle is high in pH (>7.0), there is little protein denaturation and little light scattering, and the meat is dark, firm, and dry (Mac-Dougall 1977).

Wrapping meat in an oxygen-permeable film slows the rate of fading but does not prevent it (Winstanley 1979). Oxygenation can be maintained in meat for five or six days by packing in atmospheres of 80 percent oxygen. Atmospheres used may contain 15 to 40 percent carbon dioxide for suppressing microorganisms and 60 to 85 percent oxygen, or alternatively 40 percent nitrogen (N_2) as an inert filler to prevent pack collapse, 30 percent carbon dioxide, and 30 percent oxygen (Phillips 1996). Packing in a mixture of 80 percent oxygen and 20 percent carbon dioxide allows meat to remain an attractive color for at least one week at +1°C (Taylor and MacDougall 1973). High concentrations of oxygen maintain redness but cause oxidative rancidity of fat in the meat. The use of high carbon dioxide concentrations causes gray discoloration through pH-initiated protein precipitation (Bell et al. 1996). However, mixtures of carbon dioxide and oxygen have been found beneficial in maintaining redness under chill conditions (Seideman et al. 1984). The use of pure carbon dioxide gas during the manufacture of restructured steaks was found to reduce overall discoloration both initially and during three months' frozen storage in a vacuum pack (Chu et al. 1988). Exposing fresh beef to carbon monoxide prior to vacuum packaging can yield redder steaks with extended shelf life compared with untreated, vacuum-packed steaks (Brewer and Wu 1994). Low levels of oxygen combined with 1 percent carbon monoxide seem to offer substantial improvements in storage life (Luno et al. 1998). MA packaging of pork can lead to bone discoloration. Dietary vitamin E supplementation can in some circumstances offer protection (Lanari et al. 1995).

Centralized packing demands that meat must have a useful marketing of at least one week. Metmyoglobin formation can be decreased and retail pack shelf life increased by proper control of product hygiene and storage temperature and by the use of appropriate preservative packaging (Gill 1996; Gill and Jones 1996). Retail packs are sealed into a master pack containing an anoxic atmosphere using, for example, carbon dioxide or nitrogen. Initially the product may be dulled and discolored by residual oxygen but when this has been scavenged from the atmosphere by the meat, the metmyoglobin becomes reduced to dark purple myoglobin. Alternatively, oxygen scavengers can be included in the pack (Gill and McGinnis 1995b). Meat removed from the pack will bloom to the same color as freshly cut meat (Gill and McGinnis 1995a; Gill and Jones 1996). The transient discoloration is of commercial importance only when there is little delay between preparation and display (Gill 1996). Similarly, wholesale cuts are vacuum packed for storage (Seideman and Durland 1983). They are subsequently butchered and allowed to reoxygenate for retail sale (Seideman et al. 1984). Vacuum packaging of beef is not commonly used in retail marketing because of its dull purple color, increased compression damage, and drip loss, particularly in subprimal cuts (Simard et al. 1985). Pigment oxidation occurs in salted intermediate-moisture lamb, particularly when stored in air. Vacuum packing offers some protection (Da Silva et al. 1994).

Changes in concentration of the three pigments, following non–vacuum packing in an oxygen-impermeable membrane, are illustrated in Figure 11–13. These changes to fresh beef are depicted in tristimulus measurement terms in Figure 11–14. The purple myoglobin of the freshly cut meat is converted to bright red oxymyoglobin on exposure to air. Further exposure under reduced oxygen pressure in the pack leads on oxidation to brown metmyoglobin. In this case, the growth of aerobic bacteria reduced oxygen tension at the surface, thus

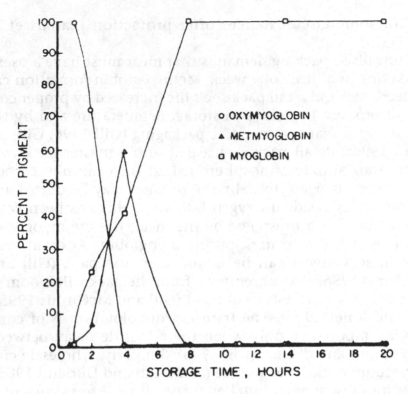

Figure 11–13 Pigment changes at 3°C of anaerobically packaged beef (Pierson et al. 1970, via Seideman et al. 1984, with permission).

enzymically reducing the metmyoglobin to the purple myoglobin (Robach and Costilow 1962). During this work, visual assessments of acceptability were also made. Figure 11–14 indicates the position of an acceptable/unacceptable boundary. This appeared to occur when the first evidence of the formation of brown metmyoglobin became visible (Hutchings 1969).

Effects of Storage Temperature and Light

Storage temperature is a major factor in the discoloration of fresh meat. Lower temperatures discourage the oxygen-consuming processes. These include respiratory enzyme activity, fat oxidation, and the promotion of oxygen dissociation from the meat pigments. At higher temperatures, more rapid reduction in oxygen tension increases the speed of pigment autoxidation, leading to brown discoloration. At lower temperatures, dissolved oxygen concentration increases, and the gas penetrates deeper into the flesh. Maintaining pigments in an oxygenated state increases their stability toward oxidation (Snyder 1964). A reduction of 5°C can halve the rate of metmyoglobin production. In aged meat, respiratory activity and oxygen consumption are lower. This leads to increased oxygen penetration and an increased depth of oxymyoglobin. Aged meat is thus a brighter red than fresh meat, but it fades more rapidly because of the loss of the reducing activity (MacDougall 1977).

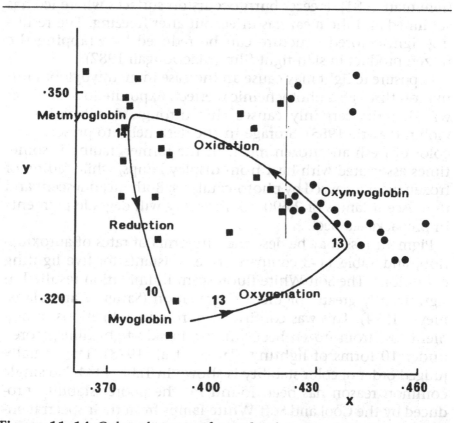

Figure 11–14 Color changes of raw beef wrapped in a permeable membrane. Illuminant C, 2° observer, 5°/d. ●, acceptable; ■, unacceptable; numbers = *Y* percent (Hutchings 1969).

Color of frozen meat is controlled by freezing rate, storage temperature, intensity of display light, and packing material. Freezing results in ice crystal formation. Slowly frozen meat is dark through pigment concentration. The faster the freeze, the smaller the crystals, the more light scattered and the paler the meat. Fresh beef allowed to bloom before freezing has a similar color to fresh meat (MacDougall 1982). Shelf life of fresh lamb chops is approximately one day. However, color deterioration in intact muscle is negligible, and a carcass can be stored for several months at −20°C. Chops cut from the thawed or frozen carcass will result in an acceptable product for display (Moore 1990). Similarly, uncured pork cuts can be successfully frozen and stored before being thawed and cut for retail display (Jeremiah 1981). Freezer burn occurs on surfaces where ice has sublimed and the meat has dried out after freezing. The resulting light-colored structure can be reduced by wrapping the frozen product in skin-tight film (MacDougall 1982).

Exposure to light may cause an increase in metmyoglobin formation through a photochemical effect. Exposure to ultraviolet wavelengths certainly causes this through denaturation of globin (Lawrie 1985). Storage in the dark helps to preserve the color of fresh and frozen meat. In the former, fading is sometimes associated with heat from display lamps, while fading of frozen meat is caused by photo-catalysis. Both incandescent and fluorescent lamps of 2,690 lux intensity will degrade pigments in prepackaged beef.

Pigment loss may be described in terms of rates of autoxidation, and Table 11–3 compares rate constants for five lighting conditions. The Soft White fluorescent illumination resulted in significantly greater pigment degradation (Satterlee and Hansmeyer 1974). This was confirmed during investigations of pigment loss from frozen beef muscles in skin-tight films, stored under 10 forms of lighting (Tuma et al. 1973). The visually judged order of color stability is shown in Table 11–4. No single common reason has been found for the poorer stability produced by the Cool and Soft White lamps from their spectral energy outputs. Soft White illumination is high in ultraviolet, and Cool White is high in yellow (Kropf 1980).

Table 11–3 Autoxidation Rate Constants for Beef at 5°C Exposed to Light at an Intensity of 2,690 lux (data from Satterlee and Hansmeyer 1974)

Light Source	Autoxidation Rate Constant (hr^{-1})
No light	3.25×10^{-3}
Cool Flood incandescent (150 watt)	5.16×10^{-3}
Incandescent (100 watt)	5.46×10^{-3}
Pink fluorescent (40 watt)	5.54×10^{-3}
Soft White fluorescent (40 watt)	8.20×10^{-3}

A good color of frozen meat can be maintained by storage in the dark. However, lighted storage at −18°C causes rapid color deterioration. The use of lower temperatures can prevent this photo-oxidation and overcome pigment degradation in frozen beef on lighted display. For example, blast-frozen bloomed beef packed in highly oxygen-permeable polyethylene has a shelf life of one to three days when stored at 1,000 lux at −18°C but over 90 days at −40°C (Lentz 1979). A maximum display intensity of 1,000 lux is recommended (Tuma et al. 1973). For long periods

Table 11–4 The Effect of Light Source on the Color Stability of Frozen Beef (Kropf 1980 quoted from Tuma et al. 1973)

Light Source	Color Stability[a]
Grolux Wide Spectrum	++
Incandescent Fluorescent	++
Deluxe Warm White	+
Standard Grolux	+
Verda-Ray	0
Incandescent Holophane (uneven intensity)	0
Cool Beam (uneven intensity)	0
Cool White	−
Soft White	−

[a] Color stability: ++, best; +, good; 0, fair; −, poor.

of frozen storage of lamb, it is better to store in the loin than as chops (Moore 1990).

Color saturation (see Chapter 7) can be used to define and determine the shelf life of raw beef. On formation of metmyoglobin (MMb), saturation (S) decreases as hue changes from bright red through red, to brown, to greenish brown. In Table 11–5, bright red to red corresponds to $S = 20$ (MacDougall 1985). The value of 18 corresponds to an MMb value of 20 percent, the amount sufficient for rejection by the customer (Hood and Riordan 1973). A distinctly brown sample (40 percent MMb) has a value of $S = 14$.

Storage lives for three beef muscle types undergoing various treatments are also given in Table 11–5. Samples were aged at 1°C before packing for display. Comparing muscle types, longissimus dorsi is more stable than semimembranosus or psoas major. (Longissimus also has a greater life than gluteus medius; Arnold et al. 1992). Aging increases rate of saturation loss. Increasing display temperature from 3 to 5°C doubles the rate of discoloration. Gas packing in 80 percent oxygen and 20 percent carbon dioxide increases display life considerably by retarding bacterial spoilage. In a comparison of initial storage regimes after oxygenation, skin packing, and freezing, meat remained red

---

Table 11–5 Changes in Saturation for Different Beef Muscles (MacDougall 1985)

Muscle Type*	Aging Treatment		Display		Color Saturation		
	Weeks	Type	°C	lux	Initial Value	Days to S = 20	Days from S = 20 to 16
LD	1	Chilled	5	1,000	2	5.5	5
	3	Chilled	5	1,000	25	5	5
PM	1	Chilled	5	1,000	25	2	3.5
	3	Chilled	5	1,000	25	1	2.5
SM	1	Chilled	5	1,000	24	3.5	4
	3	Chilled	5	1,000	24	2	3.5
	2	Chilled	8	1,000	25	<1	<1.5
	2	Chilled	1	1,000	27	10	10
		80% O_2 + 20% CO_2 Frozen, skin packed			23		
		After 2 months	−18	1,000	>20	<1	1–2
LD	2	Dark storage	−18	500	>20	<1	3

* LD, longissimus dorsi; PM, = psoas major; SM, = semimembranosus.

after three months' dark storage. Fading occurred at rates similar to those for chilled meat (MacDougall 1985). As oxygen permeability of package material increases, so frozen lean ground beef increases in brownness and decreases in redness (Brewer and Wu 1993).

Restructured Meat Products

According to the severity of deboning, meat produced contains varying quantities of hemoglobin. Oxygen mixed into the meat during processing converts most of the reduced myoglobin present to oxymyoglobin (Froning 1995). Processing under a nitrogen atmosphere has a detrimental effect (Govindarajan 1973), probably due to the low oxygen concentration accelerating pigment autoxidation (Chu et al. 1987). Tempering during processing results in a rapid increase in metmyoglobin concentration during the first month of storage. This may be due to ice crystal formation and cell disruption (Chu et al. 1987). Color of frozen lean ground meat blocks is not uniform. Surface pigments may be changed through photo- or lipid oxidation caused by the presence of oxygen at the surface (Brewer and Wu 1993).

Discoloration is a major marketing problem associated with restructured meat products. For example, eating quality of restructured pork products can be better than that of pork loin chops, but their color may not be preferred (Huffman and Cordray 1979). In restructured beef steaks, brown spots appearing among the red bloomed meat decrease acceptability (Chen and Trout 1991). Surface metmyoglobin of raw beef patties containing mechanically recovered neck bone lean first increases, then decreases during retail display. This may be due to microbial effects (Demos and Mandigo 1996). Retail-ready packs or ground beef master-packaged under an oxygen-depleted atmosphere can have a useful storage life of approximately 30 days in commercial circumstances (Gill and Jones 1994).

The presence of sodium chloride can greatly accelerate discoloration during frozen storage. It increases water-binding capacity, resulting in darker, more translucent muscle, and it denatures enzymes, thus inhibiting the production of reductants

(Fox 1987). Also, the chloride anion probably promotes myoglobin autoxidation (Huffman et al. 1981; Chu et al. 1987). Addition of tripolyphosphate partially counteracts discoloration caused by sodium chloride (Huffman et al. 1981; Matlock et al. 1984) and reduces the rate of lipid oxidation (Matlock et al. 1984).

Pork sausages sometimes discolor over the contact area between adjacent sausages. These areas are deprived of oxygen, resulting in the reduction of oxymyoglobin to the purple myoglobin. This effect is reversible in its early stages, but on storage, sufficient oxygen is available to achieve pigment conversion to metmyoglobin. This change is not reversible. If the sausages are packed in film and viewed under bright fluorescent lighting, the myoglobin color can appear green. This is not caused by bacterial action. Processing affects pork sausage color. An increase in fat content leads to an increase in lightness and a decrease in redness of the internal and external surfaces of the product. Decrease in particle size results in an increase in the external lightness of the product due to an increase in light scattering. Redness of the interior increases, possibly through a more rapid pigment oxidation of the increased surface area. On storage, this redness decreases less than the exterior value, probably due to lower oxygen availability within the product (Reagan et al. 1983).

Myoglobin and hemoglobin are the principal pigments responsible for the color of mechanically deboned meat and dark muscle in poultry. Both pigments can be extracted using water or buffered water solutions (Dawson et al. 1989) producing meat that can be used for *surimi*-type products (Froning 1995).

Cooked Meat

The range of colors found in cooked meat arises from the variation of pigment conversion rates occurring on cooking. Myoglobin completely denatures on heating between 80 and 85°C, but the conversion is slower at lower temperatures. The pigments and muscle proteins of rare cooked meat heated to a temperature of 60°C are still largely in their native state, although some denaturation takes place at lower temperatures (Wright et al. 1977). On further cooking to **medium** and **well done** (approximately 80°C), flesh becomes more opaque and the pigments more insoluble. The pigments formed are the brown, denatured globin hemichromes (Fe^{3+}) and the pink hemochromes (Fe^{2+}) (Fox 1987). Also on cooking, surface dehydration takes place, heme pigments are concentrated, soluble pigments migrate, Maillard reactions occur, thermal decomposition caramels are formed, and eventually, on roasting, charring takes place. The color of slices resulting from sides of roast beef is shown in tristimulus terms in Figure 11–15 (Hutchings 1978). A large decrease in the value of *x* occurring at cooking temperatures greater than 75°C relates to sarcoplasmic protein denaturation (Bowers et al. 1987).

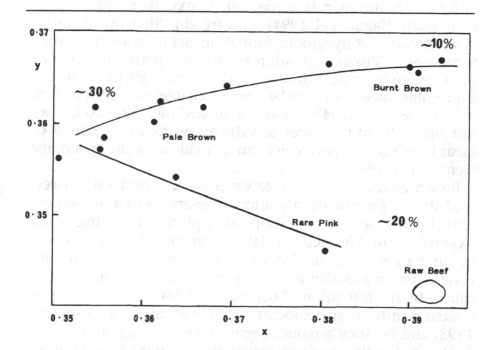

Figure 11–15 Color of in-home–prepared roast beef measured at room temperature. Lines denote increase in cooking severity. d/0°, illuminant C, 2° observer, numbers = *Y* percent (Hutchings 1978).

Internal color of ground beef patties is normally related to end-point cooking temperature. Patties cooked to 65°C (**medium rare**) have a red-pink interior, those cooked to 71°C (**medium**) are tan, and those cooked to 77°C (**well done**) have a brown interior (Marksberry et al. 1993). Cooking rate does not affect intensity of red color (Brewer and Novakofski 1996). However, deviations from this pattern can occur. Patties cooked to 71°C can retain their redness through limited myoglobin denaturation of high-pH meat or through contamination (see below) (van Laack et al. 1996). Also, there is a greater tendency for patties to retain their pinkness in products containing less than 20 percent fat (Berry 1994). Alternatively, patties with their interior myoglobin in an oxidized state immediately prior to cooking can brown prematurely at temperatures lower than 55°C. This apparent thoroughly cooked appearance of the product represents a potential safety hazard. The order of pigment stability to denaturation is myoglobin > oxymyoglobin > metmyoglobin (Hague et al. 1994). Slow freezing/thawing of patties in air favors metmyoglobin formation before cooking (Cornforth 1996). Visual evaluation of internal patty color is not therefore an accurate indication of doneness. Lack of redness in expressible juice color may be a more reliable end-point indicator (Hague et al. 1994). Storage of cooked meat at −20°C does not alter color of the juices, so valid assays for the doneness of meat in relation to juice color can be made after frozen storage (Senter et al. 1997).

Brown **medium** cooked meats can become pink on refrigerated storage because denaturation is reversible to a limited extent. In this case, renaturation takes place on cooling when oxymyoglobin is formed. The latter pigment is formed by cooking in the absence of air (Fox 1987). Pinkness can occur in well-cooked pork roasts after refrigeration and frozen storage. This is caused by the formation of denatured globin hemochromes or related non-nitrosyl hemochromes (Ghorpade and Cornforth 1993) and possibly a reduced sensitivity of myoglobin due to further processing prior to completion of rigor (Young et al. 1996). Since the pink color quickly fades to beige or gray on exposure to light and oxygen, the development of pinkness in cooked, stored pork ought not be a problem for institutions (Howe et al. 1982).

Development of a pink color in cooked poultry poses a consumer acceptance problem. Oxymyoglobin is regenerated two hours after cooking (Helmke and Froning 1971). Pinkness can also be caused by nitrates present in the chill water in which the carcasses are held (Nash et al. 1985). Discoloration can be caused by the formation of green, nitrite/gelatin reaction xanthoproteins produced through acid conditions in channels next to bones (Fox 1987). Pink discoloration occurring in cooked white meats has been reviewed by Froning (1995) and Maga (1994).

Cured Meat

Meat can be preserved by salting or fermentation. Civilizations since perhaps the Mesopotamians of 5000 B.P. have used salt from the saline deserts to preserve meat and fish. This salt contains nitrates and borax as impurities that produced the nitric oxide necessary for the formation of red complexed heme pigments. These processes were known to the Greeks, and the resulting redness was recorded later by the Romans (Binkerd and Kolari 1975).

Nitrates added to meat containing reducing agents, such as ascorbic acid and sodium ascorbate, are converted via nitrites to nitric acid. Nitrite is an antioxidant that retards spore germination and toxin formation by *Clostridium botulinum* (Cassens 1997). During the curing process, this oxidizes the meat pigments to metmyoglobin, and the brown- to tan-colored nitrosometmyoglobin is formed. Heating in the presence of reducing agents denatures the globin part of the molecule, and the pink to deep red nitrosylhemochrome and dinitrosylhemochrome are produced. It is necessary to use fresh meat because in old meat, reduction does not occur, and a brown product is obtained. Presence of fat, water, and other ingredients dilutes pinkness intensity. Presence of a high aerobic bacterial load tends to result in hydrogen dioxide production, which converts the pink pigment into a gray form. To control bacterial growth in low-salt, nonfermented cured meats, it is necessary to use direct nitrite addition. Too much nitrite can result in the formation of brown metmyoglobin or green nitrimetmyoglobin (Boles 1995; Fox 1987).

Redness of canned meats is normally developed in situ by adding curing salts to the fresh meats before processing. Lower concentrations of nitrite can be used because oxygen cannot gain access. However, the cut surface of the meat discolors more rapidly on exposure to air if light is present, the rate of the discoloration being independent of the light wavelength.

Parma ham, traditionally prepared by a lengthy seasoning, can be prepared without using nitrates. The stable red color is caused by the formation of a new myoglobin derivative, possibly formed by strains of staphylococci bacterial action (Morita et al. 1996).

Attempts have been made to reduce the need for nitrite by adding synthesized cured pigment to the comminuted product. This would obviate the need to add excess nitrite, which contributes to nitrosamine formation (Shahidi et al. 1985). Nitrite-free cured meat products, including frankfurter and salami, can be prepared using preformed cooked cured-meat pigment. This is produced from red blood cells and a nitrosating agent, such as nitric oxide, in the presence of a reductant that prevents oxidative degradation of the pigment. There is evidence that the chemical structure of the pigment produced is identical to that resulting from the nitrite-cured meat (Pegg and Shahidi 1997). The pigment can be protected through encapsulation (O'Boyle et al. 1992).

Some bacteria can survive cured-meat heat treatments, and their growth can result in a green discoloration. Bacteria such as *Lactobacillus viridescens* can grow under the anaerobic conditions existing in vacuum-packed meat products. Greening does not take place in the absence of air, but when the pack is opened, oxygen encourages the formation of hydrogen peroxide. This reacts with the pink, nitrosohemochrome, cured-meat pigment to form greenish oxidized porphyrins. Improved plant sanitation often reduces the problem, but where this fails, manufacturers are recommended to review their heat treatments (Grant et al. 1988). Bacterial growth and irradiation can lead to the formation of red peroxymetmyoglobin, green hydroperoxymetmyoglobin, or green sulfmetmyoglobin. Gamma irradiation of nitrite and nitrite-free cured meats produces no effects detrimental to the color (Shahidi et al. 1991). An undesirable bright pink stain can arise from contamination by nitrite concentrations of less than 2 mg/kg^{-1} (MacDougall and Hetherington 1992). A white film or white crystals sometimes appear in dry-cured hams. These are two different phenomena having a common origin in the presence of the amino acid tyrosine above its solubility level in the water phase (Arnau et al. 1996).

Redness of cured meats is an indicator of quality since, on aging in air or under lights, hemochrome pigments break down into brown and gray forms. Reducing agents such as ascorbate, nicotinamide, and nicotinic acid can prolong the fresh color. For example, sodium ascorbate reduces photo-oxidation of paté stored in an illuminated display cabinet (Perlo et al. 1995). However, use of reducing agents and of nitrates and nitrites themselves is subject to legislation. Interactive packaging using oxygen absorbers with concomitant development of carbon dioxide and packaging material with a low oxygen transmission rate can eliminate discoloration of pasteurized, sliced ham normally encountered as a result of photo-oxidation of nitric acid pigments during the first 24 hours of display in illuminated chill cabinets (Anderson and Rasmussen 1992). Old and faded cured meats are brown from the formation of hemichromes and bile pigments (Fox 1987).

MA packaging of cured products excludes oxygen, the most common mixtures being 20 to 50 percent carbon dioxide and 50 to 80 percent nitrogen (Phillips 1996). A layer of carbon dioxide created above the display can also prevent color change (Dages 1986). Vacuum packaging can be used, for example, for ham (Kotzekidou and Bloukas 1996), bologna (Brewer et al. 1992), and bratwurst (Ghorpade et al. 1992).

The color of cured meat has been reviewed by Pegg and Shahidi (1997), Fox (1987), and Pierson and Smoot (1982).

MARINE PRODUCTS

The major chemical groups involved in fish flesh color are heme pigments, carotenoids, and melanins. Bruising discoloration can result from catching and handling methods. Cod caught by gill net are significantly more bruised than those caught by trap, hand line, or long line (Botta, Bonnell, and Squires 1987). Fish that have struggled extensively before being brought on board (Botta, Kennedy, and Squires 1987) or have suffered delay before bleeding and gutting (Botta et al. 1986) may be given lower color grades. In freshly caught fish, blood contained in the tissue blood vessels is still red, and low temperatures are essential to prevent bacterial growth. However, holding the fish on ice before filleting results in a darker red or brown color as the hemoglobin oxidizes. Discoloration occurs in cut surfaces. This is possibly due to pH precipitation of sarcoplasmic proteins, causing a change in light-scattering properties (Church 1998a).

Depriving the flesh of oxygen retards oxidation, but neither vacuum packing nor ice glazing completely stops the deterioration. Very little oxygen is required, and the small amount of air in a can headspace is enough to discolor adjacent flesh. Air dissolved in ice glaze may also cause problems (U.S. Department of the Interior 1970). However, a properly developed, maintained, and renewed glaze is a good barrier to oxygen in the outside air and, combined with suitable packaging, will greatly retard oxidation. Yellowing of aging, raw white fish fillets is caused by oxidation and the formation of alkaline yellow anthoxanthins (Hillman 1983). Melanosis occurs in the form of enzyme-activated black spots during iced storage of some species—for example, white shrimp. The area affected increases linearly, and by 15 days, 42 percent of the shrimp becomes affected (Luzuriaga et al. 1997). "Black spot," caused by polyphenoloxidase activity, can be inhibited by addition of, for example, sulfite or by MA packaging (Church 1998a).

Normally, fish flesh is of two types, dark and light. On frozen storage, pigments in the darker meat are subject to oxidation, becoming deep yellow or brown. The color of pink or red fish is sensitive to frozen storage abuse, the pigments oxidize, light scattering increases, and the color apparently disappears. Impurities present in common salt can have an accelerating effect on the oxidative deterioration of frozen fish. Only high-quality salt should be used for the brine dipping of species particularly susceptible to oxidation (U.S. Department of the Interior 1970). Discoloration of frozen hake can be prevented by antioxidant treatment or by removing the layer of red muscle beneath the skin during filleting (Licciardello et al. 1980). The color of thawed halibut, herring, and mackerel can be dependent on thawing method, with microwaving being preferred to air or water heating at 20°C (Vorob'ev 1997).

The predominant pigment in tuna is myoglobin. This comprises 69 to 85 percent of the total heme pigments in light-colored muscle and 81 to 95 percent in dark. When first caught, the muscle varies between pink and red depending on the species. During handling and on iced or frozen storage, the color becomes progressively browner (see Chapter 8). Processors demand that as little as possible of the brown metmyoglobin color enter the canning or production line (Matthews 1983). A high myoglobin content results in the production of a dark brown color on cooking. It may also be involved in reactions with trimethylamine oxide and cysteine to form an undesirable green color in canned tuna (Francis and Clydesdale 1971).

The color of salmon and trout flesh is under genetic control (Withler and Beacham 1994) and in the wild arises from crustacean carotenoids, mainly astaxanthin. During spawning, salmonids suffer a loss of muscle carotenoid content; the female recovers lost color faster than the male (Choubert and Blanc 1993). In raw muscle, concentration of the main carotenoid is strongly related to hue, chroma, and lightness. Farmed salmon, trout, and shrimp must accumulate sufficient dietary astaxanthin to achieve market acceptability (Haard 1992). Canthaxanthin was available for use as a salmon dietary pigment in the early days of fish farming. However, astaxanthin is more strongly colored, is better deposited in the muscle, and is now the preferred pigment (Choubert et al. 1992). Ultimate flesh color depends on many factors. For example, color level induced in rainbow trout by astaxanthin is under genetic control and is sensitive to feeding rate, sex, and trout size, as well as environmental factors (Blanc and Choubert 1993). Dietary tocopherols (vitamin E) are effective in controlling pigment oxidation and result in an improved product color without affecting taste or texture (Akhtar et al. 1996; Sigurgisladottir et al. 1994). Relationships between salmon pigment, structure, and color on further canning and cooking are discussed in Chapter 9. Smoking is discussed later in this chapter. Some carotenoids change color on cooking. Blue crabs change from blue green to red orange, but redness does not indicate that the crab has been adequately cooked (Himelbloom et al. 1983). Red skin color in some species is of primary importance to the consumer. Red-skinned species of rockfish command higher prices than those having yellow or brown skins. Antioxidants can be used to delay oxidation and fading of skin carotenoids (Li et al. 1998).

Studies of the natural astaxanthin pigments occurring in freeze-dried salmon have confirmed the protective role of higher water activities in dehydrated food systems. Although pigment loss is reduced at higher water activities, nonenzymic browning increases. Free-radical mechanisms cause astaxanthin oxidation, while lipids and protein amino groups react to form browning pigments. The browning pigments in turn produce antioxidants, causing pigment oxidation to fall (Martinez and Labuza 1968).

In MA packaging, the preferred gas mixture for white fish is 40 percent carbon dioxide to 30 percent oxygen to 30 percent nitrogen. For fatty fish, gas mixtures of 40 to 60 percent carbon dioxide to 40 to 60 percent nitrogen are common (Phillips 1996); presence of oxygen encourages rancidity. Elevated concentrations of carbon dioxide inhibit growth of many spoilage organisms. However, these can result in discoloration (Parry 1993). Mackerel, for example, in an atmosphere of carbon dioxide tends to decrease in brownness and increase in yellowness. This may be due to an alteration in the balance of metmyoglobin and browning pigments (Hong et al. 1996). Greening of smoked salmon can occur in high-CO_2–packed smoked salmon. Clouding of fish eyes and other fading and browning reactions occurring in 100-percent CO_2 packs may be due to low levels of oxygen permeating into the pack during storage, 0.5 percent favoring formation of the brown metmyoglobin (Church 1998a). Fish is vacuum packed, but this can cause compression damage. Reviews of MA packaging of fish have been written by Davis (1993) and Reddy et al. (1992).

Discoloration of dried salted fish muscle during storage is caused by Maillard-type reactions and oxidative staining (Pokorny et al. 1974). Brown as well as fluorescent pigments are produced, for example, in Indonesian dried-salted mackerel (Maruf et al. 1990) and dried salted sardines (Lubis and Buckle 1990).

Irradiation can extend the shelf life of cold-stored seafood (Licciardello and Ronsivalli 1982). Its effect on appearance depends on irradiation dose and species. Work on a number of lean and oily, reef, estuarine, and ocean species revealed that no change in appearance occurred after a dose of 1 kGy. At 3 kGy, however, a noticeable change in whiting fillets was described as "gaping/dry looking" (Poole et al. 1994). No objectionable color change took place in samples of cold smoked salmon irradiated at 2 kGy. However, at doses of 4 kGy, the normal cherry red color was changed to beige white (Hammad et al. 1992).

Factors affecting the appearance of fish minces are listed in Table 6–13. Flesh yields can be increased by 20 percent by the use of mechanical separators on trimmed fish waste, but the resulting flesh can suffer from discoloration. The initial graying occurring immediately after deboning is caused by structural damage, release of black melanin pigments from the skin, presence of blood, kidney tissue, or the black lining of the belly cavity (King 1973; Dyer 1974). Water washing can improve the color (Connell and Hardy 1982), alkali or acid washing improves blood separation (Grantham 1981), and low-pH washing also reduces protein loss (Hall 1992). Whiteness can be increased by bleaching or by the addition of white light–scattering agents. Hydrogen peroxide can be used to bleach codfish mince (Grantham 1981). Chlorine in combination with MA packaging improves the whiteness of hybrid striped bass strips (Handumrongkul and Silva 1994). Ozonation results in higher whiteness and brightness in, for example, mackerel mince (King and Hung 1995) and horse mackerel mince (Chen et al. 1997). Light-scattering agents such as titanium dioxide, hydrophilic colloids such as milk, or polyphosphates (Grantham 1981) have been used to increase reflectance levels and hence act as whiteners. A suspension of titanium dioxide in xanthan gum used at a level of 1 g/kg^{-1} of titanium dioxide in cod mince, for example, yields a product stable to high-temperature cooking (Meacock et al. 1997). Yellow and brown discoloration of stored minced fish may be due to blood pigment oxidation (Jauregui and Baker 1980).

Fish oil color depends on species, but the pigments in all types can become yellow or brown through oxidation. While the whole fish or the fillet is deep frozen, oil can be forced out of the tissue to collect on the skin. These fish are "rusted." Packaging becomes noticeably discolored when the oil has oxidized. Similarly, body oils in dried fish have little protection. Autoxidation is accelerated by high temperatures, light, metals such as copper and iron, salt, and hemoproteins (Wolfe 1975).

BIVALVES

Around 66 AD, Pliny recorded the existence of red, brown, and black oysters, and since then, much attention has been paid to the appearance of bivalves. Oysters change their appearance to suit their habitat, but in general, European oysters are round and Pacific are long. The meat is a milky bisque color with dark fringes. Bivalves have been reviewed by Boon (1977).

There are at least four different causes of red or pink coloration in oysters. It may be leached from the eggs of the small oyster crab that lives in the mantle cavity of the oyster. Removal of the crab will prevent this discoloration, which is acceptable and normal. A coral pink color caused by the growth of an asporogenous yeast can develop while the bivalve is frozen. Its presence is indicative of poor sanitary conditions, and it has now been virtually eliminated. A pink discoloration can also be caused by the bacterium *Serratis marcescens*. Oysters can be rejected because of the formation of a red carotenoid contained in Gymnodinaceae phytoplankton. This is prevented by removal of the visceral material during processing, or by heating, which denatures the protein associated with the pigment.

There are two forms of green coloration in fresh bivalves. Green streaks and patches over the liver and viscera or the whole body are associated with relatively large concentrations of copper. Green mantles and gills are caused by ingestion of plankton pigments and are possibly chlorophyll degradation products. Other colorations occurring in fresh oysters include a purple color caused by ingestion of heavily pigmented seaweed spores, regional variation (for example, oysters from the South Atlantic are darker than those from further north), and blackening caused by environmental stress.

During frozen storage, the body of the oyster gradually darkens, and those parts exposed to headspace become yellow. Dark pigment from the mantle sloughs off, causing the drip to become sooty. Discoloration can be reduced by rapid freezing of the sealed cans. Antioxidants and monosodium glutamate have little or no effect, but washing in sodium mono- or dihydrogen phosphate, or in the antibiotic chlortetracycline, is more beneficial. Glazing or replacement of headspace gases has limited effect. Frozen fresh clams store well, but not if held on ice for a few days before freezing. Yellowing, caused by amino-carbonyl reactions, is prevented by antioxidants or vacuum packaging.

Iron and copper sulfides cause black discoloration in canned shellfish. Green and brown colors are caused by ingestion of degraded chlorophyll. Pigments such as copper phaeophorbide, chlorophyllin, and pheophytin have been identified. A yellow color caused by diffusion of carotenoids from the liver has also been observed (Boon 1977).

BROWNING

Two types of browning occur in foods. One is caused by enzyme activity, the other by Maillard and other chemical reactions. Browning can be desirable or undesirable, depending on the raw material and product concerned. Intensity depends on pH, temperature, water activity (a_w), and product composition. The reaction rates of the two types of browning with respect to a_w are compared with those of other food degradative mechanisms in Figure 11–16 (Labuza 1971). The reaction rate of enzymic activity increases with a_w, while nonenzymic browning commences at a lower a_w and is at a maximum at an a_w of approximately 0.7. Both types of browning can occur in a food material, although not necessarily simultaneously. For example, enzymic browning occurs in Chinese *mei* plum at the early stages of dry pickled

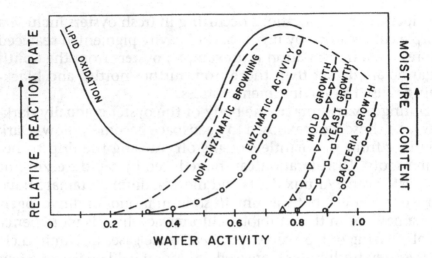

Figure 11–16 Food-degradative mechanisms as functions of water activity (Labuza 1971, reprinted with permission from *Crit Rev Food Tech*, copyright CRC Press Inc, Boca Raton, Florida).

processing, but at later stages Maillard browning reactions occur (Huang 1986). Behavior changes with cultivar. For example, in common with most fresh and minimally processed fruit and vegetables, the main browning mechanism in many *Dioscorea* species of yam is associated with polyphenoloxidase. However, in *D. rotunda*, nonenzymic browning is predominant (Omidiji and Okpuzor 1996).

Sulfur dioxide has traditionally been used to combat both types of browning. However, allergic reactions in some asthmatics have prompted much research effort into finding other effective browning inhibitor treatments.

In this chapter, enzymic and nonenzymic browning are discussed separately. Browning reaction kinetics and the use of reflectance measurements to separate natural pigment and browning changes in various foods are discussed in Chapter 8.

Enzymic Browning

Three factors restrict expansion of the fresh and minimally processed fruit and vegetable market—microbial spoilage, texture loss, and color deterioration. Careful crop and produce handling can significantly reduce loss from the first two causes. Use of refrigeration, antimicrobial and antibrowning agents, and proper packaging can reduce color degradation. Many commercially important food materials are subject to enzymic browning. Discoloration limits shelf life of fresh or minimally processed fruits, such as apples, bananas, grapes, and pears; vegetables, such as lettuce, mushrooms, and potatoes; and marine foods, such as shrimp. Bruising, slicing, and pulping of the fruit or vegetable increases enzyme activity, and when air is present, enzyme-activated oxidation occurs, producing brown and black melanin pigments. This can be a problem also in vegetables and fruits that are frozen or dehydrated. The rate of enzyme-initiated browning can rapidly increase when freezing is followed by thawing. This is caused by freezing-initiated cell damage. Careful, low-temperature washing before and after peeling or cutting helps to control microbial and enzyme action, but this is insufficient to prevent browning in, for example, apples and potatoes (Ahvenainen 1996).

Enzymic browning is not always undesirable; the quality of, for example, apple juice, dates, prunes, raisins, soy sauce, and black tea depends on it. The reactions also contribute to the quality of fermented drinks such as beer, cider, and perry. Hence, elimination of enzyme-initiated browning is necessary for some products, but control and manipulation are essential if others are to achieve their optimum desirability.

Initially, phenols are subject to enzyme oxidation, forming lightly colored quinones. These are subjected to further reactions, possibly catalyzed by enzymes, to form melanin pigments, the color of which depends upon the phenols and environmental conditions. The enzyme polyphenol oxidase (PPO) occurs widely in nature, and in the presence of atmospheric air it is responsible for most enzymic browning. The enzyme occurs in many forms and is responsible for catalyzing a number of reactions. The optimum activity of PPO occurs between pHs 5 and 7. Copper present in PPO plays an essential part in browning, but the extent can be reduced by chelation. Ascorbic acid and its derivatives temporarily control the participation of PPO in the browning reaction. Use of more stable isomers, such as erythorbic (isoascorbic) acid, has been suggested (Sapers and Hicks 1989). Action of PPO can be limited by heating and the presence of acids, sulfites, halides, reducing and chelating agents, and substrate-binding compounds. Effectiveness of PPO in, for example, potatoes can also be counteracted by genetic engineering (Thwaites 1995).

Browning reactions can be delayed or avoided by excluding oxygen from cut surfaces by immersion in water, syrup, or brine or by close wrapping or vacuum deoxygenation. Reduction of pH to below 3 inhibits enzymes. Lemon juice and citric acid are convenient additives, as they not only reduce pH but also chelate the copper present in some enzymes. Browning of lemon juice itself is caused by nonenzymic action (Fang and Ts'eng 1986).

Blanching inactivates enzymes, but heat treatment also softens the produce. Hence, fruit and vegetables that are eaten without cooking are not blanched before freezing. Fruit and vegetables destined to be cooked after a freeze/thaw process may be blanched. Blanching treatments prior to further processing are designed to inactivate enzymes at the earliest possible stage in the process. Discussions of blanching treatments required to prevent pigment destruction are included earlier in this chapter. Where protein functionality is important, as in cereals, heat treatment cannot be used for blanching.

Sulfite, in the form of sulfur dioxide or sodium bisulfite, has been widely used to keep fruit and vegetables looking fresh and to alleviate browning. A concentration of 10 ppm of sulfur dioxide is often sufficient. The sulfur dioxide reacts with quinones and prevents them becoming involved in browning reactions. However, if browning has already started, the sulfur dioxide is oxidized and prevented from acting as an inhibitor. High concentrations of salt (greater than 10 percent) can inhibit enzyme action, but this level is too high for most foods (Labuza and Schmidl 1986). Browning can be inhibited by using mixtures of ascorbic acid, citric acid, and sodium chloride in, for example, apples (Pizzocaro et al. 1993), apricots (Pizzocaro et al. 1995, cited in Forni et al. 1997), and osmodehydrofrozen intermediate-moisture apricots (Forni et al. 1997).

Extent of the browning depends on the cultivar as well as the species. Browning of fresh-cut slices of slightly underripe d'Anjou and Bartlett pears can be controlled by chemical dipping. Browning in Bosc pears cannot (Sapers and Miller 1998). Blanching immediately after maceration of pears is an effective means of juice and puree color control (Halim and Montgomery 1978). However, if browning has already started before the blanching, a browning inhibitor must be added to stabilize the product during heat treatment (Montgomery and Petropakis 1980). Ascorbic acid temporarily inhibits PPO, but quality is further improved by boiling the juice (Petropakis and Montgomery 1984).

Color and phenol content of juices from dessert apples are cultivar sensitive. Also, oxidation in the mash produces a lighter colored juice than when it takes place in the juice. However, careful control of mash oxidation can result in a closely controlled final color (Cumming et al. 1986). In clarified apple juice, a typical amber hue is desirable, and little browning is acceptable. Pulp and puree are expected to possess a yellowish or greenish color. Pulp made with unripe apples browns at a faster rate than that from ripe apples; this is related to differences in ascorbic acid and PPO activity in young fruits (Lozano et al. 1994). Malto-dextrin, particularly in the presence of potassium ions, inhibits browning in freshly ground apples (Xu et al. 1993).

Blanching of juices involves temperatures above 70 to 75°C. The time taken for complete heat inactivation of PPO in fruit juices varies widely, but at a temperature of 75°C, less than 1 minute is required for clear peach and apple juices, while more than 100 minutes is necessary for cloudy apple and plum (Labuza and Schmidl 1986). Pineapple juice effectively retards browning in fresh and dried apples (Lozano-de-Gonzalez et al. 1993).

Storage of intermediate-moisture apples at an a_w near 0.7 should be avoided because of enhanced browning activity. Storage at slightly below a_w of 0.6 is preferred for color and texture (Beveridge and Weintraub 1995). Combined mild pretreatments rather than single treatments are advocated for browning prevention of apple products to be used as components in food formulations such as ice cream (Mastrocola et al. 1996).

Browning of grape juice during processing is undesirable. Investigations involving five varieties over three years have revealed the extreme complexity of PPO development from the green stage to the mature grape. Levels of activity depend on variety, year, and climate (Sapis et al. 1983). Browning of sultana grapes is caused by PPO located mainly in the skin and by nonenzymic browning reactions (Ramshaw and Hardy 1969). Darkness of raisins produced by drying grapes is related to residual PPO activity. Light-colored raisins are traditionally produced by sulfur dioxide treatment, which prevents both types of browning. The amount of sulfur dioxide required can be reduced by increasing water diffusion rates and decreasing drying time. This can be achieved by an oil-surfactant emulsion that removes the waxy external layer and diffuses into the grape (Aguilera et al. 1987). Even lighter colored raisins can be produced using honey in sulfur dioxide–free dips (McLellan et al. 1995).

Combination treatments make it possible to hold peeled and sliced fresh potatoes in refrigerated store without browning. Prepared potatoes are dipped in a bath containing ascorbic acid to act as an oxygen scavenger and reduce the PPO initiated reactions; citric acid to lower the pH, chelate the copper, and slow the autoxidation of the ascorbic acid; and potassium sorbate to control yeast and mold growth. This is followed by draining, and vacuum packing in oxygen-impermeable bags. The product must be used straight from the bag, as the slices start browning within 10 to 30 minutes of opening (Langdon 1987). Dipping achieves a limited penetration of these chemicals compared with that of sulfite (Taylor et al. 1986), but some improvment has been gained in potato and apple plugs by pressure infiltration (Sapers et al. 1990). Dipping followed by MA packaging is a suitable combination method for fresh sliced potatoes (Sapers and Miller 1995).

The color of black tea depends on the initiation and control of enzyme action. Two conventional technologies are used in manufacture, and both involve damaging the leaf to release cell components. In the orthodox method, a rolling machine is used to twist the leaves and damage cellular membranes without greatly affecting leaf conformation. The crush-tear-curl method, on the other hand, involves shredding the leaf epidermis from the stalk. This results in the release of significant amounts of stalk cell contents. A number of reactions occur. First, acids and degradative enzymes convert the green chlorophyll pigments to yellow/gray/brown pheophytin and pheophorbide. Second, PPO catalyzes the oxidation of tea flavonols to produce yellowish-red pigments. Third, these reactions continue during the fermentation and water evaporation stages when the color changes from coppery red to brown. Conversion of the tea catechins into theaflavins and thearubigins during these second and third stages significantly contributes to the color and flavor (Mahanta and Hazarika 1985). The leaves are fired at the end of the process to terminate PPO activity. Growing conditions affect chemical makeup. For example, black teas with higher theaflavin and reduced thearubigin concentrations can be produced from bushes grown under artificial shade (Owuor and Othieno 1988). The reddish shoots appearing during the second flush and the conspicuous coloration of autumn leaves are associated with higher quality teas (Mahanta and Baruah 1992). Black teas are blended before retailing to achieve consistency in color and flavor. Reasons for this include genetic variations occurring in tea plants, the range and complexity of the reactions involved, the pigments (and flavors) formed during manufacture, and consumer sensitivity to brewed tea color (and flavor).

In whole or sliced mushrooms, browning is more intense in first-break products than in second, and in unwashed mushroom compared to washed. Colors produced on external and cut surfaces may be brown, gray, black, yellow, or purple. Purple colors are associated with bacterial lesions. Combination treatments involving dipping in 5 percent hydrogen peroxide followed by treatment with sulfite substitutes and calcium salts at pH 5.5 significantly extend shelf life (Sapers et al. 1994).

Enzymic browning has been reviewed by Lee and Whitaker (1995), Sapers (1993), McEvily and Iyengar (1992), Labuza and Schmidl (1986), and Palmer (1984); browning in apple products by Nicolas et al. (1994); grapes and wines by Macheix et al. (1991); and citrus products by Handwerk and Coleman (1988).

Nonenzymic Browning

The major nonenzymic reactions involved in food preparation and storage are, first, the Maillard carbohydrate/amino acid reactions, which cause undesirable degradation in, for example, fruit juices, condensed milk, and dried products stored at too high a temperature; second, caramelization, which involves thermal degradation of sugars without amine participation; and third, the effects of lipid oxidative decomposition products on proteins (Dworschak 1980). Most chemical changes that occur during caramelization also occur in Maillard browning. Many reactions that take place in pure sugars only at very high temperatures occur at low temperatures once they have reacted with amino acids (Mauron 1981).

Maillard browning can be found in at least three different areas of food manufacture. It has a traditional use in the development of aromas and flavors in roasting, baking, cooking, and smoking; it is used deliberately to engineer flavors in nontraditional foods; and it occurs as an undesirable byproduct of food processing, affecting color or flavor or both (Buckholz 1988).

In many processes, such as the cooking of meat or crust formation on bread, browning reactions are regarded as a benefit not only for the color development but for the aroma and flavor compounds also formed. Beef undergoes a combination of changes during roasting, and a wide color range is tolerated. All samples plotted on the roast beef chromaticity locus shown in Figure 11–15 were prepared in the home and judged **acceptable** by the cooks.

Degradative nonenzymic browning occurs on storage of a wide range of products: for example, cheese powder (Kilic et al. 1997), dried hazelnuts (Lopez et al. 1997), condensed milks (Patel et al. 1996), white and rosé wines (Gómez et al. 1995), plum pastes (Wang et al. 1995), orange juice (Johnson et al. 1995; Rassis and Saguy 1995; Maccarone et al. 1996), grapefruit juice (Cohen et al. 1994), raisins (Canellas et al. 1993), sugar solutions

(Imming et al. 1994), dried peppers (Lee et al. 1991), skim milk (Franzen et al. 1990), citrus fruit products (Handwerk and Coleman 1988), and carrots (Baloch et al. 1973).

Major factors controlling development of browning and off flavors are the ratio of reactive carbonyls to available amine groups, the pH of the system, the amount of dicarbonyls present due to prior processing, metal catalysis based on redox-couples, the amount of water or other solvent phase available (e.g., liquid fats or liquid glycols), the viscosity of the reaction phase, and the temperature of processing and storage (Labuza and Schmidl 1986).

Maillard reactions include those involving reducing sugars, aldehydes, and ketones with amines, amino acids, peptides, and proteins. In foods, the normal reactants are reducing sugars and amino acids. Reactions can be divided into three phases. The early phase consists of defined chemical reactions without browning; the second phase consists of many reactions involving the formation of volatile or soluble substances; and the final phase consists of reactions leading to the production of insoluble brown polymers (Mauron 1981). Alkaline treatment used in the production of proteins produced from new sources may also contribute to the formation of browning products (Dworschak 1980). The brown pigments themselves retard further lipid oxidation (Dworschak 1980). For example, Maillard reaction products have been shown to be effective antioxidants in model systems (Lingnert and Eriksson 1980a, 1980b), cookies (Lingnert 1980), sausages (Lingnert and Lundgren 1980), vegetable oils (Hodge and Rist 1953), and meat patties (Bedinghaus and Ockerman 1995).

Reviews and summaries of the extensive literature concerning browning reactions include those by Rizzi (1997), Yaylayan (1997), Sapers (1993), Labuza and Schmidl (1986), and Hodge (1953). Citrus browning has been reviewed by Handwerk and Coleman (1988), caramelization by Theander (1981) and Heyns and Klier (1968), and the color of smoked foods by Ruiter (1979).

Discussion can be continued conveniently under the headings of substrate and environment, including water activity, pH, and temperature.

The Substrate

Predicting the extent and rate of browning is not possible in real, or even model, food systems because of the varied nature of the process, the many factors contributing to it, and its sensitivity to the environment (Labuza and Schmidl 1986).

Reducing sugars contribute to browning reactions. Of these sugars, pentoses, such as ribose, react faster than hexoses, such as glucose, which in turn react faster than disaccharide reducing sugars, such as maltose and dextrose (Ames 1995). Although sucrose has no active carbonyl groups, it will form reducing sugars by acid-catalyzed hydrolysis even at moisture contents of less than 1 percent. It is the presence of high ribose levels that leads to browning and the rejection of many canned, dried, and frozen fish products. This is indicative of poor storage conditions prior to processing (Labuza and Schmidl 1986).

Different sugars behave differently under different conditions. Not only are the sugar/amino contents important in browning, but caramelization of the sugars can be significant even at shelf storage temperatures. Lowering the pH reduces caramelization (Buera, Chirife, Resnik, and Lozano 1987). From studies involving sugar (fructose, xylose, glucose, lactose, maltose, and sucrose) and glycine solutions, it was concluded that caramelization contributed noticeably to total color development only in the fructose/glycine system (Buera, Chirife, Resnik, and Wetzler 1987).

Sulfiting agents inhibit nonenzymic browning. In citrus products, for example, they react with carbonyl intermediates, preventing their further reaction, and also bleach the final color (Wedzicha 1987). The effect of adding sodium bisulfite to banana puree can be depicted in the form of a response surface, as in Figure 11–17 (Alzamora et al. 1995). This shows the effect of storage temperature and sodium bisulfite concentration after one month of storage.

Cysteine, calcium, disodium dihydrogen phosphate, and stable nonreducing sugars such as trehalose are effective in reducing browning rate in specific products. Ascorbic acid, citric acid, and glucose oxidase are among the additives used to prevent browning of white wines (Gómez et al. 1995). Ascorbic acid can

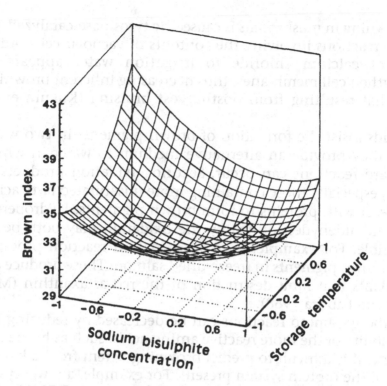

Figure 11–17 Response surface showing the effect of storage temperature and sodium bisulfite concentration on banana puree after one month of storage. *Source*: Reprinted with permission from S.M. Alzamora, P. Cerrutti, S. Guerrero, and Lopez-Malo, Minimally Processed Fruits by Combined Methods, in *Food Preservation by Moisture Control*, G.V. Barbosa-Canovas and J. Welti-Chanes, eds., p. 482, © 1995, Technomic Publishing, Co.

be a highly reactive constituent with respect to the formation of browning pigments in orange juice products. Its effect on the browning reaction is more intense in the presence of oxygen. Browning increases linearly with the concentration of amino acids incorporated but is more severe in the presence of high concentrations of ascorbic acid (Kacem et al. 1987). Calcium offers some protection to dried apples, as do oxygen scavengers in packing (Bolin and Steele 1987). It is likely that the type of packaging affects browning rate through its control of the amount of oxygen available for the reaction (Kacem et al. 1987).

Browning in mushrooms is caused by tyrosinase-catalyzed enzymic reactions involving the contents of vacuolar cells. Addition of calcium chloride to irrigation water appears to strengthen cell membranes, thus decreasing inherent browning and that resulting from postharvest bruising (Kukura et al. 1998).

Lipids assist the formation of brown pigments in two ways. First, they provide an alternative medium to water in which Maillard reactions can occur. Second, oxidation products of lipids, especially those containing more unsaturated fatty acids, can react with proteins to form brown pigments. Hydroperoxides and aldehydes formed by the oxidation may both be responsible. For example, lipid autoxidation reaction products form brown pigments in freeze-dried salmon. These produce antioxidants that limit destruction of the red astaxanthin (Martinez and Labuza 1968).

Carbonyl/amino reactions can be decreased by reducing the availability of the more reactive amino acids, such as lysine and glycine. It is difficult to predict browning extent from a knowledge of the protein system present. For example, a lower lysine content does not automatically lead to less browning, but browning can be increased by the addition of free lysine (Schnickels et al. 1976). In potato processing, glutamine contributes to browning (Rodriguez-Saona et al. 1997), and asparagine contributes to a decrease in gray color intensity in model potato systems (Khanbari and Thompson 1993). In another model system, it was found the rate of pigment production was controlled by the ratio of available reducing compound to amine. In an intermediate-moisture system, the rate of pigment formation increased linearly as the ratio of glucose to lysine was increased from 0.5 to 3. Above this, the rate did not change (Warmbier et al. 1976b).

Browning tends to occur more readily in finely ground material. For example, browning rate in dried red peppers, which contain appreciable amounts of reducing sugars and amino acids, was greater in powders from which seed had been excluded. It was assumed that active browning ingredients were in lower concentrations in the seeds. For minimum browning, storage should be at an a_w below 0.3 and in the form of whole pods or coarse powder with seeds. MA packaging with nitrogen flushing resulted in no improvement (Lee et al. 1991).

The Environment

The Effect of Water Activity. Nonenzymic browning is enhanced at intermediate values of a_w. At lower a_w, near to monolayer moisture values, solute mobility is limited, and reactions are slower because reactive groups are slower to diffuse. At higher a_w, where microbiology problems may be increased, dilution decreases reaction rates (Labuza and Schmidl 1986). This is illustrated in Figure 11–16, which compares the behavior of both types of browning. This curve shape occurs for sultana grape drying (Aguilera et al. 1987). Similarly, maximum nonenzymic browning of apple juice is governed by juice concentration and occurs at an a_w between 0.53 and 0.55 for storage times up to 100 days, as shown in Figure 11–18.

The maximum rate of browning of potatoes increases with increase in temperature of drying. Hendel's data (Hendel et al. 1955) are shown in Figure 11–19 (Mishkin et al. 1983). These data indicate that temperature should be reduced in the later stages of drying. This is especially so when browning rate is at a maximum at a moisture content between 0.1 and 0.2 g/g of solids. Alternatively, a low temperature can be used throughout the drying. These data have been used to compute a dynamic optimization model that aims to minimize browning during processing (Mishkin et al. 1983).

Common ingredients moderate reactivity at low values of a_w. Glycols can create a large phase volume, thus diluting and slowing the reaction, while sorbitol acts by increasing viscosity (Eichner and Karel 1972; Warmbier et al. 1976a). On the other hand, liquid fats can increase browning rates by dissolving the reactants. The lipid acts as a mobile phase, reducing the activation energy for the reaction (Kamman and Labuza 1985).

At lower values of a_w where reaction rates are limited by diffusion, diffusion rates are related to glass transition temperature T_g. A food matrix may exist as a viscous glass or a more liquid-like rubber, and T_g is the temperature at which the nature of a

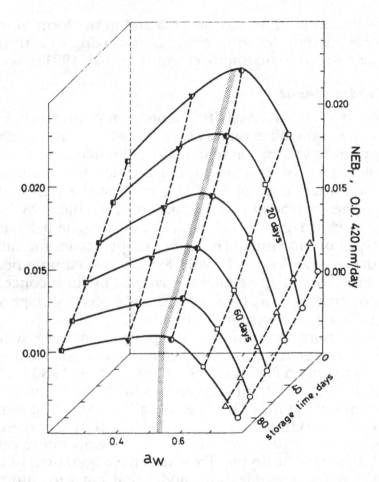

Figure 11–18 Nonenzymatic browning rate (NEB$_r$) of apple juice as a function of w_a and time of storage. °Brix 65, 70, 75, 81, 84.5, 90.5. Lines represent a polynomial equation fit. Shaded area represents NEB$_r$ maximum range (Toribio et al. 1984, with permission).

material changes from glass to rubber. The degree of browning could well be less if the product was stored below its T_g. The Williams-Landel-Ferry equations can be used to relate browning rate constants to temperature, moisture, and T_g (Buera and Karel 1993).

The Effect of Temperature. At temperatures up to 60°C, browning is normally a zero-order reaction. At higher temperatures,

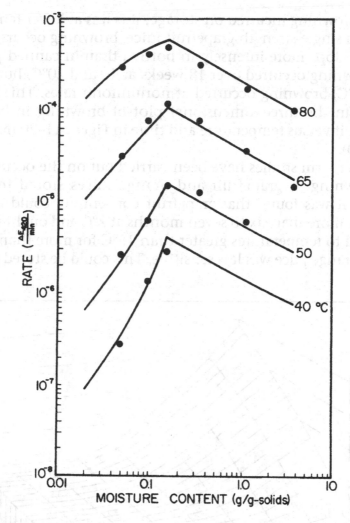

Figure 11–19 Rate of browning of potatoes as a function of moisture content and temperature (Hendel et al. 1955 via Mishkin 1983, with permission).

changes may follow a first-order reaction (Silva and Igniatidis 1996). Some materials, such as grapefruit juice, may exhibit an initial lag phase (Saguy et al. 1978). Browning reaction kinetics are discussed in more detail in Chapter 8.

Citrus juice products are affected by browning during storage. Although keeping lemon juice in the dark makes no difference to the browning incurred on storage, it is less at lower temperatures. In single-strength grapefruit juice, browning occurs faster and develops more intensely in bottled than in canned juices. No browning occurred over 18 weeks at 10 and 20°C, but at 40 and 50°C, browning occurred at nonuniform rates. This is displayed in the three-dimensional plot of browning in bottled grapefruit versus temperature and time in Figure 11–20 (Nagy et al. 1990).

Longer term studies have been carried out on the occurrence of browning in grapefruit and orange juices stored in steel drums. It was found that grapefruit concentrate could not be held for more than six to seven months at 2°C and ought not be exposed to temperatures greater than 15°C for more than a few days. Orange juice was less sensitive. This could be stored at 2°C

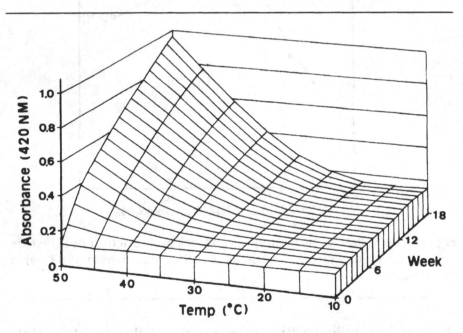

Figure 11–20 Three-dimensional plot of nonenzymic browning of bottled grapefruit juice versus storage temperature and time. *Source*: Reprinted with permission from Nagy, Lee, Rouseff, and Lin, Nonenzymic Browning of Commercially Canned and Bottled Grapefruit Juice, *Journal of Agricultural and Food Chemistry*, Vol. 38, pp. 343–346. Copyright 1990, American Chemical Society.

for 18 months with little effect, but it was suggested that ship-
ment ought not take place during hot summer temperatures
(Berk and Mannheim 1986). Maintaining citrus products at low
temperature is the only means of avoiding color and flavor deterioration through browning (Handwerk and Coleman 1988).

Thermal history is a critical factor in obtaining a lighter colored apple juice concentrate (Toribio and Lozano 1984). Variability of browning rates reported in the literature may be
caused by variety, harvest condition, and agronomic practice
(Beveridge and Harrison 1984). For example, there are cultivar
differences in apples: browning is faster in Red Delicious than in
Granny Smiths.

The Effect of pH. A typical dependance of browning rate on pH
is shown in Figure 11–21. As pH is increased, sugar fragmenta-

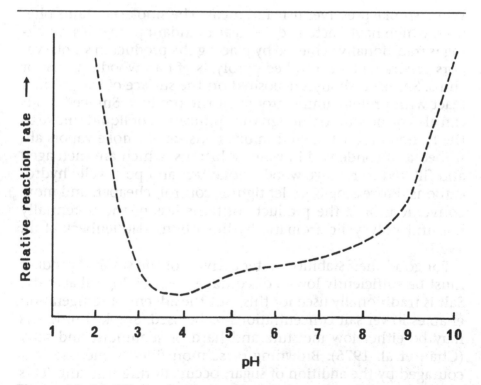

Figure 11–21 Effect of pH on the reaction rate of nonenzymatic
browning (Labuza and Schmidl 1986, with permission).

tion increases, small carbonyl compounds formed react with amino acids, ammonia, and reactive sulfur compounds, and browning increases. Carbonyl formation can be minimized by keeping the system at a pH lower than the pK_a value. This ensures that the amines are in the less reactive $-NH_3^+$ state occurring typically between pH 9.0 and 10.5. At much lower pHs, the reaction rate again increases because more of the reacting NH_2 groups are formed from $-NH_3^+$.

In processing, extremes of pH are normally avoided, although low-pH browning is used to produce pretzels and tortillas. Caramel syrups are made at high pH (Labuza and Schmidl 1986).

OTHER PRACTICAL FOOD SYSTEMS

Smoking

Wood smoke preserves fish and meat. The smoke contains phenols, which have bactericidal and antioxidant properties. Smoking is traditionally achieved by placing the product in smoke vapors generated by controlled pyrolysis of hardwood sawdust or chips. Smoke carbonyls deposited on the surface of the product react with protein amino groups in the product. Smoked foods can also be produced through the application of liquid smoke to the heated product. Liquid smoke consists of smoke vapors absorbed and condensed in water solutions, which are then aged and filtered to remove wood smoke tars and polycyclic hydrocarbons. Processing is under tighter control, cheaper, and more convenient, and the product contains less of the potentially harmful polycyclic aromatic hydrocarbons (Hollenbeck et al. 1973).

For good shelf stability, water activity of the smoked product must be sufficiently low to discourage microbiological activity. Salt is traditionally used for this, but the advent of refrigeration enables lower salt concentrations to be used. Smoked products may be either low moisture and hard or semimoist and salty (Chan et al. 1975). Browning of salmon fillet, sometimes encouraged by the addition of sugar, occurs during smoking. This leads to smoked salmon slices that are partly pink and partly brown. Whereas a great increase in light scattering occurs on cooking and canning, less change occurs on smoking. The appearance of a slice of good-quality smoked salmon is due to a delicate interplay between the natural pink pigment, the imposed browning, and the translucent nature of the flesh. The dull brown hot-smoked, or flaky, smoked salmon is cooked as it is smoked. This increases light scattering, and the translucent effect is lost.

Color of smoked cheese is also caused by smoking reactions and by the concentration of color compounds at the surface through surface dehydration. Concentration of smoke solution and contact time influence final color, but the time of subsequent heat processing is critical to the amount and hue of the color formed. Final product color is also sensitive to wood from which the smoke is obtained and to the cheese type (Riha and Wendorff 1993).

Frying

The quality of potato chips, crisps, and french fries depends to a large extent on their color. Successful manufacture requires optimum levels of reducing sugars to be present to prevent excessive darkening. Significant differences in fructose, glucose, and sucrose contents of potatoes occur according to variety, growing region, and harvest date (Leszkowiat et al. 1986). For example, potatoes grown in drier soil tend to be associated with progressively darker fry at harvest and after harvest (Eldredge et al. 1996). Russet Burbank tubers possess a significantly higher glucose-forming potential and produce darker colored chips than Norchip and Gemchip tubers, regardless of storage temperature and time (Herrman et al. 1996). Cultivars show considerable variation in the relationship between reducing sugar levels and color development. Reducing sugars do not completely explain color development in cultivars with low (<60 mg/100-g tuber) reducing sugar content (Rodriguez-Saona and Wrolstad 1997). Concentration of reducing sugars can be lowered by leaching or by allowing enzyme action to reduce them under controlled storage conditions (Labuza and Schmidl

1986). Nonuniform browning or mottling of french fries is caused by uneven distribution of sugars across the fry surface (Jankowski et al. 1997).

Addition of glucose and amino acids to potato slices by vacuum infiltration leads to increased darkening after frying. However, when filter paper disc model systems were used, sugar was the limiting factor in browning (Leszkowiat et al. 1990). Participation of amino acids in color development is marginal (Marquez and Añon 1986), and it is unlikely that determination of the amino acid content of potatoes will improve fry color predictability (Roe and Faulks 1991). Glutamine and asparagine are the major free amino acids in potato tubers. Glutamine concentration has a small effect on potato chip chroma (Rodriguez-Saona et al. 1997). Asparagine has been linked with a decrease in gray color intensity in model systems (Khanbari and Thompson 1993). Phenolic compounds can undergo nonenzymic oxidation to produce quinones that may react with amino acids to generate brown pigments (Cilliers and Singleton 1989). But chlorogenic acid, a major phenolic in tubers, has no effect on color development in model systems (Rodriguez-Saona et al. 1997). Threshold levels of amino acids and sugars for color formation in potato chips and effects of amino acids on color have been evaluated using filter paper disc technique (Chonhenchob et al. 1996). Starch structural stability factors such as crystallinity, amylose content, and starch granule size could not be used to predict or monitor chip color in stored potato tubers (O'Donoghue et al. 1996).

Extrusion Cooking

Extrusion cooking is traditionally used for processing intermediate–water activity cereals into ready-to-eat foods and expanded snack products. Materials of typical moisture content (12–18 percent) are subjected to high temperature, pressure, and shear stress (Harper 1989). Barrel temperature and screw feed affect the color of extruded products. Lightness (L^*) values and yellowness (b^*) of extrudates show marked changes and are most dependent on barrel temperature (Bhattacharya et al.

1997). During the process, there is a reduction of total sugar content (Andersson et al. 1981) and lysine (Berset 1989). These losses are due at least in part to nonenzymic browning reactions. No browning occurs when starch alone is extrusion cooked. Color development of starch-glucose-lysine mixtures depends on moisture contents and processing conditions. Using a model system, in general, color development increased with increases in die temperature, glucose, and lysine concentrations and a decrease in moisture content (Sgaramella and Ames 1993). The effect of extrusion cooking on carotene content is discussed above under "Carotenoid-Dominated Systems." To study such phenomena in the laboratory, it may be possible to use a reaction cell (Bates et al. 1994).

Also affecting the final product is pH (Figure 11–22). When a wheat flour product was used, as pH was increased from 4 to 7,

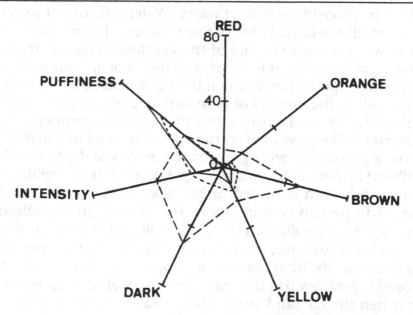

Figure 11–22 Effect of pH on appearance properties of an extrusion-cooked wheat product. *Source:* Reprinted with permission from J.M. Ames, Applications of the Maillard Reaction in the Food Industry, *Confection*, Vol. 2, No. 11, pp. 40–43, © 1995, Confection 2000.

products became more brown, red, orange, yellow, and darker, while puffiness of appearance decreased.

LIGHT SCATTERING

Light absorption and light scattering are the two physics phenomena contributing to the appearance of a material. So far, this chapter has considered mainly the light absorption behavior of food materials in terms of their natural pigments and pigment development. In most food materials, light scattering as well as absorption occurs. In combination, these give rise to basic perceptions of color and translucency. Many food processes cause changes in translucency or opacity. Cooking of starch grains causes swelling as water is absorbed, and a less opaque appearance results. Cooking of proteins leads to precipitation, and a more opaque appearance results. The color of green vegetables—for example, peas—immediately changes when the vegetables are placed into boiling water. Water replaces air in cells just beneath the outer layers of the structure. The refractive index of water is nearer to that of the vegetables than is air. As a result, light scattering is reduced, and the color appears a deeper green. The opposite occurs when the particulate size of powders is reduced. As the number of reflecting boundaries is increased, light scattering also increases, and the powder becomes paler in appearance. The physics of scattering is discussed in Chapter 3.

The degree of scattering governs the perceived depth of color exhibited by flesh foods—for example, as a result of cooking. As meat or fish flesh is heated, it becomes more opaque due to changes to protein conformation. Use of poor-quality salmon can lead to the production of smoked slices that have poor appearance. For example, poor storage conditions of the raw fish can increase its light-scattering properties. Slices from the smoked side of raw fish that have been temperature cycled during frozen storage can have a much greater light-scattering coefficient. That is, the light path within such a slice is less because the internal transmittance is decreased. This in turn leads to a decrease in light absorption; the pink tinge is no longer visible, and a slightly muddy, very undesirable appearance results. The look of the whole smoked side is not affected because the increase in light scattering within the flesh cannot be seen behind the dark brown appearance of the cured surface. The appearance of a smoked salmon slice is due to the interaction of natural pink pigment, imposed browning, and the translucent nature of the flesh. Hot-smoked salmon is cooked as it is smoked. Hence, light scattering is increased, and the translucent effect is lost. Similarly, discoloration in cut surfaces of fish is probably caused by pH precipitation of sarcoplasmic proteins, which causes an increase in light scattering (Church 1998a).

Fresh meat, particularly pork, can be unduly pale in color. Under certain pH conditions, sarcoplasmic proteins are released and precipitated; this causes an increase in light scattering. The depth to which light can penetrate is thereby reduced, so there is less opportunity for it to be absorbed. Hence, the effect of the pigment present in the muscle is reduced, and the meat looks paler. This is discussed previously in this chapter under "Meat Color" and with respect to measurement. Similarly, freezing rate affects the perceived color of foods: faster freezing leads to smaller ice crystals and paler beef.

An opalescent or iridescent sheen sometimes appears when meat is cut transversely to fiber axes. This can occur with raw, cooked, fresh, or cured meats and, when viewed from a particular direction by the naked eye, is green or less commonly red. This is an interference phenomenon believed to arise from reflections from layers within individual meat fibers (Swatland 1984).

The color of dilute skimmed milk is different in transmission than in reflection. Light from the blue end of the spectrum is preferentially scattered by casein micelles in the milk. Hence, the transmission color of the milk is red brown, and the reflected color is blue (see Chapter 3). This bluish reflected color can be changed to a more acceptable whiter color by addition of light-scattering material, such as titanium dioxide, nonfat dry milk, or milk-coagulating enzymes. The enzymes partially aggregate casein micelles, thus increasing light scatter and whiteness. Such milk is perceived as being creamier. Further enzyme treatment results in the opaqueness of cheese. Coffee can be whitened using fortified calcium carbonate whiteners or high-calcium milk proteins (Pszczola 1997). Effectiveness of whiteners depends in part on emulsion stability. This can be determined by measuring the rate of change of turbidity or light scattering with time:

$$\text{Emulsion Stability Index} = T \, \Delta t / \Delta T$$

where T is the turbidity and ΔT and Δt are the corresponding changes in turbidity and time, respectively (Parker 1987). Titanium dioxide is used to opacify and whiten confectionery, panned goods, icings, and low-fat products. Apparent creaminess and strength of coffee are related to translucency (see Chapter 9).

Tea brewed in hot water develops an unsightly surface film or scum, which increases scattering. This scum consists of calcium carbonate and an organic matrix insoluble in all solvents except highly concentrated alkali. The scum forms in water containing both calcium (or magnesium) and bicarbonate ions. It is greatly increased by the addition of milk, but no scum forms in lemon tea because of the low pH and calcium complexing (Spiro and Jaganyl 1993).

Most wines and beers must be transparent, exhibiting no signs of light scattering. Protein and pectin degradation can be controlled using appropriate enzymes (Stewart and Amerine 1982). However, these products possess subliminal turbidity that affects transmission measurements (see Chapter 9).

Turbidity in apple juice can be a positive or a negative attribute depending on the expectation of the consumer. Light scattering causing haze in juices may be biological or chemical in origin. Biological haze is associated with the growth of microorganisms. Chemical haze occurs as a result of physicochemical instability and the coagulation of particles. Opalescent apple juice has a stable cloud of soluble as well as insoluble pectin-stabilized particles. This is produced using ascorbic acid to prevent enzymic oxidation and browning, as well as rack and cloth pressing and rapid flash pasteurization to reduce enzyme activity (Beveridge 1997). A stable cloud in apple juice occurs when almost all particles have a diameter of 1 mm (Genovese et al. 1997). Some elements of the fruit can destabilize cloud. For example, addition of citrus tissue causes color instability in orange juice (Cameron et al. 1997). Great effort is put into the production of, for example, apple juice (Gökmen et al. 1998; Siebert and Lynn 1997), kiwi fruit juice (Dawes et al. 1994), plum juice (Chang et al. 1994), carrot juice (Sims et al. 1993), and blackberry juice and wine (Rommel et al. 1992). These processes generally involve heat, enzyme treatment, clarification, and filtration. This topic is considered further especially in relation to translucency and haze measurement in Chapter 9.

Light scattering also takes place from the surfaces of materials, giving rise to the basic perception of gloss. Particular gloss characteristics are associated with different fruits and vegetables, with scattering being more diffuse from rough surfaces, such as the skin of oranges, than from smooth, such as aubergine (eggplant) and unripe tomatoes (Nussinovitch et al. 1996) and bananas (Ward and Nussinovitch 1996). High-quality chocolate normally has a high gloss, and light scattered from the surface is specularly scattered. When chocolate blooms, this regular, near-mirrorlike scattering changes to diffuse scattering, and the surface becomes dull. The three types of bloom (lack of gloss) can be induced by melting and recrystallization or by storage at below–melting-point temperatures and/or by migration of non–cocoa butter fats. These treatments result in different polymorphic forms of cocoa butter from which the chocolate is made (Talbot 1995).

ADDED COLOR

Historical aspects of added color are outlined in Chapter 1. Of the many colors once available for inclusion in food, only a small proportion are now permitted. Each country has its own statutorily regulated list, and even countries within the European Economic Community (EEC) have exceptions.

Within the EEC, the directive *Colours for Use in Foodstuffs* came into effect in 1996. With two exceptions, any substance adding to or restoring color to food is regarded as a food coloring. The first exception relates to foods and flavors that are in themselves colored—for example, tomato, spinach, turmeric, and paprika—

and that are used in compound foods primarily for their flavoring or nutritional properties—that is, where their coloring effect is secondary. The second exception defines those colors used in the inedible external parts of food, such as the rind. Apart from these exceptions, only those colors listed as permitted can be used according to specific limitations specified.

The directive lists 43 permitted colors, 10 of which may be used only in particular foods. Erythrosine can be used only in candied cherries, Brown FK in kippers, and Red 2G in some sausages and burgers. Some categories of foods, including unprocessed foods, oils and fats, milks, some jams and breads, chocolate, and wine vinegar, may not contain added color. Some categories, including beer, some breads, and vinegar, may contain only certain colors within stated limits. Other foods, including butter, breakfast cereals, luncheon meat, margarine, and sausages, may contain only certain colors.

Interpretation of the directive is not straightforward, and it inevitably takes time for best practice within the directive to become established for any particular ingredient or within a particular product or product area. Among initial problems, for example, is that of differentiating primary and secondary use. For example, black currant juice is normally used as a flavor and nutritional additive, so its coloring power may be regarded as incidental. However, beetroot juice is likely to be added solely for its coloring power and hence falls within the regulations. There is doubt over the continued permission for the use of malt extracts now that they are classed among those foods that may not contain added colors.

Limits are based on the proportion of coloring in the color preparation used and also in relation to the ready-to-eat food prepared according to the instructions for use. That is, limits will refer directly to, for example, jam but not to a table jelly tablet. Then, the limits will apply to the jelly made from the product. However, this will not apply to coffee whitener. The limit will refer to the product itself—that is, the whitener—and not to the coffee that is whitened.

Correct interpretation of the regulations is the responsibility and a matter of due diligence for the food product manufacturer. Since the legislation itself and its interpretation are fluid, manufacturers are recommended to verify the interpretation placed on the additive by the supplier of the coloring (Turner 1995). Differing interpretations of regulations applying within different geographical regions make the publication of definitive, comparative lists difficult and possibly dangerous for the user (Meggos 1995). For these reasons, current regulations have not been included in this chapter.

Points to be considered before using natural (or other) colorants in a formulation include

- the nature of the possible colors to be used
- the nature of the food system, including the presence of protein, a_w, sulfur dioxide, and preservatives
- the phase to be colored, water or oil, including phase boundary problems and color migration
- the manufacturing process, including time, temperature, methods and order of mixing, oxygen exposure, and preservatives
- the package, including colorants of animal origin
- the storage conditions of the packaged product, including exposure to light, storage temperatures, and times (Freund et al. 1988; Knewstubb and Rayner 1991)

Reviews have been written on colorings in North America by Frick and Huck (1997), from the international perspective by Meggos (1995), on encapsulation of pigments by Popplewell et al. (1995), on hard candies by Chin and Frick (1995), and on current European legislation by Turner (1995). Comprehensive studies of synthetic and natural colors also can be found (Burdock 1996; Leatherhead Food Research Association 1997).

Natural Colorings

With regard to a particular food, *natural* colorings may be defined as color additives occurring naturally in the food. There is an argument that any added color, synthetic or natural, results in an artificially colored food (Pszczola 1998). However, in this chapter, *natural* colors are defined as being derived from natural sources. Included also are *nature-identical* colorants. The treatment is general; more detailed accounts include those by Meggos (1995), Collins and Hughes (1992), Freund et al. (1988), Kantha and Erdman (1987), and, in particular, Francis (1987a, 1987b, 1999).

The following natural colors are listed in terms of the major pigment types, finishing with a list of possible future commercial sources.

Betalaines

The two types of water-soluble betalaines extracted from beetroot are the strongly colored, purple-red betacyanins and the yellow betaxanthins or vulgaxanthins. Betanin is the major pigment, composing up to 95 percent of the total betacyanin content of beet (von Elbe 1975). The red to purple color of the vegetable beet appears to be controlled by the blend of the two pigment types (Timberlake and Henry 1986). A pigment apparently identical to betanin can be isolated from the pokeberry. This phytolaccanin pigment has been found to be more stable in some model food systems (Driver 1976).

Peroxidase causes discoloration of beet pigments (Wasserman et al. 1984), with endogenous polyphenoloxidase and peroxidase being responsible for the *black ring* defect in canned beetroot. This discoloration arises during time delays after treatment with live steam, which facilitates peeling prior to canning (Parkin and Im 1990). The enzymes can apparently survive freeze-drying (Kopelman and Saguy 1977). However, ascorbic acid can prevent peroxidase-catalyzed decoloration of beet homogenates occurring during the preparation of red beet powders (Wasserman et al. 1984).

As the pigment is ionic, color intensity depends upon pH, but it is reasonably stable between pH 3.0 and 7.0. The most stable range is between 4.0 and 5.0, and removal of oxygen increases stability. At slightly lower and higher pHs, it is rather bluer, but above pH 10 it becomes yellow (Huang and von Elbe 1987). The rate of betalaine pigment degradation on heating is pH dependent, but regeneration can occur, with the amount depending on the reaction pH and whether oxygen is present (Huang and von Elbe 1987).

Beet juice contains sugars and easily ferments unless protected by refrigeration or frozen storage. However, added ascorbic acid and citric acid prevent oxidation (Freund et al. 1988). The degradation of betalaines involves water, so increased pigment stability can be obtained in products that are of low water activity (von Elbe 1987). Light, metal ions, and gamma and ultraviolet radiation also cause degradation of beet pigments (von Elbe 1986a).

Common product applications for beet color include yogurt, sherbet, ice creams, frozen fruit desserts, candies, frostings, and puddings (Freund et al. 1988). A betanin concentration of 33 ppm gives marshmallow a strawberry color, but 80 ppm is needed for a cherry color. Similarly, 16 ppm in a fondant gives a strawberry color, but 32 ppm is required for raspberry (von Elbe 1986b).

Performance of beet juice powder is satisfactory for marzipan, some fruit chews and candies, cream, ice cream, and table jellies but is not recommended for inclusion in pectin and starch jellies, bakery products, and jams (Counsell et al. 1980). Betalaines have been used instead of nitrites to color cured meats such as bologna and hamburger meat (von Elbe 1975). In products such as soy-protein curd hamburgers, the fading of the betalaines during frying is an advantage in that the product behaves like real meat (Pasch et al. 1975). The chemistry of betalaines has been detailed (Mabry 1978), and their stability during food processing has been reviewed (von Elbe 1986a, 1987).

Caramels

The brown to red hues of caramel are used to color a wide variety of foods, including beer, nonalcoholic drinks, bakery products, savory sauces, sweets, cereals, syrups, soups, and pet foods. Caramels are water soluble, but solubility rates vary with type. They are melanoidan pigments, which are prepared by the controlled heat treatment of carbohydrates such as sucrose, glucose, or invert sugar. Using food-grade acids, alkalis, and salts, nitrogen-containing reactants such as ammonia, and varying processing conditions, it is possible to obtain a wide range.

Four classes of caramel are normally available to the food industry in liquid or solid forms. Class I, caustic caramels, have a comparatively low coloring power, but they are stable in high concentrations of alcohol. They are used to adjust the color of blended whiskeys and other spirits. Class II, caustic sulfite caramels, are stable in the presence of alcohol and tannins. They are therefore useful for coloring vermouth, brandies, and other alcoholic drinks containing tannin or vegetable extracts. Class III, the ammonia caramels, have a high isoelectric point of approximately 6.0 and a high coloring power. They are stable in beer and vinegar and also in the presence of high–salt concentration products such as gravies. Class IV, the ammonia sulfite caramels, have an isoelectric point of 1.0. The consequent presence of negative charges at higher pHs make them suitable for addition to cola-type drinks containing negatively charged tannins. This situation prevents agglomeration. Class IV caramels are strong emulsifiers and so are useful in drinks in which there is a risk of phase separation (Thornton 1990).

Carotenoids

Carotenoid pigments are major natural food colorant additives. This range of yellow-orange-red colors has a high tinctorial potency. Carotenoids are lipid soluble, but their emulsion and colloidal forms are water soluble. They have a natural connotation, and are noncorrosive, relatively stable under reducing conditions, and relatively stable over the acid and neutral pH range. Some possess vitamin A activity. However, they have a limited color range, and are sensitive to light and oxidative degradation. Carotenoid additives are available either as extracts from plant sources possibly containing mixtures of pigments or in a manufactured pure form known as *nature identical*.

Many carotenoids are unsuitable as colorants because they contain many other fat-soluble plant components. These are refined by extraction from plants using an organic solvent followed by solvent evaporation (Philip 1975). However, crude extracts are available of carrot, as oil-soluble and water-miscible oleoresins, and tomato. Annatto is another exception. Three methods can be used to extract this pigment from the seed pericarp of *Bixa orellana*, a small tree cultivated in Central and South America. Oil extraction removes bixin from the seed cover, but solvent extraction results in the purest form of the pigment. Alkali extraction results in the conversion of bixin to the water-soluble norbixin.

The oil-soluble pigment is used to color yellow spreads, oils, fats, cakes, confectionery coatings, and ice cream. Water-soluble norbixin is capable of binding with proteins and is therefore useful for incorporating into the cheese-making process and in other products in which color leaching is undesirable. These include breakfast cereals, salad dressings, acidic fruit drinks, smoked fish, soups, and bakery products. Butter varies in color from very pale to deep yellow depending on the breed and diet of the animal. Butter produced in the spring, when the pastures are richer in carotenoids, is a deeper color. Product uniformity can be achieved by the addition of carotenoids (Hillman 1983). The pigment is also available as mixtures, in the dried form, and as emulsions (Freund et al. 1988). In pastel coatings, the use of fractionated fats can significantly decrease color loss on storage (Hogenbirk 1987).

Consumer flavor judgments of orange juice are strongly affected by color, which is in turn subject to seasonal and varietal differences. For example, in 1965–66, 42 percent of samples were judged to be excellent or very good and 0.5 percent poor. In 1957–58, the figures were 12 percent and 22 percent respectively (Francis and Clydesdale 1975). The color of weak-colored juice produced by Florida-grown Hamlin oranges can be reinforced by pigment granules recovered from more highly colored citrus fruit (Fellers and Barron 1987). Orange peel pigments can also be used to smooth out such variations (Ting and Hendrickson 1968).

Paprika made from the chilies is valued more for its pungency than for its color. Paprika from *Capsicum annuum* is valued principally as an additive for its brilliant red color and delicate aroma (Govindarajan 1986).

Paprika extracts are available in the dried form and as oleoresins, containing flavor or deflavored, and as water or oil soluble. Although all types are stable to heat, the oleoresin colors tend to be more stable to light and are higher in pigment content than the powder (Freund et al. 1988). The color is used extensively in soup and meat products, as well as in cereals, baked goods, cheese sauces, and condiments.

Saffron has a long established use. The major source of this color and flavor is the stigmas of the *Crocus sativus*. This limited source has ensured its high cost. The major pigments are the water-soluble crocetin and its ester crocin and the fat-soluble lycopene, alpha carotene, and zeaxanthin (Sampathu et al. 1984).

Crocin and crocetin are available in greater quantities and without the saffron flavor from the genus *Gardenia*. Three pigment groups present in *G. jasminoides* are carotenoids, iridoids, and flavonoids. Colors ranging from yellow to green, red, violet, and blue can be obtained, depending on extraction time and conditions. Possible applications include fish products and substitutes, sweets, ices, and vegetable products (Francis 1987b).

Carotenoids available in the pure form are beta carotene, apocarotenal, apocarotenal ester, and canthaxanthin. These are the manufactured colorants now termed *nature identical*. They cover the color range from yellow through orange to red and are available in oil-soluble and water-dispersible forms. However, their use has not gained worldwide acceptance.

Chlorophylls

Natural magnesium chlorophyll extracts of suitable purity are not available for food use. To increase light stability, chlorophyll colors used as food colorants have been saponified to produce chlorophyllins, which may contain sodium, potassium, or copper. Copper chlorophyll is water soluble, but solutions precipitate out at low pH. They are most suitable for products with a pH greater than 7.0, when they are stable to temperatures up to 120°C (Freund et al. 1988). Copper chlorophyll has been used to replace natural chlorophyll lost during blanching (Kearsley and Katraboxakis 1980). This can be one of the more stable natural color additives, and its performance has been described as satisfactory in a wide range of products. Applications include ice cream, sweets, bakery products, jams, and jellies (Counsell et al. 1980).

Cochineal (Carmine) and Related Colorants

Cochineal is an anthraquinone-type pigment, obtained in concentrated solution after removal of the alcohol from an aqueous-alcohol extract of dried female cochineal insects (*Coccus cacti*). The principle pigment is the water-soluble carminic acid, which can be acid extracted. Below pH 7, solutions of carminic acid have little intrinsic color, but they complex with metals to produce brilliant reds (Francis 1987b). Tin and aluminium complexes produce desirable fruit colors (Lloyd 1980).

Carmine may be obtained as a water-dispersible powder insoluble in alcohol and oil or in an aqueous form such as in aqueous ammonium hydroxide. It has moderate stability to light, sulfur dioxide, and heat but is precipitated by acids. Its protein-binding property makes it a useful colorant for dairy products, but it can also be used for baked and meat products, fish, jams, and preserves. The color of carminic acid is pH sensitive. Below pH 2.0, it is orange red; between 2.0 and 5.0, a rich dark red; and above 5.0, a burgundy violet (Freund et al. 1988).

Compounds related to carmine include kermesic acid, laccaic acids, and alkannet. The last is soluble in organic solvents and has been used to color confectionery, ice cream, and wines (Coulson 1980).

Curcumin (Turmeric)

Turmeric, with its yellow coloring matter curcumin, is obtained from the roots of the dried, ground rhizome of the turmeric plant, *Curcuma longa*. Used as a cheap substitute for saffron in medieval Europe, it was called Indian saffron. It is a versatile spice for vegetarian and nonvegetarian dishes. The pigment is not soluble in water and is often used in colloidal dispersions. Various forms of the pigment are available. These include the ground powder from the root containing both color and flavor, the purified powder, and purified extracts, which may be water-miscible, oil-soluble, or oleoresins.

Curcumin is oil-soluble, has good oxygen and heat stability, but it is sensitive to light. In particular it is sensitive to air and light in combination (Souza et al. 1997). The whole spice can be used for pickles, relishes, and peppers. The deflavored forms have many applications and can be used, for example, in baked goods, frozen desserts, beverages, margarine, and salad dressings. Turmeric can be blended successfully with annatto or paprika for use in baked goods. Water- and oil-soluble or miscible versions are commercially available (Freund et al. 1988). Govindarajan (1979) has written a comprehensive review.

Flavins

Flavins are synthesized by bacteria, yeasts, and green plants. The most important member of this class is riboflavin, which is probably added as much for the vitamin B_2 content as for its yellow color. It is available as a crude extract and in two synthetic forms, riboflavin and riboflavin-5'-phosphate sodium. Neither of these latter forms is soluble in fat. The sodium salt is more soluble in water than riboflavin itself.

These pigments are relatively stable up to a temperature of 50°C but are sensitive to light, particularly at alkaline pH (Kearsley and Rodriguez 1981). They have particular application in enrobed fondants, in which a clear, vivid yellow can be achieved (Jeffries 1979). They can also be used in ice cream, water ice, lemon curd, hard candies, and fondant, except when there will be exposure to light (Counsell et al. 1980).

Flavonoids

Among the various plant sources for the pigments commercially available or suggested are grape juice lees, wine grape skins, and dried grape extract (Timberlake and Henry 1986), roselle (*Hibiscus sabdariffa*), miracle fruit (*Synsepalum dulcificum*), elderberry (Bronnum-Hansen and Flink 1985, 1986), *Zebrina pendula*, *Iomoea tricolor* (Teh and Francis 1988), berries of the *Viburnum dentatum*, and cranberries (Francis 1975). Blueberries have been used to produce attractively colored beverages, condiments, and toppings (Francis 1985a). Radish extracts have been used to color maraschino bleached cherries (Giusti et al. 1998). Acylated anthocyanins generally retain more color at higher pH values than do unmodified forms. These are fairly stable to light and thermally stable to 120°C (Freund et al. 1988). When used as additives, anthocyanins from different sources are stable to different extents. For example, those from *Tradescantia pallida* are approximately 29 times more stable than those from red cabbage (*Brassica oleracea*) when used to color a nonsugar drink model system. Fifteen times the stability was obtained in a fish protein system. Red cabbage anthocyanins are more stable than colorants from grapes (*Enocyania*) (Shi et al. 1992) as well as colorants from grape cell culture (Cormier et al. 1997). Anthocyanin extracts from *Tradescantia* are more stable than those from Concord grapes, red cabbage, and ajuga (Baublis et al. 1994).

Tannins present in anthocyanin powder additives can interact with hydrocolloids. The tannins are precipitated when used in gelatin-containing products. They can, however, be used in other gel systems. Other product applications include low-pH milkshakes and beverages, chewing gum, fruit chews, fondant, fruit toppings, and water ice. Ion-exchange purified anthocyanin pigment has been used with partial success to reinforce the color of cranberry juice cocktail (Chiriboga and Francis 1973). Enocyanin, a pigment isolated from the skins of wine grapes, is used as a colorant reinforcement for cranberry juice cocktail, and methods for its detection have been detailed (Francis 1985b). Applicable products will probably have a pH less than 3.5. These include fruit drinks, purees, and preserves.

Possible Future Natural Colorants

Other food colorants are used in different parts of the world. These may come to achieve a wider permitted status.

Monascus, a fungal pigment growing on rice, can be made water or oil soluble. It is traditionally used in the Orient for coloring foods and as a medicine. The color achieved from yellow to red depends on the growth substrate. Monascus pigments are relatively insoluble in water, but they can be reacted with compounds containing amino groups to form water-soluble colorants. The pigments are heat stable and are apparently stable at pHs between 2 and 12 (Francis 1987b). They are stable in sausage and pâté products (Fabre et al. 1993).

Biliproteins obtained from algae include the red phycoerythrins and blue phycocyanins. The function of the pigments in nature is as light absorbers, and their structures contain an open tetrapyrrole skeleton similar to chlorophyll and hemoglobin. They can be made water or alcohol soluble and are possibly stable enough for use in sweets and ices (Francis 1987b).

Many flowers are heavily pigmented and may be suitable sources of food colors. Saffron, gardenia, and lutein have already been mentioned, but other possible sources are *Butea frondosa*, hollyhocks, marigold (*Tagetes*), morning glory, poppy, and sunflower seed husks (Taylor 1980). Red and yellow carthamin pigments, dyer's saffron, can be extracted and manufactured from petals of the safflower (*Carthamus tinctorius*). Stigma of the safflower, the "bastard saffron," are sometimes passed off as the less uniform, longer saffron. Marigold blended with paprika is included in feeds for increasing egg yolk pigmentation (Fletcher and Halloran 1981). Aqueous extracts of *Clitorea ternatea* yield stable anthocyanins that retain their blue color at low pH values (Saito et al. 1985).

The roselle plant has long provided food and medicine, particularly in the West Indies (Esselen and Sammy 1973). Water extraction of the dried calyces of *Hibiscus sabdariffa* (roselle) produces a brilliant red color. This may be found suitable for addition to juices, carbonated drinks, and preserves (Al-Kahtani and Hassan 1990).

Sandalwood, which contains the natural red pigment santalin, has a historical usage in India and Canada. It is light and heat stable and is suitable for inclusion in acid products and preserves, alcoholic drinks, meats, bread, smoked fish, and vegetable oils and fats (Freund et al. 1988).

Saint John's wort is a perennial medicinal herb that is the source of a red dye. Quercitin appears to be the main pigment. The herb and fruits are sometimes used as a tea (Razhinskaite 1971).

Other colorants may be derived from, for example, squid ink, red palm oil, purple and red corn, tamarind fruit and leaves, and other grains and fruits (Wissgott and Bortlik 1996). Blue pigment can be extracted from onion and garlic (Imai et al. 1996).

In his review of lesser known colorants, Francis (1987b) listed a number of less common natural extracts and derivatives that may become useful for foods. These include red bacterial pigments, synthetic anthocyanidins, benzylpyrimidine derivatives, and berberin. All would require testing for safety.

Synthetic Colorings

Synthetic food colors are available in two forms, water-soluble dyes and lakes. The former, generally marketed in a powder or granular form, have the highest pure dye contents. Nondye content is moisture or a diluent such as common salt or dextrin. Dyes are marketed in many forms, such as pastes, plating dyes, and dispersions, and can be blended with flavors. Desirable technical features are good stability, high purity, high tinctorial value, and accurate standardization. Colors can be made insoluble by precipitation as aluminium or calcium lakes. These are weaker, containing approximately 25 percent color, and are suitable for fat-based foods and tablet coatings. Lakes tend to be more stable toward light and heat (Walford 1978).

Powder dyes can produce dust, resulting in possible health or contamination problems. However, they are marketed in larger particle–sized versions as granules and plating dyes, and also as pastes and liquid dyes. Colorants may also be microencapsulated. There are methods of controlled release that might be site, time, or stage specific. Alternatively, their release can be triggered by pH, temperature, irradiation, or osmotic shock (Kirby 1991). Many variations of colorant application are available to suit particular applications. For example, the larger particle–sized plating dyes give an increased visual appeal to dry mixes, and nonflashing blends eliminate streaking problems in products made up in the home. Pigment dispersions with titanium dioxide are used to increase product opacity. Exhibit 11–1 presents the forms, properties, and applications of FD&C dyes (adapted from Dziezak 1987).

Lakes are relatively inert in product environments that may include metal ions, pH extremes, low moisture contents, and exposure of the finished product to light. High-temperature processing considerations may lead to the choice of lake over powder form. Migration of color through leaching occurs less with lakes than with dyes. Lakes are more suitable for dark deep shades, but with light pastel shades they may result in speckling. Table 11–6 gives examples of the types of product for which dyes and lakes are suitable (adapted from Lykens 1985 and Dziezak 1987).

Exhibit 11–1 Forms, Properties and Applications of Synthetic Colorants (adapted from Dziezak 1987 with permission)

Standard Forms
Powder (pure dye content 88–93 percent)
  Pluses—easy to dissolve, uniform blending in dry mixes, broad applications, least expensive
  Minuses—potential dust problems, poor flow characteristics
  Products—dry drink mixes, extruded products

Granular (pure dye content 88–93 percent)
  Pluses—reduced dust, free flowing
  Minuses—not suitable for dry mixes, slower dissolution rate
  Products—products where color is dissolved prior to use

Liquid (pure dye content 1–8 percent)
  Pluses—ready to use, elimination of dust, ease of handling, accurate dosing
  Minuses—increased storage space needed, more costly
  Products—jelly candies, bakery and dairy products

Special-Purpose Forms, Which Have Limited Application and Are More Costly Than the Dry Color
Plating Dye (pure content 88–93 percent)
  Pluses—increased visual appeal of dry mix
  Minuses—not available in all primary colors
  Products—dry mixes, including beverage bases, puddings, cakes, gelatin desserts

Nonflashing blend (pure dye content 90 percent)
  Pluses—uniform development of color in products reconstituted by the consumer; elimination of streaking
  Products—gelatin, puddings, instant breakfast drinks

Paste (pure dye content 4–10 percent)
  Pluses—color maintenance in position while formulation is being worked
  Products—chewing gum, cake decorations, hard candy

Dispersion with Titanium Dioxide
  Pluses—opacity, cheaper than lakes
  Products—gums and sugar-coated confections

Table 11–6 Typical Product Applications for Dyes and Lakes (adapted from Lykens 1985, with permission, reprinted from The Manufacturing Confectioner, Glen Rock, NJ USA, copyright, and Dziezak 1987 with permission)

Application	Recommended Colorant Type	Comments
Hard candy	Dyes, predissolve	Dyes—brighter
Striping colors	Lakes, predisperse	Lakes—heat/light stability, late in process
Starch candies	Dyes, predissolve	Lakes minimize color
Marshmallow	Lakes, predisperse	Bleed
Liquid drinks	Dyes	
Powdered drink mixes	Lake/dye blends	Lakes powder Dyes liquid
Hard panning dextrose/sucrose	Dyes, predissolve Lakes, predisperse	Possibly use TiO_2 for sucrose opacity
Soft panning	Dyes, predissolve	Dyes—bright colors
	Lakes, predisperse	Lakes—bright colors, mottling/fading minimized
	Lakes + TiO_2	Opaque
Gum	Lakes, predisperse	No color hot spots, do not discolor mouth
Compressed candies & tablets	Lakes	
Cake mixes	Lakes	
Icing, fondant	Lakes, predisperse	Better heat/light/bleeding coatings stability
Ice cream	Dyes	Add after pasteurization
Fat-based dairy coatings	Lakes	
Pet foods	Lakes	Lake/protein affinity stabilizes color

Published studies that include detailed examinations of synthetic colorants appear infrequently. In a pharmaceutical context, liquid-solid solution interactions of colorants with gelatins are sensitive to gelatin type (Cooper 1972). In a dry food mix context, fine powders tend to be adhesive and to form structured agglomerates rather than being randomly distributed (Eagerman and Orr 1983). Powder migration between surfaces can be a problem in production, as well as affecting the appearance of the powder.

The lightness of a colored material increases as its particle size is reduced. Size reduction provides an increased number of particle boundaries and hence an increase in diffuse light scattering. Maximum scattering occurs when the particle diameter is approximately $\lambda/4$, where λ is the wavelength of the light being scattered. At this diameter, pigment particles are at their most effective. Both particle morphology and size affect the color strength of vegetable powders; hence, the method drying can be critical to performance (Javenkoski et al. 1995).

In commercial use, colored ingredients are normally accompanied by a carrier, and surface affinities between colorant and carrier differ. The rank order of affinity is citric acid (greatest), malic acid, sucrose, and sodium chloride. Within the critical range of a_w 0.33 to 0.58, powder migration almost ceases. This is probably due to the stabilization of pigment-carrier agglomerates by liquid bridges (Sapru and Peleg 1988).

When the concentration of the diluent is 10 times greater than the weight of the colorant, the lightness of the mixture is determined primarily by the diluent. A reduction in colorant particle size increases the number of light-absorbing centers and increases the chance for interaction between incident light and chromophore. That is, the lightness of the mixture will decrease. Hence, the dilution ratio and particle size of both diluent and colorant may be used to control the color of a powder mix (Saguy and Graf 1991). Industrial application of this principle of color control is illustrated by the development of an encapsulated sugar-caramel mixture to govern the extent of browning during microwave cooking. By selecting appropriate concentrations and particle sizes, it is possible to control initial and final colors as well as uniformity of color development (Graf et al. 1991).

Comparatively large amounts of natural colorants are normally needed to achieve the same color intensity as that given by synthetics. This is because the natural colors may be unpurified and not because they have inherently lower tinctorial strength (von Elbe 1986b). As has been indicated, other limitations may include increased susceptibility to oxidation, decreased solubility, and increased sensitivity to light, heat, and pH.

The stability and use of the natural colors red beet powder, copper chlorophyll, and cochineal (Kearsley and Katraboxakis 1980) and anthocyanin, beta carotene, and riboflavin (Kearsley and Rodriguez 1981) have been compared with those of selected synthetics. The natural colors were found to be relatively stable under certain defined conditions. It was concluded that these can be used with a reasonable degree of success when foods are selected so that defined conditions are met or when suitable modifications are made to the process. Many applications for which nature-identical colorants can be used have been described above (Counsell et al. 1980).

A list of suggested natural/synthetic equivalents is given in Table 11–7 (from Walford 1977; Freund et al. 1988). Specialized experience is available within the industry to reduce problems to a minimum.

Other additives affecting appearance include emulsifiers and foam stabilizers, which increase light scattering; glazing agents, such as beeswax, carnauba wax (in sugar and confectionery finishes), and shellac and lipid coatings (for fruit); bleaching agents, colorant diluents, and clarifying agents for drinks; and firming agents for jams and desserts.

Table 11–7 Potential Natural Color Substitutes for Synthetic Colors (compiled from Walford 1977, and Freund et al. 1988)

Synthetic Colorant	Natural Substitute
Tartrazine	Turmeric, turmeric/annatto blend
Sunset Yellow	Annatto, beta carotene, paprika
Allura Red	Betalaines, carmine, grape
Erythrosine	Betalaines, carmine, grape
Carmoisine, Amaranth, Red 2G	Anthocyanins, betalaines, carmine
Patent Blue, Brilliant Blue, Green S	Chlorophyll, modified carotenes

REFERENCES

Abers JE, Wrolstad RE. (1979). Causative factors of color deterioration in strawberry preserves during processing and storage. *J Food Sci* 44:75–78, 81.

Adams JB. (1973). Colour stability of red fruits. *Food Manuf* 48(2):19–20, 41.

Adams JB. (1978). The inactivation and regeneration of peroxidase in relation to the high-temperature short-time processing of vegetables. *J Food Technol* 13:281–297.

Adams PA. (1976). The kinetics and mechanism of the recombination reaction between apomyoglobin and haemin. *Biochem J* 159:371–376.

Aguerre RJ, Suarez C. (1987). Kinetics of colour change in corn, effect of temperature and moisture content. *Lebens Wiss Technol* 20:287–290.

Aguilera JM, Oppermann K, Sanchez F. (1987). Kinetics of browning of sultana grapes. *J Food Sci* 52:990–993, 1025.

Aharoni Y, Houck LG. (1980). Improvement of internal color of oranges stored in oxygen-enriched atmospheres. *Scientia Hor* 13:331–338.

Aharoni Y, Houck LG. (1982). Change in rind, flesh and juice color of blood oranges stored in air supplemented with ethylene or in oxygen-enriched atmospheres. *J Food Sci* 47:2001–2002.

Ahvenainen R. (1996). New approaches in improving the shelf life of minimally processed fruit and vegetables. *Trends Food Sci Technol* 7:179–187.

Akhtar P, Gray JI, Booren AM, Cooper TH. (1996). Effect of feed components on color and lipid stability of rainbow trout muscle during refrigerated and frozen storage. In *Book of abstracts of Institute of Food Technologists (USA) annual meeting*, 164. Chicago, IL: Institute of Food Technologists.

Al-Hooti S, Sidhu JS. (1997). Objective color measurement of fresh date fruits and processed date products. *J Food Qual* 20:257–266.

Al-Kahtani HA, Hassan BH. (1990). Spray drying of roselle (*Hibiscus sabdariffa*) extract. *J Food Sci* 55:1073–1076.

Allen MS. (1983). Sulphur dioxide and ascorbic acid: their roles in oxidation and control. *Aust Grapegrower Winemaker* (April):70–72.

Alzamora SM, Cerrutti P, Guerrero S, López-Malo M. (1995). Minimally processed fruits by combined methods. In *Food preservation by moisture control*, ed. GV Barbosa-Cánovas, J Welti-Chanes, 463–492. Lancaster, Pa: Technomic.

Ames JM. (1995). Applications of the Maillard reaction in the food industry. *Confection* 2(11):40–43.

Anderson HJ, Rasmussen MA. (1992). Interactive packaging as protection against photodegradation of the colour of pasteurized sliced ham. *Int J Food Sci Technol* 27:1–8.

Andersson Y, Hedlund B, Jonsson L, Svensson S. (1981). Extrusion cooking in a high fibre cereal product with crispbread character. *Cereal Chem* 58: 370–374.

Arnau J, Guerrero L, Hórtos M, García JA. (1996). The composition of white film and white crystals found in dry-cured hams. *J Sci Food Agric* 70:449–452.

Arnold RN, Scheller KK, Arp SC, Williams SN, Schaefer DM. (1992). Visual and spectrophotometric evaluations of beef color stability. *J Food Sci* 57:518–520.

Arya SS, Netesan V, Premaavalli KS, Vijayraghavan PK. (1982). Effect of pre-freezing on the stability of carotenoid in blanched air dried carrots. *J Food Technol* 7(1):109–113.

Bakker J, Bridle P. (1992). Strawberry juice colour: the effect of sulphur dioxide and EDTA on the stability of anthocyanins. *J Sci Food Agric* 60:477–481.

Baldwin EA, Nisoeros MO, Hagenmaier RD, Baker RA. (1997). Use of lipids in food coatings for food products. *Food Technol* 51(6):56–62, 64.

Baloch AK, Buckle KA, Edwards RA. (1973). Measurement of nonenzymatic browning of carrot. *J Sci Food Agric* 24:389–398.

Banks NH. (1984). Some effects of TAL Pro-Long coating on ripening bananas. *J Exp Bot* 35:127–137.

Barth MM, Kerbel EL, Broussard S, Schmidt SJ. (1993). Modified atmosphere packaging protects market quality in broccoli spears under ambient temperature storage. *J Food Sci* 58:1070–1072.

Basel RM. (1982). Acidified bulk storage of green beans and peas. *J Food Sci* 47:2082–2083.

Basel RM. (1983). The improvement of green colour of green beans by acidified bulk storage. *J Food Technol* 18:797–799.

Bates L, Ames JM, MacDougall DB. (1994). The use of a reaction cell to model the development and control of colour in extrusion cooked foods. *Lebens Wiss Technol* 27:375–379.

Baublis A, Spomer A, Berber-Jiménez MD. (1994). Anthocyanin pigments: comparison of extract stability. *J Food Sci* 59:1219–1221, 1233.

Bayindirli L, Sümnü G, Kamadan K. (1995). Effects of Semperfresh and Jonfresh fruit coatings on poststorage quality of Satsuma mandarins. *J Food Processing Preservation* 19:39–407.

Bedinghaus AJ, Ockerman HW. (1995). Antioxidative Maillard reaction products from reducing sugars and free amino acids in cooked ground pork patties. *J Food Sci* 60:992–995.

Bell RG, Penney N, Gilbert KV, Moorhead SM, Scott SM. (1996). The chilled storage life and retail display performance of vacuum and carbon dioxide packed hot deboned beef striploins. *Meat Sci* 42:371–386.

Bentley KW. (1960). *The natural pigments*. New York: Interscience.

Berk Z, Mannheim CH. (1986). The effect of storage temperature on quality of citrus products aseptically packed into steel drums. *J Food Processing Preservation* 10:281–293.

Berry BW. (1994). Fat level, high temperature cooking and degree of doneness affect sensory, chemical and physical properties of beef patties. *J Food Sci* 59:10–14, 19.

Berset C. (1989). Colour. In *Extrusion cooking*, ed. C Mercier, P Linko, JM Harper, 371–385. St. Paul, Minn: American Society of Cereal Chemists.

Berset C, Marty G. (1986). Utilization du beta-carotene en cuisson-extrusion. *IAA conference proceedings*, 527–532. France: Agricultural and Food Industry.

Bertola N, Chaves A, Zaritzky NE. (1990). Diffusion of carbon dioxide in tomato fruits during cold storage in modified atmosphere. *Int J Food Sci Technol* 25:318–327.

Beveridge T. (1997). Haze and cloud in apple juices. *Crit Rev Food Sci Nutr* 37:75–79.

Beveridge T, Harrison JE. (1984). Nonenzymatic browning in pear juice concentrate at elevated temperatures. *J Food Sci* 49:1335–1336, 1340.

Beveridge T, Weintraub SE. (1995). Effect of blanching pretreatment on color and texture of apple slices at various water activities. *Food Res Int* 28:83–86.

Bhattacharya S, Sivakumar V, Chakraborty D. (1997). Changes in CIELAB colour parameters due to extrusion of rice-greengram blend: a response surface approach. *J Food Eng* 32:125–131.

Bhowmik SR, Pan JC. (1992). Shelf life of mature green tomatoes stored in controlled atmosphere and high humidity. *J Food Sci* 57:948–953.

Bible BB, Suman S. (1993). Canopy position influences CIELAB coordinates of peach color. *Hort Sci* 28:992–993.

Binkerd EF, Kolari OE. (1975). The history and use of nitrate and nitrite in the curing of meat. *Food Cosmet Toxicol* 13:655–662.

Blanc J-M, Choubert G. (1993). Genetic variation of flesh color in pan-sized rainbow trout fed astaxanthin. *J Appl Aquaculture* 2:115–123.

Boles JA. (1995). *Colour of processed meat products: a literature review*. Pub. no. 951. Meat Industry Research Institute of New Zealand. Wellington, New Zealand: Meat Industry Research Institute.

Bolin HR. (1992). Retardation of surface lignification on fresh peeled carrots. *J Food Processing Preservation* 16:99–104.

Bolin HR, Huxsoll CC. (1991). Control of minimally processed carrot (*Daucus carota*) surface discoloration caused by abrasion peeling. *J Food Sci* 56:416–418.

Bolin HR, Steele RJ. (1987). Nonenzymatic browning in dried apples during storage. *J Food Sci* 52:1654–1657.

Boon DD. (1977). Coloration in bivalves. *J Food Sci* 42:1008–1012, 1015.

Botta JR, Bonnell G, Squires BE. (1987). Effect of method of catching and time of season on sensory quality of fresh raw Atlantic cod (*Gadus morhua*). *J Food Sci* 52:928–931, 938.

Botta JR, Kennedy K, Squires BE. (1987). Effect of method of catching and time of season on composition of Atlantic cod (*Gadus morhua*). *J Food Sci* 52:922–927.

Botta JR, Squires BE, Johnson J. (1986). Effect of bleeding/gutting procedures on the sensory quality of fresh raw Atlantic cod (*Gadus morhua*). *Can Inst Food Sci Technol J* 19:186–190.

Bottrill DE, Hawker JS. (1970). Chlorophylls and their derivatives during drying of sultana grapes. *J Sci Food Agric* 21:193–196.

Bowers JA, Craig JA, Kropf DA, Tucker TJ. (1987). Flavor, color and other characteristics of beef longissimus muscle heated to seven internal temperatures between 55 and 85°C. *J Food Sci* 52:533–536.

Brewer MS, Klein BP, Rastogi BK, Perry AK. (1994). Microwave blanching effects on chemical, sensory and color characteristics of frozen green beans. *J Food Qual* 17:245–259.

Brewer MS, McKeith F, Martin SE, Dallmier AW, Wu SY. (1992). Some effects of sodium lactate on shelf-life, sensory, and physical characteristics of vacuum-packaged beef Bologna. *J Food Qual* 15:369–382.

Brewer MS, Novakofski J. (1996). pH, cooking rate and endpoint temperature effects on instrumental and visual color of ground beef. *Book of abstract of Institute of Food Technologists (USA) annual meeting*, 165. Chicago, IL: Institute of Food Technologists.

Brewer MS, Wu SY. (1993). Display, packaging, and meat block location effects on color and lipid oxidation of frozen lean ground beef. *J Food Sci* 58: 1219–1223.

Brewer MS, Wu S. (1994). Carbon monoxide effects on color and microbial counts of vacuum-packaged fresh beef steaks in refrigerated storage. *J Food Qual* 17:231–244.

Britton G. (1996). Carotenoids. In *Natural food colorants*, 2nd ed, ed. GAF Hendry, JD Houghton, 197–243. London: Chapman and Hall.

Bronnum-Hansen K, Flink JM. (1985). Anthocyanin colorants from elderberry (*Sambucus nigra*): part 1. *Food Technol* 20:703–711, 713–734.

Bronnum-Hansen K, Flink JM. (1986). Anthocyanin colorants from elderberry (*Sambucus nigra*): part 2. *Food Technol* 21:605–614.

Brooks J. (1929). Postmortem formation of methaemoglobin in red muscle. *Biochem J* 23:1391–1400.

Brouillard R, Delaporte B. (1977). Chemistry of anthocyanin pigments, 2: kinetic and thermodynamic study of proton transfer, hydration and tautomeric reactions of malvidin 3-glucoside. *J Am Chem Soc* 99:8461–8468.

Buckholz LR Jr. (1988). The role of Maillard technology in flavoring food products. *Cereal Foods World* 33:547–551.

Buckle KA, Edwards RA. (1970a). Chlorophyll, colour and pH changes in HTST processed green pea puree. *J Food Technol* 5:173–186.

Buckle KA, Edwards RA. (1970b). Chlorophyll degradation and lipid oxidation in frozen unblanched peas. *J Agric Food Chem* 21:307–312.

Buera M, Chirife J, Resnik SL, Lozano RD. (1987). Nonenzymatic browning in liquid model systems of high water activity: kinetics of color changes due to caramelization of various sugars. *J Food Sci* 52:1059–1062, 1073.

Buera M, Chirife J, Resnik S, Wetzler G. (1987). Nonenzymatic browning in liquid model systems of high water activity: kinetics of color changes due to Maillard's reaction between different single sugars and glycine and comparison with caramelization browning. *J Food Sci* 52:1063–1067.

Buera M, Karel M. (1993). Application of the WLF equation to describe the combined effects of moisture and temperature on nonenzymatic browning rates in food systems. *J Food Proc and Preserv* 17:31–45.

Burdock GA. (1996). *Encyclopedia of food and color additives*. Dallas: CRC Press.

Cameron AC, Talisila PC. (1995). In *Food preservation by moisture control*, ed. GV Barbosa-Cánovas, J Welti-Chanes, 821–830. Lancaster, Pa: Technomic.

Cameron RG, Baker RA, Grohmann K. (1997). Citrus tissue extracts affect juice cloud stability. *J Food Sci* 62:242–245.

Canellas J, Rossello C, Simal S, Soler L, Mulet A. (1993). Storage conditions affect quality of raisins. *J Food Sci* 58:805–809.

Carreño J, Almela L, Martinez A, Fernandez-Lopez JA. (1995). Color changes associated with maturation of the table grape cv. *Don Mariano. J Hort Sci* 70:481–486.

Carreño J, Martinez A, Almela L, Fernandez-Lopez JA. (1995). Proposal of an index for the objective evaluation of the colour of red table grapes. *Food Res Int* 28:373–377.

Cassens RG. (1997). Residual nitrite on cured meat. *Food Technol* 51(2):53–55.

Castrillo M, Bermudez A. (1992). Post-harvest ripening in wax-coated Bocado mango. *Int J Food Sci Technol* 27:457–463.

Chai Y, Ott DB, Cash JN. (1991). Shelf-life extension of Michigan apples using sucrose polyester. *J Food Processing Preservation* 15:197–214.

Chan WS, Toledo RT, Deng J. (1975). Effect of smokehouse temperature, humidity and air flow on smoke penetration into fish muscle. *J Food Sci* 40:240–243.

Chandler BV, Clegg KM. (1970a). Pink discoloration in canned pears, I: role of tin in pigment formation. *J Sci Food Agric* 21:315–319.

Chandler BV, Clegg KM. (1970b). Pink discoloration in canned pairs, II: measurement of potential and developed colour in pear samples. *J Sci Food Agric* 21:319–323.

Chandler BV, Clegg KV. (1970c). Pink discoloration in canned pears, III: inhibition by chemical additives. *J Sci Food Agric* 21:323–328.

Chang T-S, Siddiq M, Sinha NK, Cash JN. (1994). Plum juice quality affected by enzyme treatment and fining. *J Food Sci* 59:1065–1069.

Chen CM, Trout GR. (1991). Color and its stability in restructured beef steaks during frozen storage: effects of various binders. *J Food Sci* 56:1461–1464, 1475.

Chen H-H, Chiu E-M, Huanh J-R. (1997). Color and gel-forming properties of horse mackerel (*Trachurus japonicus*) as related to washing conditions. *J Food Sci* 62:985–991.

Chervin C, Boisseau P. (1994). Quality maintenance of "ready-to-eat" shredded carrots by gamma radiation. *J Food Sci* 59:359–361, 401.

Chin M, Frick D. (1995). Formulating a color delivery system for hard candy. *Food Technol* 49(7):56–60.

Chiriboga CD, Francis FJ. (1973). Ion exchange purified anthocyanin pigments as a colorant for cranberry juice cocktail. *J Food Sci* 38:464–467.

Chonhenchob V, Cash JN, Sinha NK. (1996). Investigation on the nonenzymatic browning in potato chips based on a model system utilizing selected amino acids and sugars. In *Book of abstracts of Institute of Food Technologists (USA) annual meeting*, 92. Chicago, IL: Institute of Food Technologists.

Choubert G, Blanc J-M. (1993). Muscle pigmentation changes during and after spawning in male and female rainbow trout, *Oncorhynchus mykiss*, fed dietary carotenoids. *Aquatic Living Resources* 6:163–168.

Choubert G, Blanc J-M, Courvalin C. (1992). Muscle carotenoid content and colour of farmed rainbow trout fed astaxanthin or canthaxanthin as affected by cooking and smoke-curing procedures. *Int J Food Sci Technol* 27: 277–284.

Chu YH, Huffman DL, Egbert WR, Trout GR. (1988). Color and color stability of frozen restructured beef steaks: effect of processing under gas atmospheres with differing oxygen concentration. *J Food Sci* 53:705–710.

Chu YH, Huffman DL, Trout GR, Egbert WR. (1987). Color and color stability of frozen restructured beef steaks: effect of sodium chloride, tripolyphosphate, nitrogen atmosphere and processing procedures. *J Food Sci* 52: 869–875.

Church IJ, Parsons AA. (1995). Modified atmosphere packaging: a review. *J Sci Food Agric* 67:143–152.

Church N. (1998a). MAP fish and crustaceans: sensory enhancement. *Food Sci Technol Today* 12(2):73–83.

Church N. (1998b). Meat colour. Paper presented to the Food Research Association, Leatherhead, UK.

Cilliers JJL, Singleton VL. (1989). Nonenzymic autoxidative phenolic browning reactions in a caffeic acid model system. *J Agric Food Chem* 37:890–896.

Cisneros-Zevallos L, Saltveit ME, Krochta JM. (1995). Mechanism of surface white discoloration of peeled (minimally processed) carrots during storage. *J Food Sci* 60:320–323, 333.

Claus JR, Kropf DH, Hunt MC, Kastner CL, Dikeman ME. (1984). Effects of beef carcass electrical stimulation and hot boning on muscle display color of polyvinylchloride packaged steaks. *J Food Sci* 49:1021–1023.

Claus JR, Kropf DH, Hunt MC, Kastner CL, Dikeman ME. (1985). Effects of beef carcass electrical stimulation and hot boning on muscle display color of unfrozen vacuum packaged steaks. *J Food Sci* 50:881–883.

Clydesdale FM, Francis FJ. (1976). Pigments. In *Principles of food science*, ed. OR Fennema, 385–426. New York: Marcel Dekker.

Coghlan A. (1995). Brighter future for vegetables. *New Sci* 147:11.

Cohen E, Birk Y, Mannheim CH, Saguy IS. (1994). Kinetic parameter estimation for quality change during continuous thermal processing of grapefruit juice. *J Food Sci* 59:155–158.

Collins P, Hughes S. (1992). Natural colours: a question of stability. In *FIE conference proceedings, Dusseldorf*. Netherlands: Expo consult.

Connell JJ, Hardy R. (1982). *Trends in fish utilisation*. Farnham, UK: Fishing News Books.

Cooper JW Jr. (1972). Liquid and solid solution interactions of primary certified colorants with pharmaceutical gelatins. PhD diss, University of Georgia, Athens.

Cormier F, Couture R, Do CB, Pham TQ, Tong VH. (1997). Properties of anthocyanins from grape cell culture. *J Food Sci* 62:246–248.

Cornforth DC. (1996). Raw meat and processing effects on color changes during ground beef patty cooking. In *Book of abstracts of Institute of Food Technologists (USA) annual meeting*, 169. Chicago, IL: Institute of Food Technologists.

Coulson J. (1980). Naturally occurring materials for food. In *Developments in food colours-1*, ed. J. Walford. London: Applied Science.

Counsell JN, Jeffries GS, Knewstubb CJ. (1980). Some other natural colours and their applications. In *Natural colours for food and other uses*, ed. JN Counsell. London: Applied Science.

Cumming DB, Beveridge HJT, Gayton R. (1986). Manipulation and control of brown pigmentation in juices prepared from dessert apples. *Can Inst Food Sci Technol J* 19:223–226.

Da Silva C, Moss BW, Gault NFS. (1994). Stability of dried salted lamb. In *Proceedings of the 40th International Congress of Meat Science and Technology*. The Hague, Netherlands: International Congress of Meat Science and Technology.

Dages W. (1986). Process and equipment to stabilize the colour of sausage and meats stored in a display case. German Federal Republic Patent Application DE 35 10 313 A1.

Davis HK. (1993). Fish. In *Principles and applications of modified atmosphere packaging of food*, ed. RT Parry, 189–228. Glasgow: Blackie.

Dawes H, Struebi P, Keene J. (1994). Kiwi fruit juice clarification using a fungal proteolytic enzyme. *J Food Sci* 59:858–861.

Dawson PL, Sheldon BW, Ball HR Jr. (1989). A pilot-plant washing procedure to remove fat and color components from mechanically deboned chicken meat. *Poultry Sci* 68:749–753.

Day B. (1997). *Principles and application of modified atmosphere packaging of foods*. 2nd ed. Glasgow: Blackie Academic and Professional.

Debeaufort F, Quezada-Gallo J-A, Voilley A. (1998). Edible films and coatings: tomorrow's packagings. A review. *Crit Rev Food Sci* 38:299–313.

DellaMonica ES, McDowell PE. (1965). Comparison of beta carotene content of dried carrots prepared by three dehydration processes. *Food Technol* 19:1597–1599.

Demos BP, Mandigo RW. (1996). Color of fresh, frozen and cooked ground beef patties manufactured with mechanically recovered neck bone lean. *Meat Sci* 42:415–429.

Denny EG, Costan DC, Ballard RE. (1986). Peach skin discoloration. *J Am Soc Hort Sci* 111:549–553.

Drake SR, Moffitt HR, Eakin DE. (1994). Low dose irradiation of "Rainier" sweet cherries as a quarantine treatment. *J Food Processing Preservation* 18:473–481.

Driver MG. (1976). Comparative evaluation of phytolaccanin and betanin as food colorants. *Diss Abstr Int* 37(1):145B.

Dussi MC, Huysamer M. (1995). Severe postharvest summer pruning of mature "Forelle" pear trees influences canopy light distribution, and fruit and spur leaf characteristics in the following season. *J S Afr Soc Hort Sci* 5:57–60.

Dussi MC, Sugar D, Azarenko AN, Righetti TL. (1997). Effects of cooling by over-tree sprinkler irrigation on fruit color and firmness in "Sensation Red Bartlett" pear. *Hortechnology* 7:55–57.

Dussi MC, Sugar D, Wrolstad RE. (1995). Characterising and quantifying anthocyanins in red pears and the effect of light quality on fruit color. *J Am Soc Hort Sci* 120:785–790.

Dworschak E. (1980). Nonenzyme browning and its effect on protein nutrition. *CRC Crit Rev Food Sci Nutr* 13:1–40.

Dyer WJ. (1974). Stability of minced fish: an important factor in standard formulations. In *Second technical seminar on mechanical recovery and utilization of fish flesh*. Washington, DC: National Fisheries Institute.

Dziezak JD. (1987). Application of food colorants. *Food Technol* 41:78–88.

Eagerman H, Orr NA. (1983). Ordered mixtures—interactive mixtures. *Powder Technol* 36:117–118.

Edwards CG, Lee CY. (1986). Measurement of provitamin A carotenoids in fresh and canned carrots and green peas. *J Food Sci* 51:534–535.

Edwards EJ, Saint RE, Cobb AH. (1998). Is there a link between greening and light-enhanced glycoalkaloid accumulation in potato (*Solanum tuberosum* L) tubers? *J Sci Food Agric* 76:327–333.

Edwards RA, Reuter FH. (1967). Pigment changes during the maturation of tomato fruit. *Food Technol Aust* 19:352–357.

Eheart MS, Gott C. (1965). Chlorophyll, ascorbic acid and pH changes in green vegetables cooked by stir-fry, microwave and conventional methods and a comparison of chlorophyll methods. *Food Technol* 19:867–870.

Eichner K, Karel M. (1972). The influence of water content and water activity on the sugar-amino acid reaction in model systems under various conditions. *J Agric Food Chem* 20:218–223.

Eldredge EP, Holmes ZA, Mosley AR, Shock CC, Stieber TD. (1996). Effects of transitory water stress on potato tuber stem-end reducing sugar and fry color. *Am Potato J* 73:517–530.

Elliott RJ. (1968). Calculation and presentation of pork muscle colour from reflectance spectra. *J Sci Food Agric* 19:685–692.

Eshtiaghi MN, Stute R, Knorr D. (1994). High-pressure and freezing pretreatment effects on drying, rehydration, texture and color of green beans, carrots and potatoes. *J Food Sci* 59:1168–1170.

Esselen WB, Sammy GM. (1973). Roselle: a natural red colorant for foods? *Food Prod Dev* 7(1):80, 82, 86.

Fabre CE, Santerre AL, Loret MO, Raberian R, Pareilleux A, Goma G, Blanc PJ. (1993). Production and food applications of the red pigments of *Monascus ruber*. *J Food Sci* 58:1099–1102, 1110.

Fang T-T, Ts'eng S-F. (1986). Chemical changes in color and flavor of lemon juice during storage. In *Role of chemistry in the quality of processed food*, ed. OR Fennema, W-H. Chang, C-Y, Lii, 118–126. Westport, Conn: Food and Nutrition Press.

Farber JM, Dodds KL. (1995). Principles of modified atmosphere and sous vide product packaging. Leatherhead, UK: Pira International.

Fellers PJ, Barron RW. (1987). A commercial method for recovery of natural pigment granules from citrus juices for color enhancement purposes. *J Food Sci* 52:994–995, 1005.

Fernández-Trujillo JP, Artés F. (1998). Effect of intermittent warming and modified atmosphere packaging on color development of peaches. *J Food Qual* 21:53–69.

Fletcher DL, Halloran HR. (1981). An evaluation of commercially available marigold concentrate and paprika oleoresin on egg yolk pigmentation. *Poultry Sci* 60:1846–1853.

Forni E, Sormani A, Scalise S, Torreggiani D. (1997). The influence of sugar composition on the colour stability of osmodehydrofrozen intermediate moisture apricots. *Food Res Int* 30:87–94.

Fox JB Jr. (1987). The pigments of meat. In *The science of meat and meat products*, ed. JF Price, BS Schweigert, 193–216. Westport, CT: Food and Nutrition Press.

Francis FJ. (1975). Anthocyanins as food colors. *Food Technol* 29 (May):52, 54.

Francis FJ. (1985a). Blueberries as a colorant ingredient in food products. *J Food Sci* 50:754–756.

Francis FJ. (1985b). Detection of enocyanin in cranberry juice cocktail by color and pigment profile. *J Food Sci* 50:1640–1642, 1661.

Francis FJ. (1987a). *Handbook of food colorant patents.* Westport, CT: Food and Nutrition Press.

Francis FJ. (1987b). Lesser-known food colorants. *Food Technol* 41(4):62–68.

Francis FJ. (1999). *Handbook of food colorants.* St Paul, Minn: Eagen Press.

Francis FJ, Clydesdale FM. (1970). Orange vegetables, part I. *Food Prod Dev* 4:66–72.

Francis FJ, Clydesdale FM. (1971). Tuna. *Food Prod Dev* 5 (Oct):58, 62, 63, 66.

Francis FJ, Clydesdale FM. (1975). *Food colorimetry: theory and applications.* Westport, Conn: Avi.

Francis FJ, Harney PH, Bulstrode PC. (1955). Color and pigment changes in the flesh of McIntosh apples after removal from storage. *Proc Am Soc Hort Sci* 65:211–213.

Franzen K, Singh RK, Okos MR. (1990). Kinetics of nonenzymatic browning in dried skim milk. *J Food Eng* 11:225–239.

Freeland RD. (1970). The effects of harvest methods and storage periods on certain quality factors of southern peas, *Vigna sinesis. Diss Abstr* 31(4).

Freund PR, Washam CJ, Maggion M. (1988). Natural color for use in foods. *Cereal Foods World* 33:553–559.

Frick D, Huck P. (1997). Food colour terminology. *Food Ingredients Anal Int* 18(5):33–44.

Froning GW. (1995). Colour of poultry meat. *Poultry Avian Biol Rev* 6:83–93.

Gandul-Rojas B, Minguez-Mosquera MI. (1996). Chlorophyllase activity in olive fruits and its relationship with the loss of chlorophyll pigments in the fruit and oils. *J Sci Food Agric* 72:291–294.

García-Jares C. (1993). Research on white and red wine blending in the production of rosé by means of the partial least squares method. *J Sci Food Agric* 63:349–354.

Geeson JD, Genge PM, Smith SM, Sharples RO. (1991). The response of unripe Conference pears to modified atmosphere retail packaging. *Int J Food Sci Technol* 26:219–223.

Geeson JD, Smith SM. (1989). Retardation of apple ripening during distribution by the use of modified atmospheres. *Acta Hort* 258:245–253.

Geeson JD, Smith SM, Everson HP, Genge PM, Browne KM. (1987). Responses of CA-stored Bramley's Seedling and Cox's Orange Pippin apples to modified atmosphere retail packaging. *Int J Food Sci Technol* 22:659–668.

Genovese DB, Elustondo MP, Lozano JE. (1997). Color and cloud stabilization in cloudy apple juice by steam heating during crushing. *J Food Sci* 62:1171–1175.

George P, Stratman CJ. (1952). The oxidation of myoglobin to metmyoglobin by oxygen. *Biochem J* 51:418–425.

Ghorpade VM, Cornforth DP. (1993). Spectra of pigments responsible for pink color in pork roasts cooked to 65 or 82°C. *J Food Sci* 58:51–52, 89.

Ghorpade VM, Cornforth DP, Sisson DV. (1992). Inhibition of red discoloration in cooked, vacuum packaged bratwurst. *J Food Sci* 57:1053–1055.

Gil MI, García-Viguera C, Artés F, Tomás-Barberán FA. (1995). Changes in pomegranate juice pigmentation during ripening. *J Sci Food Agric* 68:77–81.

Gil MI, Holcroft DM, Kader AA. (1997). Changes in strawberry anthocyanins and other polyphenols in response to carbon dioxide treatments. *J Agric Food Chem* 45:1662–1667.

Gill CO. (1996). Extending the storage life of raw chilled meats. *Meat Sci* 43:S99–S109.

Gill CO, Jones T. (1994). The display of retail packs of ground beef after their storage in master packages under various atmospheres. *Meat Sci* 37:281–395.

Gill CO, Jones T. (1996). The display life of retail packaged pork chops after their storage in master packs under atmospheres of N_2, CO_2 or O_2 + CO_2. *Meat Sci* 42:203–213.

Gill CO, McGinnis JC. (1995a). The effects of residual oxygen concentration and temperature on the degradation of the colour of beef packaged under oxygen-depleted atmospheres. *Meat Sci* 39:387–394.

Gill CO, McGinnis JC. (1995b). The use of oxygen scavengers to prevent the transient discolouration of ground beef packaged under controlled, oxygen-depleted atmospheres. *Meat Sci* 41:19–27.

Giusti MM, Rodríguez-Saona LE, Baggett JR, Reed GL, Durst RW, Wrolstad RE. (1998). Anthocyanin pigment composition of red radish cultivars as potential food colorants. *J Food Sci* 63:219–224.

Gökmen V, Borneman Z, Nijhuis HH. (1998). Improved ultrafiltration for color reduction and stabilization of apple juice. *J Food Sci* 63:504–507.

Goldman M, Horev B, Saguy I. (1983). Decolorization of beta-carotene in model systems simulating dehydrated foods: mechanism and kinetic principles. *J Food Sci* 48:751–754.

Gomez E, Martinez A, Laencina J. (1995). Prevention of oxidative browning during wine storage. *Food Res Int* 28:213–217.

Gomez Cordoves C, Gonzales San Jose ML. (1995). Interpretation of color variables during the aging of red wines: relationship with families of phenolic compounds. *J Agric Food Chem* 43:557–561.

Gomez Cordoves C, Gonzales San José ML, Junquera B, Estrella I. (1995). Correlation between flavonoids and color in red wines aged in wood. *Am J Enol Viticulture* 46:295–298.

Goodburn K, Halligan AC. (1988). *Modified atmosphere packaging: a technology guide*. Leatherhead, UK: Food Research Association.

Goodman LP, Markakis P. (1965). Sulfur dioxide inhibition of anthocyanin degradation of phenolase. *J Food Sci* 30:135–137.

Gorski PM, Creasy LL. (1977). Color development in "Golden Delicious" apples. *J Am Soc Hort Sci* 102:73–75.

Govindarajan VS. (1973). Fresh meat colour. *CRC Crit Rev Food Technol* 4: 117–140.

Govindarajan VS. (1979). Turmeric: chemistry, technology and quality. *CRC Crit Rev Food Sci Nutr* 12:199–301.

Govindarajan VS. (1986). Capsicum: production, technology, chemistry and quality, part III: chemistry of the color, aroma and pungency stimuli. *CRC Crit Rev Food Sci Nutr* 24:245–355.

Graf E, Karel M, Saguy I. (1991). Color system and method of use on foods. U.S. Patent 5,002,789.

Grant GF, McCurdy AR, Osborne AD. (1988). Bacterial greening in cured meats: a review. *Can Inst Food Sci Technol J* 21:50–56.

Grantham GJ. (1981). *Minced fish today: a review.* FAO Fisheries Tech. paper no. 216. Melton Mowbray, UK: Food and Agriculture Organization.

Gray JI, Gomaa EA, Buckley DJ. (1996). Oxidative quality and shelf life of meats. *Meat Sci* 43:S111–S123.

Greene BE, Price LG. (1975). Oxidation-induced color and flavor changes in meat. *J Agric Food Chem* 23:164–166.

Grommeck R, Markakis P. (1964). Effect of peroxidase on anthocyanin pigments. *J Food Sci* 48:53–57.

Gupte SM, Francis FJ. (1964). Effect of pH adjustment and high-temperature short-time processing on the color and pigment retention in spinach puree. *Food Technol* 18:1645–1648.

Guzman-Tello R, Cheftel JC. (1990). Colour loss during extrusion cooking of beta-carotene-wheat flour mixes as an indicator of the intensity of thermal and oxidative processing. *Int J Food Sci Technol* 25:420–434.

Haard NF. (1992). Control of chemical composition and food quality attributes of cultured fish. *Food Res Int* 25:289–307.

Hague MA, Warren KE, Hunt MC, Kropf DH, Kastner CL, Stroda SL, Johnson DE. (1994). Endpoint temperature, internal cooked color, and expressible juice color relationships in ground beef patties. *J Food Sci* 59:465–470.

Halim DH, Montgomery MW. (1978). Polyphenol oxidase of d'Anjou pears (*Pyrus communis*). *J Food Sci* 43:603–605, 608.

Hall GM. (1992). *Fish processing technology.* Glasgow: Blackie.

Halpin BE, Lee CY. (1987). Effect of blanching on enzyme activity and quality changes in green peas. *J Food Sci* 52:1002–1005.

Hammad AAI, El-Mongy TM. (1992). Shelf-life extension and improvement of the microbiological quality of smoked salmon by irradiation. *J Food Processing Preservation* 16:361–370.

Hamza F, Castaigne F, Willemot C, Doyon G, Makhlouf J. (1996). Storage of minimally processed Romaine lettuce in controlled atmosphere. *J Food Qual* 19:177–188.

Handumrongkul C, Silva JL. (1994). Aerobic counts, color and adenine nucleotide changes in CO_2 packed refrigerated striped bass strips. *J Food Sci* 59:67–69.

Handwerk RL, Coleman RL. (1988). Approaches to the citrus browning problem: a review. *J Agric Food Chem* 36:231–236.

Harper JM. (1989). Food extruders and their applications. In *Extrusion cooking*, ed. C Mercier, P Linko, JM Harper, 1–15. St Paul, Minn: American Society of Cereal Chemists.

Harper KA. (1968). Structural changes of flavylium salts IV polarographic and spectrometric examination of pelargonidin chloride. *Aust J Chem* 21: 221–223.

Hartley D. (1954). *Food in England*. London: MacDonald.

Heaton JW, Marangoni AG. (1996). Chlorophyll degradation in processed foods and senescent plant tissues. *Trends Food Sci Technol* 7:8–15.

Heaton JW, Yada RY, Marangoni AG. (1996). Discoloration of coleslaw is caused by chlorophyll degradation. *J Agric Food Chem* 44:395–398.

Heiss R, Radtke R. (1968). Uber den Einfluss von Licht, Sauerstoff und temperatur auf die Haltbarkeit verpackter Lebensmittel. *Verpak Rundsch* 19(3): 17–24.

Helmke A, Froning GW. (1971). Effect of end-point cooking temperature and storage on the color of turkey meat. *Poultry Sci* 50:1832–1836.

Hendel CE, Silveira VG, Harrington WO. (1955). Rates of nonenzymic browning of white potato during dehydration. *Food Technol* 9:433–438.

Herrman TJ, Love SL, Shafii B, Dwelle RB. (1996). Chipping performance of three processing potato cultivars during long-term storage at two temperature regimes. *Am Potato J* 73:411–425.

Hetherington MJ, Martin A, MacDougall DB, Langley KR, Bratchell N. (1990). Comparison of optical and physical measurements with a sensory assessment of the ripeness of tomato fruit *Lycopersicon esculentum*. *Food Qual Pref* 2:243–253.

Hewitt L. (1995). The fruits of Provence. *Food Manuf* 70(3):51.

Heyns K, Klier A. (1968). Fragmentierungen von Kohlenhydraten. *Carbohydrate Res* 6:436–448.

Hillman H. (1983). *Kitchen science*. Boston: Houghton Mifflin.

Himelbloom BH, Routledge JE, Biede SL. (1983). Color changes in blue crabs (*Callinectes sapidus*) during cooking. *J Food Sci* 48:652–653.

Hodge JE. (1953). Dehydrated foods: chemistry of browning reactions in model systems. *J Agric Food Chem* 1:928–943.

Hodge JE, Rist CE. (1953). The Amadori rearrangement under new conditions and its significance for nonenzymatic browning reactions. *J Am Chem Soc* 75:316–322.

Hogenbirk G. (1987). Color retention in pastel coatings. *Manuf Confectioner* 68(10):66–70.

Hollenbeck CM, Moore DG, Wendorff WL. (1973). Some facets of the smoke flavoring process. In *Proceedings of the Meat Industry Research Conference* 119. Arlington, VA: American Meat Institute.

Hong LC, Leblanc EL, Hawrysh ZL, Hardin RT. (1996). Quality of Atlantic mackerel (*Scomber scrombus* L.) fillets during modified atmosphere storage. *J Food Sci* 61:646–651.

Hood DE. (1980). Factors affecting the rate of metmyoglobin accumulation in pre-packaged beef. *Meat Sci* 4:247–265.

Hood DE, Riordan EB. (1973). Discoloration in pre-packaged beef: measurement by reflectance spectrophotometry and shoppers' discrimination. *J Food Technol* 8:333–343.

Houck LG, Aharoni Y, Fouse DC. (1978). Color changes in orange fruit stored in high concentrations of oxygen and in ethylene. *Proc Florida State Hort Soc* 91:136–139.

Howard LR, Dewi T. (1995). Sensory, microbiological and chemical quality of mini-peeled carrots as affected by edible coating treatment. *J Food Sci* 60:142–144.

Howard LR, Griffin LE, Lee Y. (1994). Steam treatment of minimally processed carrot sticks to control surface discoloration. *J Food Sci* 59:356–358, 370.

Howard LR, Miller H Jr, Wagner AB. (1995). Microbiological, chemical and sensory changes in irradiated pico de gallo. *J Food Sci* 60:461–464.

Howe JL, Gullett EA, Usborne WR. (1982). Development of pink color in cooked pork. *Can Inst Food Sci Technol J* 15:19–23.

Huang AS, von Elbe JH. (1987). Effect of pH on the degradation and regeneration of betanine. *J Food Sci* 52:1689–1693.

Huang HT. (1956). The kinetics of decolorisation of anthocyanins by fungal enzyme. *J Am Chem Soc* 78:2390–2393.

Huang T-C. (1986). Pigment composition and color changes during processing of *mei* (plum). In *Role of chemistry in the quality of processed food*, ed. OR Fennema, W-H Chang, C-Y Lii, 108–117. Westport, Conn: Food and Nutrition Press.

Huffman DL, Cordray JC. (1979). Restructured fresh meat cuts from chilled and hot processed pork. *J Food Sci* 44:1564–1565, 1567.

Huffman DL, Ly AM, Cordray JC. (1981). Effect of salt concentration on quality of restructured pork chops. *J Food Sci* 46:1563–1565.

Hutchings JB. (1969). Tristimulus colour measurement in the food industry. In *Proceedings of the First International Colour Association*, 581–589. Stockholm: Swedish Colour Centre Foundation.

Hutchings JB. (1978). Psychophysics of colour and appearance in product development. In *Proceedings of a food colour and appearance symposium*, University of Surrey, Guildford, Colour Group (GB), 46–55. London: Colour Group (GB).

Hwang LS, Cheng Y-C. (1986). Pink discoloration in canned lychees. In *Role of chemistry in the quality of processed foods*, ed. OR Fennema, W-H Chang, C-Y Lii, 96–107. Westport, Conn: Food and Nutrition Press.

Imai S, Akita K, Tomotake M, Sawada H. (1996). Blue pigment formation from onion and garlic. In *Book of abstracts of Institute of Food Technologists (USA) annual meeting*, 93–94. Chicago, IL: Institute of Food Technologists.

Imming R, Bliesener KM, Buckholz KN. (1994). The fundamental chemistry of colour formation in highly concentrated sucrose solutions. *Zuckerindustrie* 119:915–919.

Irving DE, Joyce DC. (1995). Sucrose supply can increase longevity of broccoli (*Brassica oleracea*) branchlets kept at 22°C. *Plant Growth Regulation* 17:251–256.

Jackman RL, Smith JL. (1996). Anthocyanins and betalains. In *Natural food colorants*, 2nd ed, ed. GAF Hendry, JD Houghton, 244–309. London: Chapman and Hall.

Jackson JE. (1980). Light interception and utilization by orchard systems. In *Horticultural Reviews*, vol. 2, ed. J Janick. Westport, Conn: Avi.

Jankowski KM, Parkin KL, von Elbe JH. (1997). Nonuniform browning or "mottling" in French fry products associated with a heterogeneous distribution of reducing sugars. *J Food Processing Preservation* 21:33–54.

Jauregui CA, Baker RC. (1980). Discoloration problems in mechanically deboned fish. *J Food Sci* 45:1068–1069.

Javenkoski JS, Schmidt SJ, Berber-Jimenez MD, Carragher BO. (1995). A physical basis for understanding the colour strength of fruit and vegetable powders. Institute of Food Technologists Annual Meeting 1995. Chicago, IL: Institute of Food Technologists.

Jayaraman KS. (1995). Critical review on intermediate moisture fruits and vegetables. In *Food preservation by moisture control*, ed. GV Barbosa-Cánovas, J Welti-Chanes, 411–441. Lancaster, Pa: Technomic.

Jeffries G. (1979). Natural colours. *Food Processing Ind* 48 (Nov):33, 35.

Jensen C, Lauridsen C, Bertelsen G. (1998). Dietary vitamin E: quality and storage stability of pork and poultry. *Trends Food Sci Technol* 9:62–72.

Jeremiah LE. (1981). The effect of frozen storage on the retail acceptability of pork loin chops and shoulder roasts. *J Food Qual* 5:73–88.

Johnson JR, Braddock RJ, Chen CS. (1995). Kinetics of ascorbic acid loss and nonenzymatic browning in orange juice serum: experimental rate constants. *J Food Sci* 60:502–505.

Jones ID, White RC, Gibbs E. (1964). Influence of blanching and brining treatments on the formation of chlorophyllides, pheophytins and pheoborides in green plant tissue. *J Food Sci* 29:437–439.

Jones L, Roddick J, Smith D. (1996). Potatoes: the poison potential. *Food Manuf* 71(11):36–37.

Joo KJ. (1982). Effect of saccharides on anthocyanin pigments from raspberries. *J Korean Soc Food Nut* 11(2): 21–25.

Joseph K, Aworh OC. (1991). Composition, sensory quality and respiration during ripening and storage of edible wild mango (*Irvingia gabonensis*). *Int J Food Sci Technol* 26:337–342.

Jurd L, Asen S. (1966). The formation of metal and copigment complexes of cyanidin 3-glucoside. *Phytochemistry* 5:1263–1271.

Kacem B, Cornell JA, Marshall MR, Shireman RB, Matthews RF. (1987). Nonenzymatic browning in aseptically packaged orange drinks, effect of ascorbic acid, amino acids and oxygen. *J Food Sci* 52:1668–1672.

Kajuna STAR, Bilanski WK, Mittal GS. (1998). Color changes in bananas and plantains during storage. *J Food Processing Preservation* 22:27–40.

Kallio H, Pallasaho S, Karppa J, Linko RR. (1986). Comparison of the half-lives of the anthocyanins in the juice of crowberry, *Empetrum nigrum. J Food Sci* 51:408–410, 430.

Kamman JF, Labuza TP. (1985). A comparison of the effect of liquid versus plastic vegetable shortening on rates of glucose utilization in nonenzymatic browning. *J Food Processing Preservation* 9:217–222.

Kanner J, Budowski P. (1978). Carotene oxidizing factors in red pepper fruits. *J Food Sci* 43:524–526.

Kanner J, Mendel H, Budowski P. (1978). Carotene oxidizing factors in red pepper fruits. *J Food Sci* 43:709–712.

Kantha SS, Erdman JW Jr. (1987). Legume carotenoids. *CRC Crit Rev Food Sci Nutr* 26:137–712.

Kaplan HJ. (1986). Washing, waxing and colour adding. In *Fresh citrus fruits*, ed. WF Wardowski, S Nagy, W Grierson, 379–395. New York: Van Nostrand.

Karrer P. (1962). The correlation between constitution, configuration and color in carotenoids. *Palette* (Autumn): 21–27.

Katsaboxakis KZ, Papanicolaou P. (1984). The consequences of varying degrees of blanching on the quality of frozen green beans. In *Thermal processing and quality of foods*, ed. P Zeuthen et al., 684–690. London: Elsevier Applied Science.

Kearsley MW, Katraboxakis KZ. (1980). Stability and use of natural colours in foods: red beet powder, copper chlorophyll and cochineal. *J Food Technol* 15:501–514.

Kearsley MW, Rodriguez N. (1981). The stability and use of natural colours in foods: anthocyanin, beta-carotene and riboflavin. *J Food Technol* 16: 421–431.

Kent-Jones DW, Amos AJ. (1967). *Modern cereal chemistry*. London: Food Trade Press.

Khanbari OS, Thompson AK. (1993). Effects of amino acids and glucose on the fry color of potato crisps. *Potato Res* 36:359–364.

Kikuchi T, Arakawa O, Norton RN. (1997). Improving skin colour of Fuji apples in Japan. *Fruit Varieties J* 51(2): 71–75.

Kilic M, Muthukumarappan K, Gunasekaran S. (1997). Kinetics on nonenzymatic browning in cheddar cheese powder during storage. *J Food Processing Preservation* 21:379–393.

King FJ. (1973). Improving the supply of minced block for the fish stick trade: a progress report. *Marine Fish Rev* 38(8):26.

King PL, Hung MC. (1995). Effect of ozonation on color and functional properties of mackerel (*Scomber australasicus*) meal. *Food Sci Taiwan* 22:218–226.

Kirby C. (1991). Microencapsulation and controlled delivery of food ingredients. *Food Sci Technol Today* 5:74–78.

Klimczak J, Irzyniec Z, Michalowski S. (1993). Colour stability of unblanched Brussels sprouts during cold storage. *Chlodnictwo* 28:28–31.

Knee M. (1972). Anthocyanin, carotenoid, and chlorophyll changes in the peel of Cox's Orange Pippin apples during ripening on and off the tree. *J Exp Bot* 23:184–196.

Knee M. (1980). Physiological responses of apple fruits to oxygen concentrations. *Ann Appl Biol* 96:243–253.

Knewstubb CK, Rayner PB. (1991). The practicalities of using natural food colours. *Food Technol Int Eur* :207–208, 210–211.

Kopas-Lane LM, Warthesen JJ. (1995). Carotenoid photostability in raw spinach and carrots during cold storage. *J Food Sci* 60:773–776.

Kopelman IJ, Saguy I. (1977). Color stability of beet powders. *J Food Processing Preservation* 1:217–224.

Koskitalo DN, Omrod DP. (1972). Effects of sub-optimal ripening temperatures on the color quality and pigment composition of tomato fruit. *J Food Sci* 37:56–59.

Kotzekidou P, Bloukas JG. (1996). Effect of protective cultures and packaging film permeability on shelf-life of sliced vacuum-packed cooked ham. *Meat Sci* 42:333–345.

Kropf DH. (1980). Effects of retail display conditions on meat color. In Contribution no. 81-79-A. Manhattan: Kansas State University, Department of Animal Sciences and Industry, Agricultural Experiment Station.

Kukura JL, Pfeiffer M, Walsh R. (1998). Calcium chloride added to irrigation water of mushrooms (*Agaricus bisporus*) reduces postharvest browning. *J Food Sci* 63:454–457.

LaBorde LF, von Elbe JH. (1996). Method for improving the color of containerised green vegetables. U.S. Patent 5482727.

Labuza TP. (1971). Kinetics of lipid oxidation in foods. *CRC Crit Rev Food Technol* 2:355–405.

Labuza TP, Schmidl MK. (1986). Advances in the control of browning reactions in foods. In *Role of chemistry in the quality of processed food*, ed. OR Fennema, W-H Chang, C-Y Lii, 65–95. Westport, Conn: Food and Nutrition Press.

Labuza TP, Tannenbaum SR, Karel M. (1970). Water content and stability of low and intermediate moisture foods. *Food Technol* 24:543–550.

Lajolo FM, Marquez UML. (1982). Chlorophyll degradation in a spinach system at low and intermediate water activities. *J Food Sci* 47:1995–1998.

Lajolo FM, Tannenbaum SR, Labuza TP. (1971). Reactions at limited water concentrations, 2: chlorophyll degradation. *J Food Sci* 36:850.

Lanari MC, Schaefer DM, Scheller KK. (1995). Dietary vitamin E supplementation and discoloration of pork bone and muscle following modified atmosphere packaging. *Meat Sci* 41:237–250.

Lanari MC, Zaritsky NE. (1991). Effect of packaging and frozen storage temperature on beef pigments. *Int J Food Sci Technol* 26:629–640.

Lancaster JE. (1992). Regulation of skin colour in apples. *CRC Crit Rev Plant Sci* 10:487–502.

Langdon TT. (1987). Prevention of browning in fresh prepared potatoes without the use of sulfiting agents. *Food Technol* 41(5):64–67.

Lawrie RA. (1985). Chemical and biochemical constitution of muscle. In *Meat science*, ed. RA Laurie, 43–73. Oxford, UK: Pergamon Press.

Leatherhead Food Research Association. (1997). *Food colours handbook*. Leatherhead, UK: Leatherhead Food Research Association.

Ledward DA. (1983). Haemoproteins in meat and meat products. In *Developments in food proteins*, ed. BFJ Hudson, 33–68. London: Elsevier Applied Science.

Ledward DA. (1985). Post-slaughter influences on the formation of metmyoglobin in beef muscles. *Meat Sci* 15:149–171.

Lee CY, Whitaker JR, eds. (1995). *Enzymatic browning and its prevention*. Washington, DC: American Chemical Society.

Lee DS, Chung SK, Kim HK, Yam KL. (1991). Nonenzymatic browning in dried red pepper products. *J Food Qual* 14:153–163.

Lee DS, Chung SK, Yam KL. (1992). Carotenoid loss in dried red pepper products. *Int J Food Sci Technol* 27: 179–185.

Lee FA. (1958). The blanching process. *Adv Food Res* 8:63–109.

Lee HS. (1997). Issue of color in pigmented grapefruit juice. *Fruit Processing* 7(4):132–135.

Lentz CP. (1979). Effect of light intensity and other factors on the color of frozen prepackaged beef. *Can Inst Food Sci Technol J* 12:47–50.

Lesellier E, Marty C, Lebert A, Laguerre J-C, Berset C, Bimbenet J-J. (1988). Effect of different pre-drying and drying parameters of carrots. In *Proceedings of the International Drying Symposium*, vol. 2, PD1–PD6. The Hague, Netherlands: International Drying Symposium.

Leszkowiat MJ, Barichello V, Yada RY, Coffin RH, Lougheed EC, Stanley DW. (1990). Contribution of sucrose to nonenzymatic browning in potato chips. *J Food Sci* 55:281–282, 284.

Leszkowiat MJ, Yada RY, Coffin RH, Stanley DW, McKeown A. (1986). The effect of location and harvest date on the sugar content and chip colour of early harvest potatoes grown in southern Ontario. *J Inst Can Sci Technol Aliment* 19:xxxvii.

Li SJ, Seymour TA, King AJ, Morrissey MT. (1998). Color stability and lipid oxidation of rockfish as affected by antioxidant from shrimp shell waste. *J Food Sci* 63:438–441.

Licciardello JJ, Ravesi RM, Allsup MG. (1980). Extending the shelf life of frozen Argentine hake. *J Food Sci* 45:1312–1317.

Licciardello JJ, Ronsivalli LJ. (1982). Irradiation of seafoods. In *Chemistry and biochemistry of marine food products*, ed. RE Martin, GJ Flick, CE Hebard, DR Ward, 305. Westport, Conn: Avi.

Lill RE, O'Donoghue EM, King GA. (1989). Postharvest physiology of peaches and nectarines. *Hort Rev* 10: 413–452.

Lin ZM, Schyvens E. (1994). Effects of blanching on the texture and colour of sterilized vegetables. *Food Fermentation Ind* 4:8–15.

Lin Z, Schyvens E. (1995). Influence of blanching treatments on the texture and colour of some processed vegetables and fruits. *J Food Processing Preservation* 19:451–465.

Lingnert H. (1980). Antioxidative Maillard reaction products, III: applications in cookies. *J Food Processing Preservation* 4:219–233.

Lingnert H, Eriksson CE. (1980a). Antioxidative Maillard reaction products, I: products from sugars and free amino acids. *J Food Processing Preservation* 4:161–172.

Lingnert H, Eriksson CE. (1980b). Antioxidative Maillard reaction products, II: products from sugars and peptides or protein hydrolysates. *J Food Processing Preservation* 4:173–181.

Lingnert H, Lundgren B. (1980). Antioxidative Maillard reaction products, IV: application in sausage. *Food Processing Preservation* 4:235–246.

Liu Q, Scheller KK, Arp SC, Schaefer DM, Frigg M. (1996). Color coordinates for assessment of dietary vitamin E effects on beef color stability. *J Animal Sci* 74:106–116.

Lloyd AG. (1980). Extraction and chemistry of cochineal. *Food Chem* 5:91–107.

Lopez A, Pique MT, Boatella J, Romero A, Ferran A, Garcia J. (1997). Influence of drying conditions on hazel nut quality, III: browning. *Drying Technol* 15:989–1002.

Lozano JE, Drudis-Biscarri R, Ibarz-Ribas A. (1994). Enzymatic browning in apple pulps. *J Food Sci* 59:564–567.

Lozano-de-Gonzalez PG, Barrett DM, Wrolstad RE, Durst RW. (1993). Enzymatic browning inhibited in fresh and dried apple ring by pineapple juice. *J Food Sci* 58:399–615.

Lubis Z, Buckle KA. (1990). Rancidity and lipid oxidation of dried-salted sardines. *Int J Food Sci Technol* 25:295–303.

Luchsinger SE, Kropf DH, García Zepeda CM, Hynt MC, Marsden JL, Rubio Cañas EJ, Kastner CL, Kuecker WG, Mata T. (1996). Color and oxidative rancidity of gamma and electron beam-irradiated boneless pork chops. *J Food Sci* 61:1000–1005, 1093.

Lukton A, Chichester CO, MacKinney G. (1956). The breakdown of strawberry anthocyanin pigment. *Food Technol* 10:427–432.

Lundberg DE, Kotschevar LH, Ceserini V. (1973). *Understanding cooking*. London: Edward Arnold.

Luno M, Beltran JA, Roncales P. (1998). Shelf life extension and colour stabilisation of beef packaged in a low O_2 atmosphere containing CO: loin steaks and ground meat. *Meat Sci* 48:75–84.

Lurie S. (1992). Controlled atmosphere storage to decrease physiological disorders in nectarines. *Int J Food Sci Technol* 27:507–514.

Lurie S. (1993). Modified atmosphere storage of peaches and nectarines to reduce storage disorders. *J Food Qual* 16: 57–65.

Luzuriaga DA, Balaban MO, Yeralan S. (1997). Analysis of visual quality attributes of white shrimp by machine vision. *J Food Sci* 62:113–118, 130.

Lykens D. (1985). Colorants and confectionery applications. *Manuf Confectioner* 65:61–64.

Mabry TJ. (1978). The betacyanins and betaxanthins. In *Comparative phytochemistry*, ed. T Swain, 231. New York: Academic Press.

Maccarone E, Campisi S, Cataldi Lupo MC, Fallico B, Nicolosi AC. (1996). Thermal treatment effects on the red orange juice constituents. *Industrie delle Bevande* 25(144): 335–341.

Maccarone E, Maccarone A, Rapisarda P. (1987). Colour stabilization of orange fruit juice by tannic acid. *Int J Food Sci Technol* 22:159–162.

MacDougall DB. (1977). Colour in meat. In *Sensory properties of food*, ed. GG Birch, JG Brennan, KJ Parker, 59–69. London: Applied Science.

MacDougall DB. (1982). Changes in the colour and opacity of meat. *Food Chem* 9:75–88.

MacDougall DB. (1985). The influence of appearance on the shelf life of meat. *J Sci Food Agric* 36:124–125.

MacDougall DB, Hetherington MJ. (1992). The minimum quantity of nitrite required to stain sliced and homogenized cooked pork. *Meat Sci* 31:201–210.

MacDougall DB, Taylor AA. (1975). Colour retention in fresh meat stored in oxygen: a commercial scale trial. *J Food Technol* 10:339–347.

Macheix J-J, Sapis J-C, Fleuriet A. (1991). Phenolic compounds and polyphenoloxidase in relation to browning in grapes and wine. *CRC Crit Rev Food Sci Nutr* 30:441–487.

Mackinney S, Lukton A, Greenbaum L. (1958). Carotenoid stability in stored dehydrated carrots. *Food Technol* 12:164–166.

Madhavi DL, Carpenter CE. (1993). Aging and processing affect color, metmyoglobin reductase and oxygen consumption of beef muscles. *J Food Sci* 58:939–942, 947.

Maga JA. (1994). Pink discoloration in cooked white meat. *Food Rev Int* 10:273–286.

Mahanta PK, Baruah S. (1992). Changes in pigments and phenolics and their relationship with black tea quality. *J Sci Food Agric* 59:21–26.

Mahanta PK, Hazarika M. (1985). Chlorophyll and degradation products in orthodox and CTC black teas and their influence on shade of colour and sensory quality in relation to thearubigins. *J Sci Food Agric* 36:1133–1141.

Markakis P. (1982). *Anthocyanins as food colors*. New York: Academic Press.

Markakis P, Livingstone GE, Fellers RC. (1957). Quantitative aspects of strawberry-pigment degradation. *Food Res* 22:117–130.

Marksberry CL, Kropf DH, Huny MC, Hague MA, Warren KE. (1993). *Ground beef patty cooked color guide*. Manhattan: Kansas State University, Department of Animal Sciences and Industry.

Marquez G, Añon MC. (1986). Influence of reducing sugars and amino acids in the color development of fried potatoes. *J Food Sci* 51:157–160.

Martinez F, Labuza TP. (1968). Rate of deterioration of freeze-dried salmon as a function of relative humidity. *J Food Sci* 33:241–247.

Marty C, Lebert A, Lesellier E, Berset C, Bimbenet J. (1988). Color control of carrots subject to different drying experimental conditions. *Proceedings of the Congress on Progress in Food Preservation Processes*, 81–88. Brussels: CERIA.

Maruf FW, Ledward DA, Neale RJ, Poulter RG. (1990). Chemical and nutritional quality of Indonesian dried-salted mackerel (*Rastrelliger kanagurta*). *Int J Food Sci Technol* 25:66–77.

Mastrocola D, Pittia P, Lerici CR. (1996). Quality of apple slices processed by combined techniques. *J Food Qual* 19:133–146.

Matlock RG, Terrell RN, Savell JW, Rhee KS, Dutson TR. (1984). Factors affecting properties of raw-frozen pork sausage patties made with various NaCl/phosphate combinations. *J Food Sci* 49:1363–1366, 1371.

Matthews AD. (1983). Muscle colour deterioration in iced and frozen stored bonito, yellowfin and skipjack tuna caught in Seychelles waters. *J Food Technol* 18:387–392.

Mauron J. (1981). The Maillard reaction in food: a critical review from the nutritional standpoint. *Prog Food Nutr Sci* 5:5–35.

Maurisch W, DeRitter E, Bauernfeind JC. (1960). Education of carotenoid pigments for coloring egg yolks. *Poultry Sci*:1338–1345.

Mazza G. (1995). Anthocyanins in grapes and grape products. *Crit Rev Food Sci Nutr* 35:341–371.

Mazza G, Miniati H. (1993). *Anthocyanins in fruit, vegetables and grains*. Boca Raton, Fla: CRC Press.

McEvily AJ, Iyengar R. (1992). Inhibition of enzymatic browning in foods and beverages. *Crit Rev Food Sci Nutr* 32:253–273.

McLellan MR, Kime RW, Lee CY, Long TM. (1995). Effect of honey as an antibrowning agent in light raisin processing. *J Food Processing Preservation* 19:1–8.

Meacock G, Taylor KDA, Knowles MJ, Himonides A. (1997). The improved whitening of minced cod flesh using dispersed titanium dioxide. *J Sci Food Agric* 73:221–225.

Meggos H. (1995). Food colors: an international perspective. *Manuf Confectioner* 75(2):59–65.

Meheruik M, Girard B, Moyls L, Beveridge HJT, McKenkie DL, Harrison J, Weintraub S, Hocking R. (1995). Modified atmosphere packaging of "Lapins" sweet cherry. *Food Res Int* 28:239–244.

Meir S, Rosenberger I, Aharon Z, Grinberg S, Fallik F. (1995). Improvement of the postharvest keeping quality and colour development of bell pepper (cv *Major*) by packaging with polyethylene bags at a reduced temperature. *Postharvest Biol Technol* 5:303–309.

Millar SJ, Moss BW, MacDougall DB, Stevenson MH. (1995). The effect of ionising radiation on the CIELAB colour co-ordinates of chicken breast meat as measured by different instruments. *Int J Food Sci Technol* 30:663–674.

Millar SJ, Moss BW, Stevenson MH. (1996). The effect of ionising radiation on meat colour. *Meat Focus* 5:229–233.

Millar S, Wilson R, Moss BW, Ledward DA. (1994). Oxymyoglobin formation in meat and poultry. *Meat Sci* 36:397–406.

Minguez-Mosquera MI, Garrido-Fernandez J. (1989). Chlorophyll and carotenoid presence in olive fruit (*Olea europaea*). *J Agric Food Chem* 37:1–7.

Minguez-Mosquera MI, Garrido-Fernandez J, Gandul-Rojas B. (1989). Pigment changes in olives during fermentation and brine storage. *J Agric Food Chem* 37:8–11.

Mirza S, Morton ID. (1977). Effect of different types of blanching on the colour of sliced carrots. *J Sci Food Agric* 28:1035–1039.

Mishkin M, Saguy I, Karel M. (1983). Dynamic optimization of dehydration processes: minimizing browning in dehydration of potatoes. *J Food Sci* 48:1617–1621.

Mitchell GE, McLauchlan RL, Beattie TR, Banos C, Gillen AA. (1990). Effect of gamma irradiation on the carotene content of mangoes and red capsicums. *J Food Sci* 55:1185–1186.

Mitsumoto M, Cassens RG, Schaefer DM, Arnold RN, Scheller KK. (1991). Improvement of color and lipid stability in beef longissimus with dietary vitamin C dip treatment. *J Food Sci* 56:1489–1492.

Mohammed M, Wickham LD. (1995). Postharvest retardation of senescence in shado benni (*Eryng foetidum*, L) Plants. *J Food Qual* 18:325–334.

Montgomery MW, Petropakis HJ. (1980). Inactivation of Bartlett pear polyphenol oxidase with heat in the presence of ascorbic acid. *J Food Sci* 45:1090–1091.

Moore VJ. (1990). Colour stability of thawed lamb chops, II: chops from loins frozen for long periods. *Int J Food Sci Technol* 25:637–642.

Morales-Castro J, Rao MA, Hotchkiss JH, Downing DL. (1994a). Modified atmosphere packaging of head lettuce. *J Food Processing Preservation* 18: 295–304.

Morales-Castro J, Rao MA, Hotchkiss JH, Downing DL. (1994b). Modified atmosphere packaging of sweet corn on the cob. *J Food Processing Preservation* 18:279–293.

Morita H, Niu J, Sakata R, Nagata Y. (1996). Red pigment of Parma ham and bacterial influence on its formation. *J Food Sci* 61:1021–1023.

Muftugil N. (1986). Effect of different types of blanching on the color and the ascorbic acid contents of green beans. *J Food Processing Preservation* 10:69–76.

Nagy S, Lee H, Rouseff RL, Lin JCC. (1990). Nonenzymic browning of commercially canned and bottled grapefruit juice. *J Agric Food Chem* 38:343–346.

Nash DM, Proudfoot FG, Hulan HW. (1985). Pink discoloration in cooked broiler chicken. *Poultry Sci* 64:917–919.

Nicolas JJ, Richard-Forget FC, Goupy PM, Amiot M-J, Aubert SY. (1994). Enzymatic browning reactions in apple and apple products. *Crit Rev Food Sci Nutr* 34(2):109–157.

Nisperos-Carriedo MO, Shaw PE, Baldwin EA. (1990). Changes in volatile flavor components of pineapple orange juice as influenced by the application of lipid and composite films. *J Agric Food Chem* 38:1382–1387.

Nussinovitch A, Ward G, Mey-Tal E. (1996). Gloss of fruit and vegetables. *Lebens Wiss Technol* 29:184–186.

O'Boyle AR, Aladin-Kasam N, Rubin LM, Diosady LL. (1992). Encapsulated cured-meat pigment and its application in nitrite-cured ham. *J Food Sci* 57:807–812.

Ockerman HW, Cahill VR. (1977). Microbiological growth and pH effects on bovine tissue inoculated with *Pseudomonas putrefaciens, Bacillus subtilis* or *Leuconostoc mesenteroides. J Food Sci* 42:141–145.

O'Donoghue EP, Marangoni AG, Yada RY. (1996). The relationship of chip color with structural parameters of starch. *Am Potato J* 73:545–558.

Omidiji O, Okpuzor J. (1996). Time course of PPO-related browning of yams. *J Sci Food Agric* 70:190–196.

Owuor PO, Othieno CO. (1988). Studies on the use of shade in tea plantations in Kenya: effects on chemical composition and quality of made tea. *J Sci Food Agric* 46:63–70.

Palmer JK. (1984). Enzyme reactions and acceptability of plant foods. *J Chem Educ* 61:284–289.

Paradis C, Castaigne F, Desrosiers T, Fortin J, Rodrigue N, Willemot C. (1996). Sensory, nutrient and chlorophyll changes in broccoli florets during controlled atmosphere storage. *J Food Qual* 19:303–316.

Park HJ, Bunn JM, Vergano PJ, Testin RF. (1994). Gas permeation and thickness of the sucrose polyesters: Semperfresh coatings on apples. *J Food Processing Preservation* 18:349–358.

Park HJ, Chinnan MS, Shewfelt RL. (1994). Edible corn-zein film coatings to extend storage life of tomatoes. *J Food Processing Preservation* 18:317–331.

Park JM, Joo KJ. (1982). Stability of anthocyanin pigment from juice of raspberries. *J Korean Soc Food Nutr* 11(3): 67–74.

Park YW. (1987). Effect of freezing, thawing, drying and cooking on carotene retention in carrots, broccoli and spinach. *J Food Sci* 52:1022–1025.

Parker NS. (1987). Properties and functions of stabilizing agents in food emulsions. *CRC Crit Rev Food Sci Nutr* 25:285–315.

Parkin KL, Im J-S. (1990). Chemical and physical changes in beet (*Beta vulgaris* L.) root tissue during simulated processing: relevance to the "black ring" defect in canned beets. *J Food Sci* 55:1039–1041, 1053.

Parry RT. (1993). *Principles and applications of modified atmosphere packaging of food*. New York: Chapman and Hall.

Pasch JH, von Elbe JH, Dinesen N. (1975). Betalaines as natural food colorants. *Food Prod Dev* 9(11):38, 42, 45.

Patel AA, Gandhi H, Singh S, Paril GR. (1996). Shelf-life modelling of sweetened condensed milk based on kinetics of Maillard browning. *J Food Processing Preservation* 20:431–451.

Pegg RB, Shahidi F. (1997). Unravelling the chemical identity of meat pigments. *Crit Rev Food Sci Nutr* 37:561–589.

Peng CY, Markakis P. (1963). Effect of phenolase on anthocyanins. *Nature* 199:597–598.

Perlo F, Gago-Gago A, Rosmini M, Cervera-Perez R, Perez-Alvarez J, Pahgan-Moreno M, Lopez-Santovena F, Aranda-Catale V. (1995). Modification of physico-chemical and colour parameters during the marketing of paté. *Meat Sci* 41:325–333.

Perrin PW, Gaye MM. (1986). Effects of simulated retail display and overnight storage treatments on quality maintenance in fresh broccoli. *J Food Sci* 51:146–149.

Petropakis HJ, Montgomery MW. (1984). Improvement of colour stability of pear juice concentrate by heat treatment. *J Food Technol* 49:91–95.

Philip T. (1975). Utilization of plant pigments as food colorants. *Food Prod Dev* 9(3):50, 52, 54.

Phillips CA. (1996). Review: modified atmosphere packaging and its effects on the microbiological quality and safety of produce. *Int J Food Sci Technol* 31:463–479.

Piantanida L, Vallone L, Cantoni C. (1996). Classification on surface colouring of Taleggio cheese. *Industrie-Alimentari* 35:147–148.

Pierson MD, Smoot LA. (1982). Nitrite, nitrite alternatives and control of

Clostridium botulinum in cured meats. *CRC Crit Rev Food Sci Nutr* 17:141–187.

Pike M. (1974). *Catering science and technology*. London: John Murray.

Pinicelli A, Bakker J, Bridle P. (1994). Model wine solutions: effect of sulphur dioxide on colour and composition during aging. *Vitis* 33:31–35.

Pizzocaro F, Torreggiani D, Gilardi G. (1993). Inhibition of apple polyphenoloxidase by ascorbic acid, citric acid and sodium chloride. *J Food Processing Preservation* 17:21–30.

Poei-Langstron MS, Wrolstad RE. (1981). Color degradation in an ascorbic acid-anthocyanin-flavanol model system. *J Food Sci* 46:1218–1222, 1236.

Pokorny J, El-Zeany BA, Janicek G. (1974). Browning reaction of oxidized fish lipid with protein. *Proceedings of the Fourth International Congress of Food Science and Technology*, vol. 1, 217–227. Ontario: International Union of Food Science and Technology.

Pomeranz Y, Shellenberger JA. (1971). *The art and science of breadmaking*. Westport, Conn: Avi.

Poole SE, Mitchell GE, Mayze JL. (1994). Low dose irradiation affects microbiological and sensory quality of sub-tropical seafood. *J Food Sci* 59:85–87, 105.

Popplewell LM, Black J, Norris LM, Pozzio M. (1995). Encapsulation system for flavors and colors. *Food Technol* 49(5):76–82.

Pratt B Jr. (1971). Criteria of flour quality. In *Wheat: chemistry and technology*, ed. Y Pomeranz. St. Paul, Minn. American Society of Cereal Chemists.

Proctor JTA, Creasy LL. (1971). Effect of supplementary light on anthocyanin synthesis in "McIntosh" apples. *J Am Soc Hort Sci* 96:523–526.

Pszczola DE. (1997). Lookin' good: improving the appearance of food products. *Food Technol* 51(11):39–44.

Pszczola DE. (1998). Natural colors: pigments of imagination. *Food Technol* 52(6):70–82.

Purcell AE, Walter WM Jr, Tompkins GA Jr. (1969). Relationship of vegetable color to physical state of the carotenes. *J Agric Food Chem* 17:41–42.

Quaglia GB. (1988). Other durum products. In *Durum wheat: chemistry and technology*, eds. G Fabriani, C Lintas. St. Paul, Minn: American Society of Cereal Chemists.

Rahman FMM, Buckle KA. (1981). Effects of blanching and sulphur dioxide on ascorbic acid and pigments of frozen capsicums. *J Food Technol* 16:671–682.

Ramakrishnan TV, Francis FJ. (1979). Stability of carotenoids in model aqueous systems. *J Food Qual* 2:177–189.

Ramaswamy S. (1973). Coloring principles in foods. *Bombay Tech* 23:9–17.

Ramshaw EW, Hardy PJ. (1969). Volatile compounds in dried grapes. *J Sci Food Agric* 20:619–621.

Ranganna S, Govindarajan VS, Ramana KVR. (1983). Citrus fruits, part 2. *CRC Crit Rev Food Sci Nutr* 18:313–386.

Rassis D, Saguy IS. (1995). Kinetics of aseptic concentrated orange juice quality changes during commercial processing and storage. *Int J Food Sci Technol* 30:191–198.

Razhinskaite DK. (1971). Active substances of St. John's wort. *Tr Akad Nauk Lit SSR* 1:89–100.

Reagan JO, Liou FH, Reynolds AE, Carpenter JA. (1983). Effect of processing variables on the microbial, physical and sensory characteristics of pork sausage. *J Food Sci* 48:146–149, 162.

Reay PF, Fletcher RH, Thomas VJ. (1998). Chlorophylls, carotenoids and anthocyanin concentrations in the skin of "Gala" apples during maturation and the influence of foliar applications of nitrogen and magnesium. *J Sci Food Agric* 76:63–71.

Reddy NR, Armstrong DJ, Rhodehamel EJ, Kautter DA. (1992). Shelf life extension and safety concerns about fresh fishery products under modified atmospheres: a review. *J Food Safety* 12(2):87–118.

Renerre M. (1990). Review: factors involved in the discoloration of beef meat. *Int J Food Sci Technol* 25:613–630.

Rhee KS, Watts BW. (1966). Effects of antioxidants on lipoxidase activity in model systems and pea (*Pisum sativum*) slurries. *J Food Sci* 31:669.

Rickansrud DA, Hendrickson RI. (1967). Total pigments and myoglobin concentration in four bovine muscles. *J Food Sci* 32:57–61.

Riha WE, Wendorff WL. (1993). Evaluation of color in smoked cheese by sensory and objective methods. *J Dairy Sci* 76:1491–1497.

Rizzi GP. (1997). Chemical structure of colored Maillard reaction products. *Food Rev Int* 13(1):1–28.

Robach DL, Costilow RN. (1962). Role of bacteria in the oxidation of myoglobin. *Appl Microbiol* 9:529–533.

Robin JP, Lopez F, Roujou de Boubee D, Igounet O, Sauvage FX, Pradal M, Verries C. (1996). The coloration of the Shiraz grape berries during their ripening: relation between the colour descriptors, in situ dynamic and influence of some environmental factors. *J Int Sci Vigne Vin* 30: 187–199.

Rocha T, Lebert A, Marty-Audouin C. (1992). Effect of drying conditions and of blanching on drying kinetics and color of mint (*Mentha spicata* Huds.) and basil (*Ocimum basilicum*). In *Proceedings of the International Drying Symposium*. Amsterdam: Elsevier.

Rodriguez-Saona LE, Wrolstad RE. (1997). Influence of potato composition on chip color quality. *Am Potato J* 74:87–106.

Rodriguez-Saona LE, Wrolstad RE, Pereira C. (1997). Modeling the contribution of sugars, ascorbic acid, chlorogenic acid, and amino acids to non-enzymatic browning of potato chips. *J Food Sci* 62:1001–1005, 1010.

Roe MA, Faulks RM. (1991). Color development in a model system during frying: role of individual amino acids and sugars. *J Food Sci* 56:1711–1713.

Rogers AM, El-Hag NA, Shenouda YK. (1987). Preservation of the green color of blanched vegetables. U.S. Patent 4,701,330.

Rohm H, Jaros D. (1997). Colour of hard cheese, 2: factors of influence and relation to compositional parameters. *Z Lebens Forsch A* 204:259–264.

Roller S. (1996). *Guide to food biotechnology*. London: Institute of Food Science and Technology.

Romero CR, García P, Brenes M, Garrido A. (1995). Colour and texture changes during sterilization of packed ripe olives. *Int J Food Sci Technol* 30:31–36.

Rommel A, Wrolstad RE, Heatherbell DA. (1992). Blackberry juice and wine: processing and storage effects on anthocyanin composition, color and appearance. *J Food Sci* 57:385–391, 410.

Ruiter A. (1979). Color of smoked foods. *Food Technol* 33(5):54, 56, 58–60, 63.

Ruiz SR, Valenzuela BJ, Munoz SC. (1986). Association of nitrogen nutrition with colour disorders in Granny Smith apples. *Agric Tecnica* 46:369–371.

Ryall AL, Lipton WG. (1983). *Handling, transportation, and storage of fruits and vegetables*, vol. 1. Westport, Conn: Avi.

Saguy I, Goldman M, Karel M. (1985). Prediction of beta-carotene decolorization in model system under static and dynamic conditions of reduced oxygen environment. *J Food Sci* 50:526–530.

Saguy IS, Graf E. (1991). Particle size effects on the diffuse reflectance of a sucrose-caramel admixture. *J Food Sci* 56:1117–1118, 1120.

Saguy I, Kopelman IJ, Mizrahi S. (1978). Extent of nonenzymatic browning in grapefruit juice during thermal and concentration processes: kinetics and prediction. *J Food Processing Preservation* 2:175–184.

Saito N, Abe K, Honda T, Timberlake CF, Bridle P. (1985). Acylated delphinidin glucosides and flavonols from *Clitoria ternatea*. *Phytochemistry* 24: 1583–1586.

Saks Y, Waks Y, Weiss B, Franck A, Chalutz E. (1988). Light-induced postharvest regreening of pummelo fruit. *Ann Appl Biol* 113:375–381.

Salunkhe DK, Desai BB. (1984). *Post-harvest biotechnology of vegetables*, vol. 1. Boca Raton, Fla: CRC Press.

Sampathu SR, Shivashankar S, Lewis YS. (1984). Saffron (*Crocus sativus*): cultivation, processing, chemistry and standardisation. *CRC Crit Rev Food Sci Nutr* 20:123–157.

Sapers GM. (1993). Browning of foods: control by sulphites, antioxidants and other means. *Food Technol* 47(10):75–84.

Sapers GM, Garzarella L, Pilizota V. (1990). Application of browning inhibitors to cut apple and potato by vacuum and pressure infiltration. *J Food Sci* 55:1049–1051.

Sapers GM, Hicks KB. (1989). Inhibition of enzymatic browning in fruits and vegetables. In *Quality factors of fruit and vegetables: chemistry and technology*, 29. ed. JJ Jen, ACS Symposium Ser. No. 405. Washington DC: American Chemical Society.

Sapers GM, Miller RL. (1995). Heated ascorbic/citric acid solution as browning inhibitor for pre-peeled potatoes. *J Food Sci* 60:762–766, 776.

Sapers GM, Miller RL. (1998). Browning inhibition in fresh-cut pears. *J Food Sci* 63:342–346.

Sapers GM, Miller RL, Miller FC, Coke PH, Choi S-W. (1994). Enzymatic browning control in minimally processed mushrooms. *J Food Sci* 59: 1042–1047.

Sapers GM, Simmons GF. (1998). Hydrogen peroxide disinfection of minimally processed fruits and vegetables. *Food Technol* 52(2):48–52.

Sapis JC, Macheix JJ, Cordonnier RE. (1983). The browning capacity of grapes, 1: changes in polyphenol oxidase activities during development and maturation of the fruit. *J Agric Food Chem* 31:342–345.

Sapru V, Peleg M. (1988). Hierarchy in food colorants absorption on selected crystalline powders. *J Food Sci* 53:555–557.

Satterlee LD, Hansmeyer W. (1974). The role of light and surface bacteria in the color stability of prepackaged beef. *J Food Sci* 39:305–308.

Sayre CB, Robinson WB, Wishetsky T. (1953). Effect of temperature on the color, lycopene and carotene content of detached and vine-ripened tomatoes. *Proc Am Soc Hort Sci* 61:381–387.

Schnickels RA, Warmbier HC, Labuza TP. (1976). Effect of protein substitution on nonenzymatic browning in an intermediate moisture food system. *J Agric Food Chem* 24:901–903.

Schuch W. (1994). Improving tomato quality through biotechnology. *Food Technol* 48(11):79–83.

Schwartz SJ, Lorenzo TV. (1989). Chlorophyll stability during aseptic processing and storage. Paper presented at the Annual Institute of Food Technology Meeting. Chicago, IL: Institute of Food Technology.

Schwartz SJ, Lorenzo TV. (1990). Chlorophylls in food. *CRC Crit Rev Food Sci Nutr* 29:1–17.

Schwartz SJ, von Elbe JH. (1983). Kinetics of chlorophyll degradation to pheophytin in vegetables. *J Food Sci* 48:1303–1306.

Schwartz SJ, von Elbe JH, Lindsay RC. (1983). Influence of processing on the pigmentation of wild rice grain. *J Agric Food Chem* 31:349–352.

Segal B. (1971). Degradarea produselor alimenture sub actiunea luminii. *Biochim Prod Aliment*:175–181.

Segner WP, Ragusa TJ, Nank WK, Hoyle WC. (1984). Process for the preservation of green color in canned vegetables. U.S. Patent no. 4,473,591.

Seideman SC, Cross HR, Smith GC, Durland PR. (1984). Factors associated with fresh meat color: a review. *J Food Qual* 6:211–237.

Seideman SC, Durland PR. (1983). Vacuum packaging of fresh beef: a review. *J Food Qual* 6:29–47.

Senter SD, Young LL, Searcy GK. (1997). Colour values of cooked top-round beef juices as affected by end-point temperatures, frozen storage of cooked samples and storage of expressed juices. *J Sci Food Agric* 75:179–182.

Seow CC, Lee SK. (1997). Firmness and color retention in blanched green beans and green bell pepper. *J Food Qual* 20:329–336.

Sepulveda E, Saenz C, Navarrete A, Rustom A. (1996). Color parameters of passion fruit juice (*Passiflora edulis* Sims): influence of harvest season. *Ciencia Technol Alimentos Int* 2(1):29–33.

Sgaramella S, Ames JM. (1993). The development and control of colour in extrusion cooked foods. *Food Chem* 46:129–132.

Shahidi F, Pegg RB, Shamsuzzaman K. (1991). Color and oxidative stability of nitrite-free cured meat after gamma radiation. *J Food Sci* 56:1450–1452.

Shahidi F, Rubin LJ, Diosady LL, Chew V, Wood DF. (1985). Preparation of the cooked cured-meat pigment dinitrosyl ferrohemochrome from hemin and nitric oxide. *J Food Sci* 50:272–273.

Shewfelt RL, Batal KM, Heaton EK. (1983). Broccoli storage, effect of N^6-benzyladenine, packaging, and icing on color of fresh broccoli. *J Food Sci* 48:1594–1597.

Shewfelt RL, Myers SC, Resurreccion AVA. (1987). Effect of physiological maturity at harvest on peach quality during low temperature storage. *J Food Qual* 10:9–20.

Shi Z, Francis FJ, Daun H. (1992). Quantitative comparison of the stability of anthocyanins from *Brassica oleracea* and *Tradescantia pallida* in nonsugar drink model and protein model systems. *J Food Sci* 57:768–770.

Shrikhande AJ. (1986). Anthocyanins in foods. *CRC Crit Rev Food Sci Nutr* 7:193–213.

Siebert KA, Lynn PY. (1997). Haze-active protein and polyphenols in apple juice assessed by turbidimetry. *J Food Sci* 62:79–84.

Siegel A, Markakis P, Bedford CL. (1971). Stabilization of anthocyanins in frozen tart cherries by blanching. *J Food Sci* 36:962–963.

Siegelman HW, Hendricks SB. (1958). Photocontrol of anthocyanin synthesis in apple skin. *Plant Physiol* 33:185–190.

Sierra CC, Molina EB, Zaldivar CP, Flores LP, Garcia LP. (1993). Effect of harvesting season and postharvest treatments on storage life of Mexican limes. *J Food Qual* 16:339–354.

Sigurgisladottir S, Parish CC, Lall SP, Ackman RHG. (1994). Effects of feeding natural tocopherols and astaxanthin on Atlantic salmon (*Salmo salar*) fillet quality. *Food Res Int* 27:23–32.

Silva CLM, Igniatidis P. (1996). Modeling food degradation kinetics: a review. *Book of annual meeting abstracts*, Institute of Food Technologists (USA), 18. Chicago, IL: Institute of Food Technologists.

Simard RE, Lee BH, Laleye CL, Holley R. (1985). Effects of temperature and storage time on the microflora, sensory and exudate changes of vacuum or nitrogen packed beef. *Can Inst Food Sci Technol J* 18:126–132.

Sims CA, Balaban MO, Matthews RF. (1993). Optimization of carrot juice color and cloud stability. *J Food Sci* 58:1129–1131.

Sistrunk WA, Bailey FL. (1965). Relationship of processing procedure to discoloration of canned blackeye peas. *Food Technol.* 19:871–873.

Sistrunk WA, Wang RC, Morris JR. (1983). Effect of combining mechanically harvested green and ripe puree and sliced fruit, processing methodology and frozen storage on quality of strawberries. *J Food Sci* 48:1609–1612, 1616.

Skalski C, Sistrunk WA. (1973). Factors influencing color degradation in Concord grape juice. *J Food Sci* 38:1060–1062.

Skrede G, Naes T, Martens M. (1983). Visual color deterioration in blackcurrant syrup predicted by different instrumental variables. *J Food Sci* 48:1745–1749.

Skrede G, Wrolstad RE, Lea P, Enerson G. (1992). Color stability of strawberry and blackberry syrups. *J Food Sci* 57:172–177.

Smith JP, Ramaswamy HS, Simpson BK. (1990). Developments in food packaging technology. *Trends in Food Sci Technol* 1:111–118.

Smith SM, Geeson JD, Browne M, Genge PM, Everson HP. (1987). Modified-atmosphere retail packaging of Discovery apples. *J Sci Food Agric* 40:165–178.

Snyder HE. (1964). Measurement of discoloration in fresh beef. *J Food Sci* 29:535–539.

Sondheimer E, Kertesz ZI. (1953). Participation of ascorbic acid in the destruction of anthocyanin in strawberry juice and model systems. *Food Res* 18:475–479.

Souza CRA, Osme SF, Glória M. (1997). Stability of curcuminoid pigments in model systems. *J Food Processing Preservation* 21:353–363.

Speers RA, Tung MA, Jackman RL. (1987). Prediction of color deterioration in strawberry juice. *Can Inst Food Sci Technol J* 20:15–18.

Spiro M, Jaganyl D. (1993). What causes scum on tea? *Nature* 364:581.

Sprouls GK, Brewer MS. (1997). Tocopherol effects on frozen ground pork colour. *J Food Qual* 20:1–15.

Starr MS, Francis FJ. (1973). Effect of metallic ions on color and pigment content of cranberry juice cocktail. *J Food Sci* 38:1043–1046.

Steele RJ. (1987). Effect of can rotation during retorting on retention of colour in canned green beans. *CSIRO Food Res Q* 47:25–29.

Steinbuch E. (1984). Heat shock treatment for vegetables to be frozen as an alternative to blanching. In *Thermal processing and quality of foods*, ed. P Zeuthen, JC Cheftel, C Eriksson, M Jul, M Leniger, P Linko, G Varela, G Vos, 553–558. London: Elsevier Applied Science.

Stewart GF, Amerine MA. (1982). *Introduction to food science and technology.* New York: Academic Press.

Sümnü G, Bayindirli L. (1994). Effects of Semperfresh and Jonfresh fruit coatings on poststorage quality of Ankara pears. *J Food Processing Preservation* 18:189–199.

Swatland HJ. (1984). Optical characteristics of natural iridescence in meat. *J Food Sci* 49:685–686.

Sweeney JP. (1961). Process for protection of chlorophyll during cooking of green vegetables. U.S. Patent 3,044,882.

Swirski MA, Allouf R, Guimard A, Cheftal H. (1969). A water-soluble, stable, green pigment, originating during processing of canned Brussels sprouts picked before the first autumn frosts. *J Agric Food Chem* 17:799.

Talbot G. (1995). Chocolate fat: the cause and the cure. *Int Food Ingredients* (1):40–45.

Tan CT, Francis FJ. (1962). Effect of processing temperature on pigments and color of spinach. *J Food Sci* 27:232–241.

Tatsumi Y, Watada AE, Wergin WP. (1991). Scanning electron microscopy of carrot stick surface to determine cause of white translucent appearance. *J Food Sci* 56:1357–1359.

Taylor AA, MacDougall DB. (1973). Fresh beef packed in mixtures of oxygen and carbon dioxide. *Food Technol* 8:453–461.

Taylor AJ. (1980). Natural colours in food. In *Developments in food colours-2*, ed. J Walford, 159–206. London: Applied Science.

Taylor SL, Higley NA, Bush RK. (1986). Sulfites in foods: uses, analytical methods, residues, fate, exposure assessment, metabolism, toxicity and hypersensitivity. *Adv Food Res* 30:1–76.

Teh LS, Francis FJ. (1988). Stability of anthocyanins from *Zebrina pendula* and *Ipomoea tricolor* in a model beverage. *J Food Sci* 53:1580–1581.

Theander O. (1981). Novel development in caramelization. *Prog Food Nutr Sci* 5:471–476.

Thiagu R, Chand N, Habibunnisa EA, Prasad BA, Ramana KVR. (1991). Effect of evaporative cooling storage on ripening and quality of tomato. *J Food Qual* 14:127–144.

Thomas P. (1981). Radiation preservation of foods of plant origin, part 1: potatoes and other tuber crops. *CRC Crit Rev Food Sci Nutr* 19:327–379.

Thomas P. (1985). Radiation preservation of foods of plant origin, part III: tropical fruits, bananas, mangoes and papayas. *CRC Crit Rev Food Sci Nutr* 23:147–205.

Thomas P. (1986a). Radiation preservation of foods of plant origin, part IV: subtropical fruits, citrus, grapes and avocados. *CRC Crit Rev Food Sci Nutr* 24:53–89.

Thomas P. (1986b). Radiation preservation of foods of plant origin, part V:

temperate fruits, pome fruits, stone fruits and berries. *CRC Crit Rev Food Sci Nutr* 24:357–400.

Thomas P, Janave MT. (1992). Effect of temperature on chlorophyllase activity, chlorophyll degradation and carotenoids of Cavendish bananas during ripening. *J Food Sci Technol* 27:57–63.

Thorne S, Alvarez JSS. (1982). The effect of irregular storage temperatures on firmness and surface colour in tomatoes. *J Sci Food Agric* 33:671–676.

Thornton J. (1990). Caramel colours: a brief review. *Food Sci Technol Today* 4(1):9–11.

Thwaites T. (1995). Wave goodbye to discoloured fruit. *New Sci* 145:24.

Tian MS, Woolf AB, Bowen JH, Ferguson IB. (1996). Changes in color and chlorophyll fluorescence of broccoli florets following hot water treatment. *J Am Soc Hort Sci* 121:310–313.

Tichenor DA, Martin DC, Wells CE. (1965). Effects of irradiation on sweet corn. *Food Technol* 19:406–409.

Timberlake CF, Bridle P. (1983). Colour in beverages. In *Sensory quality in foods and beverages*, ed. AA Williams, RK Atkin. Chichester, UK: Ellis Horwood.

Timberlake CF, Henry BS. (1986). Plant pigments as natural food colours. *Endeavour NS* 10(1):31–36.

Ting SV, Hendrickson R. (1968). Enhancing color from orange juice with natural pigments from orange peel. *Proc Florida State Hort Soc* 81:264–268.

Toribio JL, Lozano JE. (1984). Nonenzymatic browning in apple juice concentrate during storage. *J Food Sci* 49:889–892.

Torreggiani D. (1995). Technical aspects of osmotic dehydration in foods. In *Food preservation by moisture control*, ed. GV Barbosa-Cánovas, J Welti-Chanes, 281–304. Lancaster, Pa: Technomic.

Tuma HJ, Kropf DH, Erickson DB, Harrison DL, Trieb SE, Dayton AD. (1973). Frozen meat. *Kansas Agric Exp Station Res Bull* 166.

Turner A. (1995). New controls on food colours. *Food Manuf* 70(3):58–60.

U.S. Department of the Interior. (1970). Refrigeration of fish, part 3. Fishery leaflet no. 429. Washington, DC: U.S. Department of the Interior.

van Arsdel WB. (1957). The time-temperature-tolerance of frozen foods, introduction. *Food Technol* 11:28–33.

van Laack RLJM, Berry BW, Solomon MB. (1996). Variations in internal color of cooked beef patties. *J Food Sci* 61:410–414.

Vankerschaver K, Willcox F, Smout C, Hendricks M, Tobback P. (1996). Mathematical modeling of temperature and gas composition effects on visual quality changes of cut endive. *J Food Sci* 61:613–619, 631.

Vlahov G. (1992). Flavonoids in three olive (*Olea europaea*) fruit varieties during maturation. *J Sci Food Agric* 58:157–159.

von Elbe JH. (1975). Stability of betalaines as food colors. *Food Technol* 29(5):42–44.

von Elbe JH. (1986a). Chemical changes in plant and animal pigments during food processing. In *Role of chemistry in the quality of processed food*, ed. OR Fennema, W-H Chang, C-Y Lii, 41–64. Westport, Conn: Food and Nutrition Press.

von Elbe JH. (1986b). Natural colors: where are we? *Manuf Confectioner* 66(1):43–46.

von Elbe JH. (1987). Influence of water activity on pigment stability in food products. In *Water activity in theory and applications to food*, ed. LB Rockland, LR Beuchet. New York: Marcel Dekker.

von Elbe JH. (1989). Improving the color of canned green vegetables. *Food Bev Technol Int USA*:243.

von Elbe JH, Huang AS, Attoe EL, Nank WK. (1986). Pigment composition and color of conventional and "Veri-Green" canned beans. *J Agric Food Chem* 34:52.

Vorob'ev VV. (1997). The change in fish meat colour at defrostation. *Rybnoe Khozyaistvo* 1:46–47.

Wagenmakers PS, Callesen O. (1995). Light distribution in apple orchard systems in relation to production and fruit quality. *J Hort Sci* 70:935–948.

Walford J. (1977). *Synthetic food colours: trends in future legislation.* Imperial Chemical Industries handout, Guildford, UK, April.

Walford J. (1978). The present and future scope of synthetic food colours. In *Abstracts of a food colour and appearance symposium*, University of Surrey, Colour Group (GB), 73–78. London: Colour Group (GB).

Walker GC. (1964). Color deterioration in frozen french beans (*Phaseolus vulgaris*). *J Food Sci* 29:383–388.

Wall MM, Berghage RD. (1996). Prolonging the shelf-life of fresh green chile peppers through modified atmosphere packaging and low temperature storage. *J Food Qual* 19:467–477.

Walter WM, Giesbrecht FG. (1982). Effect of lyepeeling conditions on phenolic destruction, starch hydrolysis, and carotene loss in sweet potatoes. *J Food Sci* 47:810–812.

Wang W-M, Siddiq M, Sinha NK, Cash JN. (1995). Effect of storage conditions on the chemical, physical and sensory characteristics of Stanley plum pastes. *J Food Qual* 18:1–18.

Ward G, Nussinovitch A. (1996). Peel gloss as a potential indicator of banana ripeness. *Lebens Wiss Technol* 29:289–294.

Warmbier HC, Schnickels R, Labuza TP. (1976a). Effect of glycerol on nonenzymatic browning in a solid IMF system. *J Food Sci* 41:528–531.

Warmbier HC, Schnickels RA, Labuza TP. (1976b). Nonenzymatic browning kinetics in an intermediate moisture model system: effect of glucose to lysine ratio. *J Food Sci* 41:981–983.

Warrington IJ, Stanley CJ, Tustin DS, Hirst PM, Cashmore WM. (1996). Light transmission, yield distribution, and fruit quality in six tree canopy forms of Granny Smith apple. *J Tree Fruit Prod* 1:27–54.

Wasserman BP, Eiberger LL, Guilfoy MP. (1984). Effect of hydrogen peroxide and phenolic compounds on horseradish peroxidase-catalysed decolorization of betalain pigments. *J Food Sci* 49:536–538, 557.

Watada AE, Kim SD, Kim KS, Harris TC. (1987). Quality of green beans, bell peppers and spinach stored in polythene bags. *J Food Sci* 52:1637–1641.

Watson-Smyth K. (1996). Scientists have tasted the future. *Independent* (Aug 3):3.

Wedzicha BL. (1987). The dissociation constant of hydrogen sulphite in vegetable dehydration. *Int J Food Sci Technol* 22:433–450.

White RC, Jones ID, Gibbs E. (1964). Determination of chlorophylls, chlorophyllides, pheophytins and pheoborides in plant material. *J Food Sci* 29:431–436.

Wightman JD, Wrolstad RE. (1995). Anthocyanin analysis as a measure of glycosidase activity in enzymes for juice processing. *J Food Sci* 60:862–867.

Wiley RC. (1994). *Minimally processed refrigerated fruits and vegetables*. London: Chapman and Hall.

Winstanley MA. (1979). The colour of meat. *Nutr Food Sci* 61:5–8.

Wissgott U, Bortlik K. (1996). Prospects for new natural food colorants. *Trends Food Sci Technol* 7:298–302.

Withler RE, Beacham TD. (1994). Genetic variation in body weight and flesh colour of the coho salmon (*Oncorhynchus kisutch*) in British Colombia. *Aquaculture* 119:135–148.

Wolfe ML. (1975). The effect of smoking and drying on the lipids of West African herring (*Sardinella* spp). *J Food Technol* 10:515–522.

Wright DJ, Leach IB, Wilding P. (1977). Differential scanning calorimetric studies of muscle and its constituent proteins. *J Sci Food Agric* 28:557–564.

Wrolstad RE, Skrede G, Lea P, Enerson G. (1990). Influence of sugar on anthocyanin pigment stability in frozen strawberries. *J Food Sci* 55:1064–1065, 1072.

Wu AN. (1986). Factors in the development of a Chinese fast food industry. In *Role of chemistry in the quality of processed food*, ed. OR Fennema, W-H Chang, C-Y Lii, 22–26. Westport, Conn: Food and Nutrition Press.

Xu Q, Chen Y-J, Nelson PE, Chen L-F (1993). Inhibition of the browning reaction by malto-dextrin in freshly ground apples. *J Food Processing Preservation* 16:407–419.

Yang C, Chinnan MS. (1988). Computer modeling of gas composition and color development of tomatoes stored in polymeric film. *J Food Sci* 53:869–872.

Yaylayan VA. (1997). Classification of the Maillard reaction: a conceptual approach. *Trends Food Sci Technol* 8:13–18.

Yi HD, Mitra D, Kootstra A. (1995). Postharvest stimulation of skin color in Royal Gala apple. *J Am Soc Hort Sci* 120:95–100.

Yin M-C, Cheng W-S. (1997). Oxymyoglobin and lipid oxidation in phosphatidylcholine liposomes retarded by α tocopherol and β carotene. *J Food Sci* 62:1095–1097.

Yongsawatdigul J, Gunasekaran S. (1996). Microwave-vacuum drying of cranberries: part 2, quality evaluation. *J Food Processing Preservation* 20:145–156.

Young LL, Lyon CE, Northcutt JK, Dickens JA. (1996). Effect of time postmortem on development of pink discoloration in cooked turkey breast meat. *Poultry Sci* 75:140–143.

Zagory D, Kader AA. (1988). Modified atmosphere packaging of fresh produce. *Food Technol* 42(9):70–74, 76–77.

Zhao YP, Chang KC. (1995). Sulfite and starch affect color and carotenoids of dehydrated carrots (*Daucus carota*) during storage. *J Food Sci* 60:324–326, 347.

Zhuang H, Barth MM, Hildebrand DF. (1994). Packaging influenced total chlorophyll, soluble protein, fatty acid composition and lypoxygenase activity in broccoli florets. *J Food Sci* 59:1171–1174.

Index